INFORMATION TECHNOLOGY

PROCEEDINGS OF THE 2014 INTERNATIONAL SYMPOSIUM ON INFORMATION TECHNOLOGY (ISIT 2014), DALIAN, CHINA, 14–16 OCTOBER 2014

Information Technology

Editors

Yi Wan
Lanzhou University, China

Liangshan Shao
Liaoning Technical University, China

Lipo Wang
Nanyang Technological University, Singapore

Jinguang Sun, Jingchang Nan & Quangui Zhang
Liaoning Technical University, China

CRC Press
Taylor & Francis Group
Boca Raton London New York Leiden

CRC Press is an imprint of the
Taylor & Francis Group, an **informa** business

A BALKEMA BOOK

CRC Press/Balkema is an imprint of the Taylor & Francis Group, an informa business

© 2015 Taylor & Francis Group, London, UK

Typeset by diacriTech, Chennai, India

Published by: CRC Press/Balkema
       P.O. Box 11320, 2301 EH Leiden, The Netherlands
       e-mail: Pub.NL@taylorandfrancis.com
       www.crcpress.com – www.taylorandfrancis.com

ISBN: 978-1-138-02785-5 (Hbk)
ISBN: 978-1-315-68858-9 (eBook PDF)

Information Technology – Wan et al.(Eds)
© 2015 Taylor & Francis Group, London, ISBN 978-1-138-02785-5

Table of contents

Information Technology – Wan et al.(Eds)
© 2015 Taylor & Francis Group, London, ISBN 978-1-138-02785-5

Preface

It is with great pleasure that we warmly welcome you to the 2014 International Symposium on Information Technology (ISIT 2014), held from 14-16 October 2014, Dalian, China.

ISIT 2014 aims to be an excellent platform for international exchange of state-of-the-art research and practice in information technology. This year, the conference received a large number of submissions around the world, and all papers were rigorously reviewed by many technical program committee members and peer reviewers who devoted tremendous amount of time and effort on the evaluations. After careful selection, a broad spectrum of research papers, ranging from theoretical and analytical studies to various applications, were included in the conference proceedings.

We would like to thank all those who contributed professionally to ensuring the high standards of the technical program of ISIT 2014, including the authors, technical program committee members, peer reviewers, and session chairs. We are grateful for the dedication and efforts from the Organizing Committee. Special thanks go to keynote speakers. Finally we would like to acknowledge the support from Liaoning Technical University as an outstanding sponsor.

ISIT 2014 Organizing Committee
October, 2014

Information Technology – Wan et al.(Eds)
© *2015 Taylor & Francis Group, London, ISBN 978-1-138-02785-5*

ISIT 2014 Organizing Committee

GENERAL CHAIRS

Liangshan Shao, Jinguang Sun, and Xiaowei Hui, Liaoning Technical University, China

TECHNICAL PROGRAM CHAIR

Yi Wan, Lanzhou University, China

ORGANIZING COMMITTEE CHAIRS

Xueli Shen, Liaoning Technical University, China,
Lipo Wang, Nanyang Technological University, Singapore

PROCEEDINGS CHAIRS

Jingchang Nan and Sen Lin, Liaoning Technical University, China

PUBLICITY CHAIRS

Changzheng Xing, Guangxian Xu, and Quangui Zhang, Liaoning Technical University, China

TECHNICAL PROGRAM COMMITTEE

Elias Aboutanios, *University of New South Wales, Australia*
Hamid Abrishami Moghaddam, *K.N. Toosi University of Technology, Iran*
Arvin Agah, *The University of Kansas, USA*
Metin Akay, *University of Houston, USA*
Anton Bardera, *University of Girona, Spain*
Saeid Belkasim, *Georgia State University, USA*
Nizar Bouguila, *Concordia University, Canada*
Rémy Boyer, *Université Paris-Sud (LSS-Supelec), France*
Darko Brodic, *University of Belgrade, Serbia*
Roberto Caldelli, *University of Florence, Italy*
Zhenwei Cao, *Swinburne University Technology, Australia*
Mehmet Celenk, *Ohio University, USA*
Jonathon Chambers, *Loughborough University, UK*
Juan Chen, *University of Electronic Science and Technology of China, China*
Hong Cheng, *University of Electronic Science and Technology of China, China*
Jie Cheng, *University of Hawaii, USA*
Feng-Tsun Chien, *National Chiao Tung University, Taiwan*
Wei-Ta Chu, *National Chung Cheng University, Taiwan*
Albert Chung, *The Hong Kong University of Science and Technology, Hong Kong, China*
Gouenou Coatrieux, *Institut Télécom, France*
Miguel Coimbra, *University of Porto, Portugal*

Marco Cristani, *University of Verona, Italy*
Luca De Marchi, *University of Bologna, Italy*
Yongsheng Dong, *Chinese Academy of Sciences, China*
Bogdan Dumitrescu, *Tampere University of Technology, Finland*
Mohamed El Aroussi, *Mohammed V University, Morocco*
Amar El-Sallam, *The University of Western Australia, Australia*
Boris Escalante, *National University of Mexico, Mexico*
Lionel Fillatre, *University of Nice Sophia Antipolis, France*
Luis Maria Fuentes, *Universidad de Valladolid, Spain*
Ali Gholami, *University of Tehran, Iran*
Guillaume Gravier, *IRISA, France*
Ugur Gudukbay, *Bilkent University, Turkey*
A. Ben Hamza, *Concordia University, Canada*
Samer Hanoun, *Deakin University, Australia*
Kazunori Hayashi, *Kyoto University, Japan*
Meng Hua, *City University of Hong Kong, Hong Kong, China*
Bormin Huang, *University of Wisconsin-Madison, USA*
Ting-Zhu Huang, *University of Electronic Science and Technology of China, China*
Wen-Liang Hwang, *Academic Sinica, Taiwan*
Martin Kampel, *Vienna University of Technology, Austria*
Azam Khalili, *University of Tabriz, Iran*
Seongjai Kim, *Mississippi State University, USA*
Ngai Ming Kwok, *University of New South Wales, Australia*
Edmund Lai, *Massey University, New Zealand*
Kuei-Chiang Lai, *National Cheng Kung University, Taiwan*
Ngai Fong Bonnie Law, *The Hong Kong Polytechnic University, Hong Kong, China*
Sébastien Lefèvre, *University of South Brittany, IRISA-UBS, Campus de Tohannic, France*
Xinrong Li, *University of North Texas, USA*
Xuelong Li, *Birkbeck College, UK*
Hongen Liao, *Tsinghua University , China*
Juan Liu, *Wuhan University, China*
Zhenbao Liu, *Northwestern Polytechnical University, China*
Xiaoqiang Lu, *Chiese Academy of Sciences, China*
Alessandro Marzani, *University of Bologna, Italy*
Steve Maybank, *Birkbeck College, UK*
Qurban Ali Memon, *United Arab Emirates University, United Arab Emirates*
Vasileios Mezaris, *Centre for Research and Technology Hellas, Greece*
Ji Ming, *Queen's University Belfast, UK*
Van Khanh Nguyen, *Defence Science and Technology Organisation, Australia*
Kaibao Nie, *University of Washington, USA*
Maciej Niedzwiecki, *Gdansk University of Technology, Poland*
Jean-Christophe Pesquet, *Université Paris-Est, France*
Lai-Man Po, City *University of Hong Kong, Hong Kong, China*
Jianchang Ren, *University of Strathclyde, UK*
Yong Man Ro, *Korea Advanced Institute of Science and Technology, Korea*
Saeid Sanei, *University of Surrey, UK*
Jacob Scharcanski, *Instituto de Informática, Brazil*
Javier Silvestre-Blanes, *Universidad Politécnica de Valencia, Spain*
Vaclav Smidl, *UTIA, Czech Republic*
Kai-Sheng Song, *University of North Texas, USA*
Shuli Sun, *Heilongjiang University, China*
Xue-Cheng Tai, *University of Bergen and Nanyang Technological University, Norway and Singapore*
Jiacheng Tan, *University of Portsmouth, UK*
Jijun Tang, *University of South Carolina, USA*
David Tay, *La Trobe University, Australia*
Jing Tong, *Hohai University, China*
Stefano Tubaro, *Politecnico di Milano, Italy*

Information Technology – Wan et al. (Eds)
© *2015 Taylor & Francis Group, London, ISBN 978-1-138-02785-5*

3D reconstruction and target strength calculations based on multi-photographs of the target

Xue-gang Zhang
Science and technology on underwater test and control Laboratory, Dalian, Liaoning, P.R. China;
Ocean university of china, Qing Dao, P.R. China

Gui-juan Li
Science and technology on underwater test and control Laboratory, Dalian, Liaoning, P.R. China;

Ning Wang
Science and technology on underwater test and control Laboratory, Dalian, Liaoning, P.R. China;
Ocean university of china, Qing Dao, P.R. China

ABSTRACT: The target 3D reconstruction based on multi-photographs of the target is the hotspot of the computer vision researching field, while the target strength prediction using the 3D structure of the target is the researching emphases of underwater acoustics field. Starting from 3D vision of photos, this paper has brought forward a method which makes 3D reconstruction by multi-photos of target and uses plate element method to calculate its target strength. In allusion to this method we have performed a reconstructed experiment of line-type model, and calculate the target strength of the reconstructed model. Finally, we analyze the effect of the modeling error to the calculating precision and validate the availability of this method.

KEYWORDS: target photo 3D reconstruction;target strength

1 INTRODUCTION

This paper presents a method of 3D reconstruction by many target photos on the basis of image 3D vision and carries out the reconstruction experiment of a linetype model, then the target strength of the reconstructed model is calculated and the influence of the error by this modeling method to the calculation of target strength is analyzed.

3D vision transformation and acquire the exterior linetype of the target by inserting values of the characteristic spots, use linetype in CAD tools to 3D modeling, make geometry clear to the model in meshing tools in order to get rid of the redundant inner surface and change vertical vectors, at last calculate target strength using block element method by the proper mesh (element dimension corresponding to the wavelength) divided based on the frequency.

2 BASIC IDEA

Figure 1. The flow chart of basic idea.

Fig. 1 shows the system flow of echo estimate by photos, first to acquire target photos of different angles and make camera scaling, then make inherent characteristic spots of the target matching, acquire 3D coordinate data of the characteristic spots by

3 TARGET LINETYPE INVERSION AND 3D RECONSTRUCTION METHOD BASED ON SINGLE TARGET AND MULTIPLE PHOTOS

3.1 Double eyes imaging 3D vision theory

The 3D geometry target inversion method of multiple photos is mainly based on human eyes imaging mechanism, in other words double eyes imaging relation. The geometry relation of the simplest double eyes imaging theory is as Fig. 2, it is made up of two same cameras, two image plane are at the same plane, the coordinate axles of the two cameras are parallel and superposed to *x* axle, the distance along *x* between cameras is baseline distance *b*. The imaging locations of the same characteristic spot in the scene at the two camera image plane are different in this model. We call the projection spots of the same spot in the scene to the two different images conjugated twins, one projection spot corresponding to another spot, calculating conjugated twins is just calculating corresponding problems. The distance between two conjugated twins when two images superpose (the distance between the conjugated twins) is called parallax. The plane which passes through the center of two cameras and the characteristic spots in the scene calls outer pole plane, the intersectant line between outer pole plane and image plane calls outer pole line.

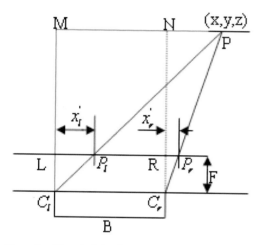

Figure 2. The principle of binocular imaging stereoscopic vision.

In the figure above, assume scene spot P is a characteristic spot of target, whose projection spots to the left and right image plane are separately P^l and P^r. Generally, assume the origin of the coordinate superposes with the center of the left lens. Compare the similar triangle PMC^l with $P^lL\ C^l$, we can get the following formula:

$$\frac{x}{z} = \frac{x_l^{'}}{F} \tag{1}$$

By the same principle, compare the similar triangle PNC_r with $P_l\ RC_r$, we can get the following formula:

$$\frac{x-B}{z} = \frac{x_r^{'}}{F} \tag{2}$$

Combine the above two formula, we can get that,

$$Z = \frac{BF}{x_l^{'} - x_r^{'}} \tag{3}$$

There into F is the focus, B is the baseline distance. As a result, the depth resume of all sorts of scene spots can be realized by calculating parallax.

3.2 Laboratory validation

To validate the precision of the acquired target linetype by this method, we make a linetype model of universal and test it in laboratory. We take photographs of different directions of the model in laboratory, make camera demarcation to the photographs and choose 9 characteristic spots of the model.

To make characteristic spots matching of photographs of different angles, assume the length of the model acknowledged and subscribe it to calculate. Assume the link line of the steering surround shell is X axle, and the link line of the steering surround shell with the model body is Y axle, the point of intersection is origin O. The matching process is as Fig. 3.

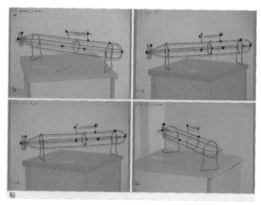

Figure 3. The matching process.

The comparition between inversion value and practical vlue of 9 characteristic spots is shown in Table I:

We can see from Table 1 that the maximum coordinate difference between the inversed value and practical value is 0.47 cm, most of the spot difference is within 0.3 cm, which shows that this inversion method can acquire the

Table 1. Compare with actual and inversed value.

Spot no.	Practical value (x, y, z) cm	Inversion value (x, y, z) cm	Difference value (x, y, z) cm
1	43.8, 6.5, 0.4	44.1, 6.8, 0	0.3, 0.3, −0.4
2	40.8, 2.9, 0.5	41.2, 3, 0	0.2, 0.1, −0.5
3	0.037, 3.4, 0.0	0, 3.2, 0	−0.04, −0.22, 0

4	13.2, 6.9, 3.8	12.7, 7.2, 3.8	−0.52, 0, 0
5	12.6, 0.0, 0.0	12.7, 0 0	0.111, 0, 0
6	0, 0, 0	0, 0, 0	0, 0, 0
7	0.12, 6.9, 3.9	0, 7.2, 3.8	−0.18, 0.33, −0.1
8	−12.08, 3.5, 0.3	−12.1, 3.15, 0	−0.024, −0.36, −0.28
9	−18.8, 7.12, 0.46	−18.8, 7.2, 0	−0.003, 0.09, −0.46

position information of each characteristic spot efficiently.

We make spline interpolation to characteristic spots in UG NX modeling software to acquire target linetype, and acquire target 3D reconstruction body by a series of curved face designing such as tension and lofting. The actual and inverted model is shown in Fig. 4.

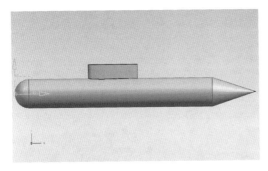

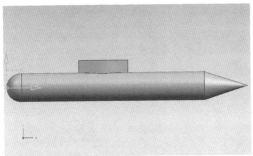

Figure 4. Actual and inversed model.

4 TARGET STRENGTH CALCULATION AND PRECISION ANALYSIS OF TARGET 3D RECONSTRUCTION BODY

4.1 *High frequency target strength estimate method of target 3D reconstruction body-plate element method*

The plate element method takes the surface of target as a series of approximate small plates; calculate the contribution to target strength of these plates separately. The scattering sound field by Kirchhoff formula is:

$$\phi_s = \frac{1}{4\pi} \int_s \left[\phi_s \frac{\partial}{\partial n} \left(\frac{e^{ikr_2}}{r_2} \right) - \frac{\partial \phi_s}{\partial n} \frac{e^{ikr_2}}{r_2} \right] ds \qquad (4)$$

As Fig. 5 shows, s is the surface of scattering body, n is the out normal line, r_2 is the radius vector of scattering spot.

To high frequency we usually assume that:

A. The contribution of the geometry shadow to the sound field can be neglected:

The practical integral area is s_0 seeing from M_1 and M_2 which locates at the bright section.

B. The surface of the object satisfies rigid boundary condition:

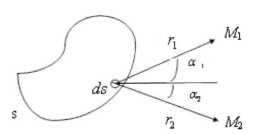

Figure 5. Integral area.

$$\begin{cases} \phi_s = \phi_i \\ \dfrac{\partial(\phi_s + \phi_i)}{\partial n} = 0 \end{cases} \qquad (5)$$

Thereinto φ_i is the potential function of incident wave (ignore $e^{-i\omega t}$), $\phi_i = (A/r_1)e^{ikr_1}$, whose derivative is:

$$\phi_s = \frac{-A}{4\pi} \int_s e^{ik(r_1+r_2)} \left[\frac{ikr_2 - 1}{r_1 r_2^2} \cos\alpha_1 + \frac{ikr_1 - 1}{r_2 r_1^2} \cos\alpha_2 \right] ds \qquad (6)$$

For receive and deliver combination case, there is

$$\phi_s = \frac{-A}{2\pi} \int_s e^{ik2r} \left(\frac{ikr - 1}{r^3} \cos\alpha \right) ds \qquad (7)$$

In principle, (6) and (7) adapt to random distance, whether near field or far field. Use far field plate

3

element method to calculate (7), we can get the far field target strength of a model.

4.2 Comparation of high frequency target strength between target 3D reconstruction body and original model

The reconstructed 3D model and original 3D model are divided into precise mesh using Hypermesh meshing tools, the dimension of the plates are equal to the wavelength of the calculating frequency. Defining incident from the stern of the target as 0° incident. To the reconstructed model and original model, the horizontal omnibearing target strength under 4.5kHz, and target strength of incident at abeam direction and 135° under different frequencies are all calculated. The comparition results are as Figs. 6–8.

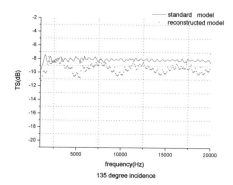

Figure 8.   The target strength of 135°incidence for varied frequncy.

incidence and top direction incidence, which has an excellent precision. Except the error between 35 and 45 degrees is relatively large, other directions are all basically agreeable, because in this range of degrees, the target strength is basically negative, whose decibel form have a larger difference, but it has a small effect on total target strength.

5   CONCLUSION

Using non-cooperating target photographs to 3D model reconstruction we can study its scattering characteristics.

This method is suitable for reconstruction and target strength calculation of regular model, which has an excellent precision. The following work should be refining this method to the inversion of unregular model.

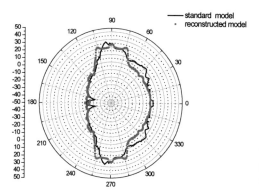

Figure 6.   The whole bearing target sthength in 4.5kHz.

From Figs. 6 and   we can see that, the target strength of reconstructed model of this method and original model agree very well at abeam direction

ACKNOWLEDGMENT

The authors acknowledge the support of research grants from Foundation of Science and technology on underwater test and control Laboratory.

REFERENCES

[1] Fan Jun, Wan Lin, Tang Weilin. Underwater target strength calculation based on 3Dmax modeling. *Acoustics Technics*, 2000, (4).

[2] Fan Jun, Tang Weilin. Plate element method of sonar target strength calculation. *Acoustics Technics*, supplement issue, 1999, 31–32.

[3] Wang Lili, Liu Rong. The geometry 3D reconstruction method based on photographs. *System Simulation Transaction*, 2001, supplement volume 13.

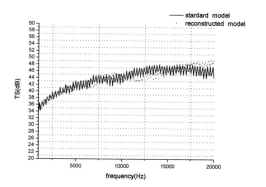

Figure 7.   The target strength of abeam incidence for varied freqency.

Information Technology – Wan et al. (Eds)
© 2015 Taylor & Francis Group, London, ISBN 978-1-138-02785-5

A 4-bit array multiplier design by reversible logic

Junzhou Qian
Shaoxing Devechip Microelectronics Co., Ltd, Shaoxing, Zhejiang, China

Junchao Wang
Department of Electrical and Computer Engineering, Illinois Institute of Techonoly, Chicago, Illinois State, USA

ABSTRACT: Low power design has attracted much attention since the energy dissipation is a significant factor in digital integrated circuit design. A certain portion of energy dissipation is caused by irreversible computing. Therefore, using reversible logic to design digital circuit is an effective way to lower decrease the energy dissipation. Through optimization of traditional array multiplier, a 4-bit array multiplier designed by reversible logic is proposed in this paper.

1 REVERSIBEL LOGIC

With the development of mobile computing, the low power design becomes increasingly important in integrated circuits. Thus, more and more concern falls on energy dissipation. However, energy dissipation in irreversible logic design is inevitable, which is explained by Landauer's principle-kTln2 joules of heat energy will be generated once erasing each bit of information [1]. One of the solutions is designing the circuits by reversible logic.

The bijective Boolean function is the basis of reversible logic. It is required that the number of input nodes must equal to the number of output nodes [2]. Therefore, each input can be reconstructed by a function of output. Generally an irreversible gate has different amounts of inputs and outputs. In order to establish reversible gate, some additional inputs and outputs are added. The additional inputs are called ancilla bit, while the extra outputs are named "garbage" bit which is abbreviated to g.

The reversible logic circuit is based on several basic reversible gates such as NOT, FEYNMAN, TOFFOLI, and PERES [3] [4].

The NOT gate has one input node and output node. Its quantum cost is 1. The FEYNMAN gate which is also called CNOT gate has two input and output nodes. Its quantum cost is 1. The TOFFOLI gate has three input and output nodes. Its quantum cost is 5. The PERES gate also has three input and output nodes. And its quantum cost is 4.

The multiplier is one of the fundamental circuits in an integrated circuit. A lot of operations in digital signal processing are multiply operation. However, the traditional array multiplier has a great energy dissipation. Besides, the area of traditional multiplier is big.

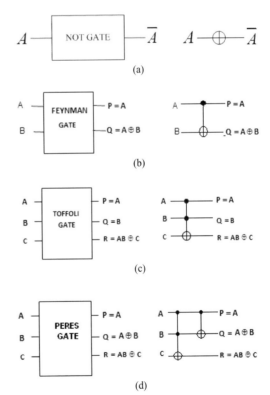

Figure 1. Basic reversible gates.

In this paper, we proposed a new design of 4 bit array multiplier which not only has a lower delay than traditional multiplier, but also declines the energy dissipated by reversible logic.

2 BACKGROUND OF ARRAY MULTIPLIER

There are many kinds of multiplier such as array multiplier, booth multiplier, and Wallace tree multiplier. Array multiplier is the most basic multiplier. In this paper, a 4 bit array multiplier will be taken for example. A 4 bit array multiplier has two 4 bit inputs which are A[3:0] and B[3:0], it also has 8 bit output C[7:0]. Traditional array multiplier consist of adder and And gate. The And gate is used to generate partial product. And then the adder is used to add partial product together. As shown in figure 2, a 4 bit multiplier need 16 And gates and three 4 bit adder.

two PERES gate is used to compose the CSA that is shown in figure 3.

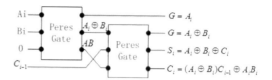

Figure 3. Reversible CSA block.

And in order to generate partial product, 16 TOFFOLI gates will be used to realize the logic function which is done by AND gate in irreversible logic. The quantum circuit figure is shown in figure 4.

Differ from the traditional array multiplier which uses 4 bit adder to add the partial product, in this paper 1 bit CSA is used to add the partial product. There are 12 CSA to form the array which will add the partial product and complete the multiplication.

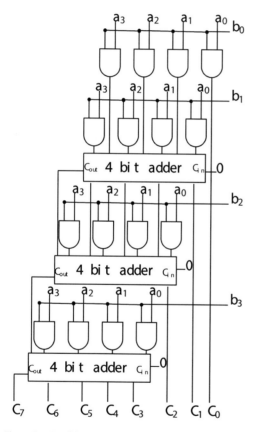

Figure 2. Traditional 4 bit array multiplier.

3 PROPOSE THE 4-BIT ARRAY MULTIPLIER

Unlike traditional array multiplier, the carry save adder (CSA) is used in this paper. The CSA has 3 inputs and 2 outputs. Its function of output SUM is $SUM = a \otimes b \otimes cin$. And its function of output Cout is $Cout = (a \otimes b) \bullet cin \otimes (a \bullet b)$. In order to redesign the array multiplier by reversible logic, the CSA block should be redesigned by reversible gate. In this paper,

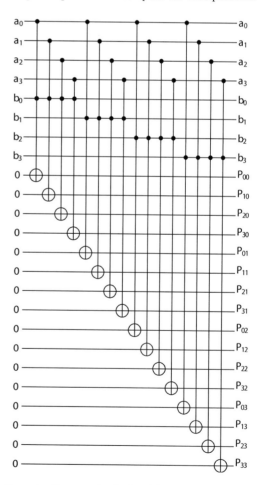

Figure 4. Quantum circuit of partial product block.

This array has three columns and four lines. All of the CSA's outputs Cout are linked to the CSA's input Cin in the next columns. This structure will decrease the delay in a line. So the CSA array multiplier's key path delay is smaller than a traditional array multiplier. The CSA array multiplier is shown in figure 5.

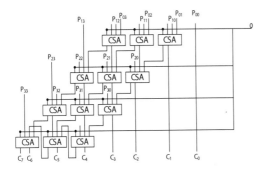

Figure 5. The CSA array of partial product add block.

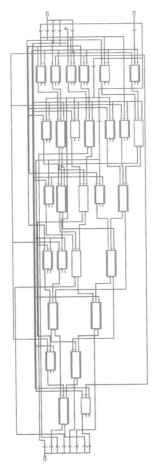

Figure 6. The simulation diagram.

4 EXPERIMENT RESULTS

To verify the function of CSA array multiplier, we use Verilog language to program the CSA array multiplier.

All the experiments are performed in Cadence IUS9.2 in an 8GB, 3.3GHZ Intel Xeon e3-1230v3 machine under Linux redhat3. 4.6-9. The verification result is shown in figure 7.

Figure 7. Verification result of CSA array multiplier.

We use five couples of 4 bit binary digits. The a[3:0] and b[3:0] represent the multiplicator and multiplicand. The product[7:0] represents the product. All the binary digits are converted into a decimal number in figure 6. All the computation is right. Therefore, the function of the reversible array multiplier is realized.

Then the delay and area of the reversible array multiplier are compared with other multiplier. The results are summarized in table 1.

Table 1. Area and delay comparison between four 4-bit multiplier.

	Area(gate)	Delay(ns)
4 bit multiplier from synopsys designware	188	3.56
Booth multiplier	124	9.67
Traditional array multiplier	152	3.95
Reversible CSA array multiplier	104	3.05

Table 1 shows the comparison of delay and area between 4-bit multiplier from the Synopsys designer IP library, booth multiplier, traditional array multiplier and reversible CSA array multiplier. Obviously, the reversible CSA multiplier has the lowest delay and smallest area in the four multipliers. So the reversible CSA multiplier will be used frequently in the high performance computing circuit.

5 CONCLUSION

In this paper, a new design of the CSA array multiplier based on reversible logic is proposed. It has the advantage of smaller area and lower delay. It consists of 16 TOFFOLI gate and 24 PERES gates with total quantum cost of 176. Because of the reversible logic, this multiplier also has the advantage of low energy dissipation. Therefore, it will have broad application prospects in the mobile computing field.

REFERENCES

[1] R. Landauer. Irreversibility and heat generation in the computing process. IBM J. of R&D, 5:183–191, 1961.

[2] Bruce J W, Thronton M A, Shivakumaraiah L, Kokate.

[3] PS, Li X (2002), "Efficient adder circuit based on a reversible conservative logic gate", IEEE Computer Society Annual Symposium on VLSI.pp.2.

[4] Yu Pang, Shaoquan Wang, Zhilong He, Jinzhao Lin, Sayeeda Sultana, Katarzyna Radecka, "Positive Davio-based Synthesis Algorithm for Reversible Logic". 2011 IEEE 29th International Conference on Computer Design (ICCD).pp. 212.

[5] Khan, M.M.M. "Quantum realization of some quaternary circuits", TENCON 2008 - 2008 IEEE Region 10 Conference.

[6] Pezaris S D. A 40-ns 17-bit by 17-bit array multiplier. Computers, IEEE Transactions on, 1971, 100(4): 442–447.

[7] Pang Y, Wang J, Wang S. A 16-bit carry skip adder designed by reversible logic, Biomedical Engineering and Informatics (BMEI), 2012 5th International Conference on. IEEE, 2012: 1332–1335.

[8] Wang J C, Pang Y, Xia Y. Sch. of Optoelectron. Eng., Chongqing Univ. of Posts & Telecommun., Chongqing, China, Wavelet Active Media Technology and Information Processing (ICWAMTIP), 2012 International Conference on. IEEE, 2012: 318–321.

[9] Bits M. TABLE I MODIFIED BOOTH'S ALGORITHM FOR A LEAST TO MOST SIGNIFICANT SCAN OF BITS[J]. IEEE Transactions on Computers, 1975.

[10] Baugh C R, Wooley B A. A two's complement parallel array multiplication algorithm. IEEE Transactions on Computers, 1973, 22(12): 1045–1047.

[11] Sato T. MOS array multiplier cell: U.S. Patent 5,151,875. 1992-9-29.

[12] Huang Z, Ercegovac M D. High-performance low-power left-to-right array multiplier design. Computers, IEEE Transactions on, 2005, 54(3): 272–283.

Information Technology – Wan et al. (Eds)
© *2015 Taylor & Francis Group, London, ISBN 978-1-138-02785-5*

A comparative research of color feature extraction for image retrieval

He Zhang & Xiuhua Jiang
Information Engineering School, Communication University of China, Beijing, China

ABSTRACT: With the rapid growth of image database, content-based image retrieval (CBIR) technology with its high value in theory and application has become a hot topic in the field of image processing. Color feature is the common characteristic of image retrieval technology. This paper discusses on the comparative algorithms based on color histograms and fuzzy color recognition. The experimental results of the two algorithms are compared and it shows that the precision of the algorithm based on fuzzy color recognition is higher than that of the traditional algorithm based on color histograms.

1 INTRODUCTION

With the development of multimedia technology and network technology, digital image databases are growing at an alarming rate. Thus, a fast and accurate algorithm is needed to pinpoint the useful images from a large pool of mass image databases. From 1970s to present, image retrieval technology were developing from text-based algorithms to content-based algorithms.

Content based image retrieval system is divided into two parts, including the image feature extraction and similarity matching. The color, texture and shape features which make up the image features library are extracted by the processing of image feature extraction. Features of the target image are used to compare with the features in the features library, and then the results of image retrieval are formed.

The color feature of images is not sensitive to the size, rotation, translation of images, so it is widely used in the field of image retrieval. Color feature extraction is mainly divided into two kinds of algorithms: the global algorithms and the local algorithms. The global algorithms include the global color histograms, the fuzzy color histograms, the color moments and the color correlogram. Furthermore, to overcome the weakness of the global feature which does not consider the spatial information of colors, researches on the extraction of local color features are more and more popular. Images are divided into sub-blocks in order to reflect the spatial information of colors. Chatzichristofis extracts the fuzzy color histogram and their texture feature of images, by using which method he obtains a higher retrieval precision.

In this paper, we try to introduce two algorithms which includes the color histogram and the fuzzy classification methods. The algorithms of color histogram are divided as the global and blocking color histograms. After both the blocking color histogram and the fuzzy classification extract the information of colors and their structures, the experimental results of the two algorithms will be compared and analyzed. It shows that precision of the algorithms which take into account the spatial information of colors is higher than the one not.

So the rest of the paper is organized as follows, Section 2 introduces the basic knowledge of color extraction. Section 3 proposed two algorithms and the processing of them. In Section 4, experimental results are discussed. Section 5 concludes the whole paper.

2 COLOR FEATURE

2.1 HSV color space

The RGB, HSV and YCbCr are commonly used color spaces. Since it is difficult to determine the proportion of each component in the RGB color space for a certain color, human's sight cannot be referenced with the RGB color space. Hue, saturation and value are the three components of the HSV color space, which are visually independent from each other. Moreover, human eye perception is directly functioned by each component, and the distance between eye perceptions of two different colors is in direct proportion in the Euclidean distance. Therefore, the HSV color space is better in accordance with the feel of human eyes than the RGB color space. Conversion from RGB to HSV color space is conducted as follows:

$$v' = \max(r,g,b), define\ r',g',b':$$

$$r' = \frac{v'-r}{v'-\min(r,g,b)}, g' = \frac{v'-g}{v'-\min(r,g,b)}, \qquad (1)$$

$$b' = \frac{v'-b}{v'-\min(r,g,b)}$$

Then $v = \dfrac{v'}{255}$ $s = \dfrac{v'-\min(r,g,b)}{v'}$

$$h' = \begin{cases} (5+b'), r = \max(r,g,b) and g = \min(r,g,b) \\ (1-g'), r = \max(r,g,b) and g \neq \min(r,g,b) \\ (1+r'), g = \max(r,g,b) and b = \min(r,g,b) \\ (3-b'), g = \max(r,g,b) and b \neq \min(r,g,b) \\ (3+g'), b = \max(r,g,b) and r = \min(r,g,b) \\ (5-g'), others \end{cases} \quad (2)$$

$$h = 60 \times h'$$

$$r,g,b \in [0,255], h \in [0,360], s \in [0,1], v \in [0,1].$$

Here if $\max(r, g, b) = \min(r, g, b)$, r', g' and b' have no meanings. So we can define $h=s=0$, $v=r/255$.

2.2   Color histogram

The HSV space is the most commonly used color space in the algorithms of color histogram. Color histogram is widely used in many methods of image retrieval. Color histogram is calculated as (3):

$$H(k) = \frac{n_k}{N}(k = 0,1,......L-1) \qquad (3)$$

N represents the total of the pixels in the image, and k represents the total of the categories of colors. The pixels belonging to the corresponding color are recorded in the same collection, so that each color corresponds to a number of pixels and the normalized color histogram is generated.

3   METHODOLOGY

3.1   The algorithms based on color histogram

3.1.1   Global color histogram
Colors need to be quantified after the selection of color space. A perfect quantified method can not only reduce the complexity of the processing, but also lose too much information of colors.

$$H = \begin{cases} 0, if(h \in [316,20]) \\ 1, if(h \in [21,40]) \\ 2, if(h \in [41,75]) \\ 3, if(h \in [76,155]) \\ 4, if(h \in [156,190]) \\ 5, if(h \in [191,270]) \\ 6, if(h \in [271,295]) \\ 7, if(h \in [296,315]) \end{cases} S = \begin{cases} 0, if(s \in [0,0.2]) \\ 1, if(s \in [0.2,0.7]) \\ 2, if(s \in [0.7,1]) \end{cases} V = \begin{cases} 0, if(v \in [0,0.2]) \\ 1, if(v \in [0.2,0.7]) \\ 2, if(v \in [0.7,1]) \end{cases} \quad (4)$$

Then these three components are synthesized into one-dimensional feature vectors.

$$G = 9H + 3S + V\ (0 <= L <= 71) \qquad (5)$$

The coefficient of each component is based on their weight in visual resolving capability of human. By this way, these three components of each pixel can be distributed in a one-dimensional vector. We can obtain the histogram of the vector for each image, and the differences of the vector between images are calculated using Euclidean distance.

However, this algorithm only describes the proportion of different colors in the entire image, but hardly caring about the spatial location of each color.

3.1.2   Blocking color histogram
To improve the above method, images are divided into blocks to express the spatial distribution of colors.

A strategy of the image overlapping division is proposed like this: Images are divided into 3×3 sub-blocks first. Then they are merged into four overlapping areas named p1, p2, p3 and p4 as Fig. 1 shows. The central block weighs 4 for it overlaps 4 times. Blocks 2, 4, 6, 8 weigh 2 for they overlap twice, and blocks 1, 3, 7, 9 weighs 1 for they overlap once. Then the differences between images are calculated using Euclidean distance. By this way, the importance of the middle block is highlighted, while the influence of the background blocks in image retrieval is suppressed.

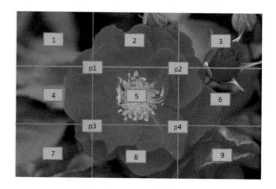

Figure 1.   Blocks divided for an image.

When the subject or our interest object is divided into different blocks, the integrity of images is destroyed. What is more, the rotation invariant of global histogram is also destroyed. Thus, image retrieval accuracy will be affected by these reasons as follows.

3.2 The algorithm based on fuzzy recognition of colors

In order to overcome the defects of color histograms and combine colors and their spatial information into a more reasonable way, a new algorithm that decomposes images into different parts according to colors which are divided into different categories is proposed. It just needs to compare the structure features of corresponding parts of different images in the similar measurement. This algorithm simplifies the complexity of retrieval and enhances the efficiency of retrieval.

3.2.1 Fuzzy recognition of colors

In HSV color space, previous research has established a mapping between basic hues and their ranges. Table 1 shows the mapping table.

Table 1. The mapping table($360° = 0°$).

Color	Low thresholds(°)	High thresholds(°)
Orange	25	46
Yellow	46	70
Green	70	165
Blue	165	270
Purple	270	340
Red	340	25

Phan found that there is a boundary between color and non-color on the plane of S and V when the H is fixed. The boundary is a fuzzy region. So color, non-color and the fuzzy region are separated using two curves as shown in Fig. 2.

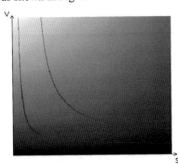

Figure 2. The dividing line of fuzzy color and non-color sets.

In Fig. 2, the curve on the left side is the right half of $(v-0.07)(s-0.03)=0.01$, on the left of which is considered to be non-color. While the other curve is the right half of $(v-0.12)(s-0.12)=0.06$, on the right of which is considered to be color. The region between the two curves is fuzzy.

The color fuzzy set and non-color fuzzy set are defined according to the boundary between the three regions. In the HSV color space, the color fuzzy set is a fuzzy subset which is defined at $[0,1]*[0,1]$. Its Membership function is (6).

$$\tilde{C}(s,v) = \begin{cases} 0, (v-0.07)(s-0.03) \leq 0.01 \\ 1, (v-0.12)(s-0.12) \geq 0.06 \\ d_l/(d_l+d_r), others \end{cases} \quad (6)$$

In (6), d_l and d_r are the distances between the two curves.

The non-color fuzzy set is a fuzzy subset which is defined at $[0,1]*[0,1]$. Its Membership function is (7).

$$\tilde{A}(s,v) = 1 - \tilde{C}(s,v) \quad (7)$$

Saturation and value are range from 0 to 1 in the HSV color space. The low, middle and high fuzzy sets of saturation and value are defined in Fig. 3 in accordance with the rules of vision. $\tilde{L}_s, \tilde{M}_s, \tilde{H}_s$ represent the low, middle and high fuzzy sets of saturation, and $\tilde{L}_v, \tilde{M}_v, \tilde{H}_v$ represent the low, middle and high fuzzy sets of value.

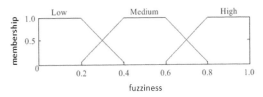

Figure 3. Trapezoidal fuzzy numbers.

As we know, color presents white, grey and black with the development of v when h cannot be expressed due to the low value of s. Meanwhile, color presents black when h and s cannot be expressed due to the low value of v. In other cases, colors present light, bright, dark and deep corresponding with the basic hues. Table 2 shows the color reasoning table.

Table 2. Color reasoning.

v	s		
	low		
High	white	light	bright
Medium	gray	dark	deep
Low	black	black	black

11

In this case, s and v from different levels can express seven different types of colors when h is fixed.

According to the attribute of color and non-color and the color reasoning table, a fuzzy color set is defined when h is fixed, including $\tilde{W}hite, \tilde{G}ray, \tilde{B}lack, \tilde{C}olor1, \tilde{C}olor2, \tilde{C}olor3, \tilde{C}olor4$. They are fuzzy subsets defined at $[0,1]*[0,1]$ of different meanings when h belongs to different basic colors, and they are fuzzy subsets which are defined at $[0,1]*[0,1]$. Their membership functions are (8).

In (8), $T(\cdot)$ presents the T norm which means take intersections.

$$
\begin{cases}
\tilde{W}hite(s,v) = T(T(\tilde{L}_s(s), \tilde{H}_v(v)), \tilde{A}(s,v)) \\
\tilde{G}ray(s,v) = T(T(\tilde{L}_s(s), \tilde{M}_v(v)), \tilde{A}(s,v)) \\
\tilde{B}lack(s,v) = T(\tilde{L}_v(v), \tilde{A}(s,v)) \\
\tilde{C}olor1(s,v) = T(T(\tilde{M}_s(s), \tilde{H}_v(v)), \tilde{C}(s,v)) \\
\tilde{C}olor2(s,v) = T(T(\tilde{H}_s(s), \tilde{H}_v(v)), \tilde{C}(s,v)) \\
\tilde{C}olor3(s,v) = T(T(\tilde{M}_s(s), \tilde{H}_v(v)), \tilde{C}(s,v)) \\
\tilde{C}olor4(s,v) = T(T(\tilde{H}_s(s), \tilde{M}_v(v)), \tilde{C}(s,v))
\end{cases} \tag{8}
$$

For a pixel $p0=(h0,s0,v0)$, the basic hue of it can be judged out according to mapping table in TABLE . Then we can set s0 and v0 into (6 and 7), getting the color fuzzy set, the non-color fuzzy set, the fuzzy sets of saturation and value. The color of pixel p0 can be calculated by (8) according to the principle of maximum degree of membership.

3.2.2 Color regions and their characteristics
We divided images into different layers which each layer contains only pixels with the same color. And the overall pixels within each layer are defined as 'single color layer (SCL)'.

Area, dispersion and centroid are three important characteristics of the SCL. Area is defined as the number of pixels of each SCL, and it will be deleted if the area of one SCL is smaller than $w*h/20$. W and h represents the width and height of images. For example, three colors of the image are extracted out after the above processing, which expressed dark red, deep red and dark green. Then the SCLs of the image can be expressed as Fig. 4.

The dispersion χ and the centroid c are defined in (9) and (10).

$$
\chi = \frac{N^2}{S} \tag{9}
$$

Figure 4.   SCR.

$$
c = (\frac{\sum_{i=1}^{n} x_i}{nw}, \frac{\sum_{i=1}^{n} y_i}{nh}) \tag{10}
$$

N represents the total number of boundary pixels and S means the area of the SCL. Dispersion reflects the distribution of pixels with the same color in images. And n means the amount of the pixels of the SCL, and x_i, y_i mean the horizontal and vertical coordinates of pixel i. Centroid reflects the location and the structure of the SCL. The feature vector of SCL with color ξ is defined like this: $F_\xi = (S_\xi, \chi_\xi, C_\xi)$.

3.2.3 Similarity measurement
Images A and B are set for example. A and B have the feature vectors of the SCL which contains the color $\xi: F_\xi^{(A)} = (S_\xi^{(A)}, \chi_\xi^{(A)}, C_\xi^{(A)})$ and $F_\xi^{(B)} = (S_\xi^{(B)}, \chi_\xi^{(B)}, C_\xi^{(B)})$. Only the features of the same SCR between different images are compared. So (11)–(13) show the difference of relative area, dispersion and centroid: $d(\cdot)$ means the Euclidean distance.

$$
\Delta S_\xi = \begin{cases} \dfrac{\left|S_\xi^{(A)} - S_\xi^{(B)}\right|}{\max(S_\xi^{(A)}, S_\xi^{(B)})}, \max(S_\xi^{(A)}, S_\xi^{(B)}) \neq 0 \\ 0, \text{others} \end{cases} \tag{11}
$$

$$
\Delta \chi_\xi = \begin{cases} \dfrac{\left|\chi_\xi^{(A)} - \chi_\xi^{(B)}\right|}{\max(\chi_\xi^{(A)}, \chi_\xi^{(B)})}, \max(\chi_\xi^{(A)}, \chi_\xi^{(B)}) \neq 0 \\ 0, \text{others} \end{cases} \tag{12}
$$

$$
d_\xi = \begin{cases} d(c_\xi^{(A)}, c_\xi^{(B)}), c_\xi^{(A)}, c_\xi^{(B)} \text{ are all exist} \\ \sqrt{2}, c_\xi^{(A)}, c_\xi^{(B)} \text{ only one exists} \\ 0, c_\xi^{(A)}, c_\xi^{(B)} \text{ no one exist} \end{cases} \tag{13}
$$

The distance between two images is defined in (14):

$$d_{AB} = \sum_{\xi \in Color} d_\xi^2 (\alpha \Delta S_\xi + \beta \Delta \chi_\xi) \tag{14}$$

In (11), α and β represent the weighted factors of area and dispersion. $\beta = 1 - \alpha$, and dAB is rotation invariance.

Fig. 5 shows the whole process of the algorithm.

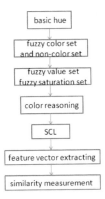

Figure 5. The process of the fuzzy recognition of color algorithm.

4 EXPEERIMENTAL RESULTS

Three experiments were made between the two methods according to the process.

The famous 'Corel' database is used as the experimental database. There are 10 categories in the database, and each category has 100 images. To evaluate the proposed descriptor we use 8 images as queries, which are extracted from 8 different categories: people, building, bus, dinosaur, flower, horse, mountain and food. Some of them are shown in Fig. 6.

We use the first 5, 10, 20, 40, 60, 80, 100 images to calculate the precisions as the results. Then the averages of them are calculated later.

Figure 6. Some of the images used as a query for the evaluation.

4.1 Experimental results of the color histogram

Table 3 shows the precision and average precision of the algorithm of the global color histogram.

The second experiment is performed by the algorithm based on the blocking color histogram. Table 4 shows the results.

Table 3. Precision of global histogram algorithm (%).

	5	10	20	40	60	80	100
People	60	70	65	67.5	63.3	60	53
Building	40	40	25	25	20	18.8	18
Bus	100	100	100	92.5	83.3	70	18
Dinosaur	100	100	95	82.5	80	71.3	61
Flower	80	70	75	57.5	50	56.7	51
Horse	100	90	90	92.5	91.7	91.3	85
Mountain	80	70	50	27.5	28.3	30	26
Food	20	20	15	27.5	33.3	36.3	34
Average	72.5	70	64.4	59	56.2	54.3	48.9

Table 4. Precision of blocking algorithm (%).

	5	10	20	40	60	80	100
People	100	90	80	75	75	66.3	62
Building	80	50	30	22.5	30	28.3	28
Bus	100	100	95	75	71.7	58.8	55
Dinosaur	100	100	95	85	80	72.5	64
Flower	100	90	70	60	53.5	46.3	44
Horse	100	90	90	95	95	92.5	88
Mountain	100	80	65	52.5	40	36.3	32
Food	60	50	65	55	53.5	47.5	46
Average	92.5	81.3	73.8	65	62.3	56.1	52.4

The first 20 results of the retrieval of the first image are shown in Fig. 7. By using the algorithm of global color histogram, five non-flower-related images appear at the first 10 position. However, three appears by using the proposed algorithm.

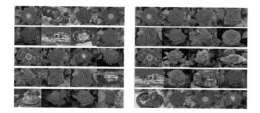

Figure 7. Results using global(left) and blocking(right) color histogram.

We can know that the algorithm of the global color histogram only describes the proportion of different colors in the entire images, but hardly caring about the spatial location of each color. Two images with different structures may have similar color histograms, which will have influence on the retrieval performance. The weighted blocking color histogram adds the spatial information of colors and makes up for the deficiency of the global

histogram. However, it may break the rotational invariance of the global histogram and the integrity of images by dividing into blocks, which will lower the efficiency of retrieval.

4.2 Experimental results of the fuzzy recognition of colors

This experiment is performed by the algorithm of fuzzy recognition of colors. Table 5 shows the experimental results. We set α =0.7, β =0.3.

Fig. 8 shows the first 20 results of the retrieval of the first image. No non-flower-related images appear at the first 20 position. The structure of colors is considered in the algorithm and it only compares the structure features of corresponding parts of different images in the similarity measurement which simplifies the complexity of retrieval and improves the efficiency of retrieval.

The average precision of the experimental results of the three algorithms are compared in Fig. 9. Obviously, the algorithms which combine the information of color spatial performance get higher precisions than the algorithm of global color histogram. Meanwhile, the algorithm based on the fuzzy color recognition with its more reasonable combination theory has obtained a higher precision than the algorithm based on blocking color histogram.

Table 5. Precision of fuzzy recognition (%)

	5	10	20	40	60	80	100
People	100	100	100	70	61.7	47.5	45
Building	60	70	60	58	55	56	59
Bus	80	60	50	65	68	59	55
Dinosaur	80	90	95	90	92	88	82
Flower	100	100	100	85	67	60	51
Horse	100	100	100	100	98	86	85
Mountain	80	80	85	90	87	85	79
Food	100	100	85	90	87	86	82
Average	87.5	87.5	84.4	81	77	71	67.3

Figure 8. Results using the algorithm of fuzzy recognition of color.

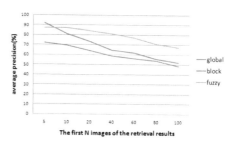

Figure 9. The comparison of the experimental results of the 3 algorithm.

14

5 CONCLUSIONS

Two algorithms based on the color feature extraction for image retrieval are proposed in this paper, including the algorithms based on color histograms and fuzzy color recognition.

The weighted blocking color histogram adds the spatial information of colors, makes up for the deficiency of the global histogram, and performs a higher precision than the global color histogram. While the algorithm of fuzzy color recognition decomposes images into different parts according to colors which are divided into different categories and compares the structure features of corresponding parts of different images in the similarity measurement. It combines colors and their spatial information in a more reasonable way than the blocking color histogram, maintaining the structure of colors. What is more, only comparing the structure of corresponding parts between different images simplifies the complexity of retrieval. Experimental results show that the algorithm of fuzzy color recognition rank the first position at the comparison.

In conclusion, the algorithm based on the fuzzy recognition of colors achieves a higher precision than the algorithm based on color histograms. And extracting the texture and shape features of the single color layers may be chosen as the next step of the work.

ACKNOWLEDGMENT

This work is supported by the National Sci-tech Support Plan of China:"Radio and television comprehensive production and management of media assets key technology research and application demonstration" (No. 2012BAH15B01)

REFERENCES

Chatzichristofis S A, Boutalis Y S. CEDD: color and edge directivity descriptor: a compact descriptor for image indexing and retrieval. Computer Vision Systems. Springer, Berlin Heidelberg, 2008, pp. 312–322.

Fang Hou, Xiuhua Jiang. Research on the Color Feature Extraction Method of the Image Retrieval. *Journal of Communication University of China*, 2013, 20(5): 35–40.

Fierro-Radilla A N, Nakano-Miyatake M, Pérez-Meana H, et al. An efficient color descriptor based on global and local color features for image retrieval. *10th International Conference on Electrical Engineering, Computing Science and Automatic Control (CCE)*, IEEE, 2013, pp. 233–238.

Huang R, Lang F, Dong S. Color image retrieval based on moments and DT-CWT. *2011 International Conference on Multimedia Technology (ICMT)*, IEEE, 2011, pp. 2998–3001.

Liu Wei. Research on some key techniques of content-based image retrieval. *Journal of Jilin University*, 2010.

Peiyuan Shi, Tingquan Deng. Fuzzy identification method of color and its application in image retrieval. *Computer engineering and Applications*, 2013, 49(18): 138–141.

Phan R, Androutsos D. Content-based retrieval of logo and trademarks in unconstrained color image databases using color edge gradient co occurrence histograms. *Computer Vision and Image Understanding*, 2010, 114:66–84.

Ying Zhao. Image retrieval system based on color distribution. *Micro Computer Information*, 2007, (18): 266–268.

Information Technology – Wan et al. (Eds)
© 2015 Taylor & Francis Group, London, ISBN 978-1-138-02785-5

A design of service-oriented gateway layer for middleware of internet of things

Jian-dong Yang
Department of Electronic Information and Control Engineering,
Beijing University of Technology, Beijing, China
CAPINFO Company Limited, Beijing, China

Pu Wang, Hui Zhao, Nong Si & Hong-bo Yu
Department of Electronic Information and Control Engineering,
Beijing University of Technology, Beijing, China

ABSTRACT: In this paper, a service-oriented middleware gateway design is proposed, which sets up the application route and control management through the identification of sensors, receives the data that send from perception layer, conducts the data cleaning and event analysis, and then transfers the data or event information that has been cleaned and filtered to the upper layers, and constructs a transparent and standard model of conversation between perception device components. The underlying device unification management is described from device management, engine operation, logic engine, event processing engines, and other technical aspects in detail, so as to when changes occur in the sensing devices or backend database software, the application client could handle the change without modifying or do simple modification, which will save the maintenance complexity problem of the many-to-many connection and solve the unified standard problem between numerous products.

1 INTRODUCTION

Internet of Things (IOT) is the network that achieves connection between any objects to carry out exchange of information and communication, in order to achieve intelligent identification, positioning, tracking, monitoring, and management of the objects based on agreed protocol by radio frequency identification devices (RFID), infrared sensors, global positioning systems, laser scanners, and other information-sensing devices (Hua-jun & Chuan-qing 2010). IOT is the latest product of information and communication technology, and the core area of the wave of a new round of information industry triggered following the computer, Internet, and mobile communication, and it has become the critical point and reached a commanding height of a new round of international information and technological competition.

The application areas of IOT can fully cover the municipal administration, public security, intelligent transportation, energy utility, smart home, finance and commerce, healthcare, industry and manufacture, agriculture, ecological environment, and other aspects of municipal administration and public management. The large and urgent demand for IOT applications will bring opportunities and strong demands for application propped-up platform of IOT,

which can achieve data transmission, filtering, and conversion of data format between a variety of tags and sensors, etc. The sensing devices and application systems have unified information access standards and have the ability for authentication and secure transmission of sense information (Hui, Zhi-gang, Shu-quan, & Zhuo 2013). The middleware of IOT is the key to its application propped-up platform and plays an intermediary role between sensing elements and applications. In the case of changes occuring in the database software or backend for storage of information of the sensing elements, such as the increase in the application that is replaced by other types of increase in software or types of read–write sensing elements, etc., the application terminal does not need modifications or can be handled through simple modifications, removing maintenance complexity of many–many connection. The module design of middleware of IOT is mainly divided into terminal device module design of perception layer, module design of middleware gateway layer, and module design of server terminal, among which the device driver of perception layer terminal device is the key to achieve standardization of the design, and the module design of middleware gateway layer is the core of operation and work of system.

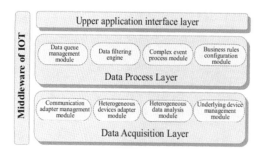

Figure 1. Application middleware for IOT hierarchical architecture diagram.

A service-oriented middleware gateway design is proposed in this paper, which sets up the application route and control management through the identification of sensors, receives the data that send from perception layer, conducts the data cleaning and event analysis, then transfers the data or event information that has been cleaned and filtered to the upper layers, and constructs a transparent and standard model of conversation between perception device components.

2 DESIGN OF APPLICATION MODEL FOR MIDDLEWARE OF IOT

The application middleware of IOT is divided into three layers from bottom to top, namely, data acquisition layer, data process layer, and upper application interface layer; the schematic diagram of conceptual model (Qi-Bo, Jie, & Shan 2010) is as shown in Figure 1.

(1) *Data acquisition layer*: Data acquisition layer mainly provides functions such as adaptation, management, and analysis of heterogeneous data for different underlying data acquisition devices. Wherein the communication adapter management module provides support for different data transmission ways of different acquisition devices; heterogeneous devices adapter module provides drive function identification for different types of data acquisition devices; heterogeneous data analysis module provides resolution services for labeling raw data generated by a variety of acquisition devices and converts it into the unified metadata of IOT and then transfers it to the data processing layer for processing; underlying devices management module provides status inquiry and control management, etc., functions for a variety of acquisition devices.

(2) *Data processing layer*: Process the metadata of IOT that is sent to the data acquisition layer to generate business event data, which can be directly used by the upper layer application in accordance with the preconfigured business rules,

and submit it for using in the upper application. Wherein the data queue management module provides caching services for the acquired data; the data filtering engine matches per event rule by the initial data layer that carries out redundant operation for the initial data, which is sent from the data acquisition layer in accordance with the matching rules of metaevents and generate metaevent queue; complex event process module can detect and handle complex events using the method based on the finite automation and active instances stack and inquiry business rules; business rules configuration module provides flexible business configuration management mechanism to achieve configurability and scalability for application middleware of IOT.

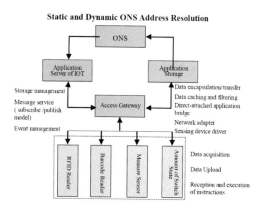

Figure 2. Technology implementation framework structure of middleware for IOT.

(3) *Upper application interface layer*: Provide unified interface support for the upper application system, such as message management, event management, business rule configuration and application management, etc., and shield the coupling dependence of external application services produced by the internal changes of application middleware of IOT.

The technology of middleware for IOT can achieve data acquisition of various sensing devices through the gateway layer module and integrate it into the gateway after processing, and carry out data encapsulation through the gateway and pass it to the upper application (Villanueva & David Villa 2012) as shown in Figure 2.

3 MODULE DESIGN OF GATEWAY LAYER OF MIDDLEWARE

Module of gateway layer of middleware is the key module for the application middleware of IOT, which

18

not only sends the data from the server terminal to the perception layer for interaction, but also transfers the data acquired from the perception layer to the server terminal. The operation mechanism data transmission of gateway layer of middleware is crucially important, among which the whole gateway application is designed based on the event-driven, and the event-driven is the key for the whole data acquisition and transmission, while the design of device management is the key to achieve unified management of the terminal devices, and the device driver provides unified device interfaces and devices IP-based standards. Therefore, the module design of gateway layer of middleware focuses on event process engine and operation engine; see the device management driver design in Figure 3:

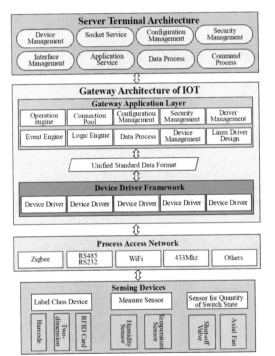

Figure 3. Gateway layer architecture of middleware for IOT.

3.1 Device management

Device management is the management of the I/O system for computer, and its main function is (Jie 2011)

(1) Select and assign I/O devices for data transmission.
(2) Control the exchanging data between I/O devices and CPU (or memory).

(3) Provide users with a friendly interface to separate the users and features of device hardware, so that the users does not need to involve specific device in preparing the application programs, and the work of device is controlled by the system according to user's requirements. Furthermore, this interface also provides an entrance connected to the core of system for the new additional user's device, allowing users to develop new device management programs.

(4) Improve the degree of parallel operation between device and device, CPU and device, and process and process to make optimal efficiency of the operation system.

Device controller is the interface between CPU and I/O device, which receives commands sent from the CPU and controls the work of I/O device. Device controller is an addressable device, it has only an unique device address in case of controlling only one device; whereas it should have more than one device addresses in case of controlling multiple devices, so that each address corresponds to one device. Device controller consists of three parts, which is shown in Figure 4.

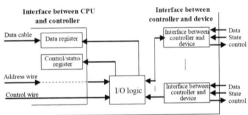

Figure 4. Composition of device controller.

- Interface between device controller and processor.
- Interface between device controller and device.
- I/O logic.

The gateway device management of IOT is the management of all devices within the gateway of IOT, including gateway devices, short-range access network devices, and sensing devices. Device management consists of device address management, device property management, and management of device registration and cancellation.

3.2 Interface management

Interface management can be achieved by setting I/O channel, which is set to make some original tasks of I/O processed by the CPU transfer to the channel to bear, so the CPU is released from the complex I/O tasks.

After setting the channel, CPU only needs to simply send an I/O command to the channel. After receiving the command, the channel will pick up the channel program to be executed from the memory and then execute the channel program. Only when the channel completes the required I/O tasks it can send an interrupt signal to the CPU.

In fact, the I/O channel is a special processor, which has the ability to execute I/O commands, and control I/O operations by executing the channel (I/O) programs. The channel has two basic types: selector channel and multiplexer channel.

I/O control mode is divided into program I/O mode, interrupt-driven I/O control mode and DMA control mode.

The interface management of gateway layer module adopts multichannel I/O system and interrupt-driven I/O control mode, so that the process can achieve parallel operation of the CPU and I/O devices in case of starting an I/O device.

3.3 Logic process engine

The main function of logic process engine is that it triggers other events of devices after the logic process, according to the data and other factors, when a device event is triggered, the logic engine will process associated logic of devices according to the preconfigured logic flow in the configuration files.

The logic flow can be configured based on business and store the logical flow in the configuration files in JSON format. The format is logic ID = {logical judgment conditions, trigger device address ...}.

The logic process of gateway of IOT is mainly for the related logic process of devices within the gateway; the complex logic process and logic process between gateways are handed over to the server terminal application layer for processing. See the logic process flow of gateway for IOT (Hui-li 2007) in Figure 5.

3.4 Operation engine

Operation engine is the core module for the operations of gateway layer module of IOT; the entire application middleware is powered by event-driven, and the task process of operation engine is the core operating system. The technical features of operation engine are

- Multithreading parallel process.
- Multitask scheduling process.
- Priority process.
- Automatic retractable thread pool.
- Interface butt.

Operation engine needs to schedule and process tasks produced by each module, specifically including event process tasks, data process tasks, logic process tasks, data transmission tasks, command process tasks, and configuration management tasks, etc.

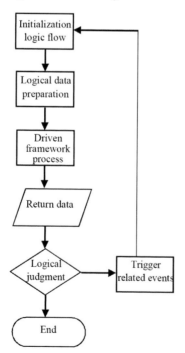

Figure 5. Process flow of logic engine.

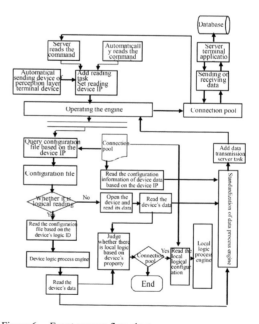

Figure 6. Event process flow chart.

20

3.5 *Event process engine*

The event is mainly divided into three types, namely, command type, timing type, and device type.

The former two types of events are similar, but have different triggering modes, although they have the same process mode. See the event process flow (Sheng-hui & Yufan 2000) in Figure 6.

The flow chart contains the entire process of event, which also has more clear description of operation engine, logic engine, data process, and device management, etc., and is the core flow of the entire gateway layer based on event-driven. The event is monitored by different threads and triggers different mechanisms. All event processes are handed over to the operation engine in task mode for unified processing.

According to different data acquisition sources, the event process method is designed into two types: (1) event process of label type of device, the label type of device such as RFID, etc.; (2) event process of measure type of device, the measure type of device such as temperature and humidity sensors and pressure sensors, etc. Based on the request type, the event type of middleware is divided into device request, application request, and setting logical request, etc.

According to the presentation forms of the business data at different stages, the middleware adopts the three layer structure such as device connection layer, event process layer, and application interactive layer, respectively. The core business is the event process, which is across the three layers. The core part of the event process system is the event process engine, and the underlying implementation mechanism adopts the most common finite-state-machine. If it is the label type of acquisition devices, the acquired data need to be filterered; if it is the sensor type of devices, the distributed architecture should be used to carry out filtering, combining, and processing of the event at the sensors network nodes, thus effectively reduces the amount of data transferred in the WSN and accelerates the response speed of the system. The complex event processing framework uses tree topological structure to carry out the hierarchical management of the network, to support the modeling, and to process the hierarchical abstraction for the complex events.

The two core technologies of event process engine are rule definition and event detection. Event process engine is the most basic module of event process system, which mainly performs the following functions:

- Generate pattern matching state machine based on the results of language parsing.
- Carry out merge and optimization for multiple state machines to improve the efficiency of pattern matching.

- Accept input of event stream, and match the event stream, and process after success or failure of the matching.

Provide real-time status inquiry service for the process system of event stream, which can be used when the user needs to inquire the processing state of the current event stream.

4 CONCLUSIONS

The module design of gateway layer of middleware for IOT is the core of work and operation for the middleware system of IOT; the module design of gateway layer of middleware for IOT is through analyzing the characteristics of the lowermost sensing devices, including label type of devices, measure type of sensors, and sensors of the amount of switch state, etc.; give full play to the role of gateway layer module, and the underlying device unification management is described from operation engine, logic engine, collection pool, and event processing engine, etc. of device management, and other technical aspects in detail, so as to when changes occur in the sensing devices or backend database software, the application client could handle the change without modifying or do simple modification, which will save the maintenance complexity problem of the many-to-many connection and solve the unified standard problem between numerous products.

REFERENCES

Hua-jun, L. & L. Chuan-qing. *Internet of Things Technology*. Beijing: Publishing House of Electronics Industry, 2010.

Hui, Y., D. Zhi-gang, Z. Shu-quan, & H. Zhuo. Design and implementation of a service-oriented middleware for iot. *Comp. Appl. Softw.*, 2013 30 (5), 65–67.

Hui-li, Z. Design and implementation of loose coupling web applications model on javaee platform. *Sci. Technol. Inform.*,2007 16, 334–335.

Jie, L. Method to integration of wireless sensor nodes and RFID data in IOT. Electron. Des. Eng., 2011 19 (7), 103–106.

Qi-Bo, S., L. Jie, & L. Shan (2010). IOC: Research overview on its concept, architecture and key technologies. *J. Beijing Univ. Posts and Telecommun.* 33(3), 1–9.

Sheng-hui, G. & S. Yu-fan. Design to information system of data dictionary library. J. Comp., 2000 23 (4), 414–418.

Villanueva, F.J. & F.M. David Villa (2012). Internet of things architecture for a RFID-based product tracking business model. *Sixth International Conference on Innovative Mobile and Internet Services in Ubiquitous Computing 249*, 235–297.

Information Technology – Wan et al. (Eds)
© *2015 Taylor & Francis Group, London, ISBN 978-1-138-02785-5*

A fast algorithm for the 4×4 discrete Krawtchouk transform

J. S. Wu, C.F. Yang & H.Z. Shu
LIST, the Key Laboratory of Computer Network and Information Integration, Southeast University, Nanjing, China
Centre de Recherche en Information Biomédicale Sino-français (CRIBs), Nanjing, China

L. Wang & L. Senhadji
INSERM U 1099, Rennes, France
Laboratoire Traitement du Signal et de l'Image, Université de Rennes 1, Rennes, France

ABSTRACT: A fast algorithm for computing the 4×4 discrete Krawtchouk transform (DKT) is proposed. By exploiting the reduced number of different values of basis functions, the proposed algorithm decreases significantly the number of arithmetic operations. It requires only 8 multiplications, 80 additions and 32 shifts, which saves about 98%, 88% and 83% multiplications compared to the direct method, the method using the properties of the DKT and the method computed by cascaded digital filters, respectively. The proposed algorithm could be used to efficiently compute the building-blocks when the process of the large image size is performed. Experiments are provided to demonstrate the efficiency of the proposed algorithm.

1 INTRODUCTION

The discrete Krawtchouk transform (DKT) is a transform method based on the discrete orthonormal Krawtchouk polynomials, it was first introduced by (Yap et al. 2003) in the field of image analysis. Since then, the DKT has found applications in image analysis and compression (See et al. 2007, Zhu et al. 2007). It was shown that the DKT has better image representation capability than the discrete Tchebichef transform (DTT), the latter transform has been extensively investigated in the recent years (Lee et al. 2001, Nakagaki & Mukundan 2006, Nakagaki & Mukundan 2007, Ishwar et al. 2008). Since the direct computation of the DKT involves a large amount of arithmetic operations, efficient algorithms for its computation are of great importance.

Some fast algorithms for computing the DKT have been reported in the literature. (Yap et al. 2003) used two properties (separability and even symmetry) of Krawtchouk polynomials to reduce the arithmetic complexity of DKT. (Asli & Flusser 2014) proposed a fast algorithm for computing the DKT by cascaded digital filters. (Raj and Venkataramana 2007) proposed a recursive algorithm for the fast computation of inverse DKT based on Clenshaw's formula. As stated in (Nakagaki & Mukundan 2006), although the recursive algorithms (Aburdene et al. 1995, Wang & Wang 2006, Raj & Venkataramana 2007) can be used to compute

various transform sizes, they even take longer time than applying directly for computation the definition with pre-computed discrete orthogonal polynomials. Therefore, more efficient DKT algorithms have to be determined for further improving the speed of image analysis application.

Recently, research effort has been paid to the fast computation of the small fixed 4×4-, 8×8-point discrete orthogonal transforms, for example (Lee & Cho 1992, Nakagaki & Mukundan 2006, Nakagaki & Mukundan 2007, Ishwar et al. 2008, Zhu et al. 2009, Bouguezel et al. 2013, Cintra et al. 2014). (Lee & Cho 1992) proposed a 4×4 DCT algorithm to serve the base case for the recursive computation of two-dimensional (2-D) DCT. Nakagaki and Mukundan derived a fast algorithm for the computation of 4×4 DTT (Nakagaki & Mukundan 2006, Nakagaki & Mukundan 2007). Inspired by the research work presented in (Lee & Cho 1992, Nakagaki & Mukundan 2006, Nakagaki & Mukundan 2007), we propose in this letter a fast 4×4 DKT algorithm which could be used as the building-block for the DKT computation.

The paper is organized as follows. In Sections 2 and 3, we introduce the definition and some properties of the DKT. The proposed algorithm is described in Section 4. The comparative analysis of arithmetic complexity and some experimental results are provided in Section 5. Section 6 concludes the paper.

2 DEFINITION OF DISCRETE KRAWTCHOUK TRANSFORM

For a given square matrix f of dimension N (i.e. a squared image), the discrete Krawtchouk transform of order $n+m; 0 \leq n \leq N-1, 0 \leq m \leq N-1$ is defined as (Yap et al. 2003)

$$Q_{nm} = \sum_{x=0}^{N-1}\sum_{y=0}^{N-1} \bar{K}_n(x; p_1, N-1)\bar{K}_m(y; p_2, N-1)f(x,y), \quad (1)$$

where $\bar{K}_n(x; p_i, N); i = 1, 2; 0 \leq x \leq N-1; 0 < p_i < 1$ are the weighted orthonormal Krawtchouk polynomials, which can be computed by the following recurrence formula (Yap et al. 2003)

$$\bar{K}_{n+1}(x; p, N) = A(Np + n - 2np - x) \\ \times \bar{K}_n(x; p, N) - B\bar{K}_{n-1}(x; p, N), \quad (2)$$

where

$$A = \frac{1}{\sqrt{p(1-p)(N-n)(n+1)}}, B = \sqrt{\frac{n(N-n+1)}{(n+1)(N-n)}}, \quad (3)$$

$$\bar{K}_0(x; p, N) = \sqrt{w(x; p, N)},$$

$$\bar{K}_1(x; p, N) = \left(1 - \frac{x}{pN}\right)\sqrt{\frac{w(x; p, N)}{\rho(1; p, N)}}, \quad (4)$$

$$w(x; p, N) = \binom{N}{x} p^x (1-p)^{N-x}, \quad (5)$$

$$\rho(n; p, N) = (-1)^n \left(\frac{1-p}{p}\right)^n \frac{n!}{(-N)_n},$$

$$(a)_k = a(a+1)(a+2)...(a+k-1) = \frac{\Gamma(a+k)}{\Gamma(a)}. \quad (6)$$

The computation of DKT can be performed directly by the definition in (1), which consists of a doubly nested summation. We call it the direct method (or naïve Method) in the following.

3 COMPUTATION OF DKT USING SEPARABILITY AND EVEN SYMMETRY

The 2-D kernel function of DKT in (1) is separable, the DKT values can thus be computed by row-column algorithm. In most applications, the probability parameters p_1 and p_2 appeared in (1) take value 0.5. This choice corresponds to the case where the origin is centered at the centroid of the image. Other choices of p_1 and p_2 permit us to extract the local information of whole image. For this reason, we take the case $p_1 = p_2 = 0.5$ into consideration in the rest of the paper. For $p_1 = p_2 = 0.5$, the weighted orthonormal

Krawtchouk polynomials satisfy the even symmetry property (Yap et al. 2003)

$$\bar{K}_n(x; p, N-1) = (-1)^n \bar{K}_n(N-1-x; p, N-1) \quad 7)$$

By using the separability and symmetry property, we can compute (1) as follows

$$Q_{nm} = \sum_{x=0}^{N/2-1} \bar{K}_n(x; 1/2, N-1) \\ \times \left(g_m(x) + (-1)^n g_m(N-1-x)\right),$$

$$g_m(x) = \sum_{y=0}^{N/2-1} \bar{K}_m(y; 1/2, N-1) \\ \times \left(f(x,y) + (-1)^m f(x, N-1-y)\right). \quad (8)$$

The above two properties can be used to reduce significantly the computational complexity of DKT compared to the direct method. In the next section, we will use the transform properties to further reduce the arithmetic complexity.

4 FAST 4 × 4 DKT ALGORITHM

Substituting $N = 4$ into (2), we can compute the values of $\bar{K}_n(x; 1/2, 3)$, $n, x = 0, 1, 2, 3$, which are shown in Table 1.

Table 1. Evaluation of the weighted orthonormal Krawtchouk polynomials for $N = 4$

	$x=0$	$x=1$	$x=2$	$x=3$
$\bar{K}_0\left(x; \frac{1}{2}, 3\right)$	$\frac{1}{2\sqrt{2}}$	$\frac{\sqrt{3}}{2\sqrt{2}}$	$\frac{\sqrt{3}}{2\sqrt{2}}$	$\frac{1}{2\sqrt{2}}$
$\bar{K}_1\left(x; \frac{1}{2}, 3\right)$	$\frac{\sqrt{3}}{2\sqrt{2}}$	$\frac{1}{2\sqrt{2}}$	$-\frac{1}{2\sqrt{2}}$	$-\frac{\sqrt{3}}{2\sqrt{2}}$
$\bar{K}_2\left(x; \frac{1}{2}, 3\right)$	$\frac{\sqrt{3}}{2\sqrt{2}}$	$-\frac{1}{2\sqrt{2}}$	$-\frac{1}{2\sqrt{2}}$	$\frac{\sqrt{3}}{2\sqrt{2}}$
$\bar{K}_3\left(x; \frac{1}{2}, 3\right)$	$\frac{1}{2\sqrt{2}}$	$-\frac{\sqrt{3}}{2\sqrt{2}}$	$\frac{\sqrt{3}}{2\sqrt{2}}$	$-\frac{1}{2\sqrt{2}}$

Since $\bar{K}_n(x; 1/2, 3)$ has two different values, there are only three different values for the product $\bar{K}_n(x; 1/2, 3)\bar{K}_m(y; 1/2, 3)$ in (1) as follows.

$$A_0 = \bar{K}_0(0; 1/2, 3)\bar{K}_0(0; 1/2, 3) = 1/8,$$
$$A_1 = \bar{K}_0(0; 1/2, 3)\bar{K}_0(1; 1/2, 3) = \sqrt{3}/8, \quad (9)$$
$$A_2 = \bar{K}_0(1; 1/2, 3)\bar{K}_0(1; 1/2, 3) = 3/8.$$

That is to say, that the number of different values of basis functions reduces from 16 to 3, which can significantly decrease the arithmetic complexity.

When $N = 4$, (1) becomes

$$Q_{nm} = \sum_{x=0}^{3}\sum_{y=0}^{3} \overline{K}_n(x;1/2,3)\overline{K}_m(y;1/2,3)f(x,y), \tag{10}$$
$$n,m = 0,1,2,3.$$

In order to facilitate the presentation, we use the notation $f_{xy} = f(x, y)$, x, $y = 0$, 1, 2, 3. For the computation of (10), we have

$$Q_{00} = \frac{\sqrt{3}}{8}[f_{01}+f_{02}+f_{10}+f_{13}+f_{20}+f_{23}+f_{31}+f_{32}]$$
$$+\frac{1}{8}[f_{00}+f_{03}+f_{30}+f_{33}]+\frac{3}{8}[f_{11}+f_{12}+f_{21}+f_{22}]$$
$$=\frac{1}{4}\left[\frac{\sqrt{3}}{2}(f_{01}+f_{02}+f_{10}+f_{13}+f_{23}+f_{31}+f_{32})\right. \tag{11}$$
$$+\frac{1}{2}(f_{00}+f_{03}+f_{30}+f_{33}+f_{11}+f_{12}+f_{21}+f_{22})$$
$$\left.+(f_{11}+f_{12}+f_{21}+f_{22})\right],$$

$$Q_{10} = \frac{1}{4}\left[\frac{\sqrt{3}}{2}(f_{00}+f_{03}+f_{11}+f_{12}-f_{21}-f_{22}-f_{30}-f_{33})\right.$$
$$+\frac{1}{2}(f_{10}+f_{13}-f_{20}-f_{23}+f_{01}+f_{02}-f_{31}-f_{32}) \tag{12}$$
$$\left.+(f_{01}+f_{02}-f_{31}-f_{32})\right],$$

$$Q_{20} = \frac{1}{4}\left[\frac{\sqrt{3}}{2}(f_{00}+f_{03}+f_{30}+f_{33}-f_{11}-f_{12}-f_{21}-f_{22})\right.$$
$$-\frac{1}{2}(f_{10}+f_{13}+f_{20}+f_{23}-f_{01}-f_{02}-f_{31}-f_{32}) \tag{13}$$
$$\left.+(f_{01}+f_{02}+f_{31}+f_{32})\right],$$

$$Q_{30} = \frac{1}{4}\left[\frac{\sqrt{3}}{2}(f_{01}+f_{02}-f_{10}-f_{13}+f_{20}+f_{23}-f_{31}-f_{32})\right.$$
$$+\frac{1}{2}(f_{00}+f_{03}-f_{30}-f_{33}-f_{11}-f_{12}+f_{21}+f_{22}) \tag{14}$$
$$\left.-(f_{11}+f_{12}-f_{21}-f_{22})\right],$$

$$Q_{01} = \frac{1}{4}\left[\frac{\sqrt{3}}{2}(f_{00}-f_{03}+f_{11}-f_{12}+f_{21}-f_{22}+f_{30}-f_{33})\right.$$
$$+\frac{1}{2}(f_{01}-f_{02}+f_{31}-f_{32}+f_{10}-f_{13}+f_{20}-f_{23}) \tag{15}$$
$$\left.+(f_{10}-f_{13}+f_{20}-f_{23})\right],$$

$$Q_{11} = \frac{1}{4}\left[\frac{\sqrt{3}}{2}(f_{01}-f_{02}+f_{10}-f_{13}-f_{20}+f_{23}-f_{31}+f_{32})\right.$$
$$+\frac{1}{2}(f_{11}-f_{12}-f_{21}+f_{22}+f_{00}-f_{03}-f_{30}+f_{33}) \tag{16}$$
$$\left.+(f_{00}-f_{03}-f_{30}+f_{33})\right],$$

$$Q_{21} = \frac{1}{4}\left[\frac{\sqrt{3}}{2}(f_{01}-f_{02}-f_{10}+f_{13}-f_{20}+f_{23}+f_{31}-f_{32})\right.$$
$$-\frac{1}{2}(f_{11}-f_{12}+f_{21}-f_{22}-f_{00}+f_{03}-f_{30}+f_{33}) \tag{17}$$
$$\left.+(f_{00}-f_{03}+f_{30}-f_{33})\right],$$

$$Q_{31} = \frac{1}{4}\left[\frac{\sqrt{3}}{2}(f_{00}-f_{03}-f_{30}+f_{33}-f_{11}+f_{12}+f_{21}-f_{22})\right.$$
$$+\frac{1}{2}(f_{01}-f_{02}-f_{31}+f_{32}-f_{10}+f_{13}+f_{20}-f_{23}) \tag{18}$$
$$\left.-(f_{10}-f_{13}-f_{20}+f_{23})\right],$$

$$Q_{02} = \frac{1}{4}\left[\frac{\sqrt{3}}{2}(f_{00}+f_{03}+f_{30}+f_{33}-f_{11}-f_{12}-f_{21}-f_{22})\right.$$
$$-\frac{1}{2}(f_{01}+f_{02}+f_{31}+f_{32}-f_{10}-f_{13}-f_{20}-f_{23}) \tag{19}$$
$$\left.+(f_{10}+f_{13}+f_{20}+f_{23})\right],$$

$$Q_{12} = -\frac{1}{4}\left[\frac{\sqrt{3}}{2}(f_{01}+f_{02}+f_{20}+f_{23}-f_{10}-f_{13}-f_{31}-f_{32})\right.$$
$$+\frac{1}{2}(f_{11}+f_{12}-f_{21}-f_{22}-f_{00}-f_{03}+f_{30}+f_{33}) \tag{20}$$
$$\left.-(f_{00}+f_{03}-f_{30}-f_{33})\right],$$

$$Q_{22} = -\frac{1}{4}\left[\frac{\sqrt{3}}{2}(f_{01}+f_{02}+f_{10}+f_{13}+f_{20}+f_{23}+f_{31}+f_{32})\right.$$
$$-\frac{1}{2}(f_{11}+f_{12}+f_{21}+f_{22}+f_{00}+f_{03}+f_{30}+f_{33}) \tag{21}$$
$$\left.-(f_{00}+f_{03}+f_{30}+f_{33})\right],$$

$$Q_{32} = \frac{1}{4}\left[\frac{\sqrt{3}}{2}(f_{00}+f_{03}-f_{30}-f_{33}+f_{11}+f_{12}-f_{21}-f_{22})\right.$$
$$-\frac{1}{2}(f_{01}+f_{02}-f_{31}-f_{32}+f_{10}+f_{13}-f_{20}-f_{23}) \tag{22}$$
$$\left.-(f_{10}+f_{13}-f_{20}-f_{23})\right],$$

$$Q_{03} = -\frac{1}{4}\left[\frac{\sqrt{3}}{2}(f_{01}-f_{02}+f_{31}-f_{32}-f_{10}+f_{13}-f_{20}+f_{23})\right.$$
$$-\frac{1}{2}(f_{00}-f_{03}+f_{30}-f_{33}-f_{11}+f_{12}-f_{21}+f_{22}) \tag{23}$$
$$\left.+(f_{11}-f_{12}+f_{21}-f_{22})\right],$$

$$Q_{13} = \frac{1}{4}\left[\frac{\sqrt{3}}{2}(f_{00}-f_{03}-f_{11}+f_{12}+f_{21}-f_{22}-f_{30}+f_{33})\right.$$
$$+\frac{1}{2}(f_{10}-f_{13}-f_{20}+f_{23}-f_{01}+f_{02}+f_{31}-f_{32}) \tag{24}$$
$$\left.-(f_{11}-f_{12}+f_{21}-f_{22})\right],$$

$$Q_{23} = \frac{1}{4}\left[\frac{\sqrt{3}}{2}(f_{00}-f_{03}+f_{30}-f_{33}+f_{11}-f_{12}+f_{21}-f_{22})\right.$$
$$-\frac{1}{2}(f_{10}-f_{13}+f_{20}-f_{23}+f_{01}-f_{02}+f_{31}-f_{32}) \tag{25}$$
$$\left.-(f_{01}-f_{02}+f_{31}-f_{32})\right],$$

$$Q_{33} = -\frac{1}{4}\left[\frac{\sqrt{3}}{2}\left(f_{01} - f_{02} + f_{10} - f_{13} - f_{20} + f_{23} - f_{31} + f_{32}\right)\right.$$
$$-\frac{1}{2}\left(f_{00} - f_{03} - f_{30} + f_{33} + f_{11} - f_{12} - f_{21} + f_{22}\right) \qquad (26)$$
$$\left.-\left(f_{11} - f_{12} - f_{21} + f_{22}\right)\right].$$

Note that in (11)-(26), we use the identity $3/8 = 1/8 + 1/4$, which can be implemented by shifts and addition. The other DKT values can be calculated in a similar way. They are listed below

$$Q_{00} = \left(E_0 + C_5\right)/4, \qquad Q_{10} = \left(E_1 + \overline{C_4}\right)/4,$$

$$Q_{20} = -\left(\overline{E_4} - C_4\right)/4, \qquad Q_{30} = \left(\overline{E_5} - \overline{C_5}\right)/4,$$

$$Q_{01} = \left(E_3 + C_3\right)/4, \qquad Q_{11} = \left(E_2 + \overline{C_2}\right)/4,$$

$$Q_{21} = \left(\overline{E_7} + C_2\right)/4, \qquad Q_{31} = -\left(\overline{E_6} + \overline{C_3}\right)/4,$$

$$Q_{02} = \left(E_4 + C_1\right)/4, \qquad Q_{12} = \left(E_5 + \overline{C_0}\right)/4,$$

$$Q_{22} = \left(\overline{E_0} + C_0\right)/4, \qquad Q_{32} = -\left(\overline{E_1} + \overline{C_1}\right)/4, \qquad (27)$$

$$Q_{03} = \left(E_7 - C_7\right)/4, \qquad Q_{13} = \left(E_6 - \overline{C_6}\right)/4,$$

$$Q_{23} = -\left(\overline{E_3} + C_6\right)/4, \quad Q_{33} = \left(\overline{E_2} + \overline{C_7}\right)/4,$$

where

$$\left(B_i, \overline{B_i}, B_{i+4}, \overline{B_{i+4}}\right)^T = \left(I_2 \otimes H_2\right)\left(f_{i0}, f_{i3}, f_{i1}, f_{i2}\right)^T, \qquad (28)$$
$$i = 0,1,2,3.$$

$$\left(C_i, \overline{C_i}, C_{i+2}, \overline{C_{i+2}}, C_{i+4}, \overline{C_{i+4}}, C_{i+6}, \overline{C_{i+6}}\right)^T = \left(I_4 \otimes H_2\right)$$
$$\times\left(B_i, B_{3-i}, \overline{B_i}, \overline{B_{3-i}}, B_{i+4}, B_{7-i}, \overline{B_{i+4}}, \overline{B_{7-i}}\right)^T, i = 0,1. \qquad (29)$$

$$\left(D_i, \overline{D_{1-i}}, D_{3-i}, \overline{D_{i+2}}, D_{5-i}, \overline{D_{i+4}}, D_{i+6}, \overline{D_{7-i}}\right)^T = \left(I_4 \otimes H_2\right)$$
$$\times\left(C_i, C_{5-i}, \overline{C_i}, \overline{C_{5-i}}, C_{i+2}, C_{7-i}, \overline{C_{i+2}}, \overline{C_{7-i}}\right)^T, i = 0,1. \qquad (30)$$

$$\left(E_4, \overline{E_4}, E_5, \overline{E_5}, E_7, \overline{E_7}, E_6, \overline{E_6}\right)^T = \left(I_4 \otimes P_2\right)$$
$$\times\left(\overline{D_0}, \overline{D_1}, \overline{D_2}, \overline{D_3}, \overline{D_4}, \overline{D_5}, \overline{D_6}, \overline{D_7}\right)^T, \qquad (31)$$

$$\left(E_0, \overline{E_0}, E_1, \overline{E_1}, E_3, \overline{E_3}, E_2, \overline{E_2}\right)^T = \left(I_4 \otimes P_2\right)$$
$$\times\left(D_0, D_1, D_2, D_3, D_4, D_5, D_6, D_7\right)^T, \qquad (32)$$

Here T denotes the transposition, "$\otimes$" is the Kronecker product (Granata et al. 1992), I_n is the identity matrix of order n. P_2 and H_2 are respectively given by

$$P_2 = \begin{bmatrix} 1/2 & \sqrt{3}/2 \\ 1/2 & -\sqrt{3}/2 \end{bmatrix}, H_2 = \begin{bmatrix} 1 & 1 \\ 1 & -1 \end{bmatrix}. \qquad (33)$$

The signal graph of the proposed algorithm is shown in Figure 1.

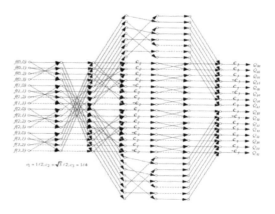

Figure 1. Signal flow graph of the 4×4 Discrete Krawtchouk Transform. Full lines represent transfer factor +1, broken lines represent transfer factor −1. $f(i, j)$, i, j=0,1,2,3, denotes image input and Q_{ij}, i, j=0,1,2,3, denotes coefficients of DKT.

5 COMPARATIVE ANALYSIS

In this section, we consider the computational complexity and the experimental analysis of the proposed 4×4 DKT algorithm and compare it with some existing algorithms.

5.1 Computational complexity analysis

The computational complexity of three methods is analyzed as follows.

1 The direct method requires $2N^4$=512 multiplications and $N^2(-1)(+1)$=240 additions, respectively.
2 The algorithm using the separability and even symmetry needs $N^3 = 64$ multiplications and $N^2(2-2) = 96$ additions, respectively.
3 It can be easily seen from (27) to (32) that the proposed algorithm requires 8 multiplications, 80 additions and 32 shifts.

The comparison results of the above three algorithms are shown in Table 2. It can be seen from this table that the proposed method can save the number of multiplications about 98%, 88% and 83% compared to the naïve method, the method using separability and even symmetry, and computed by cascaded digital filters, respectively.

5.2 Experimental analysis

In this subsection, two experiments are conducted to illustrate the efficiency of the proposed 4×4 DKT algorithm.

First, the gray-scale image 'Lena' (See Figure 2) was transformed using the proposed method.

Then, the image was reconstructed with different number of selected components along the zig-zag order used in JPEG (Wallace 1991) using the inverse

Table 2. The comparison results of the computational complexity of the proposed method compared to Naïve method and the method using separability and symmetry for 4×4 DKT. "*" denotes that, for the algorithm of (Asli & Flusser 2014), the computation of normalization factors does not take into consideration.

	Multiplications	Additions	Shifts	Saving1	Saving2
Naïve Method	512	240	0	98%	67%
Separability & Symmetry	64	96	0	88%	17%
Asli and Flusser *	48	48	0	83%	-40%
Proposed method	8	80	32	0%	0%

Figure 2. The original gray-scale image 'Lena' of size 256×256.

DKT transform. The results are shown in Figure 3. It can be seen from this figure that the reconstructed images are similar to the original image with some minor errors introduced by the zig-zag component selection.

Second, we compare the proposed algorithm with the aforementioned two methods for the 2-D DKT in terms of computer run times which tested on 128×128, 256×256, 512×512, 768×768, and 1024×1024 gray-scale 'Lena' image. The algorithms have been implemented in "C" language on a PC Intel Core2 Duo CPU with speed of 2200MHz and 3072 MB RAM. The run-times of these algorithms have been calculated using Visual C++ (VC++) Version (9). The comparison results are shown in Figure 4. It can be observed that the proposed method takes about two fifths of the time taken by the direct method and takes two thirds of the time taken by the method using the separability and symmetry properties. The results are in agreement with the arithmetic operations shown in Table 2.

Figure 3. The first row: reconstructed images of 'Lena' with 16 components, 8 components, 4 components, 2 components and 1 component from left to right; The second row: the errors between the original image and the reconstructed images.

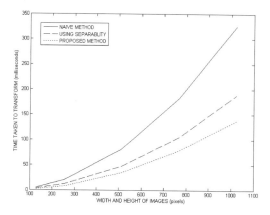

Figure 4. Time in milliseconds taken to transform image by the naïve method, the method using separability and symmetry, and the proposed method.

6 CONCLUSIONS

We have presented a fast algorithm for the efficient computation of 4×4 discrete Krawtchouk transform. By taking the advantage of a reduced number of different values of basis functions, the arithmetic complexity is reduced by the proposed algorithm compared to the direct method, the method using both the separability and even symmetry properties, and the method using cascaded digital filters. The proposed algorithm could be used to efficiently compute the base blocks when the process of the large image size is performed.

ACKNOWLEDGMENT

This work was supported by the National Basic Research Program of China under Grant 2011CB707904, by the NSFC under Grants 61201344, 61271312, 61401085, 31400842, 11301074, and 81301298, and by the SRFDP under Grants 20110092110023 and 20120092120036, the Project-sponsored by SRF for ROCS, SEM, and by Natural Science Foundation of Jiangsu Province under Grant BK2012329 and by Qing Lan Project.

REFERENCES

Aburdene, M.F., Zheng, J & Kozick R.J. 1995. Computation of discrete cosine transform using Clenshaw's recurrence formula. *IEEE Signal Process lett* 2(8): 155–156.
Asli B.H.S. & Flusser, J. 2014. Fast computation of Krawtchouk moments. *Information Science* 288: 73–86.
Bouguezel, S., Ahmad, M. O. & Swamy, M. N. S. 2013. Binary discrete cosine and Hartley transforms. *IEEE Trans Circuits Syst-I: Regular Papers* 60(4): 989–1002.
Cintra, R. J., Bayer, F. M. & Tablada, C. J. 2014. Low-complexity 8-point DCT approximations based on integer functions. *Signal Process* 99: 201–214.
Granata, J., Conner, M. & Tolimieri, R. 1992. The tensor product: A mathematical programming language for FFT's and other fast DSP operations. *IEEE Signal Process Mag* 9(1): 40–48.
Ishwar, S., Meher, P.K. & Swamy, M.N.S. 2008. Discrete Tchebichef transform-A fast 4×4 algorithm and its application in image/video compression. *IEEE ISCAS*. 260–263.
Lee, P.A., Mukundan, R. & Ong S.H. 2001. Image analysis by Tchebichef moments. *IEEE Trans Image Process* 10(9): 1357–1364.
Lee, S. U. & Cho, N. I. 1992. A fast 4×4 DCT algorithm for the recursive 2-D DCT. *IEEE Trans Signal Process* 40(9): 2166–2173.
Nakagaki, K. & Mukundan, R. 2006. A fast discrete Tchebichef Transform algorithm for image compression. *COSC460 Honours Report* 14(10): 1–29.
Nakagaki, K. & Mukundan, R. 2007. A fast 4×4 forward discrete Tchebichef transform algorithm. *IEEE Signal Process lett* 14(10): 684–687.
Raj, P.A. & Venkataramana, A. 2007. Fast computation of inverse Krawtchouk moment transform using Clenshaw's recurrence formula. *IEEE ICIP* 4: 37–40.
See, K.W., Loke, K.S., Lee, P.A. & Loe, K.F. 2007. Image reconstruction using various discrete orthogonal polynomials in comparison with DCT. *App Math Comput* 193(2): 346–359.
Wallace, G. K. 1991. The JPEG still picture compression standard. *Commun ACM* 34(4): 30–44.
Wang, G.B. & Wang, S.H. 2006. Recursive computation of Tchebichef moment and its inverse transform. *Pattern Recogn* 39(1): 47–56.
Yap, P.T., Raveendran, P. & Ong, S.H. 2003. Image analysis by Krawtchouk moments. *IEEE Trans Image Process* 12(11): 1367–1377.
Zhu, H.Q., Shu, H.Z., Liang, J., Luo, L.M. & Coatrieux, J.-L. 2007. Image analysis by discrete orthogonal Racah moments. *Signal Process* 87(4): 687–708.
Zhu, P. P., Liu, J. G. & Dai S. D. 2009. Fixed-point IDCT without multiplications based on B.G. Lee's algorithm. *Digital Signal Process* 19(4): 770–777.

Information Technology – Wan et al. (Eds)
© 2015 Taylor & Francis Group, London, ISBN 978-1-138-02785-5

A fast BEMD algorithm based on multi-scale extrema

D. Yang & X.T. Wang
Navigation Department of Dalian Navy Academy, Dalian, China

G.L. Xu
Ocean Department of Dalian Navy Academy, Dalian, China.

ABSTRACT: FABEMD is a BEMD method state of the art. This paper proposes a new algorithm to increase the decomposition Adaptivity and efficiency of FABEMD further. For this purpose, we investigate the principle of FABEMD and 1-D multi-scale extrema firstly, and then give the definition and an established method of 2-D multi-scale local extrema binary tree of an image. After that a new approach based on multi-scale extrema to determine window sizes to order-statistics and smoothing filters is presented. Lastly a improved FABEMD method is introduced. Furthermore, the experimental results demonstrate the advantages of the presented method in both decomposition Adaptivity and efficiency.

1 INTRODUCTION

Empirical mode decomposition (EMD), originally developed by Huang et al. (1998), is a powerful tool for decomposing nonlinear and nonstationary signals. Since this decomposition technique has been extended to analyze two-dimensional data by Song et al. (2001), bidimensional empirical mode de-composition(BEMD) developed rapidly (Han 2002, Nunes 2003, Liu 2005, among others). Some well-known methods are: BEMD of applying 1-D EMD directly along rows and then along columns of image (Han 2002), BEMD of considering direction of image texture, that is DEMD (Liu 2005), BEMD based on radial basis function interpolation (Nunes 2003); BEMD based on Delaunay triangulation and on piecewise cubic polynomial interpolation (Damerval 2005), BEMD based on finite-element technology (Xu 2006), neighborhood limited BEMD (Xu 2006), assisted signal BEMD (Xu 2011), fast and adaptive BEMD, that is FABEMD (Sharif 2008), BEMD based on PDE (Niang 2010), and so on. Among them, FABEMD is accepted as state of the art in both decomposition rate and effect, and has been used for image fusion (Ahmed 2010), image enhancement (Yang 2011, Zhang 2012), image registration and motion estimation (Mahraz 2012, Riffi 2013) etc.

However, we find efficiencies and adaptivity of window size determination for filters used in envelope estimation of FABEMD is fairly low. In this paper, we give a new algorithm, termed improved fast and adaptive bidimensional empirical mode decomposition (short for IFABEMD) that enhances the decomposition adaptivity and efficiency of FABEMD further.

The rest of this paper is organized as follows. In the next section, we investigate the principle of FBEMD and binary tree of 1-D multi-scale extrema (MSEx) briefly. In section 3, our method is introduced in detail. In section 4, the experimental results of our method with the other two fast BEMD approaches are presented. Finally, we make concluding remarks in Section 5.

2 FABEMD AND BINARY TREE OF MSEX OVERVIEW

2.1 *Principle of FABEMD*

Let I is an 2-D image signal, $BIMF_i$ is the ith bi-dimensional intrinsic mode function or candidate, I_{res} is the residue, R_i is an intermediate variable. The processing of FABEMD is shown in Figure 1.

We can see that, the operations to find local ex-trema, calculate adjacent extrema distance arrays and determinate window size of filters will repeat in every sifting processing and iteration processing. They are redundant and waste of time. In addition, the method to determinate window size with some extrema distance is inappropriate for 2-D signals.

In our method, the redundancies will be avoided, and a more meaningful approach is used to determinate window size of filters.

2.2 Binary tree of 1-D MSEx

Binary tree of Multi-scale extrema and its properties of 1-D signals has been studied and used for fast trend extraction in recent literatures (Yang 2013a, Yang 2013b), showing significant effects in fast analysis of 1-D signals.

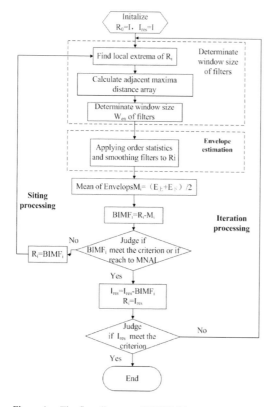

Figure 1.   The flow diagram of FABEMD.

For a given multi-component signal $f(t)$ in support $[0,T]$, find out all local maximum and local minimum to form the first level extrema set $S_1=\{S_{<1,1>},S_{<1,2>}\}$.

Take $S_{<1,1>}$ and $S_{<1,2>}$ as new series separately and find out all local maximum and local minimum of them to form sets $S_{<2,1>}$, $S_{<2,2>}$, $S_{<2,3>}$ and $S_{<2,4>}$, then the second level extrema set $S2=\{\ S_{<2,1>}, S_{<2,2>}, S_{<2,3>}, S_{<2,4>}\}$ is obtained.

Similarly, take the same operation as the above iteratively until elements in some extrema subset $S_{<i,j>}$ is less than 3. This i just equals n namely the highest level of MSEx, and then a binary tree of MSEx of $f(t)$ has been set up as Figure 2.

The characteristic scale defined by the distance between the extrema is proportion to the level number of multi-level extrema, therefore multi-level extrema was referred to as multi-scale extrema.

3   IFABEMD BASED ON 2-D MSEX

3.1   Principle of IFABEMD

To know intuitively, extrema density is more suitable to reflect the space frequency of an image than extrema distance, therefore we obtain the window size of filters w_{en} from exrema density ρ of the image by the following formula:

$$w_{en_i} = \sqrt{\frac{NumP}{\rho}} \qquad (1)$$

where $NumP$ is the pixels number of the image.

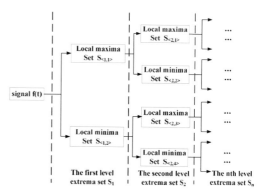

Figure 2.   Binary tree of multi-level signal extrema.

In addition, if we can obtain scale information at all levels for a composite signal once for all, without decomposite from small scale to large scale iteratively, it will save much time and has more flexibility. Binary tree of 1-D MSEx proposed a powerful tool for fast multi-scale analysis of 1-D signals. Next we extend this tool to 2-D cases.

3.2   Binary tree and its establishment of 2-D MSEx

Reference to steps in section 2.2, replace 1-D signal $f(t)$ by 2-D signal I, and find extrema in two dimensions, we can obtain the definition of the 2-D MSEx binary tree easily. An optional implementation to get 2-D MSEx of I is comparing in 8-neighborhood to form S_1, and to form higher sets by comparing in Vironi polygons constituted by extrema. However, it's not only slow but also prone to extrema sparse quickly. Here, we present a practical method:

Firstly, expand the 2-D signal I to 1-D signals f_1 and f_2 along two directions shown in Figure 3 respectively, so points in signals f_1 and f_2 are correspondence to points in signal I.

Secondly, set up a binary tree of MSEx of f_1 and f_2, denoted as S' and S'', let n' and n'' denote the highest level of S' and S'', and take $n=\min(n', n'')$ as the top level of 2-D MSEx of I.

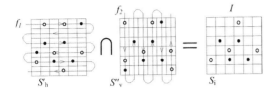

Figure 3. Set up 2-D MSEx based on 1-D MSEx. The black dots are local minima, and the circles are local maxima.

Thirdly, reallocation points of 1-D subsets $S'_{<i,j>}$ and $S''_{<i,j>}$ to 2-D extrema map according to position coordinates, to form a 2-D MSEx map sets S'_h and S''_v, where i=1,2,…,n; j=1,2,…,2^i.

Finally, take the intersection of each pair of extrema subsets corresponding in S'_h and S''_v, i.e. $S_{<i,j>} = S'_{h<i,j>} \cap S''_{v<i,j>}$, where $S_{<i,j>}$ is a 2-D MSEx subset, and finally we obtain 2-D MSEx set $S_m=\{S_i\}=\{S_{<i,j>}\}$, i=1,2,…,n; j=1,2,…,$2^i$.

The essence of this method is to find 2-D MSEx via comparison in 4-neighborhood of different scales.

3.3 Algorithm of IFABEMD

The algorithm of IFABEMD is described as follows.

Firstly, by the steps in section 3.2 to set up a binary tree of 2-D MSEx for an image I, i.e. $S=\{S_i\}=\{S_{<i,j>}\}$, i=1,2,…,n;j=1,2,…,2^i.

Secondly, compute the number N_{Sij} of extrema in a subset $S_{<i,j>}$, i=1,2,…,n; j=1,2,…,2^i.

Thirdly, sort N_{Sij} in descending order for every extrema set S_i, take the mean of the first two maximum of N_{Sij} in every extrema set Si as the average extrema number $\bar{N}_{S_i}$ of set S_i.

Fourthly, determinate window size w_{en} for filters of the ith level signal component with (1),

where $\rho = \bar{N}_{S_i}$, and then rounded to the nearest odd integer to get the final window width w_{en}.

Finally, applying order statistics and smoothing filters with window $w_{en} \times w_{en}$ to determine upper and lower envelopes of $BIMF_i$ as FABEMD.

The first four steps are the key difference between our approach and FABEMD, which could enhance the decomposition efficiency and adaptivity further as illustration in the following section.

4 EXPERIMENTAL RESULTS

In this section, we presented the experimental results of our method with the other two fast BEMD approaches (Xu 2006, Sharif 2010).

Our experiments are performed using MATLAB and a computer with a 2.13-GHz Intel Core Duo processor and 4GB RAM. The matlab code for method by Xu (2006) is available via http://www. codeforge. com/article/83809, and the matlab code for FABEMD (Sharif 2010) is available on https:// code.google. com/p/xbatdevel/source/browse/branches/heart/ Core/Util/Image/EMD/fabemd.m?r=2984.

The parameters of the three methods are as follows: both iteration numbers and maximum sifting numbers are 5 and the maximum error for sifting are 0.01 for the method by Xu (2006), iteration numbers are 15 and maximum sifting number is 1 for FABEMD, maximum sifting number is 1 and maximum iteration numbers are top level minus 1 to avoid extrema sparse for the proposed method. The decomposition effects are shown in Figure 4, and comparison of time consumed is in Table 1.

As can be seen from Figure 4, decomposition effects of the FABEMD and proposed method are

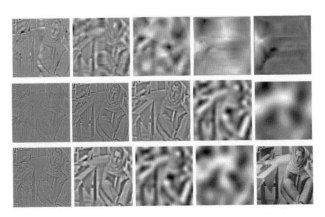

The first line is the results of method by Xu(2006), and from left to right is BIMF1~BIMF5,
The second line is the results of FABEMD, and from left to right is BIMF1~BIMF3, BIMF12, BIMF15,
The third line is the results of proposed method, and from left to right is BIMF1~BIMF4, and original of 'barbara'.
Figure 4. Decomposition results of different methods for 'barbara'.

significantly better than that of the method by Xu (2006). In addition, the proposed method reduces much redundant BIMFs than FABEMD. What more, details can be better extracted by the proposed method if we compare the BIMF2 in the third line and the second line carefully. Table 1 shows the conspicuous dominance in efficiency of the proposed approach.

Table 1. Comparison of time consumed.

	Method by Xu(2006)	FABEMD	Proposed method
Total time(s)	12.128	15.983	8.103

5 CONCLUSION

In this paper, we make full use of 2-D MSEx information, and presented a improved fast and adaptive bidimensional empirical mode decomposition (IFABEMD) that can enhance the decomposition adaptivity and efficiency for 2-D signals further. Although some advantages and disadvantages didn't discuss in detail in the limited space, the proposed method will get more study, and may be used in image enhancement, image fusion, and so on.

ACKNOWLEDGMENT

This work was supported by the National Natural Science Foundation of China under Grant 61002052, 61273262, 61250006.

REFERENCES

Ahmed, M. U., & Mandic, D. P. (2010, July). Image fusion based on fast and adaptive bidimensional empirical mode decomposition. *In Information Fusion (FUSION), 2010 13th Conference on* (pp. 1–6).

Damerval, C., Meignen, S., & Perrier, V. (2005). A fast algorithm for bidimensional EMD. *Signal Processing Letters, IEEE*, 12(10), 701–704.

Han, C, Guo H, Wang C, et al. (2002). A novel method to reduce speckle in SAR images. *International Journal of Remote Sensing*, 23(23): 5095–5101.

Huang, N. E., Shen Z. & Long S. R. (1998). The Empirical Mode Decomposition Method and the Hilbert Spectrum for Non-Stationary Time Series Analysis. *Proc. Royal. Soc. London A*, 454(A):903–995.

Liu, Z. X., Peng S. L., (2005). Directional EMD and applications in texture segmentation. *Chinese Science E Technology science*, 35(2):113–123.

Mahraz, M. A., Riffi, J., & Tairi, H. (2012). Motion estimation using the fast and adaptive bidimensional empirical mode decomposition. *Journal of Real-Time Image Processing*, 1–11.

Niang, O., Deléchelle, É., & Lemoine, J. (2010). A spectral approach for sifting process in empirical mode decomposition. *Signal Processing, IEEE Transactions on*, 58(11), 5612–5623.

Nunes, J. C., Bouaoune, Y., Delechelle, E., Niang, O., & Bunel, P. (2003). Image analysis by bidimensional empirical mode decomposition. *Image and vision computing*, 21(12), 1019–1026.

Riffi, J., Mahraz, A. M., & Tairi, H. (2013). Medical image registration based on fast and adaptive bidimensional empirical mode decomposition. *IET Image Processing*, 7 (6): 567–574.

Sharif, MA, B., Reza R, A., & Jesmin F, K. (2008). Fast and Adaptive Bidimensional Empirical Mode Decomposition Using Order-Statistics Filter Based Envelope Estimation. *EURASIP Journal on Advances in Signal Processing*, 2008. (5) :1–18.

Song, J. P., Zhang J. (2001). Application of BEMD in information seperating of remote sea images. *High-tech communications*, 9: 62–67.

Xu, G. L., Wang, X. T., & Xu, X. G. (2011). Improved bi-dimensional empirical mode decomposition based on 2D-assisted signals: analysis and application. *IET image processing*, 5(3), 205–221.

Xu, G. L., Wang, X. T., Xu, X. G., & Zhu, T. (2006). Image enhancement algorithm based on neighborhood limited empirical mode decomposition. *Dianzi Xuebao(Acta Electronica Sinica)*, 34(9), 1635–1639.

Xu, Y., Liu, B., Liu, J., & Riemenschneider, S. (2006). Two-dimensional empirical mode decomposition by finite elements. *Proceedings of the Royal Society A: Mathematical, Physical and Engineering Science*, 462(2074), 3081–3096.

Yang, D., Wang X. T., Xu G. L. (2013). A fast trend extraction for the analysis of temperature data. *In Intelligent Control and Information Processing (ICICIP), 2013 Fourth International Conference on*(pp. 338–342).

Yang, D., Wang X. T., Xu G. L. (2013). Research on 1D signal trend extracting via multi-scale extrema. *Journal of Electronics & Information Technology*, 2013 (5): 1208–1215.

Yang, H. Y., Li, S. Y., Chen, P. Y., & Shiau, Y. H. (2011, July). Efficient Color Image Enhancement Based on Fast and Adaptive Bidimensional Empirical Mode Decomposition. *In Computer Science and Society (ISCCS), 2011 International Symposium on* (pp. 319–322).

Zhang, W. Y., Hu D. H., Ding C. B. (2012). Enhancement of SAR Ship Wake Image Based on FABEMD and Goldstein Filter. *Journal of Radars*, 4:102–111.

Information Technology – Wan et al. (Eds)
© 2015 Taylor & Francis Group, London, ISBN 978-1-138-02785-5

A fixed-point algorithm for independent component analysis using AR source model

Yumin Yang
School of Mathematics and Information Science
Anshan Normal University, Anshan, China

Difei Xu
Faculty of Electronic information and Electrical Engineering
Dalian University of Technology, Dalian, China

ABSTRACT: Independent component analysis is a fundamental and important task in unsupervised learning, that was studied mainly in the domain of Hebbian learning. In this paper, the temporal dependencies are explained by assuming that each source is an autoregressive (AR) process and innovations are independently and identically distributed (i.i.d). First, we derive the likelihood of the model, unlike most likelihood the proposed method takes into account both spatial and temporal information. Next, a fixed-point algorithm is derived by using an approximative Newton method. Finally, computer simulations show that the algorithm achieves better separation of the mixed signals and mixed textures which are difficult to be separated by the basic independent component analysis algorithms. Comparison results verify the fixed-point algorithm converges faster than the existing gradient algorithm, and it is simpler to implement due to it does not need any learning rate.

1 INTRODUCTION

Blind source separation (BSS) is typically performed in a setting where the observed signals are instantaneous noise-free linear superpositions of underlying hidden source signals. Let us denote the m source signals by $s_1(t), \cdots, s_m(t)$, and the observed signals by $x_1(t), \cdots, x_m(t)$, where t is the time index. Let $\mathbf{A} \in R^{m \times m}$ is the mixing matrix. Further let us collect the source signals in a vector $\mathbf{s}(t) = (s_1(t), \cdots, s_m(t))^T$, and the observed signals vector $\mathbf{x}(t) = (x_1(t), \cdots, x_m(t))^T$. Now the mixing can be expressed as

$$\mathbf{x}(t) = \mathbf{A}\mathbf{s}(t). \tag{1}$$

The problem of the BSS is to estimate both the source signals s(t) and the mixing matrix $\mathbf{A}$ (or its inverse $\mathbf{W}$, i.e., $\mathbf{W} = \mathbf{A}^{-1}$, which is called the separation matrix), based on the observations of $\mathbf{x}(t)$ alone (Hyvärinrn & Karhunen & Oja 2001; Jutten & Hérault 1991).

The problem of BSS has been receiving wide attention in the field of signal processing in recent years. Many algorithms have been proposed for the BSS of statistically independent sources. Most of the algorithms are based on one of the following properties: nongaussianity of the sources, or their different autocorrelations. First, if all the components

(except perhaps one) have nongaussian distributions, the ensuing model refers to independent component analysis (ICA) (Common 1994), many techniques are available for estimation of the model (Hyvärinrn & Karhunen & Oja 2001). Second, many algorithms have been proposed by using temporal second-order correlations for solving a class of problems which the components are time-dependent and have stationary variances. Note that an important point is that we also need to assume that the signals have a different autocorrelation function; the mere existence of autocorrelations is not sufficient (Tong & Liu & Soon & Huang 1991; Molgedey & Schuster, 1994; Belouchrani & Abed Meraim & Cardoso & Moulines 1997; Pham 2001; Ziehe & Müller 1998; Kawamoto & Matsuoka & Oya1997; Degerine & Malki 2000).

We model each source signal by an AR model

$$s_i(t) = \sum_{\tau > 0} \alpha_i^\tau s_i(t - \tau) + n_i(t), i = 1, 2, \cdots, m. \tag{2}$$

where τ is time lag, and the innovation $n_i(t)$ is assumed to have zero-mean with a density function $q_i(n_i(t))$. In addition, innovation sequences $\{n_i(t)\}$ are assumed to be mutually independent white sequences, i.e., they are spatially independent and temporally white as well. If the zero-mean innovation term $n_i(t)$ is nongaussian, the source signal $s_i(t)$

is nongaussian as well. If the autoregressive coefficients α_i^τ are not zero, the source signals have auto-correlations, then the model (1) and (2) combines the nongaussianity and autocorrelations (Hyvärinen 2001, Kenneth & Hagai & Srikantan 2006). As always in BSS, the source signals are assumed statistically independent, and the variance of each s_i is equal to 1.

The maximum likelihood approach presented in this paper, differs from most maximum likelihood methods (Moulines & Cardoso & Gassiat 1997; Pearlmutter & Parra1996) which makes use of both the spatial and temporal structure of the sources. In this paper, a fixed-point algorithm is derived by using an approximative Newton method. Computer simulations show that the algorithm achieves better separation of the mixed signals and mixed textures which are difficult to be separated by the basic independent component analysis algorithms. Comparison results verify that the algorithm converges faster than the existing gradient algorithm and, it is more simple to implement due to its advantage of not using the learning rate.

2 LIKELIHOOD OF THE MODEL

We now compute the likelihood with respect to the innovation processes. Without loss of generality, we can derive the likelihood of the model when each source is a first-order autoregressive model. To be able to compute the innovations, we need estimates of the autoregressive coefficients, $\hat{\alpha}_i^\tau$. Note that these estimates, depend on $\mathbf{w}_i$, we write $\hat{\alpha}_i^\tau(\mathbf{w}_i)$ in the following. Let us consider latent variables $s_i(t)$ over on N-point time block. We define the vector s_i as

$$s_i = (s_i(0), \cdots, s_i(N-1))^T. \tag{3}$$

Then the joint probability density function of s_i can be written as

$$p_i(s_i) = p_i(s_i(0), \cdots, s_i(N-1))$$
$$= \prod_{t=0}^{N-1} p_i(s_i(t) \mid s_i(t-1)), \tag{4}$$

where $s_i(t) = 0$ for $t < 0$ and the statistical independence of innovation sequences was taken into account.

The conditional probability density of $s_i(t)$ given its past samples can be written as

$$p(s_i(t) \mid s_i(t-1)) = q_i(n_i(t)). \tag{5}$$

Combining (4) and (5) leads to

$$p_i(s_i) = \prod_{t=0}^{N-1} q_i(n_i(t)). \tag{6}$$

Take the statistical independent of latent variables and (6) into account, we have

$$p(s_1, \cdots, s_m) = \prod_{i=1}^{m} p_i(s_i) = \prod_{t=0}^{N-1} \prod_{i=1}^{m} q_i(n_i(t)). \tag{7}$$

Denote a set of observation data by

$$\chi = \{\mathbf{x}_1, \cdots, \mathbf{x}_m\}, \tag{8}$$

Where

$$\mathbf{x}_i = (x_i(0), \cdots, x_i(N-1))^T. \tag{9}$$

Then the normalized log-likelihood is given by

$$\frac{1}{N} \log p(\chi \mid A)$$

$$= -\log |\det A| + \frac{1}{N} \log p(s_1, \cdots, s_m) \tag{10}$$

$$= \log |\det W| + \frac{1}{N} \sum_{t=0}^{N-1} \sum_{i=1}^{m} \log q_i(n_i(t)).$$

The estimate of latent variables is denoted by $\mathbf{y}(t) = \mathbf{W}x(t)$, and $n_i(t)$ is replaced by its estimate $\tilde{y}_i(t) = y_i(t) - \hat{\alpha}_i(w_i)y_i(t-1)$, then, the objective function is given by

$$J(\mathbf{W}) = \log |\det \mathbf{W}| + \sum_{i=1}^{m} E\{\log q_i(\tilde{y}_i(t))\}, \tag{11}$$

where E is the expectation over t (sample average). The same form objective function (10) can be obtained while the source signal is expressed as (2), where $\tilde{y}_i(t) = y_i(t) - \sum_{\tau > 0} \hat{\alpha}_i^\tau(w_i)y_i(t-\tau)$.

Noted that the ensuing algorithm easily becomes unstable, more precisely, the estimates of $\log q_i(\cdot)$ must be very accurate to prevent $\mathbf{W}$ from going to zero or infinity. Therefore, we use a classic trick of stabilizing BSS algorithms: we prewhiten the data and constrain $\mathbf{W}$ to be orthogonal (Hyvärinrn & Karhunen & Oja 2001]. After this stabilization, we can obtain the objective function.

$$\max_{\|\mathbf{w}_i\|=1} J(\mathbf{w}_i) = E\{G_i(\tilde{y}_i(t))\}$$

$$= E\{G_i(\mathbf{w}_i^T(x(t) - \sum_{\tau > 0} \alpha_i^\tau(\mathbf{w}_i)x(t-\tau)))\}, \tag{12}$$

where the function G_i is the logarithm of the probability density function of the innovation process.

Theorem 1: Assume that the input data follow the linear mixture model (1) with whitened data: $\tilde{\mathbf{x}}(t) = \mathbf{VAs}(t)$, where $\mathbf{V}$ is the whitening matrix. Furthermore, assume that the innovations $n_i(t) = s_i(t) - \sum_{\tau > 0} \alpha_i^\tau s_i(t-\tau)$ $(i = 1, 2, \cdots, m)$ are mutually statistically independent and have zero mean and unit variance (assume that $\alpha_i^\tau, i = 1, 2, \cdots, m$, are the known constants), and that G_i is a sufficiently smooth even non-quadratic function. Then the local maxima of $E\{G_i(\mathbf{w}_i^T(\tilde{\mathbf{x}}(t) - \sum_{\tau>0}\alpha_i^\tau(\mathbf{w}_i)\tilde{\mathbf{x}}(t-\tau)))\}$ under the constraint $\|\mathbf{w}_i\| = 1$ include one row of the mixing matrix $\mathbf{VA}$ such that the corresponding independent innovation satisfy

$$E\{n_i g_i(n_i) - g_i{}'(n_i)\} > 0, \tag{13}$$

where g_i is the derivative of G_i, and $g_i{}'$ is the derivative of g_i.

Proof: First, assume that the input data follow the model (1) with whitened data: $\tilde{\mathbf{x}}(t) = \mathbf{VAs}(t)$, where $\mathbf{V}$ is the whitening matrix. Then the innovation processes $n_i(t) = s_i(t) - \sum_{\tau>0}\alpha_i^\tau s_i(t-\tau)$ $(i = 1, 2, \cdots, m)$ and $\bar{x}_i(t) = \tilde{x}_i(t) - \sum_{\tau>0}\alpha_i^\tau \tilde{x}_i(t-\tau)$ follow the model as well:

$$\bar{\mathbf{x}}(t) = \mathbf{VAn}(t). \tag{14}$$

Thus, the theorem stated here is a corollary of Theorem 8.1 in (Hyvärinrn & Karhunen & Oja 2001), if the innovations $n_i(t)$ are independent and have zero mean and unit variance, and G_i is a sufficiently smooth even non-quadratic function.

3 A FIXED-POINT ALGORITHM

For simplicity, we denote

$$\mathbf{z}(\mathbf{w}_i) = \tilde{\mathbf{x}}(t) - \sum_{\tau>0}\hat{\alpha}_i^\tau(\mathbf{w}_i)\tilde{\mathbf{x}}(t-\tau), \tag{15}$$

Then we can write the contrast function

$$\max_{\|\mathbf{w}_i\|^2 = 1} J(\mathbf{w}_i) = E\{G_i(\mathbf{w}_i^T\mathbf{z}(\mathbf{w}_i))\}. \tag{16}$$

To perform the optimization (16), we use a fixed-point algorithm along similar as the FastICA algorithm for maximizing nongaussianity (Hyvärinen & Oja 1997; Hyvärinen 1999). The fixed-point algorithm can be found using an approximative Newton method. According to the Lagrange conditions (Nocedal & Wright 1999), the optima of $J(\mathbf{w}_i)$ under the constraint $\|\mathbf{w}_i\|^2 = 1$ are obtained at points where

$$E\{\mathbf{z}(\mathbf{w}_i)g_i(\mathbf{w}_i^T\mathbf{z}(\mathbf{w}_i))\} +$$
$$E\{\frac{\partial\mathbf{z}(\mathbf{w}_i)}{\partial\mathbf{w}_i}\mathbf{w}_i^T g_i(\mathbf{w}_i^T\mathbf{z}(\mathbf{w}_i))\} + \beta\mathbf{w}_i = 0, \tag{17}$$

where β is a Lagrange's multiplier and the function g_i is the derivative of G_i. Because of the variance of the innovations smoothly changing, we can assume that the estimations of the $\hat{\alpha}_i^\tau(\mathbf{w}_i)$ are decoupled from the estimation of the $\mathbf{w}_i$. In other words, the $\hat{\alpha}_i^\tau(\mathbf{w}_i)$ are estimated for the fixed $\mathbf{w}_i$, and then the $\mathbf{w}_i$ are estimated for the fixed $\hat{\alpha}_i^\tau(\mathbf{w}_i)$, and so on.

Then $\frac{\partial\mathbf{z}(\mathbf{w}_i)}{\partial\mathbf{w}_i} = \mathbf{0}$. Then the equation (17) can be simplified as

$$E\{\mathbf{z}(\mathbf{w}_i)g_i(\mathbf{w}_i^T\mathbf{z}(\mathbf{w}_i))\} + \beta\mathbf{w}_i = 0. \tag{18}$$

Now let us try to solve this equation by Newton's method. Denoting the function on the left–hand side of (18) by F, we obtain its Jacobian matrix $JF(\mathbf{w}_i)$ as

$$JF(\mathbf{w}_i) = E\{\frac{\partial\mathbf{z}(\mathbf{w}_i)}{\partial\mathbf{w}_i}g_i(\mathbf{w}_i^T\mathbf{z}(\mathbf{w}_i))\}$$
$$+ E\{\frac{\partial\mathbf{z}(\mathbf{w}_i)}{\partial\mathbf{w}_i}\mathbf{w}_i g_i'(\mathbf{w}_i^T\mathbf{z}(\mathbf{w}_i))\mathbf{z}^T(\mathbf{w}_i)\} \tag{19}$$
$$+ E\{\mathbf{z}(\mathbf{w}_i)\mathbf{z}^T(\mathbf{w}_i)g_i'(\mathbf{w}_i^T\mathbf{z}(\mathbf{w}_i))\} + \beta\mathbf{I}.$$

Similar to the above analysis again, the first and the second term in (19) disappear as well and we obtain

$$JF(\mathbf{w}_i) = E\{\mathbf{z}(\mathbf{w}_i)\mathbf{z}^t(\mathbf{w}_i)g_i'(\mathbf{w}_i^t\mathbf{z}(\mathbf{w}_i))\} + \beta\mathbf{I}. \tag{20}$$

To simplify the inversion of this matrix, we decide to approximate the first term in (20). Note that $\tilde{\mathbf{x}}(t) = \mathbf{Vx}(t) = \mathbf{VAs}(t)$, then the new mixing matrix $\mathbf{VA}$ is orthogonal [15]. Then

$$E\{\mathbf{z}(\mathbf{w}_i)\mathbf{z}^T(\mathbf{w}_i)\} = \mathbf{I}. \tag{21}$$

So we have

$$JF(\mathbf{w}_i) \approx E\{g_i'(\mathbf{w}_i^T\mathbf{z}(\mathbf{w}_i))\}\mathbf{I} + \beta\mathbf{I}. \tag{22}$$

Thus the Jacobian matrix becomes diagonal, and can be easily inverted. We obtain the following approximative Newton iteration

$$\mathbf{w}_i \leftarrow \mathbf{w}_i - \frac{[E\{\mathbf{z}(\mathbf{w}_i)g_i(\mathbf{w}_i^T\mathbf{z}(\mathbf{w}_i))\} + \beta\mathbf{w}_i]}{[E\{g_i'(\mathbf{w}_i^T\mathbf{z}(\mathbf{w}_i))\} + \beta]}, \tag{23}$$

The above equation can be further simplified by multiplying both sides of by

$$E\{g_i'(\mathbf{w}_i^T\mathbf{z}(\mathbf{w}_i))\} + \beta. \tag{24}$$

This gives, after straightforward algebraic simplification

$$\mathbf{w}_i \leftarrow E\{\mathbf{z}(\mathbf{w}_i)g_i(\mathbf{w}_i^T\mathbf{z}(\mathbf{w}_i))\}$$
$$- E\{g_i'(\mathbf{w}_i^T\mathbf{z}(\mathbf{w}_i))\}\mathbf{w}_i. \tag{25}$$

Thus, the fixed-point iteration can be obtained as:

$$\mathbf{w}_i \leftarrow E\{(\tilde{\mathbf{x}}(t) - \sum_{\tau>0}\hat{\alpha}_i^\tau(\mathbf{w}_i)\tilde{\mathbf{x}}(t-\tau))$$
$$g_i(\mathbf{w}_i^T(\tilde{\mathbf{x}}(t) - \sum_{\tau>0}\hat{\alpha}_i^\tau(\mathbf{w}_i)\tilde{\mathbf{x}}(t-\tau)))\} \tag{26}$$
$$- E\{g_i'(\mathbf{w}_i^T(\tilde{\mathbf{x}}(t) - \sum_{\tau>0}\hat{\alpha}_i^\tau(\mathbf{w}_i)\tilde{\mathbf{x}}(t-\tau)))\}\mathbf{w}_i,$$

where the nonlinear functions g_i used in the algorithm are the derivatives of the log-densities G_i.

In ICA, it is well known that the exact form of the nonquadratic function G_i is not very important here either, as long as it is qualitatively similar enough. The function g_i should be chosen as in ordinary ICA, but according to the probability distribution of the estimate of the innovation process normalized to unit variance. According to (Hyvärinrn & Karhunen & Oja 2001); if the innovation is supergaussian, $g_i(u) = -tanh(au)$ is suitable, where $a \geq 1$ is a constant. For subgaussian innovations, one could use $g_i(u) = -u + tanh(u)$ or $g_i(u) = -u^3$, for example, for almost gaussian innovations, a linear g_i could be used.

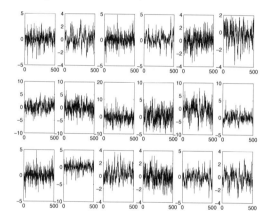

Figure 1. The first row is the original signals; the second row is the mixed signals; the third row is the estimated signals by the fixed-point algorithm in this paper.

Thus, the estimation consists of the following steps:

Step 1. Remove the mean from the data and whiten it. Denote the preprocessed data by $\tilde{\mathbf{x}}(t)$. Choose a random initial value for the matrix $\mathbf{W} = (\mathbf{w}_1, \cdots, \mathbf{w}_m)$.

Step2. Compute estimates of the source signals as

$$\tilde{s}_i(t) = \mathbf{w}_i^T \tilde{\mathbf{x}}(t). \qquad (27)$$

Step 3. Compute estimates of the autoregressive coefficients $\hat{\alpha}_i^\tau(\mathbf{w}_i)$, for example, by a classical least-squares method.

Step4. Compute estimates of the innovations as $\hat{n}_i(t) = \tilde{s}_i(t) - \sum_{\tau > 0} \hat{\alpha}_i^\tau \tilde{s}_i(t - \tau).$

Step5. Choose nonlinearity g_i for each source on the distributions of the normalized innovations $\hat{n}_i(t)$ as in ordinary ICA.

Step6. Compute (26) and orthogonalize $\mathbf{W}$ by

$$\mathbf{W} \leftarrow (\mathbf{W}\mathbf{W}^T)^{-1/2}\mathbf{W}. \qquad (28)$$

The five steps 2–6 are repeated until W has converged.

4 SIMULATIONS

In this section, to illustrate the algorithm in this paper, simulation examples on signal separation and texture separation problems are presented here.

Example 1: We created six signals using a first order autoregressive model. Signals 1, 2, 3 and 4 were created with supergaussian innovations and signals 5 and 6 with gaussian innovations; all innovations had unit variance. Signals 1, 3 and 5 had identical autoregressive coefficients (0.3) and therefore identical autocovariances; signals 2, 4 and 6 had identical coefficients (0.7) as well. The signals were mixed with a 6×6 random mixing matrix (denote by $\mathbf{A}$) as in ICA. Sample size T was 5000, and the error index defined as (Amari & Cichocki & Amari 1996):

$$PI = \sum_{i=1}^{m}(\sum_{j=1}^{m} \frac{|P_{ij}|}{\max_k |P_{ik}|} - 1) +$$

$$\sum_{j=1}^{m}(\sum_{i=1}^{m} \frac{|P_{ij}|}{\max_k |P_{kj}|} - 1), \qquad (29)$$

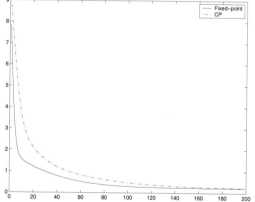

Figure 2. Convergence of the three algorithms for artificially generated data. Horizontal axis: iteration count, Vertical axis: averaged error over 100 independent runs.

where P_{ij} is the ijth element of $m \times m$ matrix $\mathbf{P} = \mathbf{W}\mathbf{A}$ (**W** is the separating matrix in ICA). For the goal of comparison, we test three algorithms (complexity pursuit (CP) (Hyvärinen 2001); Fixed-point algorithm in this paper). The nonlinearity was chosen as $g(u) = -tanh(u)$. At every trial, the three algorithms were run 200 iterations, which seemed to be always enough for convergence. The original signals, mixed signals and separated signals by the fixed-point algorithm in this paper are depicted in Figure 1. Figure 2 plots the performance indexes of the three algorithms averaged over 100 independent runs. Obviously, the fixed-point algorithm converges faster than the existing gradient algorithm.

Example 2: Separating textures has been considered a difficult case for ICA, since, the textures, considered as 1-D signals of pixel gray-scale values, are not independent (Pajunen, 1998). In this example, we show that they can be nevertheless separated if ICA is applied to their innovation processes, which seem to be independent enough. To show how dependent these signals are, we computed the correlation matrix:

$$\begin{pmatrix} 1.0000 & 0.0344 & 0.1854 \\ 0.0344 & 1.0000 & -0.0188 \\ 0.1854 & -0.0188 & 1.0000 \end{pmatrix}$$

which shows that the source textures are rather correlated. Mixed these textures using a 3×3 random mixing matrix. First, ordinary ICA was applied to the separation of the textures, using the algorithm in [15]. Secondly, the fixed-point algorithm in this paper was applied too. The results are depicted in Figure 3. Clearly, the mixed textures can be separated efficiently by the fixed-point algorithm in this paper.

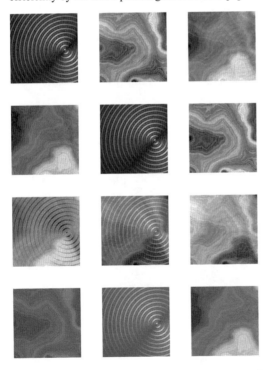

Figure 3. The first row is the original textures; the second row is the mixed textures; the third row is the estimated textures by ordinary ICA; the fourth row is the estimated textures by the fixed-point algorithm in this paper.

5 CONCLUSIONS

In this paper, we consider independent component analysis of time-dependent stochastic processes using an AR source model. First, the likelihood function of the model is developed and, it differs from most maximum likelihood methods in that it takes into account both spatial and temporal information. Secondly, the fixed-point algorithm is derived by using an approximative Newton method. Finally, computer simulations show that the fixed-point algorithm achieves better separation of the mixed signals and mixed textures which are difficult to be separated by the basic independent component analysis algorithms. Comparison results verify the fixed-point algorithm converges faster than the existing gradient algorithm and, it is more simple to implement due to it does not need any learning rate. But the case of general time-dependent signal, more sophisticated method for the signal separation need to be developed in the future.

REFERENCES

Hyvärinrn, A. Karhunen, J. Oia, E.(2001). Independent component analysis.

Common, P. (1994). Independent component analysis —a new concept?. *Signal Process. 36*, 287–314.

Tong, L. & R. -W. Liu & V. C. Soon& Y. -F. Huang(1991). Indeterminacy and identifiability of blind identification. *IEEE Trans. Circuit. Syst. 38*, 499–509.

Molgedey, L. & H. G. Schuster (1994). Separation of a mixture of independent signals using time delayed correlations. *Phys. Rev. Lett. 72*, 3634–3636.

Belouchrani, A. & K. Abed Meraim & J. -F. Cardoso & E. Moulines (1997). A blind source separation technique based on second order statistics. *IEEE Trans. Signal Process. 45 (2)*, 434–444.

Pham, D. -T.(2001). Blind separation of instantaneous mixtures of sources via the Gaussian mutual information criterion. *Signal Process. 81*, 855–870.

Ziehe, A. & K. -R. Müller (1998). TDSEP– an efficient algorithm for blind separation using time structure. *Proc. Int. Conf. Artificial Networks (ICANN'98)*, 675–680.

Kawamoto, M. & K. Matsuoka & M. Oya (1997). Blind separation of sources using temporal correlation of the observed signals. *IEICE Trans. Fundamentals E80-A(4)*, 695–704.

Amari, S. -I. (2000). Estimating functions of independent component analysis for temporally correlated signals. *Neural Comput. 12(9)*, 2083–2107.

Degerine, S. & R. Malki (2000). Second-order blind separation of sources based on canonical partial innovations. *Signal Process. 48(3)*, 629–641.

Hyvärinen, A. (2001). Complexity pursuit: separating interesting component from time-series. *Neural Comput. 13(4)*, 883–898.

Kenneth, E. H. & T. A. Hagai & S. N. Srikantan (2006). An EM method for spatio-temporal blind source separation using an AR-MOG source model. *Proc. ICA 2006.*, 98–105.

Moulines, E. & J. F. Cardoso & E. Gassiat (1997). Maximum likelihood for blind source separation and deconvolution of noisy signals using mixture methods,. *Intl. Conf. on Acoustics, Speech, and Signal Process. 5*, 3617–3620.

Pearlmutter, B. A. & L. C. Parra (1996). Maximum likelihood blind source separation: A context-sensitive generalization of ICA. In *Advances in Neural Information Proc. Systems 9*, 613–619. Stockholm, Sweden, pp. 39–64.

Hyvärinen, A. & E. Oja (1997). A fast fixed-point algorithm for independent component analysis. *Neural Comput. 9(7)*, 1483–1492.

Hyvärinen, A. (1999). Fast and robust fixed-point algorithms for independent component analysis. *TEEE Trans. Neural Networks. 10(3)*, 626–634.

Nocedal, J. S. J. Wright(1999). Numerical Optimization.

Jutten, C. & J. Hérault (1999). Blind separation of sources, Part I: an adaptive algorithm based on neuromimetic architecture. *Signal Process. 24*, 1–10.

Amari, S. -I.& A. Cichocki & H. Yang (1996). A new learning algorithm for blind source separation. In *Advances in Neural Information Proc. Systems 8*, 757–763.

Pajunen, P. (1998). Blind source separation using algorithmic information theory. *Neurocomputing. 22*, 35–48.

Information Technology – Wan et al. (Eds)
© *2015 Taylor & Francis Group, London, ISBN 978-1-138-02785-5*

A hybrid system for skin lesion detection: Based on gabor wavelet and support vector machine

Mohamed Khaled Abu Mahmoud & Adel Al-Jumaily
Faculty of Engineering and Information Technology, University of Technology,
Sydney Ultimo NSW Australia

ABSTRACT: Severe melanoma is potentially life-threatening. A novel methodology for automatic feature extraction from histo-pathological images and subsequent classification is presented. The proposed automated system uses a number of features extracted from images of skin lesions through image processing techniques which consisted of a spatially winner and adaptive median filter then applied Gabor filter bank to improve diagnostic accuracy. Histogram equalization to enhance the contrast of the images prior to segmentation is used. Then, a wavelet approach is used to extract the features; more specifically Wavelet Packet Transform (WPT). This article introduces a novel melanoma detection strategy using a hybrid particle swarm - based support vector machine (SVM–WLG–PSO) technique. The extracted features are reduced by using a particle swarm optimization (PSO), this was used to optimize the SVM parameters as a feature selection and finally, the obtained statistics are fed to a support vector machine (SVM) binary classifier to diagnose skin biopsies from patients as either malignant melanoma or benign nevi. The obtained classification accuracies show better performance in comparison to similar approaches for feature extraction. The proposed system is able to achieve one of the best results with classification accuracy of 87.13%, sensitivity of 94.1% and specificity of 80.22%.

KEYWORDS: Lesion, Histo-pathological images, Winner, Adaptive median filter (AMF), Gabor filter bank, Histogram Equalization, Particle swarm optimization (PSO), Support vector machine (SVM).

1 INTRODUCTION

Australia is one of many countries in which skin cancer is simply "wide spread" in comparison to other types of cancer [1] [2]. Researchers have found that Australian melanoma rates are the highest globally at almost four times the rates seen in Canada, the United Kingdom and the United States[3]. Skin cancer costs the health system around $300 million Australian dollars annually, the highest cost of all cancers. Melanoma has near 95% cure rate if detected and treated in its early stages [4]. This study proposes an automated system for discrimination between melanocytic nevi and malignant melanoma. The general approach of developing a Computer Aided Diagnostic system for the diagnosis of skin cancer is to find the location of a lesion and also to determine an estimate of the probability of a disease. As mentioned in the literature the digital images are often corrupted during acquisition and transmission [5]. Filters are then very important as pre-processing tools.

Wiener, Gabor and adaptive median filters are used to remove unwanted features like the fine hairs, noise and air bubbles on the skin [6]. Median filter is used to remove noise without blurring edges. In addition these filters have been shown to possess optimal localization properties in both the spatial and frequency domains and thus are well suited for quality segmentation problems[7].

Segmentation is one of the most widely investigated research areas in pathological image analysis. In swarm-based SVM, a particle swarm optimization (PSO) was used to optimize the SVM parameters. A fitness function was defined in the optimization to find a high performance of the melanoma detection, especially in the sensitivity. The support vector machine (SVM) classier is used widely in bioinformatics, due to its high accuracy, ability to deal with high-dimensional data and in this syntax diverse sources of data[8,9,10].

This article introduces a novel melanoma detection strategy using a hybrid particle swarm-based support vector machine (SVM–WLG – PSO) technique. It is possible that the data found from the RBF mapping is more likely to be correctly classified by SVM than from the other kernel functions mappings. Furthermore, SVM-linear could be a special case of SVM-RBF[9,11], which means that SVM-RBF has more possibilities to obtain a better performance than SVM-linear. The better performance of the SVM based system which employs RBF is confirmed in another application [8,12].

This paper is organized as follows: its next section describe the computer-aided diagnosing (CAD) which consists of the (a) pre-processing, (b) segmentation, (c) features extraction and selection, and (d) the automated procedure using SVM. The followed section presents the result and discussion, and the final section presents conclusions.

2   COMPUTER-AIDED DESIGN (CAD) FOR SKIN LESIONS

Automated diagnostic of medical images analyzing digital images has become one of the major medical research areas, and a dynamic area in several applications [13]. It has been shown through research that Melanoma if detected in the early stages improves survival rates. Fig. 1 displays the four stages of the CAD system for skin lesion images.

2.1   Pre-processing stage

This stage includes image resizing, masking, cropping, hair removal, and conversion from RGB color to intensity grey image. It is meant to facilitate image segmentation by filtering the image and enhancing its important features [5].

2.1.1   Wiener2 filter
Wiener filtering gives the optimal way of narrowing off the noisy components, so as to give the best reconstruction of the original signal. It can be applied in spatial basis, Fourier basis (frequency components), wavelet basis, etc. Wiener2 is low pass-filters to a gray scale image that has been corrupted by constant power additive noise. The Wiener method is based on statistics estimated from a local neighborhood of each pixel. The additive noise (Gaussian white noise)

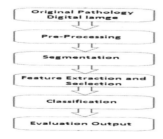

Figure 1.   Displays the four stages of the CAD system for skin lesion images.

is assumed to be noise. Wiener filtering is best when applied to the difference between an image and a smoothed image [14,15,16].

2.1.2   Gabor filter
The Gabor filter is basically a Gaussian (with variances Sx and Sy along x and y-axes respectively) modulated by a complex sinusoid (with center frequencies U and V along x and y-axes respectively). Gabor filters have been used in many applications, such as quality segmentation, target detection, fractal dimension management, document analysis, edge detection, retina identification, image coding and image representation[17]. A Gabor filter can be viewed as a sinusoidal plane of particular frequency and orientation, modulated by a Gaussian envelope.

Thus the 2-D Gabor filter can be written as in equation 1:

$$h(x,y) = e^{-0.5(\frac{x^2}{\sigma x^2}+\frac{x^2}{\sigma y^2})} e^{-j2\omega(u_0 x+v_0 y)}$$
$$= g(x,y)e^{-j2\pi(u_0 x+v_0 y)} \qquad (1)$$

The frequency response of the filter is:

$$H(u,v) = G(u-u_0, v-v_0)2\pi\sigma_x\sigma_y$$
$$= \left[e^{-2\pi^2\left[(u-u_0)^2\sigma_x^2+(v-v_0)^2\sigma_y^2\right]} \right]$$
$$= \frac{1}{2\pi\sigma_x\sigma_y} e^{-0.5\left[\frac{(u-u_0)^2}{\sigma_u^2}+\frac{(v-v_0)^2}{\sigma_v^2}\right]}$$

$$\text{Ωηερε} \quad \sigma_u \frac{1}{2\pi\sigma_x}, \sigma_v \frac{1}{2\pi\sigma_y} \qquad (2)$$

this is equivalent to translating the Gaussian function by (u_0, v_0) in the frequency domain. Thus the Gabor function can be thought of as being a Gaussian function shifted in frequency to position (u_0, v_0) i.e. at a distance of $\sqrt{u_0^2+v_0^2}$ from the origin $\tan^{-1}\frac{u_0}{v_0}$. In the above equations (1) & (2), (u_0, v_0) are referred to as the Gabor filter spatial central frequency. The parameters σ_x, σ_y are the standard deviation of the Gaussian envelope along X and Y directions and determine the filter bandwidth.

2.1.3   Median filter
Median filtering is a nonlinear operation often used in image processing to reduce noise and not preserve edges. The median filter considers each pixel in the image in turn and looks at its nearby neighbors to decide whether or not it is representative of its surroundings [5]. It replaces it with the median of those values. The median is calculated by first sorting all the pixel values from the surrounding neighborhood into numerical order and then replacing the pixel being considered with the middle pixel value. A 3×3 square neighborhood is used here; larger neighborhoods will produce more severe smoothing[5].

The performance of the filter was evaluated by computing the two criteria, classically used in the literature[18] [19], the mean absolute error (MAE) and the mean square error (MSE), respectively.

2.2 Image processing

Segmentation and classification are important steps in the medical image analyses for radiological evaluation or (CAD). One of the early steps in this stage is image enhancement. The purpose of image enhancement methods is to process a picked image for better contrast and visibility of features of interest for visual examination as well as subsequent computer-aided analysis and diagnoses. As described in [14,20] different medical imaging modalities provide specific characteristics information about internal organs or biological skins. Image contrast and visibility of the features of interest depend on the imaging modality as well as the anatomical regions [7].

The simplest form of intensity transformation function is written as shown in Eq. (3):

$$s = T(r) \tag{3}$$

where r denotes the intensity of f and s the intensity of g, both at any corresponding point (x, y) in the images.

2.2.1 Histogram processing

Histogram is the estimation of the probability distribution of a particular type of data. The histogram in Fig. 2 shows that the image contains only a fraction of the total range of grey levels. Therefore this image has low contrast. We focus on using histograms for image enhancement.

2.2.2 Histogram equalization

This application describes a method of imaging processing that allows medical images to have better contrast.

2.3 Image segmentation

Segmentation is one of the most difficult tasks in image processing. Image segmentation methods can be broadly classified into three categories: (1) Edge-based methods, (2) pixel-based direct classification methods, (3) region-based methods.

2.3.1 Edge detection operations

The gradient magnitude and directional information from the Sobel horizontal and vertical direction masks can be obtained by convolving the respective Gx and Gy masks with the image. The magnitude (M) of the gradient that can be approximated as the sum of the absolute values of the horizontal and vertical gradient

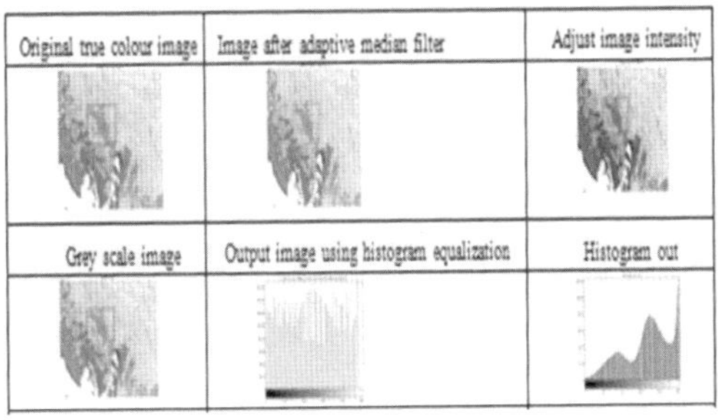

Original true colour image	Image after adaptive median filter	Adjust image intensity
Grey scale image	Output image using histogram equalization	Histogram out

Figure 2. The results showed, original image, image after median filter, gray scale image, histogram out; adjust image intensity, and output image using histogram equalization

images is obtained by convolving the image with the horizontal and vertical masks, Gx and Gy[21,22].

2.3.2 Thresholding

This has used the gradient to improve the histogram by combining intensity and gradient information for better separation of objects and background, as displayed in Fig. 3.

Figure 3. Selecting a threshold by visually analysing a bimodal histogram. (Principle of histogram peak separation).

2.4 Optimal global thresholding

To determine an optimal global grey value threshold for image segmentation, parametric distribution based methods can be applied to the histogram of an image [23]. Let us assume that the histogram of an image to be segmented has two normal distributions belonging to two respective classes such as background and object. The overall probability of error in pixel classification using the threshold T is expressed in Eq. (4):

$$E(T) = P_2(T)E_1(T) + P_1(T)E_2(T) \tag{4}$$

$$E_1(T) = \int_{-\infty}^{T} P_2(z)\,dz \tag{5}$$

$$E_2(T) = \int_{-\infty}^{T} P_1(z)\,dz \tag{6}$$

where E1 (T) in Eq. (5) and E2 (T) in Eq. (6) are the probability of incorrectly classifying a class 1 pixel to class 2 and a class 2 pixels to class 1 respectively.

For image segmentation, the objective is to find an optimal threshold T that minimizes the overall probability of error in pixel classification. The optimization process requires the parameterization of the probability density distributions and possibility of both classes. These parameters can be determined from a model or set of training images [19].

Let us assume σi and μi to be the standard deviation and mean of the Gaussian probability density function of the class i (i=1,2 for two classes) as explained in equation 7:

$$p(z) = \frac{P_1}{\sqrt{2\pi}\sigma_1} e^{-(z-\mu_1)^2/2\sigma_1^2} + \Pi 2/\sqrt{2\pi}\,\sigma_2$$
$$e - (z - \mu_1)^2/2\sigma_2^2 \tag{7}$$

The optimal global threshold T can be determined by finding a general solution that minimizes Eq. (4), with the mixture distribution in Eq. (7) and thus satisfies the following quadratic expression [2] as in Eq. (8):

$$AT^2 + BT + C = 0, \tag{8}$$

where Eqs. (9)–(11) display the components of Eq. (10)

$$A = \sigma_1^2 - \sigma_2^2 \tag{9}$$

$$B = 2(\mu_1\sigma_2^2 - \mu_2\sigma_1^2) \tag{10}$$

$$C = \sigma_1^2\mu_2^2 - \sigma_2^2\mu_1^2 + 2\sigma_1^2\sigma_2^2 \ln(\sigma_2 P_1/\sigma_1 P_2) \tag{11}$$

If the variances of both classes can be assumed to be equal to σ^2, the optimal threshold T can be determined as in equation 12:

$$T = \frac{\mu_1 + \mu_2}{2} + \frac{\sigma^2}{\mu_1 - \mu_2} \ln\left(\frac{P_2}{P_1}\right) \tag{12}$$

It should be noted that in the case of equal probability of classes, the above expression for determining the optimal threshold is simply reduced to the average of the mean values of two classes. In this study, Edge Detection Operations were used as the segmentation method.

2.5 Feature extraction

This paper used six features extracted from histogram, 21 features were extracted from co-occurrence matrix for each channel and 255 features were extracted from wavelet packet transform (WPT). WPT was used to implement the feature extraction process. Variability appears to be what most separates malignant melanoma from benign nevi, therefore, the best approach at feature extraction would retain be to retain as much of the data variability as possible[24]. Wavelet analysis looks at these changes over different scales which should detect whole lesion changes such as texture, color, and local changes like granularity. The WPT is a generalized version of the wavelet transform: the high-frequency part also splits into a low and a high frequency part, this produce a decomposition tree as shown in Fig.4.

The WPT provides a high dimensional feature vector thus providing more information about the images. However, the WPT complicates the analysis process as the high dimensionality of the feature vector causes an increase in the learning parameters of the pattern classifier, and the convergence of

The learning error deteriorates. Consequently, dimensionality reduction will play an important role before applying the feature vector to the pattern classifier.

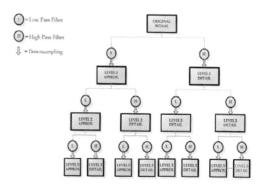

Figure 4. A Wavelet Packet decomposition tree.

The extracted features were reduced by using a particle swarm optimization (PSO)[25] . This was used to optimize the SVM parameters as a feature selection and finally, the obtained statistics were fed to a support vector machine (SVM) binary classifier to diagnose skin biopsies from patients as either malignant melanoma or benign nevi. The obtained classification accuracies show better performance in comparison to similar approaches for feature extraction.

2.6 Support vector machine based classification

The SVM classifier is widely used in bioinformatics, due to its high accuracy, ability to deal with high-dimensional data such as genetic factor expression, and edibility in these special context diverse sources of data[8] [26] . The SVM is used in the classification of histopathology images which is often the final goal in image analysis, particularly in cancer applications. Features derived from segmented nuclei from histopathology are usually a prerequisite to extracting higher level information regarding the state of the disease.

3   RESULTS AND DISCUSSION

Feature Process: In our algorithm we used a 5-fold cross validation test. The LIBSVM [8] [27] is a process library for support vector machines, where the linear, (RBF) radial basis function kernel [8] and polynomial kernel are each under magnification. The melanoma images that are confirmed by the pathologist are treated as negative images, while the nevus ones are treated as positive images. For instance, the classifying of breast tissue by [18] found 67.1% accuracy in labeling nuclei as benign or malignant, the classification of histopathology images presented in [28] resulted in 88.7% accuracy in the diagnosis of lung cancer, 94.9% accuracy in the typing of malignant mesothelioma, and 80.0% to 82.9% accuracy in the prognosis of malignant mesothelioma for Fulgent stained lung sections. Keenan et al [13] reported accuracies of 62.3–76.5% in the grading of H&E stained cervical tissue. Skin cancer is a fast developing disease of modern society, reaching 20% [13] increase of diagnosed cases every year. As reported in [13], only experts arrive at 90% sensitivity and 59% specificity in skin lesion diagnosis, while for less trained doctors, these figures shown a significant drop to around 62–63% for general practitioners , as mentioned in Table 1.

Table 1.   Described Sensitivity And Specificity.

	Sensitivity	Specificity
Experts	**90 %**	**59 %**
Dermatologists	**81%**	**60 %**
General	**85%**	**36 %**
practitioners	**62%**	**63 %**

The presented work has been implemented using MATLAB R2013b, and tested using 79 images (29 benign images and 50 melanoma images). The images are sampled and split into training set (mixed 64 images) and test set (mixed 15 images). These 79 images were fed into the proposed SVM network. The performance of the algorithm was evaluated by computing the percentages of sensitivity (SE) equation (13), specificity (SP) equation (14) and accuracy (AC) equation (15): the respective definitions are as follows:

$$SE = \frac{TP}{(TP+FN)} \times 100 \tag{13}$$

$$SP = \frac{TN}{(TN+FP)} \times 100 \tag{14}$$

$$AC = \frac{(TP+FN)}{(TN+TP+FN+FP)} \tag{15}$$

where TP is the number of true positives (expects malignant as malignant), TN is the number of true negatives (expects benign as benign), FN is the number of false negatives (expects malignant as benign), and FP is the number of false positives (predicts benign

as malignant). Since it is interesting to estimate the performance of classifier based on the classification of benign and malignant skin cell nuclei, sensitivity, specificity and accuracy of prediction have been calculated according to the above equations for all of the testing data. Table 2 shows the average resulting SE, SP and AC of 5 fold validation of the proposed networks by using LIBSVM. This study used RBF kernel with gamma=0.0084732 and C=3525.0051. We did many trials to improve our experimental results; Table 3 shows the results without using (SFS) technique. These low results can be compared with our proposed results displayed in Table 2 by using (SFS) sequential feature selection technique.

Table 2.   The result after training of the (svm) network. with using (SVM+WLG+PSO).

No of images	Sensitivity	Specificity	Accuracy
79	94.1	80.2	87.1

Table 3.   The Result After Training Of The (Svm) Network, Without Using (Sfs)Technique.

No of images	Sensitivity	Specificity	Accuracy
79	83.6	70.7	77.4

4 CONCLUSION

In this paper, SVM has been implemented for classification of benign from malignant skin tumor. 79 images sampled from microscopic slides of skin biopsy have been used in the current work. MATLAB software is used to implement the proposed work; these features were carried out to generate training and testing of the proposed SVM. The present work is a new application based on histo-pathological images of skin lesions that required finding out new features getting the reduced numbers of features and getting better accuracy. It was required to modify many of the mentioned techniques to make them work for such an application. The higher accuracy of diagnosis of the proposed work is calculated and displayed. The results show that using the (SVM+WLG+PSO) technique produces one of the best results compared with expertise/physicians results displayed in Table 1. The obtained accuracy of the system is 87.1%, whereas the sensitivity and specificity were found to equal 94.1% and 80.2% respectively. By comparing the results in Tables 1 and 2, it can be seen that the proposed system produced more accurate classification results than physicians did. This paper concludes that there are some possible factors to improve the accuracy of detecting malignant melanoma by having a higher number of images for training of the SVM network. Future work directions will be to use a hybrid approach of genetic algorithms and another different method to improve feature extraction and feature selection. Moreover, we are planning to use different classification techniques such as Neuron-Fuzzy algorithms to improve the accuracy.

5 REFERENCES

[1] A. B.O.S., "Causes of death 2010," Commonwealth of Australia, 2012.

[2] A. I. o. H. a. Welfare, "Cancer in Australia: An Overview," A. I. O. H. A. Welfare, Ed., ed. Canberra, Australia: Australian Institute of Health and Welfare, 2006.

[3] P. Baade and M. Coory, "Trends in melanoma mortality in Australia: 1950–2002 and their implications for melanoma control," Australian and New Zealand Journal of Public Health, 2007, 29:383–386.

[4] AIHW, "Cancer in Australia 2001," Australian Institute of Health and Welfare, Canberra, Australia Cat. No. CAN 23.), 2004.

[5] M. Jayamanmadharao, M. S. Ramanaidu, and K.Reddy, "Impulse Noise Removal from Digital Images-A Computational Hybrid Approach," Global Journal of Computer Science and Technology Graphics & Vision, vol. 13, 2013.

[6] F. Robert-Inacio, Dinet, E., "An adaptive median filter for colour image processing," Proc. 3rd CGIV, pp. 205–210, 2006.

[7] A. K. Jain, "Fundamentals of Digital Image Processing," Englewood Cliffs, 1989.

[8] B. Scholkopf, Tsuda, K., Vert, J.P., "Kernel Methods in Computational Biology," MIT Press, 2004.

[9] V. Kecman, Learning and Soft Computing: Support Vector Machines, Neural Networks, and Fuzzy Logic Models (Complex Adaptive Systems). Cambridge, MA, 2001: The MIT Press, 2001.

[10] L. P. Wang and X. J. Fu, Data Mining with Computational Intelligence. Berlin: Springer, 2006.

[11] A. Sloin and D. Burshtein, "Support vector machine training for improved hidden markov modeling," Signal Processing, IEEE Transactions, 2008, 56: 172–188.

[12] F. Chu and L. P. Wang, "Applications of support vector machines to cancer classification with microarray data," International Journal of Neural Systems, 2005, 15: 475–484.

[13] R. H., C. H. Chan, and M. Nikolova, "Salt-and-Pepper Noise Removal by Median-Type Noise Detectors and Detail-Preserving Regularization," IEEE Transactions On Image ProcessinG, 2005, 14.

[14] A. Criminisi, P. Perez, and K. Toyama, "Region filling and object removal by exemplar-based image inpainting," IEEE Transactions on Image Processing, 2004, 13: 1200–1212.

[15] M. OpenCourseWare and http://ocw.mit.edu, "Introduction to Communication, Control, and Signal Processing For Wiener Filtering," S. 2010S, Ed., ed, 2010.

[16] W. Page. (2014). Wiener Filtering and Image Processing. Available: https://www.clear.rice.edu/elec431/projects95/lords/wiener.html

[17] J. Pavlovicova, M. Oravec, and M. Osadský. (2010, An application of Gabor filters for texture classification ELMAR, 2010 PROCEEDINGS 23–26.

[18] QinzLi, "Dark line detection with line width extraction," in 15th IEEE International Conference on Image Processing, 2008. ICIP 2008, ed, 2008.

[19] R. M. Hodgson, Bailey, D.G., Naylor, M.J., Ng, A.L.M., McNeill, S.J., "Properties, implementations and applications of rank filters," image and vision computing, vol. 3, pp. 1–14, 1985.

[20] M. R. Sweet, "adaptive and recursive median filtering ".

[21] A. Merritt, M. Mehta, and A. Warke, "Parallel Edge Detection," 2011.

[22] Q. Abbas," Lesion border detection in dermoscopy images using dynamic programming," Skin Research and Technology, vol. 17, pp. 91–100, 2011.

[23] S. Wu and A. Amin, Eds., Automatic Thresholding of Gray-level Using Multi-stage Approach (Proceedings of the Seventh International Conference on Document Analysis and Recognition (ICDAR 2003). IEEE computer Society, 2003, p.^pp. Pages.

[24] J. Sikorski, "Identification of malignant melanoma by wavelet analysis," in Student/Faculty Research Day, CSIS, Pace University., 2004.

[25] Y. del Valle, G. K. Venayagamoorthy, S. Mohagheghi, J.-C. Hernandez, and R. G. Harley, "Particle Swarm Optimization: Basic Concepts, Variants and Applications in Power Systems," IEEE TRANSACTIONS ON EVOLUTIONARY COMPUTATION, vol. 12, pp. 171–195, 2008.

[26] L. P. Wang, B. Liu, and C. R. Wan, "Classification using support vector machines with grading resolution," IEEE International Conference on Granular Computing, vol. 2, pp. 666–670, 2005.

[27] A. H. Khandoker, M. Palaniswami, and C. K. Karmakar, "Support Vector Machine for Automated Recognition of Obstructive Sleep Apnea Syndrome From ECG Recordings," IEEE Trans. Information Technology in Biomedicine, vol. 13, pp. 37–48, 2009.

[28] J. C. Caicedo, A. Cruz, and A. Gonzalez, "histopathology image classification using Bag of Features and Kernel Functions," Bioingenium

Information Technology – Wan et al. (Eds)
© 2015 Taylor & Francis Group, London, ISBN 978-1-138-02785-5

A new effective method of image enhancement for license plate location

Zhen Jiang, Yule Yuan & Yong Zhao
Key Laboratory of Integrated Microsystems School of Electronic and Computer Engineering, Peking University Shenzhen, China

ABSTRACT: This paper introduces a novel framework to enhance the car image in extreme lighting conditions for license plate location. First, our method decides the plateau threshold value of the histogram from the input image and modified this histogram to get a preliminary enhanced image. Then divide this image into 32*32 windows. Finally, we calculate the mean luminance and standard deviation of the pixels in each window that determine the gray value of pixels in the output image. Experimental results show that this method can successfully enhance images both in under-exposed and over-exposed conditions.

1 INTRODUCTION

As a result of population growth and changes in the way of people's travel, ITS (intelligent transportation system) technologies have been the key research issue to solve the traffic congestion, and license plate location is an essential important stage in vehicle license plate recognition for ITS. Researchers have found many diverse methods for license plate location. However, it is difficult to precisely locate the license plate under certain lighting conditions. It is necessary to enhance low contrast images before plate location, and the edge details of an image captured at different lighting conditions using an industrial camera will always be obscured when the image is under-exposed or over-exposed.

One of the common methods used for digital image enhancement is Global Histogram Equalization (GHE). The idea of GHE is to redistribute the gray levels of an image and produces an image with a uniform probability density function. This method is generally successful in improving the contrast of an image and simple to be implemented. GHE can increase the entropy of the image, therefore, more information can be obtained from the image.

However, GHE still has some shortcomings, some unwanted effects are usually produced by GHE mapping, such as over enhancement for intensity levels with high probability value, and loss of contrast in levels with low probability value. As a result, the background area will be enhanced more intensely compared with the object of interest because of occupying a large part of an image. So, GHE is not really the most effective method to enhance license plate images.

Since car image always has low contrast between background and license plate, in order to recognize license plate area correctly, gray stretch is another widely used enhancement algorithm. Yuan proposed a method in [8], which is given by (1):

$$I'_{i,j} = \begin{cases} \dfrac{c}{a} * P_{i,j}, & I_{x,y} < a \\[2mm] \dfrac{(d-c)}{(b-a)} * (I_{x,y} - a) + c, & a \le I_{x,y} < b \\[2mm] \dfrac{(255-d)}{(255-b)} * (I_{x,y} - b) + d, & I_{x,y} \ge b \end{cases} \quad (1)$$

where $I_{x,y}$ is the gray value of the source image pixel and its gray-scale range is [a, b], the expected dynamic range of the enhanced image $I_{x,y}$ is [c, d], the values of c, d are set as 30, 225 respectively. This method can be very efficient and simple to be implemented for car image captured in normal lighting conditions. But when the image is under-exposed or over-exposed, the details of the license plate area will be lost.

Zheng devised a new method. They divided the car image into several windows, and modified the gray values of pixels according to the window's mean luminance and standard deviation (MLSD). This method will be explained in the next section. This method of image enhancement works well for under-exposed image, but for image in over-exposed condition, it leads to an overly enhanced output image. The experimental results are shown in section 4, this paper further researches the issue and improve the method through a preprocessing algorithm based on a modified histogram equalization method. The details of the method are presented in section 3.

2 RELATED WORK

MLSD was introduced by Zheng in 2005, In this scheme, the car image was first divided into many blocks equably, then it computed out the mean luminance of pixels in each block to make sure that the output image has the same luminance with the input image. The standard deviation of pixels in each block should also be calculated to measure the contrast of the blocks. The algorithm of MLSD is described below.

Firstly, this method chooses an example car image with 384*288 pixels and 256 gray levels, then $I_{i,j}$ is used to denote the luminance of the pixels $P_{i,j}$ in the car image and $I'_{i,j}$ is used to denote the luminance in the enhanced image (row: $0<i<288$, column:$0< j<384$). Finally $I_{i,j}$ and $I'_{i,j}$ are set to satisfy (2), where $W_{i,j}$ is a window centered on pixel $P_{i,j}$, $\bar{I}_{Wi,j}$ and $\sigma w_{i,j}$ are the mean luminance and standard deviation of the pixels in the window $W_{i,j}$,

$$I'_{i,j} = f(\sigma W_{i,j}) \bullet (I_{i,j} - \bar{I} W_{i,j}) + \bar{I} W_{i,j} \qquad (2)$$

$f(\sigma W_{i,j})$ is an enhancement coefficient. Here we select a 48*36 rectangle as the window $W_{i,j}$ and the car image is separated into 8*8 windows. Considering the computational complexity, it is not advisable to compute out the values of $\bar{I}_{Wi,j}$ and $\sigma w_{i,j}$ at each pixel, we can use the bilinear interpolation algorithm to get them, it is shown in (3) and Figure.1:

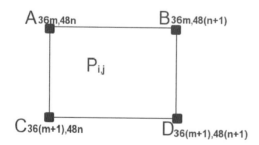

Figure 1.   Bilinear interpolation to compute $\sigma w_{i,j}$ and $\bar{I}_{Wi,j}$ in the rectangle block ABCD.

$$\bar{I}_{W_{i,j}} = (1-c_y)[(1-c_y)\,\bar{I}_{W_A} + c_x \bar{I}_{W_B}]$$
$$+ c_y[(1-c_x)\,\bar{I}_{W_C} + c_x \bar{I}_{W_D}] \qquad (3)$$
$$\sigma_{W_{i,j}} = (1-c_y)[(1-c_x)\sigma_{W_A} + c_x \sigma_{W_B}]$$
$$+ c_y[(1-c_x)\sigma_{W_C} + c_x \sigma_{W_D}]$$

where $36m< i <36(m+1)$, $48n< j <48(n+1)$, m, n=1, 2, 3..., $c_x = (j-48n)/48$, and $c_y = (i-36m)/36$. After $\sigma w_{i,j}$ is computed out, $f(\sigma W_{i,j})$ is given by (4):

$$f(\sigma W_{i,j}) = \begin{cases} \dfrac{3}{\frac{2}{400}(\sigma W_{i,j}-20)^2 +1} & if\ 0 \leq \sigma W_{i,j} < 20 \\[2mm] \dfrac{3}{\frac{2}{1600}(\sigma W_{i,j}-20)^2 +1} & if\ 20 \leq \sigma W_{i,j} < 60 \\[2mm] 1 & if\ \sigma W_{i,j} \geq 60 \end{cases} \qquad (4)$$

The car images of under-exposed and over-exposed enhanced by MLSD are shown in Figure. 2 and Figure. 3, we can see that the license plate region in under-exposed image has been strengthened, but in over-exposed condition, the result was even worse than the original gray scale image of the car.

The limitation of the principle should be analyzed in order to improve this method. Briefly, the purpose of the algorithm is to enhance the contrast between license plate image and the background for locating the license plate, and at the same time enhance the contrast between characters and license plate region's background for character recognition. In under-exposed condition, the license plate has a relatively high luminance, characters and background of the car image belong to dark section. After the enhancement by MLSD, it is easy to distinguish the license plate and characters on it. However, in over-exposed condition, characters and the license plate have a higher luminance compared with the background of the car image, so the whole region of the license plate is overly enhanced as experimental result shows.

3 METHODOLOGY

3.1 *Image enhancement*

To improve the performance of MLSD, we introduce a method called SAPHE (Self-Adaptive Plateau Histogram Equalization) as preprocessing algorithm which is mainly to optimize the contrast of the whole image. SAPHE is based on Histogram Equalization (HE), it uses a threshold value to modify the original histogram, and then histogram equalization based on the modified histogram is carried out. Unlike global histogram equalization, SAPHE will not overly enhance the background area, which normally occupies a large portion of an image. The procedure of SAPHE is described in the next paragraph.

Firstly, we calculate the histogram $h(x)$ of the input image, then we found the median value of $h(x)$ and set it as the plateau threshold value T [11], then we create the modified histogram $h_{mod}(x)$ based on the threshold value T, by using (5), and the cumulative function $c(x)$ is also determined.

$$h_{\mathrm{mod}}(x) = \begin{cases} h(x), & h(x) \le T \\ T, & otherwise \end{cases} \qquad (5)$$

Finally, the gray scale of pixels in the output image is given by (6):

$$f(x) = \left\lfloor \frac{c(x)}{T} \right\rfloor \qquad (6)$$

where $\lfloor\ \rfloor$ represents truncation to the next lower integer. After steps described above, we get the image enhanced by SAPHE, then we use MLSD to enhance the image to make it easy for license plate location.

3.2 License plate location

Here we use MSER (Maximally Stable Extremal Region) to detect the license plate. MSER was proposed by J. Matas, O. Chum, M. Urban, and T. Pajdla [10]. Extremal regions are connected components of an image binarized at a certain threshold. In an image I, extremal regionsare defined by (7) and (8):

$$\forall p \in R_i, \forall q \in boundary(R_i) \rightarrow I(p) > I(q) \qquad (7)$$

$$\forall p \in R_i, \forall q \in boundary(R_i) \rightarrow I(p) < I(q) \qquad (8)$$

where p and q are pixels in image I. (7) Defines the maximum intensity region and (8) defines the minimum intensity region. In license plate images, every character can be treated as an extremal region. We perform the standard MSER algorithm on the enhanced image. In our method, we use rectangles to denote the regions. For every detected MSER, we can construct a rectangle. However, many MSERs, which are not characters, are also detected. These noise regions seriously affect the extraction of the license plate.

As we all know, the aspect ratio, width to height of a character is less than 1. In our algorithm, we set this as a threshold to filter out non-character MSER rectangles. After that, the MSER rectangles of plate characters conspicuously lay in a row and are close to each other. We set a threshold to eliminate the rows whose rectangle number is below 5, the character rectangles which are very close to each other are connected, which forms a bigger rectangle, this bigger rectangle is a candidate plate region.

However, in some images, some regions similar to license plate are also extracted as candidates.

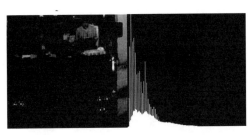

(a). GSI and HI of input image.

(b) SI and HI of output image enhanced by MLSD.

(c) GSI and HI of output image enhanced by the proposed method.

Figure 2. A car image in under-exposed condition, and we got the images enhanced by MLSD and the proposed method. The gray scale images (GSI) and histogram images (HI) of those images can be compared in this figure. After enhanced by the proposed method and MLSD, the details of the license plate region in the image are strengthened.

(a). GSI and HI of input image.

(b). GSI and HI of output image enhanced by MLSD.

(c). GSI and HI of output image enhanced by the proposed method.

Figure 3. A car image in over-exposed condition was selected, and we got the image enhanced by MLSD and the proposed method. Characters on the car image enhanced by MLSD can hardly be recognized.

(a). Candidate license regions of input GSI (left), result of location (right).

(b). Candidate license regions of GSI enhanced by MLSD (left), result of location (right).

(c). Candidate license regions of GSI enhanced by the proposed method (left), result of location (right).

(d). Candidate license regions of input GSI (left), result of location (right).

(e). Candidate license regions of GSI enhanced by MLSD (left), result of location (right).

(f). Candidate license regions of GSI enhanced by the proposed method (left), result of location (right).

Figure 4. License plate location of images enhanced by MLSD and the proposed method.

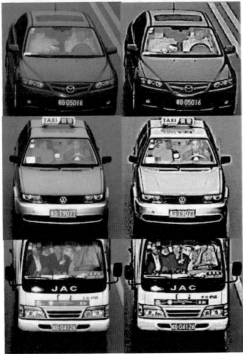

Figure 5. The gray scale images (left) and images enhanced by the proposed method (right).

Since characters in license plate have almost the same height, therefore the variance of a real plate is small and the variance is big for a fake one. So the candidate with the smallest height variance is selected to be the last real license plate.

Since the enhancement of car images that captured in normal lighting situation is easy to be

implemented, the work of the paper is mainly focused on images in extreme lighting situation. However, in order to illustrate that the method composed in this paper is available in a wide range, we compare the gray images and images enhanced by the proposed method below.

4 EXPERIMENT RESULTS

We evaluate the proposed method on images which were captured from a real highway using a Basler industrial camera, we choose two images in respective under-exposed and over-exposed lighting condition with 384*288 pixels and 256 gray levels, and select a 32*32 rectangle as the rectangular window. Experiments have been implemented to test the method proposed by the paper, in addition, we include results from MLSD and original gray scale images for comparison.

The results are shown in Figure. 2, the input images have been successfully enhanced by the proposed method and MLSD method compared with gray scale image. The enhancement of car images captured in under-exposed condition has always been a key issue, and MLSD is a very effective method, this is why we choose MLSD method to improve, we get an even better result from the proposed method and the histogram image shows that pixels with high gray levels are more clear compared with histogram images from MLSD.

However, as shown in Figure. 3, in over-exposed condition, the car image is overly enhanced by MLSD, and details of license are almost lost, this can be seen in histogram image. In the proposed method, car image is first optimized to enhance the contrast between characters on license and license plate to avoid being overly enhanced, as a result, the proposed method has proven to be an effective way to enhance images in over-exposed lighting condition. Nevertheless, in MLSD method, the license plate, including characters is considered as a high gray scale section. After the enhancement, we can hardly recognize the characters on the license plate. Experiments also show that, in the proposed method, the accuracy of locating license plate is greatly improved, as Figure 4.

5 CONCLUSION

This paper proposed a method based on MLSD to enhance car images captured by industrial cameras, SAPHE is used as a preprocessing method to make sure that the image is not over enhanced. Results show that the proposed method is much better than MLSD, especially in over-exposed condition. The details of the license plate region are completely saved. Experiments demonstrate that the proposed method can successfully enhance car images both in under-exposed and over-exposed conditions.

ACKNOWLEDGMENT

We would particularly like to thank my research lab, which has sponsored us to do research on our paper. We would also thank our teammates. We got many inspirations, and they have given us so many valuable suggestions.

REFERENCES

Abdullah-Al-Wadud, M & Kabir, M.H & Dewan, M.A.A. "A dynamic histogram equalization for image contrast enhancement," *Consumer Electronics*, IEEE Transactions on, 2007, 53(2): 593–600.

Ibrahim, H & Kong, N.S.P. "Brightness preserving dynamic histogram equalization for image contrast enhancement," *Consumer Electronics*, IEEE Transactions on, 2007, 53(4): 1752–1758.

J. Matas & O. Chum & M. Urban & T. Pajdla. "Robust wide baseline stereo from maximally stable extremal regions," In BMVC02, volume 1, pages 384–393, London, UK, 2002.

Kong,N.S.P & Ibrahim, H & Ooi, C.H. "Enhancement of microscopic images using modified self-adaptive plateau histogram equalization[C]," *Computer Technology and Development*, 2009. ICCTD'09. International Conference on. IEEE, 2009, 2: 308–310.

R.C.Gonzalez & R.E.Woods. *Digital Image Processing*, 2nd ed., Prentice-Hall of India, New Delhi, 2002.

Stark.J.A. "Adaptive image contrast enhancement using generalizations of histogram equalization" *Image Processing*, IEEE Transactions on, 2000, 9(5): 889–896.

Wang, B & Liu, S.Q & Li,Q. "A real-time contrast enhancement algorithm for infrared images based on plateau histogram," *Infrared Physics & Technology*, 2006, 48(1): 77–82.

Wang, C & Ye, Z. "Brightness preserving histogram equalization with maximum entropy: a variational perspective," *Consumer Electronics*, IEEE Transactions on, 2005, 51(4): 1326–1334.

Zhao, Y & Yuan, Y & Bai, S. "Voting-based license plate location" 2011 14th International IEEE Conference on. IEEE, 2011: 314–317.

Zheng, D & Zhao, Y & Wang, J. "An efficient method of license plate location," *Pattern Recognition Letters*, 2005, 26(15): 2431–2438.

Zhicai, L & Zhi-guang, L. "A Review on Image Process Technique of Thermal Imager," *Infrared Technology*, 2000, 6: 005.

Information Technology – Wan et al. (Eds)
© 2015 Taylor & Francis Group, London, ISBN 978-1-138-02785-5

A new method for HRRP motion compensation based on twice velocity estimating

Shuguang Jian & Tao Sun
National Key Laboratory of Science and Technology on Blind Signal Processing,Chengdu, China

ABSTRACT: High-resolution range profile (HRRP) can be obtained through the stepped-frequency radar, keeping of the wideband receiver in wideband radars. However, HRRP is troubled by distortions, such as range profile shift, peak reduction, and point spread function spreading, which are caused by target radial motion, especially for the space targets with a high velocity. In this paper, a new motion compensation method based on twice velocity estimating is proposed for the stepped-frequency radar. Use the minimum entropy criterion and maximum pulses sum criterion to do twice velocity estimate, then the accurate velocity will be obtained. Simulation and practical echo processing results have proved the proposed method is able to reduce the range profile distortions effectively.

KEYWORDS: stepped frequency radar; motion compensation; twice velocity estimating; HRRP

1 INTRODUCTION

In radar systems, synthetic high-resolution range profiles (HRRPs) can be obtained by using stepped-frequency waveforms (SFWs) [1,2]. SFW consists of a few chirp signals which has same narrow bandwidth and continuous frequency, which has advantages of both synthetic wide bandwidth and moment narrow bandwidth. The frequency signal needs lower bandwidth of the radar receivers and A/D sampling rate, so it has been applied comprehensively. However, range profile shift, peak reduction, and point spread function (PSF) spreading are produced by target radial velocity, especially for a space target that has a higher speed. Motion compensation is necessary in this situation.

To compensate the influence of the target motion, parameter estimation is usually used. The researches have shown that the range profile distortions are majorly caused by the space target's radial velocity and the radial acceleration can be ignored [3]. An analysis of distortions induced by the radial velocity can be found in [4]. Specific SFWs can be designed for minimizing HRRP distortions due to the Doppler effect (radial velocity). In [5], Doppler processing is used to estimate the target velocity based on range-profile sharpening. A new approach has been presented in [6] where the phase errors caused by motion are removed by using a new waveform composed of two SFW pulse trains. In addition, in this case, the technique proposed in [7] only allows compensating phase terms due to the target velocity.

But these documents are not suited to space targets, which has a lower signal noise ratio (SNR) and a higher speed.

2 SFW RANGE PROFILE

2.1 Received signal model

A SFW waveform consists of a sequence of N narrow band pulses centered at increasing carrier frequencies, namely $f_n = f_0 + n\Delta f$ with $n = 0,1,2,...,N–1$. The expression of the analytic transmitted signal is as follows:

$$u(t) = \sum_{i=0}^{N-1} u_1(t - iTr)\exp[j2\pi(f_0 + i\Delta f)(t - iTr)]$$

$$= \sum_{i=0}^{N-1} rect(\frac{t - iTr}{T_1})\exp[j\pi K(t - iTr)^2]$$

$$\bullet \exp[j2\pi(f_0 + i\Delta f)(t - iTr)] \tag{1}$$

where $u(t)$ is the complex envelope of the narrow band transmitted pulse, Tr is the pulse repetition interval, f_0 is the carrier frequency, Δf is the frequency step, and N is the number of transmitted frequencies. The total time length of the SFW is $(N-1)PRI + Ti$ where Ti is the time length of the transmitted pulse $u(t)$. The time interval NTr is called ramp repetition interval (PRI), and it represents the processing time required to form the HRRP. The signal return from a moving target is given by the following:

$$u_r(t) = Au(t - \tau(t))$$

$$= A\sum_{i=0}^{N-1} u_1(t - iTr - \tau(t))$$

$$\bullet \exp[j2\pi(f_0 + i\Delta f)(t - iTr - \tau(t))] \qquad (2)$$

where A is a complex constant, which accounts for the attenuation and the phase displacement introduced by the scatter, while the time varying delay $\tau(t)$ is as follows:

$$\tau(t) = 2R(t)/c \approx \tau_0 + 2vt/c \qquad (3)$$

where $\tau_0 = 2R(0)/c$ and v is the space target's velocity.

2.2 Range profile distortions

One of the most critical problems in synthetic range profile reconstruction is the distortions induced by the target motion. The target velocity generates first and second order phase terms in the received signal, which produce shifts and distortions in the PSF. The problem exists in typical radars, especially in the scenarios of the low Pulse Repetition Frequency (PRF) radars and/or the fast maneuvering target detection. The technique used to remove the motion-induced phase errors is called "motion compensation" [8].

The phase of received signal can be written as follows, with only translational motion considered.

$$\varphi_i(t) = \varphi_1 + \varphi_2 + \varphi_3 + \varphi_4 \qquad (4)$$

where

$$\varphi_1 = \frac{4\pi f_0}{c}\left[R(0) - v(\frac{T}{2} + \frac{2R}{c})\right]$$

$$\varphi_2 = \frac{4\pi k\Delta f R(0)}{c}$$

$$\varphi_3 = \frac{4\pi k\Delta fv}{c}(\frac{T}{2} + \frac{2R(0)}{c}) + \frac{4\pi kf_0 vT_r}{c}$$

$$\varphi_4 = \frac{4\pi k^2\Delta fvT_r}{c}$$

It can be observed that the first order phase item does not affect the range profile forming, and the second order phase item causes the constant shift of the range profile in different PRIs. The third order phase item is the linear phase change caused by target motion while the fourth phase item is the nonlinear one. Both of them can cause an apparent range shift Δr that maps the target into a wrong position in the range profile, and play the principal role in the signal mismatch.

The main radar and target parameters are shown in Table 1.

Table 1. Radar and target parameters.

f_0(GHz)	N	Δf(MHz)	v (m/s)	R(0)(Km)
3.5	128	5	80	1.8

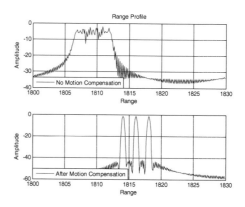

Figure 1. Typical Range Profiles.

Figure 1 compares the range profiles before and after the motion compensation. In this case, a three-point target is considered, and each point is located 2 m from the others.

It is shown that the range shift, PSF spreading, and peak reduction occur without the motion compensation.

3 A NEW METHOD OF VELOCITY ESTIMATE

A space target's velocity is much higher than common targets, which can achieve 7000 m/s. To make matters worse, the SNR from the echoes from a space target is much lower, because of the very large distance. These problems make it necessary to find a better way to estimate the velocity of the space target. An effective velocity estimation method based on minimum entropy and the maximum pulse sum criterions is brought up in this paper.

In this part, positive SFW and negative SFW are launched alternately, $S_1(k)$ is the sampling value of a positive received waveform whose frequency is from f_0 to $f_0 + (N-1) \Delta f$ and the interval is Δf, $S_2(k)$ is the negative one. N is the total number of stepped frequency.

3.1 Minimum entropy criterion

Entropy is the measurement for the disorder, and it decreases with the disorder or quality of the range profile improved. The entropy definition of the HRRP is as follows:

$$En = \sum_{i=0}^{N-1} \overline{H(i)} \log(\overline{H(i)}) \tag{5}$$

It can be observed that the entropy is able to reach least when the influence of the target motion is effectively compensated.

The main radar and target parameters in simulations are shown in Table 2.

Table 2. Radar and target parameters.

f_0(GHz)	N	Δf (MHz)	v(m/s)	$R(0)$ (Km)	SNR(dB)
3.5	128	5	500	1.8	0

Through Equation (4), the results of entropy in the situation can be obtained when the radial velocity of target changes from 0 to 1000 m/s. It is shown that the radial velocity of the target is 506 m/s when the entropy reaches least, which is very close to the true value (500 m/s) in the simulation.

3.2 Maximum pulses sum criterion

Through 3.1 we get 506 m/s as the suspected velocity, now using the maximum pulses sum criterion we can make the velocity estimate more accurate.

In this part, we launch positive SFW and negative SFW alternately, $S_1(k)$ is the sampling value of a positive received waveform whose frequency is from f_0 to $f_0 + (N-1) \Delta f$ and the interval is Δf, $S_2(k)$ is the negative one. N is the total number of stepped frequency.

We use the pulses sum function as the convergence criterion, which defines as follows:

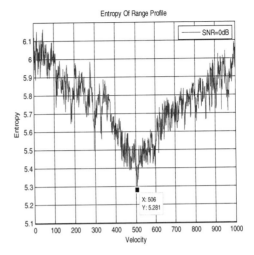

Figure 2. Entropy Of Range Profile.

$$Ps(v) = \left| \sum_{k=0}^{N-1} S_1(k) S_2(k) \right| \tag{6}$$

For an SFW radar, the error function can be written as follows:

$$R(\Delta v) = \left| \frac{\sin(2\pi N \Delta v T_r (2f_0 - \Delta f)/c)}{\sin(2\pi \Delta v T_r (2f_0 - \Delta f)/c)} \right| \tag{7}$$

when $\Delta v = 0$, which means the velocity estimate equals the real one, Equation (7) can get the maximum.

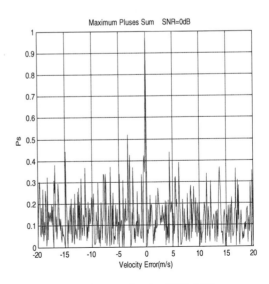

Figure 3. The pulses sum function (SNR=0dB).

In Figure 3, we can conclude that the maximum pulses sum criterion has a good ability of cancelling noise.

3.3 *Motion compensation steps*

a Use the exterior system or known information to get first estimate velocity of the space target.
b Employ first estimate to get a search scope of minimum entropy criterion, and then use the mean of estimates as the second estimate.
c Do motion compensation by using the second estimate, through the maximum pulses sum criterion, a more accurate velocity estimate can be got.
d Use the final estimate to do motion compensation, so that a clear range profile will be obtained.

4 SIMULATION AND CONCLUSION

The compared range profiles are shown in Figure 4. The target's velocity is 500 m/s. Through the estimate method, the final estimate velocity is 500.47 m/s, which is really close to the real velocity. In the range profiles, there are no clear points in the range profile before motion compensation, but, after motion compensation, the three points have been shown clearly just as expected.

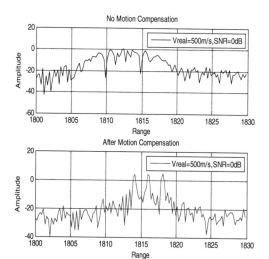

Figure 4. Range Profile Comparison.

Through the simulations, we can see that the method also has a good ability of cancelling noise, which is very important to a space target. While

the state of the technology has some defects, for example, the speed of space targets always is higher than 7000 m/s, while the traditional methods can not estimate the high velocity accurately. However,

Table 3. Radar and target parameters.

f_0(GHz)	N	Δf(MHz)	v(m/s)	R(0)(Km)
3.1	100	2	0	1.8

in this way, the range profile can be obtained clearly even though the velocity is so high.

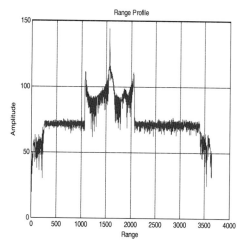

Figure 5. Range Profile Of Real Signals.

5 REAL SIGNALS PROCESSING

The main radar and target parameters in real signals are shown in Table 3.

The real signals are received from a plane, which has a velocity of 180 m/s, after motion compensation using the twice velocity estimation, range profile has been clearly obtained, as Figure 5 shows.

REFERENCES

[1] A.Scheer and J.L.Kurtz, "Coherent Radar Performance Estimation," Artech House, 1193.
[2] D.R.Wehner, High Resolution Radar. Norwood, MA:Artech House, 1995.
[3] Wen xiang, "The influence of a space target's HRRP caused by target motion," Research On

Telecommunication Technology, Vol. 382, No. 6, 2013.

[4] N.Levanov and E.Mozeson, "Nullifying ACF grating lobes in stepped-frequency train of LFM pulses," IEEE Trans. Aerosp. Electron. Syst. vol. 39, no. 2, pp. 694–703, Apr. 2003.

[5] F.Berizzi, M.Martorella, and A.Cacciamano, "Synthetic range profile focusing via contrast optimization," unpublished.

[6] N.Levanov, "Stepped-frequency pulse train radar signal," IEE proceedings on Radar, Sonar and Navigation, Volume 149, Issue 6,Dec. 2002 Page(s): 297–309.

[7] M.A.Temple, K.L.Sitler, R.A.Raines, and J.A.Hughes, "High Range resolution (HRR) improvement using synthetic HRR processing and stepped-frequency polyphase coding," IEE Proc. Radar Sonar and Navigation, Vol.151, No.1, February 2004.

[8] F.Berizzil and M.Martorellal, A.Cacciamano, "A Contrast-Based Algorithm For Synthetic Range-Profile Motion Compensation," IEEE, 2008, 46(10).

Information Technology – Wan et al. (Eds)
© 2015 Taylor & Francis Group, London, ISBN 978-1-138-02785-5

A new species identification method of freshwater fish based on machine vision

Jing Zhang & Qihua Wang

School of Information Science and Technology, Dalian Maritime University
Educational Technology & Computing Center, Dalian Ocean University, Dalian, china

ABSTRACT: China is the major country that farms and produces freshwater fish in the world. The accurate identification of different species is the prerequisite for deep processing of freshwater fish. Aiming at the current problems emerged in our freshwater fish identification; a new species identification method of freshwater fish based on machine vision was put forward. The ratio of long axis and short axis of the whole fish body and the ratio of long axis and short axis of the segment fish body, as well as the texture features of the fish body surface, were extracted and connected in series to construct the feature vector respecting the fish species. The linear support vector machine classifier was utilized for fish species identification and the randomly bought crucians and cyprinoids undertook duty of experimental materials. The proposed method was tested and compared with other methods. The experimental results show that our method obtains satisfactory species identification results and is robust to light, angle and other details. It provides methods and ideas for the prerequisite species identification before deep processing of freshwater fish.

KEYWORDS: Freshwater fish; Species identification; Identification accuracy

1   INTRODUCTION

Freshwater fish have many distinguishing characteristics, such as tender meat, delicate taste and rich nutrition. They are one of the most favorite foods and people's most important source of protein in daily life [1,2,3]. China is the major country which farms and produces freshwater fish in the world. The annual output accounts for more than 70% of the world's total output. The common freshwater fish include herring, grass carp, silver carp, bighead, cyprinoid, crucian and bream, etc. [4,5]. The farming and producing of freshwater fish are developed quickly in China, but the processing industry lags to some extent and the market of freshwater fish is still dominated by fresh sales. So, during the seasons in which freshwater fish are sent into market heavily, it is difficult to sell the freshwater fish and the enthusiasm of producers are dampened consequentially. In order to solve this problem, it is necessary to process freshwater fish. Because in China, the farming form of freshwater fish is polyculture, the accurate species identification for different freshwater fish is the premise of deep processing [6,7].

At present, species identification of freshwater fish mainly relies on manual work in China. The shortage of this form is because of poor working environment, high labor intensity and low efficiency [8,9]. In view of this situation, Zhang [10] proposed a species identification method of freshwater fish based on color

components and ratio of long axis and short axis (called RLS in this paper). They first extracted the blue, green and red components and statistically analyzed the threshold in each color component for all fish species. Then they extracted the contour of the fish body, and calculated RLS, which reflected the proportion of body length and body width. And lastly, the classification was achieved through multiple conditions judgment, and the species identification of silver carp, bream, cyprinoid and crucian was successfully accomplished. In this method, the effect of color component is dependent on the image acquisition hardware. Different acquisition equipments may lead to different color component features. And RLS only takes the overall characteristics of the fish body in account, so, considering the different fish which own little variety in body contours, such as crucian and cyprinoid, RLS cannot obtain ideal identification effect. Wan [11] put forward a species identification method of crucian and cyprinoid based on RLSs of the segment fish body. They divided the fish body into 5 segments in the length direction of freshwater fish after the body contour was extracted. Then the radio of length and averaged width of each body segment were employed to be the features which represent the fish body. Species identification for crucian and cyprinoid was finally executed by using BP neural network. In the method, RLSs of the segment fish body overcome the shortage of RLS of the whole fish body. But they still consider only the features based on fish contour, and have not fully utilized the more abundant characteristics of the fish surface texture.

Based on the above consideration, a species identification method of freshwater fish comprehensively considering the fish body contour and the fish surface texture was proposed in this paper. First, we calculated the fish body contour, and extracted RLS of the whole fish body which served as the whole body contour feature. Second, we segmented the fish body contour and extracted RLSs of each segment fish body which served as the segment body contour feature. Finally, we intercepted part of fish body surface and performed Gabor transform on the gotten sub image, and extract Gabor features which served as fish surface texture features. All the gotten features were connected in series to form a feature vector and the linear support vector machine (SVM) classifier was utilized for fish species identification. The proposed method obtained satisfactory species identification results in view of crucian and cyprinoid. It lays a good foundation for the prerequisite species identification before deep processing of freshwater fish.

2 IMAGE ACQUISITION

We adopted two species, crucian and cyprinoid, to verify our method. Experimental materials were randomly purchased from the farmers market. We bought a total of 100 fish, including 50 crucians and 50 cyprinoids, whose weights were in the range of 300–1000 g. In them, 70 randomly selected fish body samples, with 35 samples for each species, served as training set, and the remaining 30 samples served as test set.

Image acquisition system was comprised of hardware and software. The hardware mainly included lights, camera, image acquisition card and computer. The software mainly included image collection program and image processing program. Image collection program was in charge of the collection and storage of image, and image processing program was in charge of fish body contour extraction, fish body segmentation,

Figure 1.   The collected image of crucian sample.

Gabor transform and feature extraction. The collected crucian image after cutting was shown in Fig. 1.

3   FEATURE EXTRACTION

3.1   *Feature of fish body contour*

Feature extraction based on fish body contour is the extraction of image features from the fish body contour image. Features of fish body contour mainly include long axis, short axis, projection area, perimeter and RLS, etc. In this paper, we mainly took the whole body contour feature and the segment body contour feature into account.

First, we extracted RLS of the whole fish body as the whole body contour feature. Binaryzation for the collected fish image was carried out to extract target image from original image. The principle of binaryzation was to divide the image into two subsets through a certain threshold θ. The subset in which pixel values were larger than θ was considered as the background and noise, and was set to 255, which represents white. The other subset in which pixel values were smaller than θ was considered as target image and was set to 0, which represents black. The value of threshold θ was obtained through a lot of experiments, and was set to 230. The image after binaryzation was shown in Fig. 2.

Figure 2.   The image after binaryzation.

We utilized neighborhood boundary extraction algorithm to calculate the fish body contour. If there was a pixel whose value equaled to 0, and the values of its 8 adjacent pixels all equaled to 0, we identified this pixel as the interior point, and its value was set to 255. Fig. 3 showed the gotten fish body contour.

RLS of the whole fish body was calculated by dividing long axis by short axis. Long axis reflected the length of the fish body, and was calculated by the following steps. The contour image was scanned from left to right and the scanning was stopped when the first pixel whose value equaled to 0 was found. The

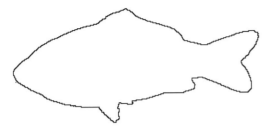

Figure 3. The gotten fish body contour.

abscissa of this pixel was recorded as P_{left}. Then the contour image was scanned from right to left in the same way, and P_{right} was found. Finally, long axis was set to the difference of P_{right} minus P_{left}. Short axis reflected the width of the fish body, and was calculated by the following steps. The contour image was scanned from top to bottom and the scanning was stopped when the first pixel whose value equaled to 0 was found. The ordinate of this pixel was recorded as P_{top}. Then the contour image was scanned from bottom to top in the same way, and P_{bottom} was found. Finally, short axis was set to the difference of P_{bottom} minus P_{top}.

Second, we extracted RLSs of the segment fish body as the segment body contour feature. The gotten contour image was segmented in the direction of the long axis of fish body. The fish body was divided into 5 segments, and in each segment the average body width was calculated in accordance with the short axis direction. RLSs of all segments were extracted as the segment body contour feature.

3.2 Feature of fish surface texture

RLS only considers the characteristics of fish body contour. For the fish which owns little variety in body contours, such as crucian and cyprinoid, we cannot obtain ideal identification results only by using RLS. Because there exists great difference between the fish scale of crucian and cyprinoid, we extracted Gabor features of fish surface image as the fish surface texture feature.

Gabor filter is a bandpass filter. It is selective in direction and frequency and is widely used in research fields of computer vision and image processing. Gabor filter has excellent distinguish ability and is robust to light, angle and other details. Kernel function of Gabor transform is defined as follows:

$$\psi_{\mu\nu}(z) = \frac{\|k_{\mu\nu}\|^2}{\delta^2} \cdot \exp\left(-\frac{\|k_{\mu\nu}\|^2\|z\|^2}{2\delta^2}\right)\cdot\left(\exp(ik_{\mu\nu}z)-\exp\left(-\frac{\delta^2}{2}\right)\right) \quad (1)$$

where $z=(x,y)$ represents the coordinate of pixel. $k_{\mu\nu}$ is defined as $k_{\mu\nu}=k_\nu e^{i\phi\mu}$, in which $k_\nu=k_{max}/f^\nu$

and $\phi_\nu=\frac{\mu\pi}{8}$, k_{max} is the maximum frequency, μ represents orientation and ν represents scale, δ determines the ratio of width and length of the Gauss window. In this paper, we set 6 orientations and 4 scales, i.e. $\mu = 0, 1, ..., 5$, $\nu = 0, 1,..., 3$. $z=(x,y)$ represents fish surface image. The intercepted part of fish surface image was shown in Fig. 4. After Gabor transform, we got 24 Gabor features.

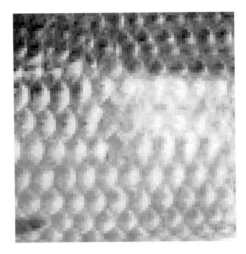

Figure 4. The intercepted part of fish body surface.

If all Gabor features were connected in series simply, we would obtain a high dimensional feature vector. Therefore, for all Gabor features in every orientations and scales, we calculated mean and variance of all point values. And the gotten 24 means and 24 variances served as features represented fish surface texture.

The above gotten whole body contour feature, segment body contour feature and fish surface texture feature were connected together to constitute a feature vector representing freshwater fish.

4 SPECIES IDENTIFICATION

4.1 Identification result

Linear SVM classifier was adopted for species identification of freshwater fish. SVM works are based on the structural risk minimization theory, and arrives at global optimization by constructing optimal hyperplane in the feature space. The expected risk of SVM in the whole sample space is less than a certain upper bound. For the situation of nonlinear sample space, SVM transforms the nonlinear low dimensional input space into a linear high dimensional feature space through nonlinear mapping algorithm, so that it becomes possible to make linear analysis on the nonlinear characteristics of the samples.

We randomly selected 70 samples from the above 100 freshwater fish, including 35 crucians and 35 cyprinoids, to constituting training set. The linear SVM classifier was trained by training set and was then tested by test set constituted by the remaining 30 fish samples. Identification results were listed in table 1.

Table 1. Species identification results.

Species	Test numble	Accuracy number	Identification accuracy (%)
Crucian	15	14	93.33
Cyprinoid	15	15	100
Average			96.67

It can be seen from Table 1, our method obtained satisfactory identification accuracy, which indicated the effectiveness of the method. In order to further investigate its superiority, we compared our method with the previous species identification ones that only considered fish body contour. Table 2 showed the comparison results.

Table 2. Comparison of three species identification methods.

Method	Test numble	Accuracy number	Identification accuracy (%)
Whole body Contour feature	30	27	90.00
Segment body Contour feature	30	28	93.33
Our method	30	29	96.67

As we can see from Table 2, because it comprehensively considered features of fish body contour and features of fish surface texture, our method avoided the limitation of methods which only using fish body contour feature. In view of crucian and cyprinoid, two species owning little variety in body contours, we got satisfactory identification accuracy. In addition, the adoptive Gabor transform was robust to light, angle and other details, so our method could achieve eligible results in the situation with weak changes of above details.

4.2 Effect of training sample number

In the above experiment, we used 70 fish samples for training the SVM classifier. Sequentially, in order to investigate the influence of training sample number on identification accuracy, we randomly selected 30, 50, 70, 80 fish samples respectively for training the SVM classifier, and the remaining samples for testing. The gotten identification accuracies were shown in Fig. 5.

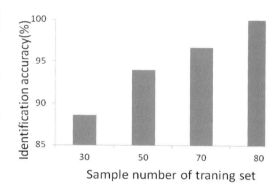

Figure 5. Identification accuracies of different training sample number.

As can be seen, because of the outstanding ability of SVM classifier to solve the problem of few samples, the proposed method could also obtain eligible identification results with less fish body samples. It provided the experimental basis for the practical application.

5 CONCLUSION

The accurate identification of different species is the precondition for deep processing of freshwater fish. Aiming at the current problem emerged in freshwater fish identification in our country, a new species identification method of freshwater fish based on fish body contour feature and fish surface texture feature was put forward. Whole body contour feature, segment body contour features and fish surface texture features were extracted and constituted the feature vector which represented the fish sample. Linear SVM classifier was utilized to accomplish species identification of freshwater fish. The proposed method was verified by using randomly bought crucian and cyprinoid samples and was compared with the traditional identification methods. The experimental results show that our method can obtain satisfied identification effect and has certain robustness. It provides methods and ideas for the pretreatment of freshwater fish processing.

REFERENCES

[1] T. Aurele, B. Olivier, O. Thierry, etc. Historical assemblage distinctiveness and the introduction of widespread

non-native species explain worldwide changes in freshwater fish taxonomic dissimilarity. *Global Ecology and Biogeography*, 2014, 23(5): 574–584.

[2] E. Borzym, M. Matras, J. Maj-Paluch, etc. First isolation of hirame rhabdovirus from freshwater fish in Europe. *Journal of Fish Diseases*, 2014, 37(5): 423–430.

[3] J. Fletcher. Ecology of north american freshwater fishes. *Freshwater Science*, 2014, 33(1): 374–375.

[4] Z. Q. Zhang, Z. Y. Niu, S. M. Zhao, etc. Weight grading of freshwater fish based on computer vision. *Transactions of the CSAE*, 2011, 27(2):350–354.

[5] P. Wan, H. B. Pan, C. J. Long, etc. Design of the on-line identification device of freshwater fish species based on machine vision technology[J]. Food & Machinery, 2012, 28(6):164-167.

[6] S. Eder Marques, W. Marina Sek Lien, M. Cesar, etc. Screening and characterization of sex-specific DNA fragments in the freshwater fish matrincha, Brycon amazonicus (Teleostei: Characiformes: Characidae.

[7] Z. A. Muchlisin, Z. Thomy, N. Fadli, etc. DNA barcoding of freshwater fishes from lake Laut Tawar, Aceh Province, Indonesia. *Acta Ichthyologica et Piscatoria*, 2013, 43(1): 21–29.

[8] Z. L. Liu, H. X. Fan. Study on cost efficiency measurement of intensive fresh water fish cultivation and its influence factor analysis. *Journal of Anhui Agriculture Science*, 2007, 35(8): 2204–2206.

[9] Z. L. Liu, H. X. Fan. Deterministic frontier model and its application in calculating of technical efficiency of intensive freshwater fish producing in China. *Fisheries Economy Research*, 2006, (4): 2–7.

[10] Z. Q. Zhang, Z. Y. Niu, S. M. Zhao. Identification of freshwater fish species based on computer vision. *Transactions of the CSAE*, 2011, 27(11): 388–392.

[11] P. Wan, H. B. Pan, L. Zong, etc. Study on crucian and cyprinoid species identification method based on machine vision. *Guangdong Agricultural Sciences*, 2012, (17): 184–187.

Fish Physiology and Biochemistry, 2012, 38(5): 1487–1496.

Information Technology – Wan et al. (Eds)
© 2015 Taylor & Francis Group, London, ISBN 978-1-138-02785-5

A no-reference image blur metric based on artificial neural network

Xiaoyu Ma & Xiuhua Jiang
Information Engineering School, Communication University of China, Beijing, China

ABSTRACT: This paper presents an efficient no-reference blur metric for images based on an artificial neural network (ANN) implemented by the back-propagation (BP) algorithm, named as BP-ANN. Three image features are extracted from spatial domains as the input vectors of the BP-ANN. The three features are No-Reference Structural Sharpness (NRSS), edge width and neighborhood correlation respectively. Experiments using the LIVE blur database demonstrate that the proposed algorithm correlates well with subjective quality evaluations.

1 INTRODUCTION

Image quality assessment plays a very important role in image processing, image analysis and other related fields. During the coding and transporting of images, a lot of perceptual distortions come about. Blurriness is one of the most common perceptual distortions. So it is very meaningful to propose an accurate blur metric which can be applied in many applications such as image analysis and image monitoring.

Many researchers have proposed a lot of novel blur metrics to estimate the image blur, so we firstly presents an overview of existing high-performance no-reference blur metrics, then we will propose our metric that based on artificial neural network(ANN) based on back-propagation (BP), named as BP-ANN. The input vector of the BP-ANN is three of the features we mentioned below.

The blurriness is caused by the attenuation of the high spatial frequencies (Ferzli & Karam, 2007). And the attenuation of the high spatial frequencies result in the spread of the sharp edges (Marziliano et al., 2002). So we can estimate the image quality by calculating the edge width. This measurement does not make any assumption on the blurring process and just calculate the average edge widths of all edges found, so it is very fast and is near real-time. Edge width metrics are very effective ways to measure image quality, and it is more accurate than Derivative-based metrics and variance metrics.

Considering that with the reduction of the high frequency component, neighboring pixels tend to be similar because of the smoothing effect. So we can calculate the neighborhood correlation of the pixels which can reveal the degree of image blur. As the blur is more likely to be perceived in the edge of the image, we can calculate the neighborhood correlation of the pixels in the wavelet domain. Neighborhood correlation metric is as fast as the edge width metric,

it is a very popular algorithm in image blur assessment domain.

Xiaofu Xie and Jian Zhou proposed a no-reference image blur metric based on structural similarity index metric (SSIM), named as no-reference structural sharpness (NRSS). This method constructed a reference image by a low-pass filter and assessed the image quality by computing the SSIM between the original image and the reference. The NRSS combines the traditional image blur assessment method with the SSIM and obtained good performance.

As the image quality assessment is actually a classification process, using artificial neural networks (ANN) to "understand" the distortion information is a reasonable approach (Jiang et al., 2008). Generally an ANN should be trained by the image features and the different mean opinion score (DMOS) of images to find out implicit relationship between the image features and the DMOS. After the training, we can use this ANN for the assessment of the degree of image blur like human observers do.

In this paper, we combined the three metrics which we mentioned before together by the ANN to obtain a higher consistency with the subjective evaluation. So we extract the edge width, neighborhood correlation and NRSS of images as the input vectors of the ANN. Experiments using the LIVE blur database demonstrate that the output scores of the ANN correlates well with subjective quality evaluations.

2 FEATURE EXTRACTION

In this section we will give a schematic description of the feature extraction. We define three features to describe the image blur. They are edge width, neighborhood correlation and NRSS respectively.

2.1 The extraction of edge width

An image appears blurred when the high spatial frequency components are attenuated. This will lead to the spreadability of the edges. So the edge width can estimate the image blur to some extent. In this paper we calculate the edge width based on the method proposed by Pina Marziliano (Marziliano et al., 2002). We improved this method and the procedure of the edge width metric, which is summarized in Fig. 1.

First, we used an edge detector such as the Canny operator to figure out the edges in the image (Wang et al., 2012). Then we iterates through every edge point in the image to calculate its width. The global measure for the whole image is calculated by averaging the width of all the edge points.

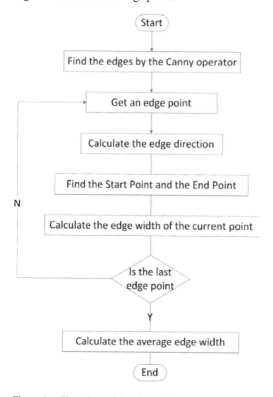

Figure 1.   Flow chart of the edge width extraction.

The calculation of the width of the edge point is as follows. First, we get the positions of the two edge points whose distance to the current edge point is two pixels. Then we calculate the slope of the two points to describe the edge direction of the current edge point. The start position and end position of an edge point are defined as the local extrema locations closet to the edge along the edge direction. The edge width of the current point is then given by the distance between the start position and end position.

2.2 The extraction of neighborhood correlation

Considering that the blur is actually a process of convolution, which will cause every pixels contain the information of neighboring pixels (Wang et al., 2004). This leads to the increase of the correlation. Therefore, we can estimate the image blur in terms of neighborhood correlation.

The process of neighborhood correlation calculation is as follows. First, we use a Sobel filter to remove the influence of the flat regions of the image because the correlation of the flat regions is very high no matter the image is sharp or blurred. Then we sampled the filtered image into 9 samples, named as sample 1 to sample 9. The vertical and horizontal sampling intervals are both 3. After this we calculated the correlations between each two samples. The correlation between sample i and sample j is named as R_{ij} and R_{ij}, which is calculated by (1). The neighborhood correlation of the image is the sum of all the R_{ij} values.

$$R_{ij} = \frac{1}{n-1} \sum_{k=1}^{n} (\frac{Y_{ik} - \overline{Y_i}}{S_{Yi}})(\frac{Y_{jk} - \overline{Y_j}}{S_{Yj}}) \quad (1)$$

The n is the number of the elements in a sample. The Y_i is the average of the sample i and the Y_j is the average of the sample j. The S_{Yi} is the standard deviation of sample i and the S_{Yj} is the standard deviation of sample 5. Y_{ik} and Y_{jk} are the elements in sample i and sample j. The the procedure of the neighborhood correlation metric is summarized in Fig. 2.

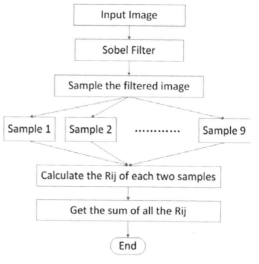

Figure 2.   Flow chart of the neighborhood correlation extraction.

This metric has got good performance in LIVE image database and TID2008 image database (Wang et al. 2004). But this metric is not efficient when the scene in the image is shrouded in haze or the scene in the image has a very large flat area.

2.3 The extraction of neighborhood correlation

The structural similarity index measurement (SSIM) is widely used in full-reference (FR) image quality assessment (Zhang, 2013). This metric has good performance and its computational complexity is very low. The main process of SSIM calculation is summarized as follows. The luminance factor, contrast factor and structural similarity factor of the source image and distorted image should be compared at first. Then we compare the three factors between source image and distorted image. At last we can get the SSIM value by integrating the comparison results of three factors together.

Xiaofu Xie and Jin Zhon applied the SSIM metric into no-reference blur assessment (Xie et al., 2007). They construct a reference image (I_r) of the input image (I_i) by a low pass filter at first. Then extract the gradient information of I_r and I_i by the Sobel operator. The gradient images of the I_r and I_i are named as G_r and G_i, respectively. Then divide the G_r and G_i into 8×8 blocks and find 30 blocks with the maximum variance in Gi named as x_i^ϵ, i[1,30]. Then get the corresponding blocks in G_r, named as y_i, i$^\epsilon$[1,30]. At last we can get the NRSS of the input image by (2), SSIM(x_i,y_i) stands for calculating the SSIM value between x_i and y_i.

$$NRSS = 1 - \frac{1}{30}\sum_{i=1}^{30} SSIM(\mathbf{x}_i, \mathbf{y}_i) \qquad (2)$$

The the procedure of the NRSS metric mentioned above is summarized in Fig. 3.

3 BP-ANN BASED IMAGE BLUR METRIC

The BP algorithm is used widely in artificial neural network based application because it has good performance to find out the implicit relationship between the variables. And that is just the method we need to integrate the three features together. So we use BP-ANN to integrate the three metrics together to obtain a metric with more excellent performance.

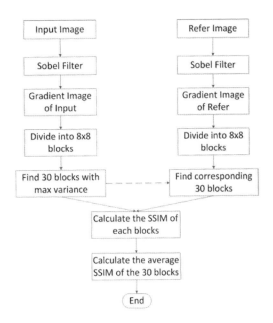

Figure 3. Flow Chart of the NRSS Extraction.

3.1 Architecture of the BP-ANN we used

In this paper, we apply a fully connected three-layer feed-forward network to estimate image blur. The Network is trained by using error back-propagation algorithm. Our Network consists of the following three layers.

- Input layer: As we extract three features to estimate image blur in this paper, we assign three neurons at this layer, corresponding to the edge width, neighborhood correlation and NRSS respectively.
- Hidden layer: We assign 8 neurons at this layer for the tradeoff between the system robustness and efficiency (Jiang et al., 2008).
- Output layer: The output of our network is on behalf of the degree of image blur. So we just assign one neuron at this layer.

To establish an artificial neural network with good performance, the following parameters are of much importance, such as learning rate, initial value and stopping criteria. Our parameter settings in this paper are as follows. The learning rate is set as 0.05, the stopping criterion is 0.00065 and the initial value of each neuron is sett with a random value.

3.2 Architecture of the BP-ANN we used

We use 45 images in the LIVE blur database to train the BP-ANN as mentioned above. First, we calculate the three features of each image and get 3 vectors whose lengths are all 45. The 3 vectors are on behalf of the edge width, neighborhood correlation and NRSS of the 45 images, respectively. Then we normalize the vectors individually according to their respective ranges. Then we apply these vectors into the input layer.

After 45 training images passes through the BP-ANN, we can get an output vector from the output layer. Then we can get the mean square errors of each video by the output vector and the different mean opinion scores of the 45 images. And we can use this error value to regulate the connecting weights among neurons. Then we calculate the output vector of the network with new weights and repeat the above steps until the stopping criterion are met.

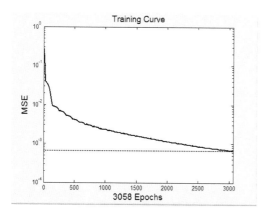

Figure 4.   Learning curve of our BP-ANN.

Every time we calculate the output vector of the network with new weights, we can get a smaller MSE. So we can draw a learning curve to show how the MSE reduces as the training times grows. The learning curve is shown in Fig. 4.

From Fig. 4 we can see that the output vector match the stopping criteria after 3058 times of training. After the training, we can use this BP-ANN based metric to evaluate image blur with a excellent performance.

4   EXPERIMENTS AND RESULTS

In this section, we show results of the BP-ANN based blur metric which we had proposed in the previous section. We testify the proposed metric using LIVE blur database. The dataset contains 174 images: 29 of them are original images and the other 145 images are Gaussian blurred. We use 45 images of the Gaussian blurred image to train the BP-ANN as we mentioned above, these images are called training set

together. The other images are used to testify the performance of the BP-ANN based metric we proposed, these images are called testing set together.

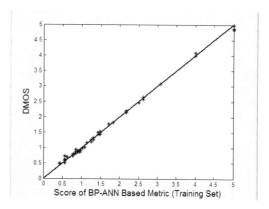

Figure 5.   The scatter plot of DMOS versus the BP-ANN output of the training set.

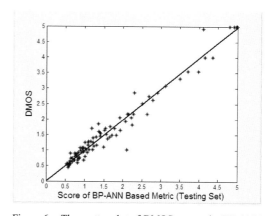

Figure 6.   The scatter plot of DMOS versus the BP-ANN output of the testing set.

The scatter plot of different mean opinion score (DMOS) versus the BP-ANN output of the training set is shown in Fig. 5. And the scatter plot of DMOS versus the BP-ANN output of the testing set is shown in Fig. 6.

From Figs. 5 and 6 we can infer that by training the BP-ANN, the output of the Network in the training set can get a extremely high consistence with the DMOS. What's more, the output of the Network in the testing set is has also a high consistence with the DMOS. The results demonstrate that integrating the three features together based on BP-ANN to get a metric with better performance is feasible.

In order to show the superiority of our metric, we also testify the performance of the edge width metric, neighborhood correlation metric and the NRSS metric using the LIVE blur database. The scatter plots of the three metric are shown in Figs. 7–9. The Pearson correlation coefficient (CC) and Spearman rank order correlation coefficient (SROCC) of our proposed metric and other reference metrics are shown in Table 1.

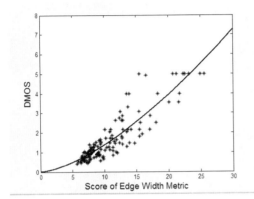

Figure 7. The scatter plot of DMOS versus the Score of Edge Width Metric.

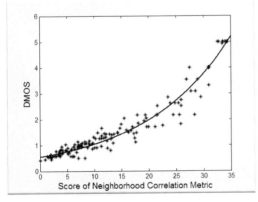

Figure 8. The scatter plot of DMOS versus the Score of Neighborhood Correlation Metric.

From the figures and tables we can infer that by integrating the three metrics together, out proposed metric has obtained a much more better performance than the performance of the single metric we mentioned above. In detail, the CC and ROCC of the edge width metric are only 0.8530 and 0.8723 respectively. The CC and ROCC of the NRSS metric are only 0.7564 and 0.8962 respectively. We can find that the two metrics above are not very effective. The CC and ROCC of the neighborhood correlation metric are about 0.92 and 0.94. The neighborhood correlation metric is much more effective than the two metrics

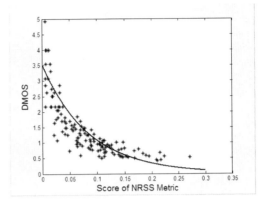

Figure 9. The scatter plot of DMOS versus the score of neighborhood correlation metric.

Table 1. Performance comparison.

Setting	CC	ROCC
BP-ANN based	0.9826	0.9610
Edge width	0.8530	0.8723
NC*	0.9202	0.9410
NRSS	0.7564	0.8962

* NC stands for the Neighborhood Correlation.

above. However, the efficiency of the neighborhood correlation metric still can be improved. By integrating the three methods together using the BP-ANN, the CC and ROCC of the proposed method are as high as 0.9826 and 0.9610.

5 CONCLUSIONS

In this paper, a no-reference blur metric is proposed based on the BP-ANN. We first extract three features of images and then integrate them into an image blur score by the trained BP-ANN. The results shown in previous section demonstrate that our metric has a significant potential to be used in practical blur assessment applications.

ACKNOWLEDGMENT

This work is supported by the Specialized Research Fund for the Doctoral Program of Higher Education: " Research of Visual Perception for Impairments of Color Information in High-Definition Images ". (No. 20110018110001).

REFERENCES

Ferzli, R. & Karam, L.J. 2007. A no-reference objective image sharpness metric based on just-noticeable blur and probability summation. *IEEE International Conference on Image Processing (ICIP 2007), San Antonio, US, 2007*: 445–448.

Jiang, X. et al. 2008. No-reference perceptual video quality measurement for high definition videos based on an artificial neural network. *International Conference on Computer and Electrical Engineering (ICCEE 2008), Phuket, Thailand, 2008:* 424–427.

Marziliano, P. et al. 2002. A no-reference perceptual blur metric. *IEEE International Conference on Image Processing, New York, US, 2002:* 57–60.

Wang, Y. et al. 2012. A no-reference perceptual blur metric based on complex edge analysis. *International Conference on Network Infrastructure and Digital Content (IC-NIDC 2012), Beijing, China, 2012:* 487–491.

Wang, Z. et al. Quality Assessment: From Error Visibility to Structural Similarity. *IEEE Transactions on Image Processing, ,* 2010, 3(4): 600–612.

Xie, X. et al. No-reference quality index for image blur. *Journal of Computer Applications* 2010, 30(4): 921–924.

Zhang, D. et al. 2013. Applications of natural image Statistics in image processing. *Master Thesis in Zhejiang University.* Hangzhou: Zhejiang University.

Information Technology – Wan et al. (Eds)
© *2015 Taylor & Francis Group, London, ISBN 978-1-138-02785-5*

A novel cell model to detect the carcinogenic potential of polycyclic aromatic hydrocarbons

Shuang Wu, Ping Gong, Youxi Zhao, Dan Jiang & Baining Liu
College of Biochemical Engineering, Beijing Union University, Beijing, China

Rugang Zhong & Yi Zeng
College of Life Science and Bioengineering, Beijing University of Technology, Beijing, China

ABSTRACT: The aim of this study is to detect the carcinogenic potential of three polycyclic aromatic hydrocarbons (PAHs) employing the novel cell model. The results indicated that the novel cell model (an improved cell transformation assay using Balb-E6E7 cells) could detect the initiating and promoting activities of three PAHs. The initiating and promoting activities of Benz[a]anthracene (B[a]A) were positive and positive respectively, whereas that of Benzo[ghi]perylene (B[ghi]P) were negative and positive, respectively. Pyr didn't show any carcinogenic potential. The Balb-E6E7 cell transformation assay possessed increases in transformed foci as many as 1-10 times and one-third reduction of time consumed comparing with conventional Balb/c 3T3 cell system. According to the standard of agents classified by the IARC, the data demonstrated the sensitivity and effectiveness of the Balb-E6E7 cell transformation assay were confirmed in predicting performance of PAHs carcinogenicity with the more focus transformation, less time consumed and better performance of cost and labor. The results might be of significance for improving the overall process of safety and risk assessment of chemical carcinogens and contributing to the cancer prevention.

Chemical carcinogens can be divided into two categories, i.e., initiators and promoters, based on the two stage model of carcinogenesis. Most initiators can be detected by various genotoxicity tests, the results of which are used for carcinogenicity prediction (IARC/NCI/EPA Working Group, 1985). In the case of tumor promoters, several methods have been proposed (Fitzgerald, D.J. et al. 1989), but none of them have been routinely used for regulatory purposes. Therefore, developing a method for the detection of genotoxic and non-genotoxic carcinogens is a major challenge for the safety evaluation of chemicals. The utilization of additional screening tests covering a wide range of carcinogens has an advantage before contemplating in vivo long-term carcinogenicity experiments for chemical safety assessment.

The cell transformation assays using BALB/c 3T3 cells could simulate the process of two-stage carcinogenesis. The original procedure of Balb/c 3T3 transformation assay was described by Kakunaga (Kakunaga, T. 1973), the development history of which had been reviewed by Schechtman (Schechtman, L.M. 1985). A variety of modifications were introduced to the standard transformation assay including changes of exposure and post-exposure conditions for highly cytotoxic and short-lived chemicals, the addition of various forms of endogenous metabolic activation and the two-stage protocol (Combes, R. et al. 1999, Hayashi, K. 2008, Heidelberger, C. et al. 1983, Isfort, R.J. et al. 1996, Kajiwara, Y. et al. 2003, Tsuchiya, T. 2010). Formation of transformed foci is the consequence of the complex process of cells from normal to malignant state. Since the conventional Balb/c 3T3 two-stage cell transformation assay could detect both initiating and promoting activities (Tsuchiya, T. 2010), it was anticipated as a screening tool to be useful for detection of tumor initiators and tumor promoters. In spite of this expectation, it has not been accepted as a routine screening method, because of the laborious and time-consuming procedure and the unsatisfied sensitivity and effectiveness (IARC/NCI/EPA Working Group, 1985).

The novel cell model which involved a new cell line (Balb-E6E7 cells) and an optimized cell transformation assay was performed. First, a recombinant plasmid pcDNA3.1 containing HPV 16 E6 and E7 DNA was constructed and then transfected into the Balb/c 3T3 cells. After screening by G418, the cells obtained were named Balb-E6E7 cells (Shuang, W. 2012). Furthermore, an improved Balb-E6E7 cell transformation assay was established by selecting an appropriate cell culture medium and the concentration of FBS, and optimizing the chemicals administration and the density of cell inoculation (Shuang, W. unpubl.).

The purpose of this paper is to detect the carcinogenic potential of three polycyclic aromatic hydrocarbons (PAHs), including both initiating and promoting activities employing the novel cell model, compare predicting performance of PAHs carcinogenicity with that of conventional Balb/c 3T3 cell system, validate the reliability and sensitivity of novel cell model and increase the responsiveness of the novel cell model to a broader range of chemical carcinogens.

1 MATERIALS AND METHODS

1.1 Chemicals

Chemicals employed were Benz[a]anthracene (B[a]A), Pyrene (Pyr), Benzo[ghi]perylene (B[ghi]P), 3-methylcholanthrene (MCA), 12-O-tetradecanoyl-phorbol-13-acetate (TPA), dimethyl sulphoxide (DMSO) and ITES (consisting of bovine pancreas insulin, human transferrin, ethanol-amine and sodium selenite) purchased from Sigma Chemical Co.. The chemicals were dissolved in DMSO, the final concentration of which in the medium was adjusted to 0.5% v/v.

1.2 Cell culture

Balb/c 3T3 clone A31 cells provided by the Institute for Viral Disease Control and Prevention (Beijing, China) and Balb-E6E7 cells were grown in DMEM (Hyclone) supplemented with 10% FBS (Hyclone), 4 mM L-glutamine and 1% antibiotic mixture (Sigma) under standard culture conditions (95% air/5% CO_2 at 37°C in a humidified atmosphere).

1.3 Detection of initiating activity of PAHs employing Balb-E6E7 cells

1.3.1 Initiation assay to examine the initiating activity of PAHs

Exponentially-growing Balb-E6E7 cells were seeded at a density of 10^4 cells per 25 cm^2 culture flask in 8 replicates for determination of morphological transformation. After a 24 h incubation period, the cells were exposed to various concentrations of PAHs for 3 days. After that, cells were subsequently cultured in chemical-free DMEM supplemented with 5% FBS. The medium was renewed every 5 days. On Day 21, the cells were fixed with methanol and stained with Giemsa solution. Each assay was performed, including MCA as the positive control and 0.1% DMSO as the negative control.

1.3.2 Cytotoxicity assay to examine the initiating activity of PAHs

Cell viability and cytotoxicity were determined by Cell Counting Kit-8 (CCK-8, Dojindo Laboratories, Japan). The cell suspension containing 10^3 cells was plated onto 96-well sterile plates and allowed to attach and grow for 24 h. The cells were exposed to various concentrations of PAHs for 72 h. After the culture media was aspirated, 10 μL CCK-8 solution was added to each well of the plate and the plate was incubated for 2 h. The absorbance was measured at 450 nm and 630 nm by a spectrophotometic plate reader (Bio-Rad instruments, Model 680). All values were corrected by cell-free controls. All experiments were repeated five times. The percentage of viability calculated was the average absorbance of PAHs at different concentrations divided by that of the corresponding solvent control group.

1.4 Detection of promoting activity of PAHs employing Balb-E6E7 cells

1.4.1 Promotion assay to examine promoting activity of PAHs

Exponentially-growing Balb-E6E7 cells were seeded at a density of 10^4 cells per 25 cm^2 culture flask in 8 replicates for determination of morphological transformation. After a 24 h incubation period, the culture medium was replaced with a fresh DMEM medium supplemented with 5% FBS for 72 h. And then the cells were exposed to various concentrations of PAHs dissolved in the DMEM/F12 supplemented with 2% FBS and 0.2% ITES. The medium was renewed every 5 days. On Day 14, all treatment media were removed and the cells were subsequently cultured in chemical-free DMEM/F12 supplemented with 2% FBS and 0.2% ITES. On Day 21, the cells were fixed with methanol and stained with Giemsa solution. Each assay was performed, including TPA as the positive control and 0.1% DMSO as the negative control.

1.4.2 Cytotoxicity assay to examine promoting activity of PAHs

The cell suspension containing 2×10^2 cells was plated onto 96-well sterile plates and allowed to attach and grow for 24 h. The cells were cultured in the DMEM supplemented with 5% FBS for 72 h. And then the cells were exposed to various concentrations of PAHs dissolved in the DMEM/F12 supplemented with 2% FBS and 0.2% ITES for 72 h. After the culture media was aspirated, 10 μL CCK-8 solution was added to each well of the plate and the plate was incubated for 2 h. The absorbance was measured at 450 nm and 630 nm. All values were corrected by cell-free controls. All experiments were repeated five times.

1.5 Detection of initiating and promoting activities of PAHs employing balb/c 3T3 cells

The protocols of initiating and promoting assays and the relevant cytotoxicity assays using Balb/c 3T3 cells were described in detail (Tsuchiya, T. 2010).

1.6 Counting of transformed foci and statistical analysis

According to the basis of the morphological characteristics, the criteria of transformed foci was described (IARC/NCI/EPA Working Group, 1985). All data were analyzed using the SPSS statistical software (version 13.0, SPSS), with $P<0.05$ values considered as statistically significant. Comparisons between the groups were assessed by paired-samples t test.

2 RESULTS

2.1 Normal and transformed cell morphology

As shown in Figure 1, the micrographs of normal and transformed cell morphology were obtained. Exposed to different concentration of PAHs for several days, both Balb/c 3T3 cells and Balb-E6E7 cells showed the morphological cell transformation.

According to criteria described (IARC/NCI/EPA Working Group, 1985), the scoring of transformed foci was performed by the following morphological characteristics: deep basophilic staining, dense multilayering of cells, random orientation of cells at the edge of the foci, invasion growth into the monolayer of surrounding contact-inhibited cells, and a diameter of more than 2 mm.

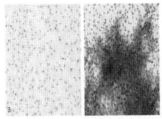

a: Normal cell   b: Transformed cell

Figure 1. Micrographs of cell morphology (Giemsa staining, ×100).

2.2 The initiating and promoting activities of PAHs

The results of the initiating and promoting activities of B[a]A, B[ghi]P and Pyr were summarized in Figure 2 and 3.

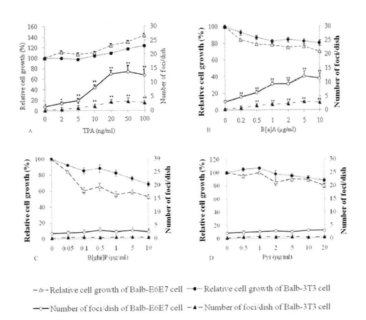

Figure 2. The initiating activity of PAHs in Balb-E6E7 cell transformation assays and Balb/c 3T3 cell transformation assays.

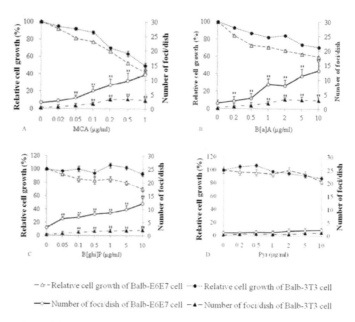

- △ - Relative cell growth of Balb-E6E7 cell   ⋯•⋯ Relative cell growth of Balb-3T3 cell

—○— Number of foci/dish of Balb-E6E7 cell   – ▲ – Number of foci/dish of Balb-3T3 cell

Figure 3.   The promoting activity of PAHs in Balb-E6E7 cell transformation assays and Balb/c 3T3 cell transformation assays.

2.2.1   The initiating and promoting activities of B[a]A

The results of the initiating and promoting activities of B[a]A in Balb-E6E7 cell transformation assays are shown in Figure 2B and 3B, respectively. Significant increase of focus formation was observed in the initiation assay of B[a]A at the concentrations of 0.2 μg/mL and above (P<0.05). B[a]A showed a similar response in the promotion assay, which induced increased transformed foci at a concentration of 0.2 μg/mL and higher (P<0.05). Compared with cytotoxicity assay to examine initiating activity, the B[a]A displayed a weaker toxic effect to Balb-E6E7 cell. While the concentration of B[a]A was 10 μg/mL, the relative cell growth percent observed in promoting assay was approximately 70%, which was higher than that of in initiating assay, approximately 60%.

On the basis of epidemiological and experimental evidence of carcinogenicity, mechanistic and other relevant data, agents were assigned to Group 1 (carcinogenic to humans), Group 2A (probably carcinogenic to humans) or Group 2B (possibly carcinogenic to humans), Group 3 (not classifiable as to its carcinogenicity to humans) and Group 4 (probably not carcinogenic to humans) (WHO, 2005). The data demonstrated B[a]A possessed both tumor-initiating activity and tumor-promoting activities, which was

up to the standard of B[a]A classified by the IARC (Group 2B) (WHO, 2005).

2.2.2   The initiating and promoting activities of B[ghi]P

Results on the initiation and promotion assays for B[ghi]P are shown in Figure 2C and 3C. No increase in focus number was observed in the initiation assay when the concentration was up to 100 μg/mL. Severe cytotoxicity was observed in the Balb-E6E7 cell. While the concentration of B[ghi]P was 0.1 μg/mL, the relative cell growth percent was approximately 60%. And then it changed gently as the concentration above 0.1 μg/mL (Figure 2C). The lower relative cell growth percent meant the more intense cytotoxicity, which displayed a weaker tumor-initiating activity. Because the carcinogenesis was known to be a multi-step and long-term process, it needed a persistent accumulation of DNA damage with low cytotoxicity concentrations to induce tumor.

In contrast, a significant increase (P<0.05) of focus formation was observed in the promoting assay at all concentrations tested up to 10 μg/ml (Fig. 3C). The foci was increased with increasing concentrations of B[ghi]P. Therefore, the data demonstrated B[ghi]P could be a tumor-promoting agent, not a tumor-initiating agent, and it was classified to Group 3 by the IARC (WHO, 2005).

2.2.3 The initiating and promoting activities of Pyr

As shown in the Figure 2D and 3D, Pyr exhibited negative results in both initiating and promoting assays. In the relevant cytotoxicity assays, Pyr did not show obvious cytotoxicity at the concentration ranges tested.

According to the IARC (WHO, 2005), Pyr didn't show any carcinogenic potential, both the initiating and promoting activities. The results were the same as that of Balb-E6E7 cell transformation assays.

2.3 The comparison of sensitivity to PAHs between the novel cell model with conventional cell system

The results of the initiating and promoting activities of PAHs in both Balb-E6E7 cell transformation assays and Balb/c 3T3 cell transformation assays are shown in the Figures 2 and 3.

In the initiating activity assays, three PAHs exhibited the higher cytotoxicity and more focus formation in Balb-E6E7 cell transformation assays compared with that of in BALB/c 3T3 cell transformation assays. To detect the initiating activity of MCA, it induced a statistically significant increase ($P<0.05$) of transformed foci at the concentration ranges from 0.05 μg/mL to 1.0 μg/mL in Balb-E6E7 cell transformation assays, whereas the positive results was observed at the concentration of 0.1 μg/mL and higher in the BALB/c 3T3 cell transformation assays. Similarly, in the Balb-E6E7 cell transformation assays, significant increase of focus formation was observed in the initiation assay of B[a]A at the concentrations of 0.2 μg/mL and above ($P<0.05$). While, a response similar to that in BALB/c 3T3 cell transformation assay at the concentrations of 1.0 μg/mL to 10 μg/mL. The Balb-E6E7 cell transformation assays possessed increases in transformed foci as many as 3-8 times and one-third reduction of time consumed compared with Balb/c 3T3 cell transformation assays.

In the case of the promotion assay, the similar transformed response was observed in Balb-E6E7 cell transformation assays and Balb/c 3T3 cell transformation assays. At the six concentration dose tested of B[a]A and B[ghi]P (from 0.2 μg/mL to 10 μg/mL, and from 0.05 μg/mL to 10 μg/mL, respectively), the significant increase ($P<0.05$) of focus formation were observed in Balb-E6E7 cell transformation assays, whereas positive results were observed at five concentration dose tested in Balb-E6E7 cell transformation assays. The number of transformed foci was about 1-10 times greater in Balb-E6E7 cell transformation assays than that of in the Balb/3T3 cell transformation assays. Furthermore, there was a one-third reduction of time consumed in Balb-E6E7 cell transformation assays compared with Balb/c 3T3 cell transformation assays.

With the more focus transformation, less time consumed and more satisfied performance of cost and labor, Balb-E6E7 cell transformation assays were the more sensitive and efficient method to predict the carcinogenicity of PAHs than Balb/c 3T3 cell transformation assays.

3 DISCUSSION

The novel cell model (Balb-E6E7 cell transformation assays) was performed to detect the initiating and promoting activities of three PAHs, which are classified as environmental carcinogens for human by the IARC. In order to estimate the sensitivity of Balb-E6E7 cell transformation assays and its applicability to the prediction of chemical carcinogenicity, we compared the results obtained in Balb-E6E7 cell transformation assays with BALB/c 3T3 cell transformation assays.

Based on our previous results (Shuang, W. unpubl. & 2012), the Balb-E6E7 cell, as a new cell line, was performed by transfecting a recombinant plasmid pcDNA3.1 containing HPV 16 E6 and E7 DNA into the Balb/c 3T3 cells and screening by G418. After selecting appropriate cell culture medium and optimizing the density of cell inoculation, an improved Balb-E6E7 cell transformation assay was established. In the BALB/c 3T3 two-stage cell transformation assay (Tsuchiya, T. 2010), the initiating activity was examined by treating the cultures with a test chemical in the initiation phase and then with the known tumor promoter TPA in the promotion phase. To test promoting activity the cells were exposed to the known tumor initiator, the sub-threshold dose of MCA (0.2 μg/mL) in the initiation phase and then to a test chemical in the promotion phase. However, in Balb-E6E7 cell transformation assays, there was no redundant addition of MCA or TPA in promotion assay or initiation assay, which could provide multiple protection of researchers' healthy and friendly environment. Furthermore, Balb-E6E7 cell transformation assays possessed increases in transformed foci as many as 1-10 times and one-third reduction of time consumed compared with Balb/c 3T3 cell transformation assays. The novel cell model was more sensitive and efficient system than the conventional BALB/c 3T3 cell system.

Human papillomaviruses (HPVs) are small, double-stranded DNA viruses that induce proliferative lesions in epithelial tissues. One of the primary activities of E6 was to target p53 for degradation, which was mediated by complex formation with the cellular ubiquitin ligase E6-AP. E6 had been shown to activate telomerase via upregulation of hTERT expression (Münger, K. 2004). Furthermore, the E6 proteins had been shown to interact with a number of other cellular proteins, such as the putative calcium-binding protein E6-BP, the focal adhesion rotein paxillin, p300/CBP, and PDZ domain proteins (Münger, K. 2004). E7 acted by binding to members of the Rb tumor suppressor protein family and inhibiting their ability to modulate the function of E2F transcription factors (Song, S. 2000). HPV16 E6 and E7 could disrupt the normal regulation of the cell proliferation, differentiation and apoptosis, and contribute to the immortalization and transformation of cells. Therefore, the Balb-E6E7 cell exhibited the high sensitivity for both genotoxic and non-genotoxic chemicals to perform the morphological transformation and malignant progression.

According to the standard of agents classified by the IARC, the sensitivity and effectiveness of the novel cell model were confirmed in predicting performance of PAHs carcinogenicity compared with that of conventional Balb/c 3T3 cell system. After the reliability and sensitivity was validated, the further study is increasing the responsiveness of the novel cell model to a broader range of chemical carcinogens. The further development, evaluation, routine use of the novel cell model, in conjunction with other research results and relevant information of cytology, toxicology, molecular biology and biochemistry, would be of significance for improving the overall process of safety and risk assessment of chemical carcinogens and contributing to the cancer prevention.

ACKNOWLEDGMENT

Funding for this work was provided by the "New Start" Academic Research Projects of Beijing Union University (Zk10201306).

REFERENCES

Combes, R., Balls, M. & Curren, R. 1999. Cell transformation assays as predictors of human carcinogenicity. *Atla-Alternatives to Laboratory Animals*, 27: 745–767.

Fitzgerald, D.J., Piccoli, C. & Yamasaki, H. 1989. Detection of nongenotoxic carcinogens in the BALB/c 3T3 cell transformation mutation assay system. *Mutagenesis*, 4: 286–291.

Hayashi, K., Sasaki, K. & Asada, S. 2008. Technical modification of the BALB/c 3T3 cell transformation assay: The use of the serum-reduced medium to optimise the practicability of the protocol. *Atla-Alternatives to Laboratory Animals*, 36: 653–665.

Heidelberger, C. Freeman, A.E. & Pienta, R.J. 1983. Cell transformation by chemical agents-a review and analysis of the literature. *Mutation Research*, 114: 283–385.

IARC/NCI/EPA Working Group. 1985. Cellular and molecular mechanisms of cell transformation and standardization of transformation assays of established cell lines for the prediction of carcinogenic chemicals: Overview and recommended protocols. *Cancer Research*, 45: 2395–2399.

Isfort, R.J. & LeBoeuf, R.A. 1996. Application of in vitro cell transformation assays to predict the carcinogenic potential of chemicals. *Mutation Research-Reviews in Genetic Toxicology*, 365: 161–173.

Kajiwara, Y. & Ajimi, S. 2003. Verification of the BALB/c 3T3 cell transformation assay after improvement by using an ITES medium. *Toxicology in Vitro*, 17: 489–496.

Kakunaga, T. 1973. A quantitative system for assay of malignant transformation by chemical carcinogens using a clone derived from BALB/c 3T3. *International Journal of Cancer*, 12: 463–473.

Münger, K., Baldwin, A. & Edwards, K. M. 2004. Mechanisms of human papillomavirus-induced oncogenesis. *Journal of Virology*, 78: 11451–11460.

Schechtman, L.M. 1985. Balb/c 3T3 cell transformation: protocols, problems and improvements. Transformation Assay of Established Cell Lines: Mechanisms and Application. *IARC Scientific Publications* No. 67.

Shuang, W., Jintao, L. & Rugang, Z. 2012. A synergistic transformation response induced by HPV 16 E6E7 gene and MCA/TPA in Balb/c 3T3 cells. *Shandong Medical Journal*, 52: 35–37.

Shuang, W. unpublished. Synergistic effect on carcinogenesis of polycyclic aromatic hydrocarbons and HPV 16 E6E7.

Song, S., Liem, A. & Miller, J. A. 2000. Human Papillomavirus Types 16 E6 and E7 Contribute Differently to Carcinogenesis. *Virology*, 267: 141–150.

Tsuchiya, T., Umeda, M. & Tanaka, N. 2010. Application of the improved Balb/c 3T3 cell transformation assay to the examination of the initiating and promoting activities of chemicals: the second interlaboratory collaborative study by the non-genotoxic carcinogen study group of Japan. *Atla-Alternatives to Laboratory Animals*, 38: 11–27.

WHO. 2005. IARC Monographs on the Evaluation of Carcinogenic Risks to Humans. *Meeting of the IARC Working Group on the Evaluation of Carcinogenic Risks to Humans*. Lyon.

Information Technology – Wan et al. (Eds)
© 2015 Taylor & Francis Group, London, ISBN 978-1-138-02785-5

A novel image decompression technique using image restoration and coefficient fusion

Zhiyong Zuo, Jing Hu & Lihua Deng
Institute for Pattern Recognition and Artificial Intelligence, Huazhong University of Science and Technology, Wuhan, China

Zhiyong Zuo & Xia Lan
The 10th Institute of China Electronics Technology Group Corporation, Chengdu, China

Xiaoping Wang
Institute of Radar and Electronic Warfare, Equipment Academy of Air Force, Beijing, China

Zhao Cheng
Institute of Manned Space System Engineering, China Academy of Space technology, Beijing, China

ABSTRACT: Due to the increasing traffic cased by multimedia information and digitized form of representation of images; image compression has become a necessity. Wavelet transform which has the ability to analyze signals in different scales is widely used in lossy image compression, while more and more image detail will be thrown away with the increase of the compression ratio (CR). In order to overcome this drawback, in this paper we incorporate image restoration and coefficient fusion techniques into the decompression procedure, and propose a novel image decompression technique. We restore part of high-frequency coefficient by using image restoration with the low-frequency component, new high-frequency components are next retrieved by fusing the restored high-frequency and the corresponding original high-frequency, and then the reconstructed image is last obtained by applying the inverse wavelet transform to the new coefficient map which is generated through combing the original low-frequency with the retrievable high-frequency components. Finally, a set of experiments and performance evaluation are provided to assess the effectiveness of the proposed method.

1 INTRODUCTION

As digital imagery becomes more commonplace and of higher quality, there is the need to manipulate more and more data. Thus image compression is needed to reduce the necessary storage and bandwidth requirements [1]. During the past decade, a variety of image compression methods have been proposed [2]. Among them, one of the most popular techniques is wavelet transform, which has the ability to analyze signals in different directions and scales [3].

Image compression can be classified as lossy and lossless. In lossless compression methods, the images are encoded in full quality and can be reconstructed without any change in pixel intensity but the compression ratio (CR) achieved is low. In the lossy compression schemes, which are implemented in frequency domains, encoding is achieved with losing of the frequency coefficients' precision to obtain higher compression ratio [4]. It is well known that the low frequency components constitute the base of an image, and the high frequency components express the texture information of the image, that is to say, the low frequency is more important than the high frequency. Therefore the quantization step-size in the high frequency is much bigger than that in the low frequency [2].

According to the above analysis, the detail of the reconstructed image is thrown away because the high frequency is lost in the compression procedure. In order to improve the quality of the reconstructed image, in this paper image restoration and coefficient fusion techniques are introduced to retrieve the high-frequency in the decompression procedure. First, we restore part of high-frequency coefficient by using image restoration with the low-frequency component whose precision is much more reserved by coefficient quantization. Next, new high-frequency components are retrieved by fusing the restored high-frequency and the corresponding original high-frequency. Then, a new coefficient map is generated by combing the original low-frequency with the retrievable high-frequency components. Finally, the reconstructed image is obtained by applying the inverse discrete wavelet transform (IDWT) to the combined coefficient map.

This paper is structured as follows. In Section 2, we review some related work. The proposed image decompression technique is presented in detail in Section 3. Section 4 gives some experimental results and discussion. Finally, conclusions are drawn in Section 5.

2 PREVIOUS WORK

In this section, we review some works of the wavelet-based lossy image compression method and the blind deconvolution method.

2.1 Wavelet-based lossy image compression

Most natural images have smooth colour variations, with the fine details being represented as sharp edges in between the smooth variations. Technically, the smooth variations in colour can be termed as low frequency variations and the sharp variations as high frequency variations [5]. The low frequency components constitute the base of an image, and the high frequency components add upon them to refine the image, thereby giving a detailed image [6]. Hence, the smooth variations are demanding more importance than the details. Separating the smooth variations and details of the image can be done by decomposition of the image using a Discrete Wavelet Transform (DWT) [5-7], and the two-layer decomposition structure is illustrated in Fig.1.

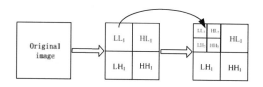

Figure 1.　Two level decomposition of an image.

The human eye is fairly good at seeing small differences in brightness over a relatively large area, but not so good at distinguishing the exact strength of a high frequency brightness variation. This fact allows one to reduce the amount of information required by ignoring the high frequency components according to the compression ratio's requirement. After the transform coefficients have been quantized into a finite set of values, it can be encoded using an entropy coder to give additional compression.

The general block diagram of the wavelet-based encoder is illustrated in Fig. 2(a). A discrete wavelet transform is first applied on the original image.

The transform coefficients are then quantized and entropy coded before forming the output bit-stream. The decoder process is the inverse procedure of the encoder, and the block diagram is shown in Fig. 2(b).

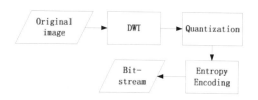

Figure 2(a).　General block diagram of the encoder.

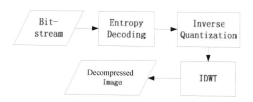

Figure 2(b).　General block diagram of the decoder.

2.2 Blind deconvolution algorithm

Blind image deconvolution is an ill-posed problem that requires regularization to solve. Then many blind deconvolution methods have been proposed to simultaneously estimate the true image and blur kernel (or the point spread function). Recently, Levin et al. pointed out that this way is ill-posed, while estimating the blur kernel alone is better conditioned because the number of parameters to estimate is small relative to the number of image pixels measured [8]. As a result, Krishnan et al. [9] introduced a normalized sparsity measure to accurately estimate the blur kernel, and then the image restoration problem is turned into the non-blind deconvolution problem.

The point spread function (PSF) is first performed on the high frequencies of the image. More precisely, consider $\nabla_x = [1, -1]$ and $\nabla_y = [1, -1]^T$ the discrete filters. Let $y = [\nabla_x g, \nabla_y g]$ and x be the high-frequency version of the degraded image g and the unknown sharp image f, respectively. The cost function for estimating the blur kernel is:

$$\min_{x,k} \left\{ E(x,k) = \lambda \|x \otimes k - y\|_2^2 + \frac{\|x\|_1}{\|x\|_2} + \beta \|k\|_1 \right\}$$

subject to: $k \geq 0, \sum k = 1$ (1)

where k is the unknown blur kernel, $\otimes$ is the 2D convolution operator, λ and β are the regularization parameters.

Eqn. 1 is highly nonconvex. The standard approach to optimizing such a problem is to start with an initialization on x and k, and then alternate between x and k updates. Once the blur kernel k has been estimated, the unknown sharp image f can be recovered by the following l_p -norm cost function:

$$\min_f \left\{ E(f) = \mu \| f \otimes k - g \|_2^2 + \| \nabla_x g \|_p + \| \nabla_y g \|_p \right\} \quad (2)$$

where μ is the regularization parameter, and p is chose as 0.8 as recommended in [9]. The nonblind deconvlution problem (2) can be easily solved and produce very satisfactory results.

3 PROPOSED METHOD

In image compression based on wavelet analysis, the quantization step-size in the high frequency is much bigger than that in the low frequency, and the low-frequency component has litter precision loss [10], which may result in blurring effect in the reconstructed image. In order to improve the quality of the reconstructed image, we need make the best of the low-frequency component to retrieve the lost high-frequency domain coefficients. In this paper we introduce image restoration and coefficient fusion techniques into the decompressed algorithm, and propose a novel image decompression technique.

Here we use orthogonal wavelet filter to ensure the reversibility of transformation. As the wavelet filters are orthogonal, the processes in x direction and y direction can be equated to one 2-D transformation [11]. So, the LL component's acquirement is equated to the process shown in Fig.3.

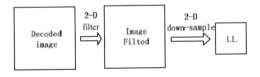

Figure 3.   Procedure of acquiring the LL component from decoded Image.

In our experiment, we choose the low-pass 1-D filter as follows:

$$L = \frac{1}{8} [-1, \ 2, \ 6, \ 2, \ -1] \quad (3)$$

and its 2-D filter can be expressed as:

$$D = \frac{1}{64} \begin{bmatrix} 1 & -2 & -6 & -2 & 1 \\ -2 & 4 & 12 & 4 & -2 \\ -6 & 12 & 36 & 12 & -6 \\ -2 & 4 & 12 & 4 & -2 \\ 1 & -2 & -6 & -2 & 1 \end{bmatrix} \quad (4)$$

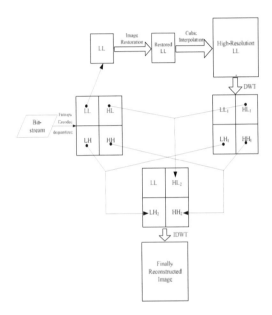

Figure 4.   The flow chart of the proposed decoder.

The proposed decompressed approach is described below, and the flow chart is shown in Fig. 4:

Step 1: The bit-stream is first entropy decoded and dequantized, thus result in the decoded wavelet coefficients;

Step 2: Extract the LL component from the wavelet coefficients, then restore the LL component with blind deconvolution algorithm, and the point spread function is initialized to D (described in Section II.B), retrieve the lost precision of the LL component, and obtain the restored LL;

Step 3: Bicubic B-spline interpolation is taken to generate a high-resolution LL with double size as follow:

$$F(u,v) = \sum_{i=0}^{3} \sum_{j=0}^{3} a_{ij} F(u + w_1, v)^i (u, v + w_2)^j \quad (5)$$

$$w_1, w_2 \in \{0,1\}$$

Step 4: Wavelet transform the high-resolution LL with the same filter, and acquire the new LL_1, LH_1, HL_1, HH_1 components.

Step 5: Fuse LH_1, HL_1, HH_1 with LH, HL, HH respectively to produce new retrievable LH_2, HL_2 and HH_2 components by simply take the corresponding coefficients which have larger absolute values of the wavelet coefficients in each subband [12], which is defined as:

$$WL_2(i,j) = \begin{cases} WL_1(i,j), & |WL_1(i,j)| > |WL(i,j)| \\ WL(i,j), & otherwise \end{cases} \quad (6)$$
$$and \quad WL \in \{LH, HL, HH\}$$

Step 6: Combine the LL with the retrievable LH_2, HL_2 and HH_2 to generate the combined coefficient map, and inverse discrete wavelet transform (IDWT) is then applied to the combined coefficient map to produce the finally reconstructed image.

4 EXPERIMENTAL RESULTS AND DISCUSSION

In this section, we present two simulated data experiments under different compression ratios to illustrate the performance of the proposed algorithm (because of page limitation, test images are omitted here). To assess the relative merits of the proposed methodology, we compare the proposed algorithm with the traditional image decompression [2] and the image compression algorithm using image restoration based on wavelet analysis [4] in the experiments. In order to give a visual impression about performances of the methods included in the comparison, the detailed regions cropped from the results of the test images are presented in Fig. 5, respectively.

It can be observed from Fig. 5 that the proposed decompression method gives better reconstructed results, compared to the decompression results of the other two approaches. The good performance of the proposed method can also be illustrated by the PSNR values shown in Table 1. It is shown that the proposed method produces the highest PSNR value, which illustrates that our method produces a better reconstructed result, close to the original image.

5 CONCLUSION

Image compression is unavoidable due to large amount of storage space or high bandwidth for communication in its original form. Wavelet transform is widely used in lossy image compression, while

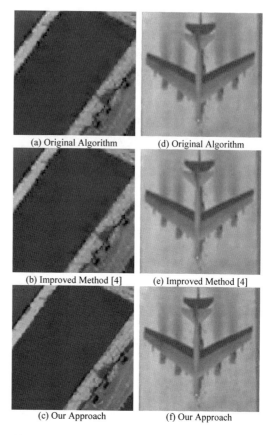

(a) Original Algorithm   (d) Original Algorithm

(b) Improved Method [4]   (e) Improved Method [4]

(c) Our Approach   (f) Our Approach

Figure 5. The decompression results of different methods in the experiments (CR = 20).

Table 1. The PSNR Values of Different Methods in the Experiemts.

Image	CR	Original Algorithm	Improved Method [4]	Our Approach
Aircraft Carrier	20	33.774	34.024	34.171
	30	32.481	32.767	32.898
	40	30.986	31.185	31.249
	50	30.387	30.593	30.618
Airplane	20	37.547	37.874	37.967
	30	35.237	35.469	35.519
	40	34.312	34.419	34.442
	50	33.285	33.457	33.488

more and more image detail will be thrown away with the increase of the compression ratio. In order to overcome this problem, we proposed a novel image decompression method by introducing image restoration and coefficient fusion techniques to the decompression procedure. Some high-frequency coefficients are firstly restored by using image restoration with the low-frequency; new high-frequency components are next retrieved by fusing the restored high-frequency and the corresponding original high-frequency, and then a new coefficient map is generated by combing the original low-frequency with the retrievable high-frequency components. Finally, the reconstructed image is obtained by applying the inverse discrete wavelet transform to the combine coefficient map. Experimental results illustrate that the proposed method outperforms the traditional image decompression methods, both in visual effects and in quantitative terms.

ACKNOWLEDGMENT

This work was supported by the Project of the key National Natural Science Foundation of China under Grant No.60736010, No.60902060, No.61227007, Innovation Research Fund Committee of HUST (2011QN073) and Innovation Fund of CASC (CASC 201104).

REFERENCES

[1] M. W. Marcellin, M. J. Gormish, A. Bilgin, M. P. Boliek, An Overview of JPEG-2000, in Proc. Data Compression Conf., 2000; 523–541.

[2] K. Sayood, Introduction to data compression, 3rd Ed, Academic Press, Morgan Kaufmann Publishers, 2006.

[3] D. S. Taubman and M. W. Marcellin, JPEG2000: Image Compression Fundamentals, Standards and Practice. Norwell, MA: Kluwer, 2002.

[4] Z. Cheng, T. X. Zhang, H. F. Lu, Image Compression Algorithm using Image Restoration based on Wavelet Analysis, Remote Sensing and GIS Data Processing and Other Applications, Oct. , 2009, doi:10.1117 /12.832555.

[5] G. Padmaja, P. Nirupama, Analysis of Various Image Compression Techniques, ARPN Journal of Science and Technology 2012, 2(4): 371–376.

[6] R. Sudhakar, R. Karthiga, S. Jayaraman, Image compression using coding of wavelet coefficients –a survey, ICGST-GVIP Journal 2005; 5(6):25–38.

[7] S. Mallat, A theory for multiresolution signal decomposition: the wavelet representation, IEEE Trans. Pattern Anal. Mach. Intell. 1989; 11(7): 674–693.

[8] A. Levin, Y. Weiss, F. Durand and W. Freeman, Understanding and evaluating blind deconvolu tion algorithms, IEEE Conference on Computer Vision and Pattern Recognition 2009; 1964– 1971.

[9] D. Krishnan, T. Tay and R. Fergus, Blind Deconvolution Using a Normalized Sparsity Measure, IEEE Conference on Computer Vision and Pattern Recognition 2011; 233–240.

[10] J. M. Shapiro, Embedded Image Coding Using Zerotrees of Wavelet Coefficients, IEEE Trans. Signal Process 1993; 41(12): 3445–3462.

[11] A. Lewis, and G. Knowles, Image compression using the 2-D wavelet transform, IEEE Trans. Image Process 1992; 1(2):244–250.

[12] A. Petrosian, F. Meyer, Wavelets in Signal and Image Analysis: From Theory to Practice (Chapter 8), Kluwer Academic Pub, 2001.

Information Technology – Wan et al. (Eds)
© *2015 Taylor & Francis Group, London, ISBN 978-1-138-02785-5*

A novel zero-watermarking algorithm using fractional Fourier transform

Zhen-yu Jiang, Jun Lang & Zhen-ding Shi
College of Information Science and Engineering, Northern University Shenyang, China

ABSTRACT: In this article, a novel method is proposed to make use of digital watermarking for digital image copyright protection. It uses the digital watermarking algorithm based on fractional Fourier transform (FRFT). Meanwhile the letter proposes an improved scheme in which that can achieve the zero-watermarking. And record the pixels of the fractional Fourier transform (FRFT) as the secret key to strength the security of the image. After the experiments, results show that the proposed method is not only of good imperceptibility and security and very robust to JPEG compression and to noise attacks, but also can provide protection under the cropping and the filtering.

1 INTRODUCTION

With the use of the Internet and the development in computer industry widely, the digital content, are easily copied, which need an urgent solution for protecting the copyright of digital image against piracy. Digital watermarking is proposed as an effective solution for the copyright problem.

Digital watermarking has been demonstrated to be very useful in identifying the source, creator, owner, distributor, or authorized consumer of a document or an imager (Yi-Ta Wu & Shih Frank Y 2004). According to the embedding domain, they can be divided into spatial domain and transform domain based techniques. Generally speaking, the spatial domain based schemes have lower computational complexity than the other methods, such as, the least significant bit plane modification is a kind of classical spatial domain of the watermarking method. The watermark in the transform domain based schemes usually exhibits good robustness. Several watermarking techniques based on discrete cosine transformation (DCT) domain (Xiao-Chen Bo 2001, Jiwu Huang 2000), discrete wavelet transforms (DWT) (Yaw Low Cheng et al. 2008, Bao P & Xiaohu M 2005), discrete Fourier transform (DFT) (Vassilis Solachidis & Ioannis Pitas 2001) and the Fractional Fourier Transform (FRFT) (Xia mu Niu & Shenghe Sun 2004, Delong Cui 2009) have been presented in Refs.

In this paper, a zero-watermarking technique is introduced by using Fourier transform and addressing the robustness of a watermarking system for the requirement of the medical image's safety and robustness. The watermarking is encrypted in the spatial domain and making two-dimensional discrete fractional Fourier transform (2D-FRFT). Then it is embedded into the mid-frequency coefficients of the FRFT of the original image.

2 PRELIMINARIES

2.1 The Fractional Fourier Transform (FRFT)

In view of the importance of the frequency domain, the Fourier transform (FT) has become one of the most widely used tools, and the FRFT is a kind of time-frequency, joint representation of a signal, it has found extensive applications in image processing (Haldum. M. Ozaktas 2000). The FRFT of order α of any signal $s(t)$ is defined as:

$$F^{\alpha}[s](\xi) = \int_{-\infty}^{+\infty} s(t)K_{\alpha}(t,\xi)dt \qquad (1)$$

The transformation kernel $K_{\alpha}(t,\xi)$ is given by:

$$K_{\alpha}(t,\xi) = \begin{cases} \sqrt{1-j\cot\alpha}\,\exp\left[j\pi\left(u^2\cot\alpha - 2ut\csc\alpha + t^2\cot\alpha\right)\right], & \alpha \neq 2\pi \\ \delta(u-t), & \alpha = 2n\pi \\ \delta(u+t), & \alpha = (2n+1)\pi \end{cases} \qquad (2)$$

where the t can be thought as the fractional domain. n is an integer and α can be interpreted as a rotation angle in the phase plane.

The fractional Fourier operator F^{α} has the following properties for the special cases:

1 $F^{0}[s](\xi) = s(\xi)$
2 $F^{\pi/2}[s](\xi) = s(\xi)$
3 $F^{\pi}[s](\xi) = s(-\xi)$
4 $F^{2\pi}[s](\xi) = s(\xi)$

let $s(t)$ be a sampled periodic signal with a period Δ, the α order discrete fractional Fourier transform of $s(t)$ can be obtained by using an equation. The formula (1) can be written as:

$$F^\alpha = \sum_{K=-N/2}^{N/2-1} s(k\Delta/N) \sum_{-\infty}^{+\infty} K_\alpha(t,(n+k/N)\Delta) \qquad (3)$$

The forward and inverse two-dimensional discrete fractional Fourier transform (2D-DFRFT) (Soo-Chang Pei & Min-Hung Yeh 1998) of the image signal are computed as:

$$F_{\alpha,\beta}(m,n) = \sum_{p=0}^{M-1}\sum_{q=0}^{N-1} f(p,q)K_{\alpha,\beta}(p,q,m,n) \qquad (4)$$

$$f(p,q) = \sum_{p=0}^{M-1}\sum_{q=0}^{N-1} F_{\alpha,\beta}(m,n)K_{-\alpha,-\beta}(p,q,m,n) \qquad (5)$$

where (α,β) is the order of 2D-DFRFT, $K_{\alpha,\beta}(p,q,m,n) = K_\alpha \times K_\beta$ is the transform kernel, K_α and K_β are the discrete fractional Fourier transform kernels. The two orders of the fractional Fourier transform can be the same, or may not be the same. In this paper, in order to enhance the security of watermark, the order in the fractional Fourier transform is different.

3 PROPOSED WATERMARK ALGORITHM

In the paper, in order to avoid image distortion and obtain a better robustness performance, middle frequency coefficients of the FRFT of the original images are chosen to be embedded watermark, to make a compromise between invisibility and robustness.

3.1 Watermark embedding

Assume $I_1 = \{w(i,j), 0 \leq i \leq N, 0 \leq j \leq M\}$ as the original image, $I_2 = \{z(u,v), 0 \leq u \leq P, 0 \leq v \leq Q\}$ is the original watermark image, $w(i,j)$ and $z(u,v)$ represent the pixel gray level located on coordinates (i,j) and (u,v). The watermark is embedded as follows:
Process:

1 The images W and Z are transformed by the fractional Fourier transform with two transform orders (α,β). As shown in Fig.1. Then we get the transformed image W' and Z'.
2 Convert the image matrix W' into a vector A_1, and then resort the vector A_1 in ascending order to $B_1(x_1,x_2,...,x_m), m = 256 \times 256$. The image matrix Z' makes the same transform as W', we can get $B_2(y_1,y_2,...,y_n), n = 32 \times 32$.
3 $\{Z_n = X_m \mid \min(|X_m - Y_n|)_{m=1,2,...256\times256}\}$ n represents the element's number in the vector. Then compare the elements in B_2 with the elements in

B_1, find the elements in B_1 close to the elements in B_2, and record the location $C_1(z_1,z_2,...,z_n)$ of elements in B_1.
4 Save transform orders (α,β), embedding location, the size of original watermark and the scrambling index. We can get the watermarked image.
5 The process of embedding is shown in Fig.2.

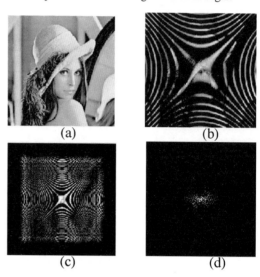
(a)   (b)

(c)   (d)

Figure 1. Results of the image transformed with different orders. (a) Original image Lena (b) The image transformed with transform orders (0.1, 0.1) (c) The image transformed with transform orders (0.6, 0.6) (d) The image transformed with transform orders (1, 1).

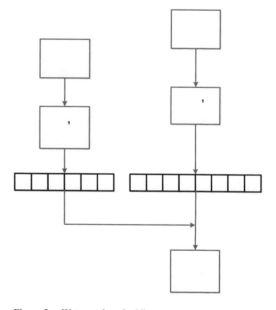

Figure 2. Watermark embedding.

3.2 Watermark extracting

In this proposed method, an original image is required for extracting process, and the extracting process is just the inverse process of the embedding one. The watermarks are detected by using the embedded location and fractional orders (α, β) after the FRFT decomposition of the watermarked image, as follows:
Process:

1 The watermarked image is transformed by the FRFT transform with orders (α, β), and then converts the transformed im. matrix W' into $(256 \times 256) \times 1$ vector A_3.

2 The vector A_3 is rewritten as the process of embedding. Return the location of embedding in vector A_3, and take these elements to constitute a new vector A_4 from smallest to largest.

3 Use the permutation index to shuffle the items of vector A_4, and then reshape the new vector into a matrix Z_2 whose size is 32×32 as watermark.

4 The matrix Z_2 is transformed Z_3 two dimensional fractional Fourier transform with two transform orders $(-\alpha, -\beta)$. We can get the extracted watermark at last.

5 The process of extracting is shown in Fig.3

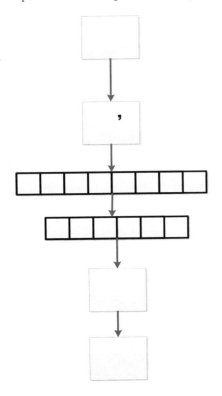

Figure 3. Watermark extracting.

4 EXPERIMENT RESULT

The algorithm is tested on various standard test images and attacks. In the examples, standard test image "Lena" (256×256) is used, and the watermarking is a binary image with (32×32) the pixels [Fig. 4].

The watermarked image and the extracted watermark are shown in Fig. 5

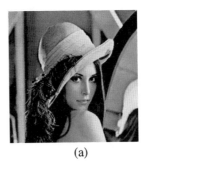

(a) (b)

Figure 4. (a) original image. (b) original watermarking.

(a) (b)

Figure 5. (a) watermarked image. (b) watermark.

4.1 Attack simulation

In order to protect the image, the embedding watermark must be imperceptible and do not distort the watermarked image in comparison with the original image. So the imperceptibility of the algorithm is a factor to qualify an algorithm's performance. We measure the imperceptibility of our algorithm by determining PSNR between a 256 gray scale original image $I_{m,n}$ and watermarked image $I'_{m,n}$ as:

$$PSNR = 10 \log_{10} \left(\frac{255^2 MN}{\sum_{m,n}(I_{m,n} - I'_{m,n})^2} \right) \quad (6)$$

Normalized correlation (NC) (Chuan-ching Wang 2000) is another factor to qualify an algorithm's performance. It determines correlation between an embedded and extracted binary watermark image as:

$$NC = \frac{\sum_{m,n} I_{m,n} I'_{m,n}}{\sum_{m,n} I^2_{m,n}} \tag{7}$$

$I_{m,n}$ shows the ranks of the coordination in the original watermark image pixels, $I'_{m,n}$ indicates the extracted watermark image that coordinates of the pixel, the greater NC value indicates that the extracted watermark are similar to the original watermark.

In addition, the watermark should be able to resist against various attacks. Such as JPEG compression, Gaussian noise, Salt and Pepper Noise, low pass filtering. In order to confirm the validity of the proposed method, we do a lot of experiments; the various attacks results and extracted watermark are shown in Fig. 6.

4.2 Table show

Table 1 shows the corresponding extracted watermarking results, it indicates the proposed algorithm possesses a great robustness against all kinds of attacks.

Compared to other approaches, our algorithm has a more comprehensive performance in robustness. As mentioned in Section 1, Niu and Sun embedded a gray-level watermark into images using discrete fractional Fourier transform (DFRFT). Experimental results show that the algorithm is only against lossy compression attacks. The NC value is equal to 0.458 at JPEG compression quality factor 50. Our algorithm, the NC value is 0.9993 at JPEG compression quality factor 50.

Table 2 shows the algorithm in this paper compared with those in (Jun Lang et al 2008). We use the extracted watermark image and original image's NC value as compared results. And the results indicate that our algorithm has better robustness against various attacks.

Compared with the algorithm in reference (Jun Lang et al. 2008), we can see that our algorithm has a better performance in robustness. No matter what the attacks image suffers, The NC value is all above ninety percentages higher than the Reference (Jun Lang et al 2008). Especially the NC value of Gaussian noise 0.01, JPEG compression factor 20 and Median filter all nearly approaches 1. In all, our algorithm shows a better performance in resisting attacks. And our algorithm has a good quality in image's robustness.

a) JPEG compression and extracted watermark

b) Gaussian noise and extracted watermark

c) Gaussian low-pass filter and extracted watermark

d) Salt and pepper noise and extracted watermark

e) Median filter and extracted watermark

Figure 6. Extract watermarking after several attacking operations.

Table 1. The watermarking robustness test results.

Attacks	Against parameters and NC			
	standard deviation	0.003	0.009	0.03
Gaussian noise	Watermarks			
	NC	0.9996	0.9958	0.9908
	noise intensity	0.04	0.06	0.08
Salt and Pepper Noise	Watermarks			
	NC	0.9985	0.9936	0.9906
	quality factor	70	20	5
JPEG Compression	Watermarks			
	NC	0.9997	0.9988	0.9955
	Standard deviation	0.5	0.8	1.2
Gaussian low-pass filter	Watermarks			
	NC	0.9965	0.9919	0.9903

Table 2. Compared with the algorithm in (Jun Lang et al. 2008).

	NC	
Attacks	Proposed method	Reference [12]
Shearing	0.9358	0.89
Gaussian noise 0.01	0.9952	0.94
JPEG Compression (quality factor: 20)	0.9988	0.97
Median Filter	0.9914	0.95

Table 3. Compared with the algorithm in (Rui-qing Tian 2006).

	NC		
Attacks	Proposed method	Reference [13] I	Reference [13] II
JPEG Compression (quality factor: 90)	1	0.9707	0.9386
JPEG Compression (quality factor: 70)	0.9997	0.9439	0.9332
JPEG Compression (quality factor: 50)	0.9993	0.9070	0.9303
Gaussian lowpass filter (Standard deviation: 0.1)	1	0.9738	0.9797
Gaussian lowpass filter (Standard deviation: 0.3)	1	0.9482	0.9819
Gaussian lowpass filter (Standard deviation: 0.5)	0.9965	0.8696	0.9696

For showing our algorithm's quality more clearly, we compared our algorithm with the algorithm in reference (Rui-qing Tian 2006). From the table, we can find the NC value that the method we propose nearly keeps 1, even though we change the JPEG compression quality factor and Gaussian low-pass filter's standard deviation. By contrast, the results in reference (Rui-qing Tian 2006) are not as good as us. So we can see the good quality of our algorithm in resisting different level's attacks and robustness against the different kinds of attacks.

5 CONCLUSION

A blind image watermarking scheme based on FRNT is presented in this paper. The experimental results show that the proposed algorithm possesses preferable non-visibility, security and robustness resistant against the attacks of JPEG compression, noise, cropping, and filtering, so it can be applied to the image copyright protection effectively.

ACKNOWLEDGMENT

The authors would like to thank the anonymous reviewers for their great efforts and valuable comments that are greatly helpful to improve the clarity and quality of this manuscript. This work was supported by "985 Project" of Northeastern University (No. 985-3-DC-F24), the Fundamental Research Funds for the Central Universities (N130504009), the National Natural Science Foundation of China (No. 61202446), and National Key Technologies R&D Program of China (2012BAK24B0104).

REFERENCES

Yi-Ta Wu, Shih Frank Y. (2004). "An adjusted-purpose digital watermarking technique." Pattern Recognition, 37: 2349–2359.

Xiao-chen Bo et al. (2001). "Sign correlation detector for blind image watermarking in the DCT domain," Lecture Notes in Computer Science, Spring Ver Lag, Berlin Heidelberg, 780–787.

Ji-wu Huang et al. (2000). "Image watermarking in IXT domain: countermeasure and algorithm." Chinese Journal of Electronics, 28: 386–389.

Yaw Low Cheng et al. (2008). "Fusion of LSB and DWT Biometric Watermarking for Offline Handwritten Signature," Proceedings of Congress on Image and Signal Processing, CISP'08, Vol. 5, 702–708.

Bao P & Xiaohu M. (2005). "Image adaptive watermarking using wavelet domain singular value decomposition." IEEE Trans Circuits Syst Video Technol 15(1):96–102.

Vassilis.Solachidis & Ioannis.Pitas. (2001). "Circularly Symmetric Watermark Embedding in 2D-DFT domain," IEEE Transactions on Image Processing, vol. 10, No. 11, pp. 1741–1753, Nov.

Xia mu Niu & Shenghe Sun (2004). "A digital watermarking for still image based on discrete fractional fourier transform," Journal of Harbin Insitude of Technology, vol. 8, no. 3, pp. 302–311.

Delong Cui. (2009). "Dual Digital Watermarking Algorithm for Image Based on Fractional Fourier Transform," IEEE trans. Web Mining and Web-based Application, pp. 51–54.

Haldum.M. Ozaktas et al. (2000). "The fractional fourier transform with applications in optics and signal processing" New York, John Wiley & Sons Ltd, p. 117.

Soo-Chang Pei & Min-Hung Yeh. (1998). "Two dimensional discrete fractional Fourier transform," Signal Processing, vol. 67, pp. 99–108.

Chuan-ching Wang et al. (2000). "Repeating image watermarking technique by the visual cryptography," IEEE Trans on Fundamentals, 36(13): 1108–1110.

Jun Lang et al. (2008). "The Multiple-parameter fractional Fourier Transform," Sci China Ser F-Inf Sci, 51(8), 1010–1024.

Rui-qing Tian. (2006). "Image Digital Watermarking Using the Fractional Fourier Transform" Beijing University of Chemical Technology

Information Technology – Wan et al. (Eds)
© 2015 Taylor & Francis Group, London, ISBN 978-1-138-02785-5

A Qt platform-based automatic train diagram computerization tool

Yan Li
NARI Technology Development Co., Ltd., Nanjing, P. R. China

Xing Zhao
College of Civil and Transportation Engineering, Hohai University, Nanjing, P. R. China

Wei-wei Zhu
NARI Technology Development Co., Ltd., Nanjing, P. R. China

ABSTRACT: Based on display of train diagram, this research provides a convenient computer-aided design tool to execute visualized operation of train diagram compilation combining with requirement of train operation. It makes train diagram drafting easier by providing simple manipulation to achieve data inputting, outputting, and a variety of editing functions automatically. The proposed tool is programmed on Qt platform. It enriches contents of diagram display by providing zooming, picture saving, and printing functions for users. Combining with Qt packaging technology, train diagram drafting tool provides complete and convenient interfaces of importing and exporting, which makes this tool more applicable and open.

1 INTRODUCTION

The drawing and application of train diagram has a great significance on enhancing circuit efficiency in urban rail transit system. In the construction of domestic urban rail transit, facilities of signaling system are mostly manufactured by foreign companies. It leads to lots of inconvenience and high costs in project fields such as facilities installation, commissioning, fault detection, and integration with other software systems. Therefore, the localization of urban rail transit signaling system will make a great help in domestic urban rail transit construction, and the diagram part is an important component of ATS in signaling system. Based on train traction calculation, this research is aimed to develop a train diagram drawing tool with convenient maneuverability and good portability.

Qt is a cross-platform application, and also a C++ graphical user interface library, which was produced by TrollTech Company of Norway at the end of 1995. It provides rich libraries and a unique signal/slot mechanism for programmers to achieve the graphic interface development work conveniently and efficiently. These advantages of Qt platform ensure the quality of interface display and extend the range of the software application.

Therefore, this research develops the train diagram drawing tool based on Qt platform. This tool provides auxiliary display functions and relative conventional editing functions, including operation parameters input, basic diagram output, trip adding and deleting, train run-type adjustment, departure and arrival setting and turn-around setting. It also provides data exchange channel with other relative applications to enhance the applicability of the tool. And on the base of diagram display, functions such as zooming, printing, picture saving (PNG format), and display style switching can be easily called to match different requirements of train dispatching staffs.

2 SYSTEM STRUCTURE

The system of train diagram drawing tool is designed as four modules, including fundamental data collecting module, data processing module, display module, and output module. Thereunto, fundamental data collecting module achieves the input process of data for whole system. The process is designed in two ways: manual input parameters or interact with other program. Then, data processing module verifies the data acquired in fundamental data collecting module by algorithm specified, and completes all calculating and packaging of data which display module requires. Display module shows the operating result to users, and accomplishes human-computer interaction accordingly. This module is the core part of application. Finally, output module stores all the data during diagram drawing procedure to support communication interface with other program.

3 SYSTEM MAIN FLOW

The train diagram drawing tool is compiled on Qt platform. Qt can match the drawing requirement of software interface well with the powerful painting engine, and ensure the compatibility of software between different operation systems on the cross-platform character. The main flow of the system is shown in Figure 1.

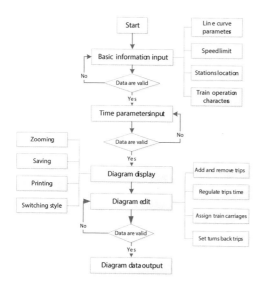

Figure 1. System main flow.

4 SYSTEM INTERFACE DESIGN

4.1 Data input

The data input includes station mileage information, flat and vertical curve parameters of line, train motion parameters, and speed limit information of line. Conduct the standard train traction calculation with the input data above to acquire the speed-distance-time relationship of standard train run curve. Besides manual input pattern, these data can be acquired by communicating with other process that can be provided by a train traction tool established by our research team before.

4.2 Diagram display

After acquiring the external data, it is necessary to verify their validity. In the context of getting reasonable data, specify the train operation time range and grid spacing. A group of equal-time-interval time-displacement curve of train can be drafted.

The train diagram drawing tool is designed for main line passenger vehicles, and is focused on description for regional train running status and corresponding turn back relationship in main line. There are different staffing patterns of diagram in different situation, and combined with the characteristics of urban rail transit, this research adopts one minute time per cell as horizontal axis, scaled line mileage distance that separates into different direction sites at each station as vertical axis. Record the arrival and departure time at each station and compute the turn back time of each trip on the graph established. The diagram display interface is shown as Figure 2, including stations, trip IDs, and train vehicle IDs. Time tips represented by horizontal axis can be shown when cursor moves to corresponding positions, as shown in Figure 2. In addition, whether to display different direction sites at each station can be switched by click on corresponding station name, as the double line pattern in Figure 2.

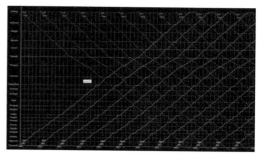

Figure 2. Train diagram display.

In the diagram display interface, several effective assistant functions are provided to enrich the contents of images, including diagram zooming, images saving, printing, and display style switching.

4.2.1 Diagram zooming function
Zooming function is supplied to show various kinds of elements more accurately for users in diagram display interface. The function can be triggered for a single axis, it means that time axis and the distance axis can be zoomed in or zoomed out separately.

4.2.2 Diagram saving function
The diagram can be saved as a picture in conventional formats such as PNG, JPG, BMP.

4.2.3 Diagram printing function
The diagram can be printed in regular procedure according to the corresponding connected printer.

4.2.4 Interface style switching function
Several operating system rendering styles can be chosen for interface display, including CDE, plastique, and so on.

4.3 Diagram edit

On the basis of train diagram generated described in Section 4.2, more detailed edit function is supplied to adjust to the operating habits of users. The functions include adding and removing trip curves, regulating trip arrival and departure time between stations, assigning train carriages to trips, setting turns back trips, and so on. Diagram editing interface is shown as Figure 3; in the left part of the picture is train list window, in the right part of the picture is diagram display window, and the bottom left zone in a label is used for edit function triggering.

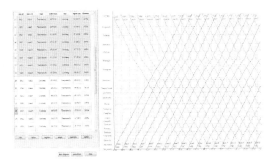

Figure 3.   Train diagram edit interface.

4.4 Adding and removing trip curves

Adding a trip curve can be implemented by assigning departure time, trip number, train carriage number, and the direction. Adapt to the standard curve style and the current user settings, a new trip curve can be added at corresponding position in diagram automatically.

Deleting a trip curve is implemented by identifying specified trip number. The corresponding trip curve can be deleted from diagram automatically.

4.5 Regulating trip arrival and departure time between stations

In a single trip, the operational time between stations can be adjusted, including train run time and train park time at stations. As shown in PIC, scheme arrival time, scheme departure time, or park time can be modified by clicking left mouse button, and at the same time, the three parameters are coordinated in relations:

$$t_{arrive} + t_{stay} = t_{depart} \qquad (1)$$

After the input of time, the new value needs to be verified in the method below. According to the standard train run curve, it is assumed that the interval time value between train departure at station 1 and train arrival at station 2 is as follows:

$$t_{nes}(s1, s2) = t_{depart}(s2) - t_{arrive}(s1) \qquad (2)$$

where $s1 < s2$, and the new value which user inputs to identify train arrival time at station 2 is $t_{m\_depart}(s2)$, if $t_{m\_depart}(s2) - t_{arrive}(s1) < t_{nes}(s1, s2)$, then the value input will be denied. The regulations to other parameters are verified in similar principles, which ensure the regulations are effective and reasonable. The regulation interface is shown as Figure 4.

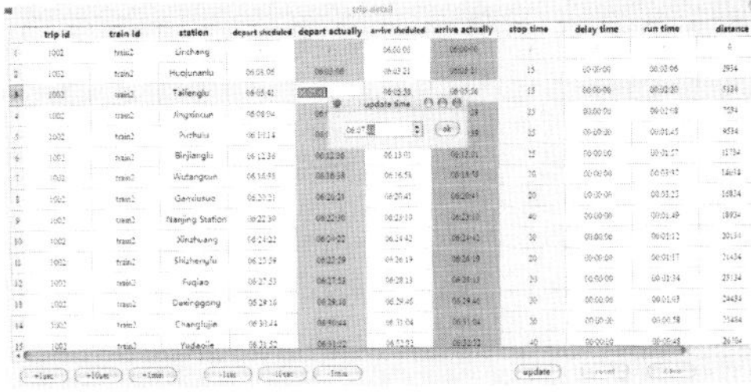

Figure 4.   Regulating trips time function interface.

4.6 Assigning train carriages to trips

In the initialization process of diagram settings, every train is assigned a train carriage number and a unique trip number. Assigning train carriages to trips function can realize this assignment manually to re-associate train carriage numbers and trip numbers. Trips delegated by a same train carriage number are connected in the diagram automatically.

4.7 Setting turn back trips

Turn back train can be set by indicating trip number and turn back position. The starting and ending position of the relative trips will be adjusted in diagram.

In addition, all functions above can be realized in right click menu in diagram showing interface or tool bar at top, which improve the convenience of the diagram plan preparation.

In the whole diagram plan preparation process, it is important to verify user's all operations, which include physical verification and logical verification. The physical verification is against the validity of user input characters, as well as the logical verification is against unreasonable data effect to diagram. All invalid operations should be excluded to improve the utility of the tool.

4.8 Data output

Some relative interfaces in data processing module are packaged to support data recording in output module. All data generated in diagram editing process will be recorded, including trip numbers, train carriage numbers, operation directions, positions for each moment, train departure time, and train arrival time, and so on.

Data exchange interfaces are open to other program processes. The system stores data on standard xml files, and packets them with QDomDocument class. The operation data include train trip numbers, train carriage numbers, directions, and so on, and can be used by other applications for further research.

5 CONCLUSION

This article describes the developed train diagram drawing tools from three aspects including system structure, flow design, and interface design. It mainly introduces the basic information input process, various editing functions, the auxiliary display function, the output model of map information, and the realization of the data exchange with other related application programs. In the future, work will be devoted to enrich display content of the train diagram and to improve operation convenience of this software.

ACKNOWLEDGMENTS

This research is supported by the National Natural Science Foundation of China (No. 51408190).

REFERENCES

Zhang. Z. Y. (2009). Train Traction Calculation. *China Railway Press,* 5–10.

Wu, L. N., et al. (2010). Operating Indicator Design of Inner Route and Stop for Bus. *Journal of Heilongjiang Institute of Technology,* 24, 14–17.

Zou, Z. J. & Yang, D. Y. (2001). Software Design of Urban Road Traffic Simulation System. *Journal of Traffic and Transportation Engineering,* 1, 86–88.

Liu, J. F., et al. (2005). Multi-train Movement Simulation System for Urban Rail Transit. *Journal of Transportation Systems Engineering* and *Information Technology,* 5, 79–82.

Mao, B. H., et al. (2000). A General-proposed Simulation System on Train Movement. *Journal of the China Railway Society,* 22, 1–6.

Wang, B., et al. (2007). The Research on the Train Operation Plan of the Beijing-Tianjin Inter-city Railway Based on Periodic Train Diagrams, *Journal of the China Railway Society,* 29, 8–12.

Xu, H., et al. (2007). Research on the Model and Algorithm of the Train Working Diagram of Dedicated Passenger Line, *Journal of the China Railway Society,* 29, 1–6.

Peng, Q. Y., et al. (2001). Study on a General Optimization Model and Its Solution for Railway Network Train Diagram, *Journal of the China Railway Society,* 23, 1–8.

Zhang, Y. S. & Jin, W. D. (2009). Iterative Repair Method of Single-track Railway Rescheduling based on Discrete Event Topologic Diagram Model. *Journal of System Simulation,* 21, 7003–7007.

Information Technology – Wan et al. (Eds)
© 2015 Taylor & Francis Group, London, ISBN 978-1-138-02785-5

A semi-supervised local linear embedding method in kernel space

Cuiping Jing & Jiulong Zhang

Xi'an University of Technology, Xi'an, China

ABSTRACT: Motivated by the fact that traditional local linear embedding algorithm does not use label information well, an improved semi-supervised local linear embedding method in kernel space is proposed. For the unlabeled sample faces, we need to approximate nearest neighborhood distance by adjusting distance matrix. Gaussian kernel space distance is used here instead of Euclidean distance, therefore manifold structure will not be destroyed in low dimension space. In the ORL and Yale B face database, experiment results and comparison charts show that our approach has a better performance.

1   INTRODUCTIONS

A major kind of face recognition methods is appearance based; taking the whole image as a high-dimensional random vector, and applying linear subspace transform method or manifold learning method for feature extraction. There are many linear subspace methods applied in dimension reduction of face image such as principle component analysis (PCA), linear discriminate analysis (LDA), independent component analysis (ICA), and much more improved nonlinear algorithms are developed in recent years. The linear structure of data can be found by the linear subspace method, but the problem is that there is more nonlinear information contained in the data. Studies have shown that the high dimensional face images exist substantially in a lower dimensional manifold [1–2]. Facial images under expression, illumination and pose variation can be reduced to corresponding low-dimensional features with concrete physical meaning, which show good result, so we consider using manifold learning algorithms for face recognition.

To make use of the few available labeled samples, it is necessary to introduce the supervised learning algorithms [3], which can enhance the clustering and classification ability of different human face data. Semi-supervised algorithm was introduced to the manifold learning by Xin Yang [4]. Junping Zhang [5] introduced the kernel function to LLE, solving the lack of projection matrix for traditional manifold learning algorithm. We introduce a method for choosing kernel function for LLE and demonstrate how it works, and provide several examples. First, preprocess the input face image; select the training samples and testing samples, and then use the SS-KLLE for dimension reduction, the final step is three-nearest neighborhood algorithm for recognition.

This paper is organized as follows: first, we present manifold learning principle and gives a brief description of main algorithms; then an improved semi-supervised local linear embedding method in kernel space are introduced, and have a discussion about the steps of SS-KLLE algorithm; it is the experimental results and analysis, finally conclusions are given.

2   MANIFOLD LEARNING

Manifold learning is a machine learning method based on differential geometry theory, To discover the intrinsic manifold structure of the data, many methods were proposed, which can be categorized into two classes, global structure preserving and local structure preserving. Global structure preserving methods include isometric mapping (Isomap) and local structure preserving methods include local linear embedding (LLE), local projection preserving (LPP), Laplace Eigen maps (LE), local tangent space alignment (LTSA), etc.

Isomap algorithm is based on classical multi-dimensional scaling (MDS), which seeks to preserve the global geodesic distance between pair of linear sample points, and gains the shortest distance by constructing adjacency graph, and uses shortest path graph to approximate geodesic distance. Fig. 1 shows some of sculpture face dataset, 64 pixel by 64 pixel raw images would be processed as 4096 dimensional vectors. Fig. 2 illustrates the existing of low dimensional manifold. Left-right pose and up-down pose was separated by Isomap ($K = 7$) dimensional reduction method. However, computational complexity of isomap is high to being a global structure preserving method.

By LLE method, each point can be represented by k-nearest neighbors with linear relation. LLE minimizes the rebuilding error matrix, and obtain the embedded data in low dimension. Thus, the neighborhood in high dimension can influence embedded result in low dimension.

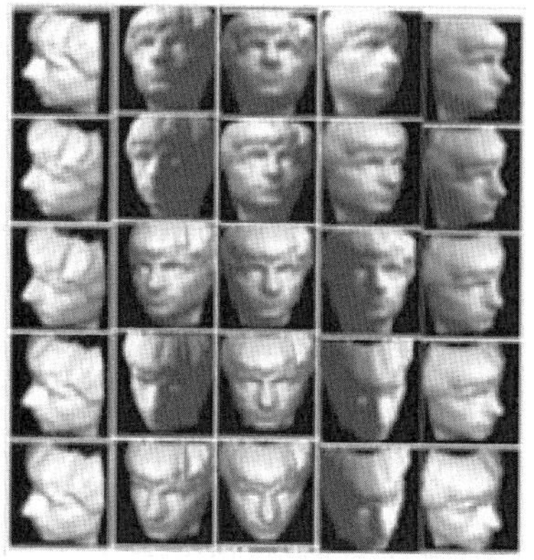

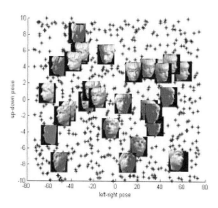

Figure 1. Some examples of sculpture face dataset.

Figure 2. Visual result by Isomap method.

Suppose original dataset X is consisted of N vectors x_i, which is a high dimensional vector of D, that is

$$X = \{x_1, x_2, \cdots, x_N\}, x_i \in R^D ;$$

y_i is the low dimensional vector, which is embedded from x_i of d, where $D \gg d$, that is

$$Y = \{y_1, y_2, \cdots, y_N\}, y_i \in R^d ;$$

LLE maps dataset X globally to embedded dataset Y by the assumption of retaining the neighborhood relations: after mapping, the original near neighbor points are also near. For example, in high-dimensional space, w_{ij} = construction coefficient of point x_i by its neighbor points x_{ij}; w_{ij} should remain the same relationship in low dimension space after dimension reduction.

3 SEMI-SUPERVISED LOCAL LINEAR EMBEDDING IN KERNEL SPACE

Semi-supervised learning is an important technology of unlabeled learning, which can automatically take advantage of a large number of unlabeled data to improve learning generalization capability without outside intervention. SS-KLLE are mainly applicable to high-dimensional datasets, which can exhibit clearly manifold structure by Kernel function in higher dimension space.

3.1 Steps of SS-KLLE algorithm

3.1.1 Compute the neighborhood for each data point

The Euclidean distance of the neighborhoods between x_i and x_j is defined:

$$d(x_i, x_j) = \|x_i - x_j\| = \left\{ \sum_{i=1}^{k} (x_i - x_j)^2 \right\}^{1/2} \quad (1)$$

We not only use kernel space distance to calculate K-nearest neighbors for each point x_i, but also combine it with the label information, so the distance matrix will be adjusted as follows:

$$D(x_i, x_j) = \begin{cases} \left(1 - \exp(-d^2(x_i,x_j)/\beta) + \sqrt{\exp(d^2(x_i,x_j)/\beta)}\right)/2 &, x_i \text{ or } x_j \text{ is unlabeled} \\ 1 - \exp(-d^2(x_i,x_j)/\beta) &, c_i = c_j \\ \sqrt{\exp(d^2(x_i,x_j)/\beta)} &, \text{otherwise} \end{cases} \quad (2)$$

where the parameter β is used to avoid large distance $d(x_i, x_j)$ to grow too fast, and β = mean of $d(x_i, x_j)$, so we can calculate K-nearest neighbors of each point by $D(x_i, x_j)$.

3.1.2 Construct weights matrix W_{ij}
Construct the weight matrix W_{ij} in the neighborhood of x_i, and the reconstruction error matrix is as below:

$$J_1(W) = \sum_{i=1}^{N} \left\| x_i - \sum_{j=1}^{N} W_{ij} x_j \right\|^2 \quad (3)$$

Reconstruction weight W_{ij} must be subjected to two constraints:

- If j are not in the same neighbor of i, W_{ij} will be zero,

$$W_{ij} = 0 \qquad (4)$$

- The sum of the rows of the weight matrix is one, that is,

$$\sum_i W_{ij} = 1 \qquad (5)$$

The optimal weights can be obtained by solving a least squares problem. Finally, the sparse weights matrix W was formed.

3.1.3 Computing the low-dimensional embedding

Maintain the weight W_{ij}, calculate y_i, which is the mapping of x_i, making minimum reconstruction error rate in low-dimensional space, where reconstruction error rate is:

$$\sum_{i=1}^{N} \left\| y_i - \sum_{j=1}^{N} W_{ij} y_j \right\|^2 = trace\left(Y(I-W)(I-W)^T Y^T \right) \qquad (6)$$

By solving this optimization problem, we can get that low-dimensional embedding vector matrix M, which comprised of the minimum d eigenvectors of the matrix from second to $d+1$, where $M=(I-W)(I-W)^T$. Therefore, we can calculate the low-dimensional embedding vector y_i from the optimal reconstruction matrix.

3.2 Visual result by SS-KLLE algorithm

To make the data be visual, we reduced the data to be two-dimensional by the SS-KLLE algorithm, and then figure out the scatter chart of some Yale B face image in two dimensions, according to the comparison result between the two-dimensional coordinates after face embedding and the raw image. In Fig. 3, we can see that most of the same person get together, the right side is darker for each image. It shows that we can find variables which control some essential characteristics of human face by SS-KLLE algorithm, which is the potential manifold for face input space.

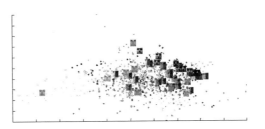

Figure. 3 Yale B scatters chart by the SS-KLLE when reduced to two-dimensional and the corresponding face image.

4 EXPERIMENTS AND ANALYSIS

In this experiment, we used ORL face database and a subset of public Yale B face database. Yale B face database is under different lighting and expressions of face images, we take 38 people, 60 images each person for example, resolution rate is 168×192. In the preprocessing stage, the size of each individual face database image is cropped to 128×128, and then histogram equalization processing is done.

In ORL and Yale B face database, one half of images were randomly selected for training, and the remaining is as testing set. SS-KLLE parameters choosing are as follows: the number of neighbors $K=13$, coefficient of information utilization degree between different class $\alpha = 0.1$. Figs. 4 and 5 are recognition results in different identifying dimensions for a variety of algorithms, they are, respectively KLLE, SKLLE, SS-KLLE in ORL face database and Yale B face database.

As it shows in ORL face database, LLE, SLLE and SSLLE had almost same recognition effect. But in Yale B face database, SKLLE obtained much better performance, the recognition rate of SSKLLE had reached above 90%, and it is much higher than KLLE algorithm.

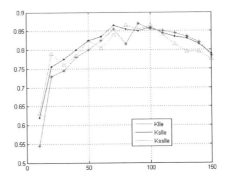

Figure 4. Experimental results under ORL database.

93

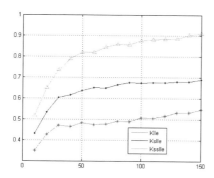

Figure 5. Experimental results under Yale B database.

5 CONCLUSION

In this paper, we had introduced a semi-supervised kernel local linear embedding manifold learning method for face recognition, the light sub manifold was found in Yale B dataset. First, rationality of this method was analyzed, and the validity was verified by experiments. Then, Guassian kernel function and known label information were incorporated into LLE method. Owing to labeling every sample artificially is not accurate and flexible, SLLE method is discarded on the occasion of big dataset. In addition, further experiments in the two datasets proved that SS-KLLE algorithm has a higher recognition rate than unsupervised and supervised LLE in human face recognition under light.

ACKNOWLEDGMENTS

This paper was sponsored by the Science Plan Research Project of Beilin District (under Grant No. GX1410), Natural Science Foundation of China (No. 61272284), the Nature Science Foundation of Shaanxi province (Grant No. 2014JQ8299).

REFERENCES

[1] Roweis S T, Saul L K. Nonlinear dimensionality reduction by locally linear embedding. *Science*, 2000, 290(5500):2323–2326.
[2] Tenenbaum J B, de Silva V, Langford J C.A Global geometric framework for nonlinear dimensionality reduction. 5, 2000, 290 (5500):2319–2323.
[3] Ridder D, Olga K. Supervised Locally Linear Embedding. Artificial Neural Networks and Neural Information Processing, New York, USA. 2003, pp. 333–341.
[4] Xin Yang, Haoying Fu, Hongyuan Zha, Jesse Barlow, Semi-Supervised Nonlinear Dimensionality Reduction. Proceedings of the 23rd International Conference on Machine learning,, New York, USA, 2006, pp. 1065–1072.
[5] Junping Zhang, Stan Z.Li, JueWang. Manifold Learning and Applications in Recognition. Intelligent Multimedia Processing with Soft Computing. Springer-Verlag, Heidelberg. 2005, 281–300.

Information Technology – Wan et al. (Eds)
© 2015 Taylor & Francis Group, London, ISBN 978-1-138-02785-5

A star crude extraction algorithm based on shearlets and top-hat transform

Jie Li
College of Electronic Information Engineering, Changchun University, Changchun, China

Pan Guo
College of Electronic Information Engineering, Changchun University of science and Technology,
Changchun, China

Chunzhe Wang
College of Electronic Information Engineering, Changchun University of Science and Technology,
Changchun, China

ABSTRACT: A star crude extraction algorithm is based on shearlets transform and morphology. Top-hat transform is proposed for that the star extraction is the key to realize the high precision and accuracy of star pattern recognition and attitude determination. The algorithm filters the noise in the star image by Shearlets transform through different threshold processing to Shearlets coefficients in different directions and reconstructing the denoising image. Then Top-hat transform is performed on shearlets reconstructed star image to suppress the background noise and using the adaptive threshold method processing to remove the residual noise after background suppression. The algorithm completes the star crude extraction effectively. The experimental results show that the algorithm can remove the noise effectively and make a good foundation for sub-pixel star extraction.

1 INTRODUCTION

Starlight attitude determination generally includes four parts of pre-processing star, star target extraction, star pattern recognition and attitude solution. The star target extraction is an important part among them, and its extraction precision and accuracy will affect the performance of star sensor. Star target extraction consists mainly of two steps: star crude extraction and sub-pixel extraction. Because the stars represent the spot generally in the dark background on the CCD plane and the diameter of spot is 3–5 pixels, star point targets of star image are easily affected by noises. So it is an important factor to influence the accuracy of centroid localization that whether the star crude extract can maximize the reduction of the actual distribution of pixels and gray value of star point or not [1].

At present, there are three kinds of crude extraction algorithms, such as high pass filtering method[2], local entropy method[3] and mathematical morphology algorithm,[4] etc. High pass filtering method has a limitation of the template and it is not applicable; for removing noise with local entropy method, because of its accuracy affected by the size of local window, when the target is located in multiple local windows, there will be greater error; mathematical morphology algorithm based on Top-hat transform is influenced by the background. So only in the bright background and complicated background, denoising effect is obvious and its accuracy is low.

Based on the research of star crude extraction algorithms at the present, a crude extraction algorithm by using shear wave transform and combined with Top-hat transform is proposed in this paper, which can suppress the noise and background noise effectively, and use the adaptive threshold processing to remove the residual noise. The experiments show that the algorithm has high antinoise performance, and effectively ensures the accuracy of sub-pixel extracting.

2 RESEARCH ON STAR CRUDE EXTRACTION ALGORITHM

2.1 The theory of shearlets transform algorithm

Shearlets transform is proposed by Guo and Easley et al. through the using of affine systems with composite dilations, which can be used for image representation and at the same time can be with the advantages of multiresolution, locality and directionality[5–7]. In this paper, the star image with noises is processed by Shearlets transform to get coefficients of different scales and different directions, which are performed by different thresholds, and then we can get the star image removed the noise through Shearlets inversing transform to the processed coefficients.

The affine systems are expressed in Eq. (1) as,

$$A_{AB}(\psi) = \{\psi_{j,l,k}(x) = |\det A|^{j/2} \psi$$
$$(B^l A^j x - k) : j,l \in Z, k \in Z^2\} \quad (1)$$

where both and are 2×2 invertible matrices and $|\det B| = 1$. If $A_{AB}(\psi)$ forms a Parseval tight framework for L2(R2), the elements of the system are called composite wavelets.

Shearlets are defined as follows:

$$SH_\psi f(a,s,t) = \langle f, \psi_{a,s,t} \rangle, a > 0, s \in R, t \in R^2 \quad (2)$$

$\psi_{a,s,t}(x)$ is defined as follows,

$$\psi_{a,s,t}(x) = |\det M_{as}|^{1/2} \psi(M_{as}^{-1}(x-t)) \quad (3)$$

Note that

$$M_{as} = B_s A_a \quad A_a = \begin{bmatrix} a & 0 \\ 0 & \sqrt{a} \end{bmatrix} \quad B_s = \begin{bmatrix} 1 & s \\ 0 & 1 \end{bmatrix}, \quad (4)$$

where A_a is an anisotropic dilation matrix, and B_s is a shear matrix.

As the formula (2) shows, the Shearlets are a set of functions with three parameters which are scale, direction and position. The tiling of the frequency plane induced by Shearlets and the comparison of frequency support of Shearlets and wavelets are illustrated in Fig. 1. It can be seen from the figure that the frequency support of Shearlets transform is a trapezoidal region which is symmetric relatively for the origin along the direction of the slope in different scales. Compared with the wavelet transform, the Shearlets can be better to realize the positioning in each of scale, direction and position.

An adaptive threshold method based on Bayesian estimation[8] is used for threshold processing to Shearlets coefficients. The estimate $\hat{\omega}$ to the true Shearlets coefficient is given by

$$\hat{\omega}(y) = \text{sign}(y) \times (|y| - T)_+ \quad (5)$$

where y is the Shearlets coefficient with noise and is a threshold which is given by

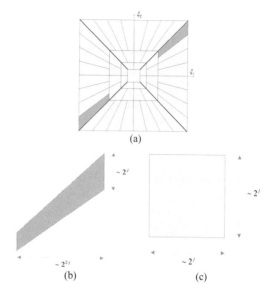

(a)

(b)   (c)

Figure 1.   (a) The tiling of the frequency plane induced by Shearlets. (b) frequency support of Shearlets. (c) frequency support of wavelets.

$$T = \sqrt{2}\sigma_k^2 / \sigma \quad (6)$$

where σ_k^2 is the variance of noise, and σ is standard deviation of Laplace edge distribution. The symbol $(g)_+$ is expressed as follows,

$$(g)_+ = \begin{cases} 0, if\ g < 0 \\ g, otherwise \end{cases} \quad (7)$$

2.2   The theory of morphological Top-hat transform

The gray scale opening operation in morphological can remove some small bright details and keep the whole gray level of image and the large bright region invariant at the same time[9]. Top-hat transform can remove the background noises of the image through using the gray opening operation to get the estimation to the background and then making the estimation and the original image cancel. Assume that $f(x,y)$ is the gray value of the image, and (x,y) is the coordinate point of the gray value. and are translation parameters, and is a matrix of structural element and D_b is the domain of definition of . So the gray dilation $f \oplus b$ $f$ois defined as follows,

$$(f \oplus b)(x,y) = \max\{f(x-x',y-y')$$
$$+ b(x',y')|(x',y') \in D_b\} \qquad (8)$$

The gray erosion $f \ominus b$ is defined as follows:

$$(f \ominus b)(x,y) = \min\{f(x+x',y+y')$$
$$- b(x',y')|(x',y') \in D_b\} \qquad (9)$$

The gray opening operation $f \circ b$ is defined as follows:

$$f \circ b = (f \ominus b) \oplus b \qquad (10)$$

Top-hat transform $H_{top-hat}$ is defined as follows:

$$H_{top-hat} = f - f \circ b \qquad (11)$$

2.3 The threshold processing

After shearlets denoising and Top-hat transformation, most noises can be removed, then the threshold processing can be used to remove the residual noise, which can realize the star crude extraction. Assume that $f(x,y)$ is the gray value of the image, and T is the threshold of the background of the image, then the threshold processing is expressed as

$$F(x,y) = \begin{cases} f(x,y) & f(x,y) > T \\ 0 & f(x,y) < T \end{cases} \qquad (12)$$

The threshold T is given as

$$T = \mu + \alpha \cdot \sigma \qquad (13)$$

where μ is the mean of the image and σ is the variance of the image, and α is a constant which is taken one of 3–15 according to general experience[10-11].

3 THE SIMULATION OF STAR IMAGE

The actual gray distribution of star sensor imaging pixel approximately accords with Gauss distribution, therefore, we need to do gray diffusion in accordance with the Gauss rule to generate simulated star

spots[12–13]. The Gauss distribution of the star spots simulation as shown in Fig. 2, and the formula is given as follows:

$$g_{i,j} = \frac{A}{2\pi\sigma^2} \exp(-\frac{(x_i - x_0)^2 + (y_j - y_0)^2}{2\sigma^2}) \qquad (14)$$

where is a total energy of stars, and (x_0, y_0) is a mapping coordinate in the CCD star point which is the true star coordinate, and $g_{i,j}$ is gray value of the simulation star point (x_i, y_j). When σ <0.671, 95% of the energy of Gauss distribution is in 3pixel×3pixel[14], which is consistent with the actual distribution of star imaging.

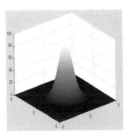

Figure 2. the energy distribution of the star spots

4 THE REALIZATION OF THE STAR CRUDE EXTRACTION ALGORITHM

The realization process of star crude extraction algorithm based on the shearlets transform and morphological Top-hat transform is, first, using the shearlets transform to filter the noises map by using different threshold processing to shearlets coefficients to suppress noises, and then using Top-hat transform for the reconstructed star image to estimate background of star image to suppress background noises, finally using adaptive threshold processing to remove residual noises, which can complete the star crude extraction algorithm. Concrete steps are as follows:

Step 1: use star simulation method of the Gauss distribution and add noises to the simulation star image;

Step 2: the star image with noises is processed by shearlets transform with 4 layer shearlets decomposition to get the low frequency coefficients and the high frequency coefficients in different scales and directions;

Step 3: obtain the estimation of real coefficient $\hat{\omega}$ through the formula (5);

Step 4: take the shearlets inverse transform to get the reconstructed star image;

Step 5: Top-hat transform is performed on shear-lets reconstructed star image using the size of 5×5 square and diamond as structural elements;

Step 6: in accordance with the formula (12), take the threshold processing to the transformed star image of step 5 to obtain the image after star crude extraction algorithm.

5 EXPERIMENTAL RESULTS AND ANALYSIS

5.1 Star image simulation experimental results

The Gauss gray pervasion method is used for accurate simulation of star image, in accordance with the star standard schemes in Ref. [15]. For any of the selected a pointing, right ascension (66.7352) and declination (88.26437), assume that the star sensor field is 8 × 8 and CCD is 512 pixel×512pixel. We choose 10 stars as shown in table 1. According to the formula (14),

Table 1. Star simulation coordinates

| Numbers | Mapping coordinates | |
	x_0	y_0
1	34.2815	492.6473
2	179.2209	461.3400
3	328.4338	394.3868
4	379.3006	427.4299
5	121.1035	267.9541
6	73.9042	300.3970
7	240.1077	360.1191
8	398.4109	271.5984
9	212.8773	190.8897
10	387.0235	77.9893

simulate stars in Table 1, and add Gauss noise and random noise to the simulation star image, respectively. The simulate and added noise star images are shown in Fig. 3(a)–(c), and Fig. 3(d)–(f) are 3D star image of (a)–(c), respectively.

5.2 Star image de-noising based on Shear Wave transform

According to the steps of 1–4, the simulate star images added random noise and Gauss noise are processed by Shear Wave de-noising respectively, which is used with wavelets de-noising for contrast. The star image with random noise processed after shear wave and wavelets de-noising are shown in Fig. 4(a) and (b) and the 3D images of Fig. 4(a) and (b) are shown in Fig. 4(c) and (d), respectively. The star image with Gauss noise processed after shearlets and wavelets de-noising are shown in Fig. 5(a) and (b) and the 3D images of Fig. 5(a) and (b) are shown in Fig. 5(c) and (d), respectively. Table 2 shows MSE value of the star images before and after denoising processing.

By comparing on the wavelet de-noising and shear wave de-noising, from the visual effect, the shear wave de-noising is obviously superior to the traditional wavelets de-noising; from the contrast of MSE values, the effect of shear wave de-noising is obviously better than the wavelets de-noising.

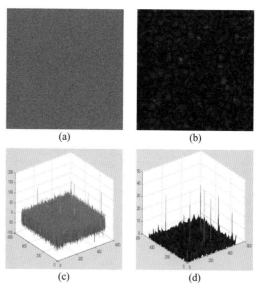

Figure 4. (a) The star image with random noise processed after wavelets denoising; (b) the star image with random noise processed after shearlets denoising; (c) the 3D image of (a); (d) the 3D image of (b).

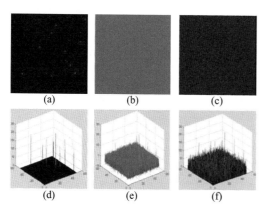

Figure 3. (a) Simulation star image; (b) star image with random noise of =19;(c) star image with Gauss noise of = 0.1;(d) 3D star image of (a); (e) 3D star image of (b); (f) 3D star image of (c).

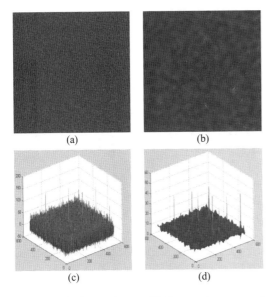

(a)　　　　　　(b)

(c)　　　　　　(d)

Figure 5. (a) The star image with Gauss noise processed after wavelets denoising; (b) the star image with Gauss noise processed after shearlets denoising; (c) the 3D image of (a); (d) the 3D image of (b).

Table 2.　Comparison of MSE values.

The types of noise	Noise	After wavelets denoising	After shearlets denoising
random noise	360.993163	242.387778	2.654689
Gauss noise	379.853214	299.914566	134.056090

5.3 Background noise suppression based on Top-hat transform and thresholding processing of star image

In accordance with the steps of 5 and 6, Top-hat transform and threshold processing are performed on the reconstructed image of shear wave de-noising. Fig. 6(a) and (b) shows that the images (with random noise) processed by Top-hat transform and then processed by threshold processing. Fig. 6(c) and (d) shows that the 3D images of Fig. 6(a) and (b), respectively. The images (with Gauss noise) processed by Top-hat transform and then processed by threshold processing are shown in Fig. 7(a) and (b). Fig. 7(c) and (d) shows that the 3D images of Fig. 7(a) and (b), respectively. shows the comparison of MSE value

for the image before and after Top-hat transform and thresholding processing.

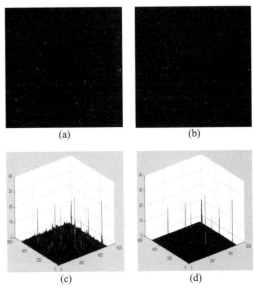

(a)　　　　　　(b)

(c)　　　　　　(d)

Figure 6. (a) The image (with random noise) processed after Top-hat transform; (b) the image (with random noise) processed after threshold processing; (c) the 3D image of (a); (d) the 3D image of (b).

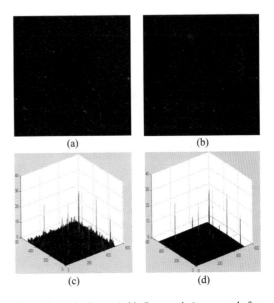

(a)　　　　　　(b)

(c)　　　　　　(d)

Figure 7. (a) the image (with Gauss noise) processed after Top-hat transform; (b) the image (with Gauss noise) processed after threshold processing; (c) the 3D image of (a); (d) the 3D image of (b).

Table 3. Comparison of MSE values.

The types of noise	shearlets denoising	After Top-hat transform	After threshold processing
random noise	2.654689	1.671234	1.063116
Gauss noise	134.056090	1.555886	1.010031

From the visual effect and the comparison of MSE values, Top-hat transform can complete the background noise suppression well. After threshold processing, the image is basically identical with the primitive simulate star image. The star crude extraction algorithm proposed in this paper is very effective for both the random noise and Gauss noise.

6 CONCLUSION

A star crude extraction algorithm based on shear wave transform and Top-hat transform is proposed in this paper. The algorithm firstly uses shear wave de-noising, and then Top-hat transform for the background noise suppression of star image, and adaptive threshold processing for removing the residual noise of star image are performed to obtain the star image after crude extraction. The experimental results show that the shear wave de-noising is obviously better than the wavelets de-noising, and the effect of background noise suppression based on Top-hat transform is also very effective. And the star image after adaptive threshold processing is consistent with the star image without noises.

ACKNOWLEDGMENT

The authors would like to thank for the developers of the shearlet toolbox and wavelet toolbox. This work is supported by Jilin Province Science and Technology Department (No. 201215107) and Jilin Province Education Department (No. 2013264).

REFERENCES

Wang Haiyong, Wu Wenqing, Xue Xiaofeng, and Zhao Yanwu, "Local block peak growth of star extraction," *Optics and Precision Engineering*. China, 2012, 20: 2507–2515.

Tian Jinwen, Ouyang Hua, Zheng Sheng, and Zhang Jun, " A method of extracting star," *Journal of Huazhong University of Science and Technology, China*, 2005, 33: 38–40.

Tian Yulong, Wang Guangjun, Fang Jiancheng, and Tian Jinwen, "A new method of star image high precision extraction based on local entropy," *Journal of Harbin Institute of Technology, China*, 37: 1068–1069.

Li Zhenzhen and Wei Honggang, "The algorithm of star extraction based on mathematical morphological in the bright-background," *Opto Electronic Engineering*. China, 2011, 38: 23–27.

K Guo and D Labate, "Optimally sparse multidimensional representation using shearlets," *SIAM J. Math. Anal. United States*, 2007, 39: 298–318.

K Guo, D Labate and W Q Lim, "Wavelets with composite dilations and their MRA properties," Appl. Comput. Harmon. Anal. United States, 2006, 20: 231–249.

K. Guo, D. Labate and W. Q. Lim, "Wavelets with composite dilations," *Electron. Res. Announc. Of AMS, United States*, 2004, 10: 78–87

Gong Junliang, He Xin, Wei Zhonghui and Wang Fangyu, "Local adaptive image denoising based on Bayesian estimation Shearlet domain," *Liquid Crystals and Displays, China*, 2013, 28:799–804.

Wang Lingling and Xin Yunhong, "A small IR target detection and tracking algorithm based on morphological and genetic-particle filter," *Acta Photonica Sinica, China*, 2013, 42:849–856.

Wang Hongtao, Luo Changzhou, Wang Yu, Wang Xiangzhou and Zhao Shufang, "Algorithm for star extraction based on self-adaptive background prediction," *Optical Technique, China*, 2009, 35:412–414.Yuan Yulei and Zheng Yong, "A large field of view star extraction threshold algorithm," *Marine Surveying and Mapping, China*, 2011, 31:41–43.

CARL C. L, "Accuracy performance of star tracker-atutorial," *IEEE. United States*, 2002, 38:587–597.

Wang Haiyong, Fei Zhenghong and Wang Xinlong, "Precise simulation of star spots and the centroid calculation based on Gaussian distribution," *Optics and Precision Engineering, China*, 2009, 17:1672–1677.

Zhang Hui, Zhong Jianyong, Yuan Jiahu and Liu Enhai, "The circuit noise effects on position accuracy of star sensor," *Optics and Precision Engineering. China*, 2006, 14: 1052–1056.Fan Xiaoyu and Jiang Hao, "The research of fast centroid locating for star image," *Aerospace Control, China*, 2011, 29:48–52.

Information Technology – Wan et al. (Eds)
© *2015 Taylor & Francis Group, London, ISBN 978-1-138-02785-5*

A video communication platform based on service-oriented architecture

Zhongbo Li, Yongqiang Xie, Weiguo Zhang & Jin Qi
Institute of the Chinese Electronic Equipment System Engineering Company, Beijing, China

ABSTRACT: Different video communication systems are difficult to communicate with each other due to diverse session control protocol, media processing mechanism and application logic. In this paper we propose a video communication platform based on service-oriented architecture. The design principle, architecture and implementation are introduced in detail. The architecture of the proposed communication system is flexible enough to support a diverse set of application and users in heterogeneous networks.

KEYWORDS: Service-oriented architecture; Video communication platform

1 INTRODUCTION

Video communication systems enable people in remote places to communicate and cooperate. And most video communication systems are not designed in the approach of an open system and cannot communicate with each other. In call control aspect, different networks have their own call control protocols, e.g., H.323 [1] or SIP [2]. In transmission aspect, different networks have different channel characteristics, e.g., different packet loss rate or delay jitter. Moreover, terminals of different networks have different processing capability. For example, the processing capability of PC in wired network is high, and that of mobile phone in wireless network is low. In the face of heterogeneous networks the current video communication system with monolithic design methodology is very difficult to meet users' needs. Therefore, it will bring substantial benefits to end users if we can build a video communication platform, which combines conference, surveillance, collaboration as well as other video applications into a single easy-to-use, integrated environment.

Service-oriented architecture (SOA) has been widely regarded as a promising method in large-scale heterogeneous communications networks [3,4]. SOA is defined as "an application architecture within which all functions are defined as independent services with well defined invokable interfaces which can be called in defined sequences to form business processes" [5]. From the standpoint of SOA, the entire communication system can be decomposed into many different services provided by one or more service providers.

Some work has been done to design the communication system based on SOA principle. Literature [6,7] presents a solution to interoperate SIP and H.323 systems. Literature [8] provides a general session management protocol to accommodate more protocols and applications with an easy to use web interface. Literature [9] proposes a quality-driven decision engine for real-time video transmissions. Literature [10-13] proposes a Global Multimedia Collaboration System (Global-MMCS) based on the XGSP web-services framework. In this paper, we propose a more general service-oriented video communication platform which can provide services in different aspects, e.g., session control, media processing and application logic.

The remainder of this paper is organized as follows. In Section 2, the proposed service-oriented video communication platform is described. The experimental results are presented in Section 3. Section 4 concludes the paper.

2 PROPOSED VIDEO COMMUNICATION PLATFORM BASED ON SOA

The proposed service-oriented video communication platform is described in this section.

2.1 *Platformc architecture*

The proposed video communication platform can provide different types of video communication services such as video conference, video surveillance, video collaboration, etc. And the service-oriented architecture makes it flexible enough to integrate various video services to satisfy the ever changing application.

The architecture of the video communication platform is shown in Fig. 1. There are six main components in this architecture: infrastructure service, service bus, platform service, application service, development and management.

Infrastructure service is a set of base service loosely coupled with application. It consists of data management service, resource management service, protocol adaptation service, device management service, etc. Data management service provides a unified interface for the platform to inquire and manage data. Resource management service is responsible for distributing resource to platform and application

service. Protocol adaptation service makes different kinds of terminals and devices compatible to the platform.

Service bus is the core of the whole platform. It contains the function of service discovery, service composition, service execution and broker network. It is a unified middleware for all communications between services, including synchronous communication, asynchronous communication, broadcast and multicast. Moreover, the service bus is low delay to meet the real-time requirements.

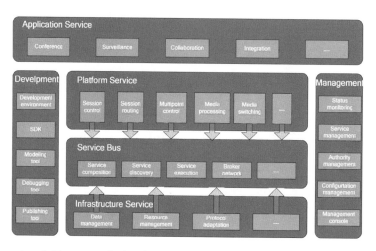

Figure 1.   Service-oriented video communication platform architecture.

Platform service is a kind of application related atomic service. It contains session control service, session routing service, multipoint control service, media processing service, and media switching service. All these services are reusable and can be composed to form a new application service.

Application service provides services for application directly. It is an encapsulation of application logic and has clear business value. It is often composed of different infrastructure services and platform services. Our platform provides video conference service, video surveillance service, video collaboration service and video integration service, etc.

Development provides multilevel developing tools for users and developers. It consists of development environment, SDK, application modeling tool, debugging tool and publishing tool. Management includes status monitoring, service management, authority management and configuration management, etc.

2.2   Service bus

service bus is responsible for all communication between service providers (e.g., infrastructure

service, platform service and application service) and service users. It contains the function of service discovery, service composition, service execution and broker network. We will describe them respectively as below.

3   SERVICE COMPOSITION

Service composition is concerned with designing a solution based on a set of existing services. Its role is to specify a list of services involved in a composition, and the way they interact and the composition topology. It is based on BPEL (Business Process Execution Language) engine to accomplish the service composition and orchestration in order to meet the ever changing application need.

3.1   Service discovery

Service discovery help service users find their required services. In our approach, a service user sends an inquiry message to the service provider. When service providers receive this message, they respond by sending a service description message, in which they include the current status of that service provider. The information provided in this service description message depends on the nature of the service being provided. But it must be helpful for the service user to decide from which service provider to ask for the service. The service user waits for a period of time for responses to arrive, and evaluates the received messages.

3.2 Service execution

When the service user selects the service provider on which it intends to run its service, it sends a request message to the service provider for the execution of the service. If the service provider can handle this request, it sends a success message. Otherwise, it sends a fail message. In case of failure, the service user either starts this process from the beginning or tries the second best option. A service can be terminated by the service user by sending a stop message.

3.3 Broker network

Broker network is responsible for routing event messages. The broker network is organized in a cluster-based hierarchy. Each broker keeps a broker network map of its own perspective to efficiently route the messages to their destinations. A message can be transported over HTTP while traversing a firewall but later TCP or UDP can be used to deliver it to its final destinations.

The broker network model is shown in figure 2. It is based on publish-subscribe messaging model. MessageDispatcher is responsible for distributing message based on the subscription condition. Two message queues can realize various messaging mechanism including synchronous communication, asynchronous communication, broadcast and multicast. MessageHandler supports local processing of specific message. MessageRelay receives the message subscriptions and relay them to corresponding services. MessagePostHook is used for delay diagnosis and status examination.

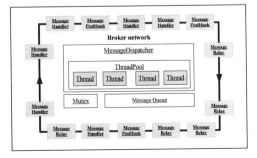

Figure 2. Broker network model.

3.4 Platorm service

Platform service is the base of application service. It contains session control service, session routing service, multipoint control service, media processing service, and media switching service, etc. Here we only introduce some important ones.

3.5 Session control service

Session control service provides the capability of creation and deletion of an audio session. Thus two endpoints can be connected/disconnected to each other or join/leave an audio conference. Two main session control services-- H.323 and SIP are provided in the proposed system.

H.323 is an audio/video conferencing recommendation from International Telecommunications Union (ITU) for packet based communication system. It defines a complete regulation for session control, e.g., H.225 and H.245. And these session control functions can be encapsulated as services for H.323 terminals.

SIP is a session management protocol from Internet Engineering Task Force (IETF). It is designed to discover other terminals and to negotiate the characterics of a real time session. And these session control functions can be encapsulated as services for SIP terminals.

3.6 Media processing service

Media processing service provides different capabilities of video and audio coding responding to different network bandwidth and client capabilities. Some clients have high network bandwidth and advanced processing and display capacity. They can receive process and display multiple concurrent audio and video streams. Therefore, they can receive all audio and video streams in a video conference. Some other clients have limited network bandwidth, processing and display capacity. Either they can not receive multiple audio and video streams or they can not process and display them. For different application services, different media processing services should be provided. The video conference service usually needs the high definition media processing service, and the video surveillance service usually needs the low definition media processing service. H.264, AVS and MPEG4 are the mainstream video coding services. G.711, G.729, AMR-WB are the commonly used audio coding services.

3.7 Application service

Application service is an encapsulation of application logic and often composed of

103

different infrastructure services and platform services. It includes video conference service, video surveillance service, video collaboration service, video integration service, etc.

Video conference service contains the application logic of video conference. It is coarse-grained and realized by the composition of fine-grained service below. Based on the video conference service, H.323 terminals, SIP endpoints and other video conferencing terminals can take part in a same meeting. Chairman mode and lecture mode are supported in our platform.

Video surveillance service contains the application logic of video surveillance. It mappings the surveillance logic to the corresponding session control service, media processing service, media switching service, etc. Different from the video conference service, it adopts a one-way transmission way.

Video collaboration service contains the application logic of video collaboration. Text, picture and map can be shared among multi persons. And the interoperability is supported by using the protocol adaptation service.

Video integration service is a recombination of video conference service, video surveillance service and video collaboration service. For example, the user can access a surveillance video when he is in a video conference, and he is able to label on map with other people at the same time.

4 EXPERIMENTAL RESULTS

We have developed a prototype video communication platform based on the service-oriented architecture. The prototype hardware contains an ATX mainboard, Intel C602J chip set, Intel Xeon E5-2600 CUP of 2.1GHz, 8GB memory and 600G hard disk. It supports H.323, SIP and other specific protocols. The video coding format includes H.264, AVS and MPEG4. The video resolution is up to 1080p. The video service template is more than 20. It can support 1000 concurrent users. The service bus switching time is smaller than 400ms.

Fig. 3 shows a single video conference service which mainly consists of session control service, multipoint control service and media switching service. Fig. 4 shows the integration of video conference and surveillance service by adding a video surveillance service. Fig. 5 shows the integration of video conference, surveillance and video on demand service. Fig. 6 shows the integration of all the video communication service. Fig. 7 shows the video service composition and changing with application. All these figures indicate that the service-oriented architecture makes the video communication platform so flexible that it can adapt to the ever changing application.

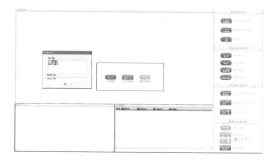

Figure 3. A single video conference service.

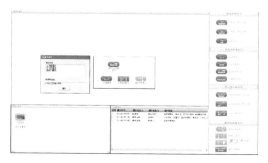

Figure 4. The integration of video conference and surveillance service.

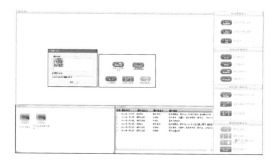

Figure 5. The integration of video conference, surveillance and video on demand service.

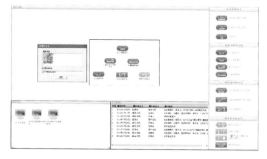

Figure 6. The integration of all the video communication service.

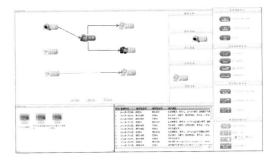

Figure 7. The video service composition and changing with application.

5 CONCLUSION

In this paper we propose a service-oriented video communication platform where different types of services are jointly considered. In this framework, all the components of video communication system are regarded as service entities. And they can be coupled together using service bus. Experimental results show that the proposed service-oriented communication system can support various video applications, e.g., video conference, video surveillance, video collaboration and video integration. And it can provide a much better end-user experience than existing approaches. In the next step, we will try to add more video applications, and extend the capacity of the platform to 5000 users.

REFERENCES

[1] Recommendation H.323, Packet-base multimedia communications systems, ITU, September, 1999.

[2] J.Rosenberg, H.Schulzrinne, et al. "SIP: Session Initiation Protocol", RFC 3261. Internet Engineering Task Force, http://www.ietf.org/rfc/rfc3261.txt.

[3] W. Wu et al., "Service Oriented Architecture for VoIP Conferencing," Special Issue on Voice over IP — Theory and Practice, *Int'l. J. Commun. Sys.*, 2006, 19(4): 445–461.

[4] I. Brunkhorst, S. Tonnies, and W. Balke, "Multimedia Content Provisioning using Service Oriented Architectures," *Proc. IEEE ICWS,* Beijing, China, October, 2008.

[5] K. Channabasavaiah, K. Holley, and E. Tuggle, "Migrating to a Service-Oriented Architecture," IBM Developer-Works, 16 December. 2003.

[6] K. Singh and H. Schulzrinne. Interworking between SIP/SDP and H.323. In *Proceedings of the 1st IPTelephony Workshop (IPtel 2000)*, Berlin, Germany, April, 2000.

[7] Jiann-Min Ho, Jia-Cheng Hu, Peter Steenkiste, A conference gateway supporting interoperability between SIP and H.323, *Proceedings of the ninth ACM international conference on Multimedia,* 2001, Ottawa, Canada.

[8] W. Wu, G. C. Fox, H. Bulut, A. Uyar, H. Altay, "Design and Implementation of A Collaboration Web-services system", *Journal of Neural, Parallel & Scientific Computations*, 2004, 12.

[9] Dalei Wu, Song Ci, Haiyan Luo, Haohong Wang, Katsaggelos, A, "A quality-driven decision engine for live video transmission under service-oriented architecture". *IEEE Wireless Communications*, 2009 (4): 48–54.

[10] Geoffrey C. Fox and Shrideep Pallickara. "The Narada Event Brokering System: Overview and Extensions", proceedings of the 2002 International Conference on Parallel and Distributed Processing Techniques and Applications (PDPTA'02), 2002.

[11] Geoffrey Fox, Ozgur Balsoy, Shrideep Pallickara, Ahmet Uyar, Dennis Gannon, and Aleksander Slominski, "Community Grids" *Invited talk at The 2002 International Conference on Computational Science,* April 21–24, 2002 Amsterdam, The Netherlands.

[12] Fox,G.C., Bulut, H., Kim, K., et al.. Collaborative Web Services and Peer-to-Peer Grids, *Keynote speech 2003 Collaborative Technologies Symposium (CTS'03)*, 2003.

[13] Hasan Bulut, Geoffrey Fox, Shrideep Pallickara,Ahmet Uyar and Wenjun Wu, Integration of NaradaBrokering and Audio/Video Conferencing as a Web Service, *IASTED International Conference on Communications, Internet, and Information Technology*, November 2002, US Virgin Islands.

Information Technology – Wan et al. (Eds)
© 2015 Taylor & Francis Group, London, ISBN 978-1-138-02785-5

Acute toxicity of zinc oxide nanoparticles and bulk ZnCl$_2$ to rats

Huijuan Kuang, Lin Yang, Wanyi Zhang & Hengyi Xu
State Key Laboratory of Food Science and Technology, Nanchang University, Nanchang, China

ABSTRACT: The present study was aimed to determine the significance of dose by comparing acute toxicological potential of ZnO nanoparticles (NPs) (30 nm) with its bulk materials ZnCl$_2$. Sprague Dawley rats, 8 to 9 weeks old, were administered with 20 mg/kg body weight (bw) of ZnO nanoparticles (30 nm) with its bulk materials ZnCl$_2$ suspended in distilled water once via tail vein injection. The liver, lung, and spleen could be possible target organs for accumulation of NPs and their bulk counterparts. Results showed that the nanoparticles-form material was more toxic than the bulk materials, which may due to the properties of nanoparticles.

1 INTRODUCTION

Nanoparticles are defined as small objects that have at least one dimension in the range of 1-100 nm. Compared to particles in micro scale, Nanoparticles (NPs) due to their small size and large-specific surface area, show different degrees of biological effect (Esmaeillou 2013). NPs offer a great possibility for biomedical application, not only to deliver pharmaceutics, but also to be used as novel diagnostic and therapeutic approaches (Caruther 2007). The small sizes NPs imply that they could get close to a biological target of interest. Furthermore, metallic NPs can be made to resonantly respond a time-varying magnetic field, with advantageous results relater the transfer of energy to the particles (Pissuwan 2006, Kogan 2007). This leads to its use as a hyperthermic agent, thereby delivering toxic amounts of thermal energy to targeted bodies such as tumors (El-Sayed 2006, Kogan 2006, Zharov 2006). In additional, NPs was also used in the food industry, paints, electronics, sports, environmental cleanup, cosmetics, sunscreens, etc (Asharani 2008). At the same time, the novel and useful properties possessed by these engineered nanomaterials can lead to unpredictable outcomes in terms of their interactions with biological systems. Therefore, it is necessary to understand and assess the potential toxicity of NPs to avoid their adverse effects on human health.

Among the varieties of NPs being used today, zinc oxide nanoparticles (ZnO-NPs) are one of the most widely used in consumer products. They are extensively used in cosmetics and sunscreens because of their efficient UV absorption properties without scattering the visible light. This makes them transparent and more aesthetically acceptable as compared to their bulk counterpart (Schilling 2010). ZnO-NPs are being used in the food industry as additives and in packaging due to their antimicrobial properties (Gerloff 2009, Jin 2009). They are also being explored for their potential use as fungicides in agriculture (He 2011) and as anticancer drugs and imaging in biomedical applications (John 2010).

Along with extensive application of ZnO-NPs in the industrial field, it is conceivable that the human body may be intentionally or unintentionally exposed to nanoparticles via several possible routes, including oral ingestion, inhalation, intravenous injection, and dermal penetration (Baek 2012). Some studies evaluated the acute oral toxicity of ZnO-NPs at a high dose range (1–5 g/kg body weight) and found a decrease in damage to the liver, spleen and pancreas with an increase in nanoparticle dose. The spleen showed enlargement of the splenic corpuscle while an infiltration of inflammatory cells was observed in the pancreas interstice (Wang 2008). However, to date, almost all in vivo studies of the toxicity of nanoparticles have been focused on acute toxicity or repeated-dose toxicity evaluation via oral ingestion or inhalation. A few studies have investigated the acute toxicity and distribution in the body of ZnO-NPs via intravenous injection.

Therefore, the present study was undertaken to investigate the acute toxicity of ZnO-NPs including distribution in the body via intravenous injection, and also to determine if the toxicity is solely due to dissolution of metal oxide NPs by comparing to their reference toxicants: soluble metal ions and bulk metal oxides of non-nanoparticulate form. For this we exposed mice to ZnO-NPs and ZnCl$_2$ via tail-vein injections and the concentration of zinc in different tissues was investigated. The effects of zinc on serum biochemical parameters were also examined. Moreover, the histopathological changes had also been examined.

2 MATERIALS AND METHODS

2.1 Nanoparticles and suspension preparation

ZnO-NPs (30 nm) at 99.983% purity were purchased from Xiya shijiCo. Ltd, (Chendu, China), $ZnCl_2$ at 99.983% purity were purchased from Sinopharm Chemical Reagent Co. Ltd, (Shanghai, China). Suspensions of ZnO-NPs and bulk $ZnCl_2$ were prepared with single-distilled water.

2.2 Animals

Twelve adult male KunMing (450-500 g each) were purchased from the experimental animal center of Nanchang University, China, and raised in an animal facility under 25°C with a 12 h light/dark cycle and supplement with food and water ad libitum. All procedures involving animals were approved by the Animal Care Review Committee (approval number 0064257), Nanchang University, Jiangxi, China and care for institutional animal care committee guidelines. Studies have investigated the acute toxicity and distribution in the body of ZnO-NPs via intravenous injection.

2.3 Dosing and sample collection

After 7 days acclimation, 9 mice were divided into three groups with three mice in each group. Mice were ethically treated by all mean during the procedure. ZnO-NPs were diluted using ultrapure water resulting in the same mass concentration of 8 mg/mL. Two groups of male rats received a single dose of 20 mg/kg of ZnO nanoparticles by tail-vein injections, and one additional group of three rats received an equivalent volume of physiological saline as controls for all the experiments one day after administration, the mice were sacrificed after being anesthetized for appraising the biochemical factors.

Blood samples are collected from mice, including three treated groups and one control group for 1 days after injection with various. Blood samples were collected via the ocular vein. The serum was obtained by centrifugation of the whole blood at 3000 rpm for 15 min, and analyzed for ALB (albumin), alkaline phosphatase (ALP), creatinine (CRE), alanine aminotransferase (ALT), aspartate aminotransferase (AST), blood urea nitrogen (BUN) , total bilirubin (TBIL), uric acid (UA), and ratio of albumin to globulin (A/G). These biochemical parameters were determined by an automated biochemical analyzer. The blood serum was examined at the First Affiliated Hospital of Nanchang University.

For the tissue distribution study, tissue samples of the brain, heart, kidney, liver, lung, spleen, intestine and tests were collected.

2.4 Sample preparation for AAS analysis

The total zinc content was determined using atomic absorption spectroscopy (Thermo Scientific iCE 3500, AAS). A small fraction of each sample (0.2-0.5 g) were tailored and dissolved in digest solution (HNO_3 : $HClO_4$=5:1) with the volume of 12 mL and were heated to 230°C. The temperature could be increased to 280°C when the reaction reached to stability. Then the samples were diluted with ultrapure water to 25 mL after reaction stopping, and subsequently to determine the iron concentrations using AAS.

2.5 Histopathology

Heart, liver, spleen, kidneys, lung, intestine, tests and brain were collected from all animals and preserved in 10% neutral buffered formalin, These samples were dehydrated and embedded in paraffin wax. Subsequently, 5-μm -thick tissue sections were cut. These sections were transferred onto clean slides, dried overnight and stored at room temperature until use. The mounted tissue sections were stained with hematoxylin and eosin for histologic examination.

2.6 Statistical analysis

Data were analyzed using the statistical package SPSS (Ver. 16.0, SPSS Inc., Chicago, IL, USA). Data for biochemical tests are expressed as the means ± SE from three independent experiments. Differences were considered statistically significant when $p < 0.05$.

3 RESULT

3.1 Serum biochemical parameters

Table 1 shows the changes in biochemical parameters in the serum of mice induced by ZnO-NP or $ZnCl_2$. There was a significant increase ($p<0.01$) in ALT and AST for rats treated with ZnO-NP. A significant decrease ($p<0.01$) in UA was found in the rats treated with ZnO-NP. No significant changes in the serum biochemical parameters were noted except for a significant increase ($p<0.05$) in ALT, ALB, and GLB. There were no significant differences in the other serum biochemical parameters between the groups.

Table 1. Serum biochemical values for rats (Mean ± SD).

paramrter	control	ZnO-NP	$ZnCl_2$
ALT	34.5±9.19	108±1.41**	26±2.82*
AST	152±16.97	329.5±2.12**	174.5±3.53
TBIL	0.75±0.35	0.95±0.07	0.6±0.28
ALB	26.95±0.92	25±0.14	29.25±1.06*
GLB	48.15±3.88	46.45±0.63	53.8±3.11*
A/G	0.56±0.02	0.53±0.01	0.78±0.32
CRE	27.3±9.33	28.5±0.70	33.2±2.55
BUN	6.15±0.50	5.55±0.07	8.55±3.74
UA	378.5±34.64	247.5±27.57**	398±35.35

**Significant difference vs. control, $p < 0.01$,
*Significant difference vs. control, $p < 0.05$

3.2 Znic content

The tissue distribution of ZnO nanoparticles was determined in the heart, brain, kidney, liver, lung, spleen, and testis by measuring the total zinc level with AAS (Figure 1). When the concentrations of total zinc levels were compared between the experimental groups and the control, no statistically significant increases were observed in the kidney, heart, and testis. However, there was a significant increase ($p<0.05$) in heart for rats in the experimental groups. A significant increase ($p<0.01$) in liver was also found in the experimental groups. The concentrations of total zinc levels in the spleen and lung were highest for rats treated $ZnCl_2$.

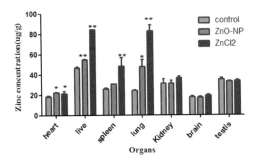

Figure 1. Tissue distribution of Zn in rats after single-dose tail-vein injections of 20 mg/kg 30 nm ZnO-NP and ZnCl2 in male rats, respectively. **Significant difference vs. control, $p < 0.01$, *Significant difference vs. control, $p < 0.05$.

3.3 Histopathologic examination

To investigate the toxicity, Histopathological examination of organs was also conducted to determine the potential tissue damage, lesions, or inflammation from toxic exposure.

There were dramatic morphological alternations to the hepatic lobules in livers from rats treated with 30 nm ZnO-NP, as indicated by disordered hepatic cords and reduced central veins (Fig.2), compared to those from the vehicle control rats. No noticeable alternation was detected in the livers from rats treated with $ZnCl_2$.

Rats exposed to ZnO-NP or $ZnCl_2$ showed no tissue damage in any of the sections obtained from the spleen, kidney, and testis. (Fig. 2).

4 DISCUSSION

The current study investigated the toxicity response of nano-size ZnO in Sprague Dawley rats in comparison with their bulk particles.

After intravenous injection, a significant increase in AST and ALT levels was in the experimental groups, especially the rats treated with ZnO-NP, which suggested that the damage to the liver in the ZnO-NP group when compared to control. AST and

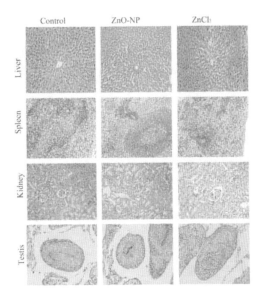

Figure 2. Histopathological findings in organs. The original magnification was 200×.

ALT are the two liver enzymes whose serum level gets elevated during necrosis, degeneration, hepatitis and inflammatory condition (Pasupuleti 2012). The incidence of histopathology lesions of liver correlated with the elevated liver enzyme levels. Although the levels of AST and ALT were increased, no lesions were observed in the liver from the group of $ZnCl_2$. In contradictory to our findings, previous studies indicated the toxicity of ZnO-NP and bulk ZnO were identical (Hussain 2005, Wang 2009, Zhu 2008, Zhu 2009), the toxicity caused by ZnO-NP and bulk ZnO was the dissolved Zn^{2+} ions. Generally, ZnO nanoparticles are classified as poorly water soluble, $ZnCl_2$ was easily released Zn^{2+} into a culture medium, the concentration of Zn^{2+} in ZnO-NP suspensions in this study would be limited, therefore, Zn^{2+} could not have been the main mechanism of the lethal toxicity of ZnO-NPs and bulk $ZnCl_2$ suspensions observed here. The tissue damage may cause by nanoparticles.

In order to use ZnO-NP in more fields, it is essential to characterize the bioaccumulation and toxicity associated. In this study, we evaluated the bioaccumulation and acute toxicity of ZnO-NP after intravenous injection. After injection, the zinc concentration in biological samples was determined by AAS. In all organs studied, the liver was the target organ and the tissue distribution of zinc. Although the level of zinc in the lung was higher in rats treated with $ZnCl_2$, histopathologic examination of lung tissue did not show any treatment-related effect. It indicated that the tissue damage may not originate from the Zn^{2+} ions.

Particularly interesting is the case of brain and testis: considering the relatively constant level of zinc in tissues after injection, which indicated that zinc is difficult across biological barriers, such as the blood-brain barrier or blood-testis barrier.

The tissue distribution between nanoparticles and bulk material were similar. Although the concentration of zinc in organs of rats treated with $ZnCl_2$ was higher than those treated with ZnO-NP, the histopathology evaluation of liver revealed damaged in rats treated with ZnO-NP, the damage may be due to the generation of nanoparticle. It is well known that most of the novel properties of nanomaterials relate to their size. Several studies have reported that nanosized particles always showed more serious toxicity than bulks (Kasemets 2009) and suggested that size was one of the key factors influencing the toxic effect of NPs (Chang, 2012).

5 CONCLUSION

Toxicity and behavior of ZnO-NP and bulk materials $ZnCl_2$ to rats in an aqueous exposure medium via intravenous injection after single-dose administration was characterized in the present study. Results indicated that metal oxide (ZnO) NPs are toxic to rats, and the toxicity effect might possibly come from the form of nanoparticles.

ACKNOWLEDGMENT

This work was supported by Training Plan for the Young Scientist (Jinggang Star) of Jiangxi Province (20142BCB23004).

REFERENCES

Asharani, P.V., Y. Wu, Z. Gong, & S. Valiyaveettil (2008). Toxicity of silver nanoparticles in zebrafish models. *Nanotechnology. 19*, 5102–5109.

Baek, M., H. Chung, J. Yu, J. Lee, T. Kim, J. Oh, W. Lee, S. Paek, J. Lee, J. Jeong, J. Choy, & S. Choi (2012). Pharmacokinetics, tissue distribution, and excretion of zinc oxide nanoparticles. *Int J Nanomed. 7*, 3081–3097.

Caruther, S.D., S. Wickline, & G. Lanza (2007). Nanotechnological applications in medicine. *Curr Opin Biotech. 18*, 26–30.

Chang, Y., M. Zhang, L. Xia, J. Zhang, & G. Xing (2012). The Toxic Effects and Mechanisms of CuO and ZnO Nanoparticles. *Mater. 5*, 2850–2871.

El-Sayed, I. H., X. Huang, & M. El-Sayed (2006). Selective laser photo-thermal therapy of epithelial carcinoma using anti-EGFR antibody conjugated gold nanoparticles. *Cancer Lett. 239*, 129–135.

Esmaeillou, M., M. Moharamnejad, R. Hsankhani, A. Tehrani, & H. Maddi (2013). Toxicity of ZnO nanoparticles in healthy adult mice. *Environt Toxiciol Phar. 35*, 67–71.

Gerloff, K., C. Albrecht, A. Boots, I. Förster, & R. Schins (2009).Cytotoxicity and oxidative DNA damage by nanoparticles in human intestinal Caco-2 cells. *Nanotoxicology. 3*, 355–364.

He, L., Y. Liu, A. Mustapha, & M. Lin (2011). Antifungal activity of zinc oxide nanoparticles against Botrytis cinereaand Penicillium expansum. *Microbiol Res. 166*, 207–215.

Hussain, S., K. Hess, J. Gearhart, K. Geiss, J. Schlager (2005). In vitro toxicity of nanoparticles in BRL 3A rat liver cells. *Toxicol in vitro. 19*, 975–983.

Jin, T., D. Sun, J. Su, H. Zhang, & H. Sue (2009). Antimicrobial efficacy of zinc oxide quantum dots against Listeria monocytogenes, Salmonella enteritidis, and Escherichia coliO157:H. *J Food Sci. 74*, 46–52.

John, S., S. Marpu, J. Li, M. Omary, Z. Hu, & Y. Fujita, A. Neogi (2010). Hybrid zinc oxide nanoparticles for biophotonics. *J Nanosci Nanotechnol. 10*, 1707–1712.

Kasemets, K., A. Ivask, H. Dubourguier, & A.Kahru (2009). Toxicity of nanoparticles of ZnO, CuO and TiO2 to yeast Saccharomyces cerevisiae," *Toxicol in Vitro. 23*, 1116–1122.

Kogan, M.J., N. Bastus, R. Amigo., D. Grillo-Bosch, E. Araya, A. Turiel, A. Labarta, E. Giralt, & V. Puntes (2006). Nanoparticle-mediated local and remote manipulation of protein aggregation. *Nano Lett, 6*, 110–115.

Kogan, M.J., I. Olmedo, L. Hosta, A. Guerrero, L. Cruz., & F. Albericio (2007). Peptides and metallic nanoparticles for biomedical applications. *Nanomedicine. 2*, 287–306.

Pasupuleti, S., S. Alapati, S. Ganapathy, G. Anumolu, N. Pully, & B. Prakhya (2012). Toxicity of zinc oxide nanoparticles through oral route. *Toxicol Ind Health. 28*, 675–686.

Pissuwan, D., S. Valenzuela., & M. Cortie (2006). Therapeutic possibilities of plasmonically heated gold nanoparticles. *Trends Biotechnol. 24*, 62–67.

Schilling, K., B. Bradford, D. Castelli, E. Dufour, J. Nash, W. Pape, S. Schulte, I. Tooley, J. Bosch, & F. Schellauf (2010). Human safety review of nano titanium dioxide and zinc oxide. *Photochem Photobiol Sci. 9*, 495–509.

Wang, B., W. Feng, M. Wang, T. Wang, Y. Gu, M. Zhu, H. Ouyang, J. Shi, F. Zhang, Y. Zhao, Z. Chai, H. Wang, & J. Wang (2008). Acute toxicological impact of nano- and submicro-scaled zinc oxide powder on healthy adult mice. *J Nanopart Res. 10*, 263–276.

Wang, H., R. Wick, & B. Xing (2009). Toxicity of nanoparticulate and bulk ZnO, Al_2O_3 and TiO_2 to the nematode Caenorhabditis elegans. *Environ Pollut. 157*, 1171–1177.

Zharov, V.P., K. Mercer, E. Galitovskaya, & M. Smeltzer (2006). Photothermal nanotherapeutics and nanodiagnostics for selective killing of bacteria targeted with gold nanoparticles. *Biophys J. 90*, 619–627.

Zhu, X., J. Wang, X. Zhang, Y. Chang, & Y. Chen (2009). The impact of ZnO nanoparticle aggregates on the embryonic development of zebrafish (*Danio rerio*). *Nanotechnology. 20*, 5103–5111.

Zhu, X. S., L. Zhu, Z. Duan, R. Qi, Y. Li, & Y. Lang (2008). Comparative toxicity of several metal oxide nanoparticle aqueous suspensions to Zebrafish (Danio rerio) early developmental stage. *J Environ Sci Heal A. 43*, 278–284.

Information Technology – Wan et al. (Eds)
© 2015 Taylor & Francis Group, London, ISBN 978-1-138-02785-5

Adaptive radar STC based on RBF neural network prediction sea clutter

Yunfeng Liu, Jidong Suo & Xiaoming Liu
College of Information Science and Technology, Dalian Maritime University, Dalian, China

ABSTRACT: Sea clutter is the main interference for marine radar. Sea clutter waveforms are generally quite complicated. Radar receiver will overload for close signal caused by sea clutter. By analyzing the correlation of sea clutter, we change traditional STC method and make most range of A/D show fast changing component of echo for our interest. We proposed an adaptive STC based on RBF neural network prediction sea clutter. It increases the dynamic range of radar receiver and prevent overload in close range.

1 INTRODUCTION

Sensitivity time control (STC) is an important unit of radar receiver [1]. Its performance has a direct impact on the performance of receiver system. Typically, marine radar uses STC control the echo intensities whose disadvantage is that it cannot adapt to the clutter environment [2]. There is a better result only to the clutter with certain regularity using traditional STC.

Sea clutter refers to the backscattered returns from a patch of the sea surface illuminated by a transmitted radar pulse [1]. Since the complicated sea clutter signals depend on the complex wave motions on the sea surface, it is reasonable to study sea clutter from nonlinear dynamics, point of view, instead of simply based on random processes [3]. The neural network, with strong nonlinear fitting capabilities, is an analysis method which has strong robustness to parameters variation. It can get accurate output by training samples without building precise model. Moreover, it is suitable for establishing the model which the parameters present nonlinear changes [4-8].

Sea clutter is the main interference for marine radar. The intensity of sea clutter changes with radar parameters, sea state and the like. Typically, sea clutter is described as a combination of two components, slow changing component and fast changing component, in the statistical sense. Traditional STC does not distinguish slow and fast changing component. It generates attenuation control signal based on sea surface reflection intensity with distance changing. However, we are more concerned with fast changing of sea clutter. Therefore, our aim is to find out slow changing component of sea clutter and control the gain of radar receiver to amplify fast changing component of sea clutter by limited A/D resources. The

paper proposed adaptive radar STC based on RBF neural network predict sea clutter to prevent signal distortion and make radar receiver work well.

The remainder of this paper is organized as follows. In Section 2, we present traditional STC. Section 3 summarizes sea clutter correlation and Section 4 describes RBF neural network. The effectiveness of the proposed method is discussed in Section 5 using measured data from IPIX. Finally, we give the conclusion.

2 TRADITIONAL STC

STC technology is often used to control intensity of short range target. Due to the high amplitude of sea clutter, it could easily cause false alarm. So the primary role of STC is to detect close range target that obscured by sea clutter echoes. Based on echo intensity, you can clearly distinguish between different target echo signals, especially the short-term objective [9]. According to the radar equation [1], radar echo intensity is inversely proportional to the fourth power of range. STC is a time-varying attenuator added to the receiver that reduces the effect of distance on echo power [2]. STC's basic principle is after transmitting a pulse signal by transmitter, the receiver outputs a control voltage that matched echo power variation. The gain of receiver accords to this pattern changes. Echo power variation is shown in Fig. 1. As can be seen from Fig. 1, STC makes echoes (include sea clutter and target) attenuation by the same proportion. That makes amplitude attenuation of little target larger. It is of no use for detection of small target in close range. Fig. 2 shows a traditional STC diagram of radar receiver.

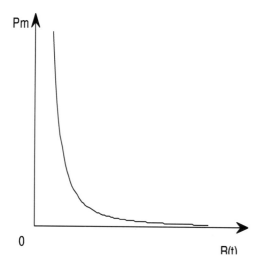

Figure 1. Echo power variation.

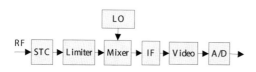

Figure 2. Traditional STC diagram.

3 TEMPORAL AND SPATIAL CORRELATION OF SEA CLUTTER

Sea clutter characteristic analysis and modeling are the basic premise for target detection and track in sea clutter. Its research aims to fully grasp the characteristics information of sea clutter, providing necessary prior information for the algorithm of target detection. Correlation of sea clutter is divided into temporal and spatial correlation.

Spatial correlation includes radial spatial correlation and azimuth spatial correlation. Usually clutter models used to describe the magnitude of sea clutter distribution have Rayleigh distribution, Log-normal distribution, Weibull distribution and K-distribution. Spatial correlation of sea clutter amplitude is related to sea state. In the literature [10], Watts thinks the correlation length of sea clutter depends on the correlation length of the sea surface, and gives an empirical formula between correlation lengths and wind speed, acceleration and other parameters. In addition, spatial correlation of sea clutter is also affected by the radar range resolution, the sea conditions, the polarization, and the impact of grazing angle. In literature [3], Guanjian gives conclusion of sea clutter amplitude correlation: sea clutter includes anti-continuing long-range spatial correlation, and its anti-continuing degree is related to sea conditions.

Recently, scholars studied temporal correlation extensively, and gave mathematical model of power spectrum [11–13]. Temporal correlation usually means correlation of multiple echo continuous exposure to the sea on the same radar cell. Compound k-distributed model is closer to the actual situation. In compound K-distributed model, clutter is described as a product of two factors, one part is basic amplitude modulation components (slow changing component), with a long temporal correlation, following Gamma distribution. The second part is speckle (fast changing component), it was formed by stacking a large scattering of reflection, in line with Rayleigh distribution.

4 RBF NEURAL NETWORK

RBF neural network, namely, is radical basis function neural network. It is a kind of three layer forward network consisting of input layer, hidden layer, and output layer. As shown in Fig. 3. RBF neural network has many applications in function approximation, time series analysis, nonlinear control, pattern recognition, image processing, and so on. RBF Neural network algorithm can acquire RBF neural network by training based on the training data. Corresponding calculation can be quickly completed by trained RBF neural network.

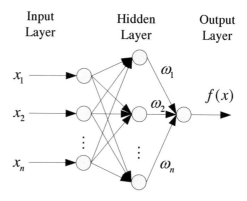

Figure 3. RBF network diagram.

5 EXPERIMENTAL RESULTS AND DISCUSSION

5.1 Real sea clutter data and prediction

The database used for the present study was collected by using instrument quality radar called the IPIX radar. The sampling rate is 1000 Hz. Because of sea clutter's non-stationarity [14] and RBF neural network specializes in time series analysis, we use RBF neural network predict sea clutter. Prediction values using RBFNN for IPIX 54# data, vv polarization are shown in Fig. 4. As can be seen from Fig. 4, we can predict sea clutter for its correlation. For this reason, we get prediction error by prediction sea clutter based on RBFNN. As shown in Fig. 5.

Section 3, we can get prediction of sea clutter for spatial and temporal correlation of sea clutter. Next, we adjust STC using prediction result and reduce the impact of slow changing component of sea clutter. We make most dynamic range of receiver for the fast changing component we interested in. An improved STC diagram is shown in Fig. 6.

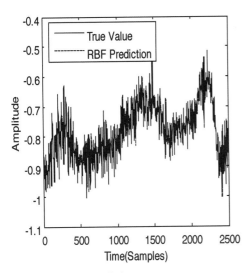

Figure 4.   RBFNN prediction.

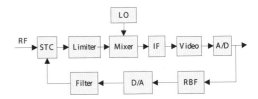

Figure 6.   Improved STC diagram.

6   CONCLUSIONS

Due to the correlation of sea clutter, it provides possibility to prediction of sea clutter. We are interested in fast changing component of sea clutter through predicting sea clutter using RBF neural networks. Traditional STC does not distinguish fast and slow parts of sea clutter. The defect is likely to cause overload and saturation in close range. Dynamic range of receiver is used to amplify fast changing component which minus slow changing of sea clutter by prediction sea clutter. It proposes a method to prevent overload and saturation. We can optimize RBF algorithm by K-means cluster, GA and PSO algorithm and further reduce prediction error. However, whether error decrease is beneficial to target detection is the future research topic.

ACKNOWLEDGMENT

This work is supported by the National High Technology Research and Development Program of China (863 Program, No. 2012BAH36B02).

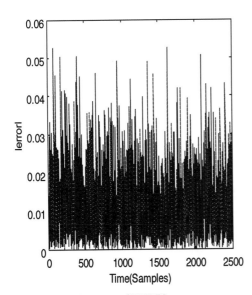

Figure 5.   Absolute error of RBFNN.

5.2   Discussion

From Section 2, shortcoming of traditional STC is making whole attenuate by same proportion. From

REFERENCES

[1] Skolnik, M.I. 1990. Radar Handbook. New York: McGraw-Hill.
[2] Foreman, T. L. A model to quantify the effects of sensitivity time control on radar-to-radar interference. IEEE Trans.on Electromagnetic Compatibility, 1995, 37(2): 299–301.
[3] Guan J. Spatial correlation property with measured sea clutter data, Chinese Journal of Radio Science, 2012, 27(5): 943–953.
[4] Xu H.B., Li C.J., et al. Application of RBF neural network in the prediction of yeast, Manufacturing Automation, 2002, 24(6): 44–47.

[5] Yao Y.H., Niu Y.Y. CPI prediction based on RBF neural network, Computer Applications and Software, 2010, 27(10): 92–94.

[6] Simon H. Neural Networks, Beijing: Machinery Industry Press, 2004.

[7] Oudjana, S.H., Hellal, A., Mahamed, I.H. Short term photovoltaic power generation forecasting using neural network. 11th International Conf. on Environment and Electrical Engineering, 2012, pp. 706–711.

[8] Wu K.H. & Yuan Y., Cheng B.H. Research of wind power prediction model based on RBF neural network. 5th International Conference on Computational and Information Sciences, 2013, pp. 237–240.

[9] Meena, D. & Prakasam, L. FPGA based real time solution for sensitivity time control, , 2008, pp. 244–248.

[10] Watts, S. 1996. Cell-averaging CFAR gain in spatially correlated K-distributed clutter. IEEE Proc. Radar Sonar Navig., 1996, 143(5): 321–327.

[11] LEE, P. BARTER J.D. & LAKE, B.M. Lineshape analysis of breaking wave doppler spectra. IEE Proc. Radar Sonar Navig., 1998, 145(2): 135–139.

[12] WALKER, D. Doppler modeling of radar sea clutter, IEE Proc Radar Sonar Navig., 2001, 148(2): 73–80.

[13] LAMONT, S.T. Azimuth dependence of doppler spectra of sea clutter at low grazing angle. IEE Proc Radar Sonar Navig. 2008, 2(2): 97–103.

[14] Ward, K.D., Barker, C.J. & Watts, S. 1990. Maritime surveillance radar part 1: radar scattering from the ocean surface. IEE Proceedings, 1990, 137(2): 51–62.

Information Technology – Wan et al. (Eds)
© 2015 Taylor & Francis Group, London, ISBN 978-1-138-02785-5

Adaptive truncation coding for computed tomography images

Zhenzhen Zhai & Huiqin Jiang
College of Information Engineering, Zhengzhou University, Henan Province, China

Liping Lu
Information Engineering Department, Henan College of Finance and Taxation, Henan Province, China

Yumin Liu
Engineering Technology Research Center for Medical Information, Zhengzhou University, Henan Province, China

ABSTRACT: In order to improve the storage and transmission efficiency of CT images, this paper proposes a new adaptive truncation coding compression algorithm. First, the CT images are de-noised in wavelet domain. Second, the processed wavelet coefficients are block coded by truncating adaptively the bit stream, where the channel numbers are decided according to the relative importance of wavelet sub-bands. Finally, the reconstructed image is obtained by decoding and inverse wavelet transform. The experimental results show that the proposed algorithm can not only reduce the computational complexity, but can also improve the quality of the reconstructed image. The peak signal to noise ratio (PSNR) and the mean structural similarity (MSSIM) increase by 17% and 2%, respectively compared to the EBCOT algorithm with the same compression ratio, and the execute time reduces by 30%.

1 INTRODUCTION

With the application of multi-row spiral CT and the medical picture achieving and communication system (PACS) based on JPEG2000 are widely used, the compression for CT images has a considerable significance (Ge Wang & Hengyong Yu, 2008).

Medical image compression applied to PACS system has become a hotspot. Classical image compression algorithms in current PACS system include the tree structure-based, the block structure-based method (Rabbani et al., 2002), (Zandi, A et al., 1995), (Ting Zhang & Ji'an Yang, 2006), (Lei Xu, 2009) and etc. The tree structure-based method such as embedded zero tree (EZW) and the set partitioning in hierarchical trees (SPIHT) are used widely. The EZW uses the tree structure to scan the wavelet coefficients, which leads to low compression efficiency because of incomplete utilization of the wavelet transform characteristics. To deal with the scheduling of wavelet coefficients, A. Said and W.A. Pearlman proposed the SPIHT algorithm by adopting spatial orientation Trees structure. Zandi and Allen developed compression method with reversible embedded wavelets to decrease the quantization error introduced by wavelet transform, which emphasized the importance of wavelet transform.

The representative block structure-based methods have the set partitioned block coder (SPECK) and block coding with optimized truncation (EBCOT). The SPECK performed the refinement coding both in the high frequency and the low frequency sub-band, which leads to a lower encoding efficiency. The EBCOT use the post compression rate distortion optimization to optimize the encoding efficiency, which is accepted as the core coding algorithm by JPEG2000.

For medical images, lossless compression is widely utilized. The compression ratio is only 2–4 times, which cannot satisfy the need of effective storage and high speed transmission for massive medical images (David et al., 2006). In our previous study, we proposed shearlet-based nearly-lossless compression algorithms (Ping Li & Huiqin Jiang, 2013). Although the proposed methods can obtain preferable effect, it still will take a long time for applying to actual PACS system.

In this paper, we propose a new compression algorithm for CT images. Our method combines denoising and the EBCOT method. The main contribution is to decide the channel numbers and truncating the bit stream adaptively. Experimental results show that the proposed method has better results than the existing methods.

2 INVESTIGATION OF THE EXISTING METHORDS

In order to confirm the compression effect between typical tree-structure coding SPIHT algorithm and block structure coding EBCOT algorithm in JPEG2000 for CT images, we perform a series of experiments on the same medical images with setting the same compression ratio (CR), PSNR and MSSIM (Wang Zhou et al., 2004) are chosen as quantitative evaluation standard.

The PSNR and MSSIM are calculated respectively as

$$PSNR = \lg(S^2 / MSE) \qquad (1)$$

where S is the maximum value of the pixel and the unit of PSNR is decibels (dB). MSE is mean square error between the reconstructed image and the original image.

$$MSSIM = (\sum SSIM) / K \qquad (2)$$

where K is the number of local window in the image and $SSIM$ is the similarity structure of the image.

Results of quantitative evaluations are presented in Table.1.

Table 1. Simulation results of two algorithms.

Setting	PSNR		MSSIM	
CR	SPIHT	EBCOT	SPIHT	EBCOT
4	31.9608	43.6614	0.9497	0.9806
8	31.7121	39.0362	0.9433	0.9700
15	31.1871	35.6107	0.9316	0.9572
20	30.9296	34.8734	0.9268	0.9443
30	30.2177	33.6861	0.9135	0.9413

As we can see from Table 1, with CR increases from 4 to 30, the PSNR and MSSIM of the two algorithms have a tendency of declining, but EBCOT is better than SPIHT. Therefore, we study the improved algorithm based on the EBCOT.

3 PROPOSED ALGORITHM

In the EBCOT algorithm, each wavelet sub-band is divided into small rectangular blocks, defined as code-blocks. The bit planes are associated with the coefficients of the code-block, where the bit plane numbers are decided by the maximum of wavelet coefficients. Then the coefficients are entropy-coded by bit plane coding and binary arithmetic coding in MQ coder. Finally the bit streams are organized into the compressed streams using post compression rate distortion.

Because the bit plane number is decided by using the maximum value of wavelet coefficients in low frequency sub-band, it is usually surplus for high frequency sub-band. On the basis of the wavelet theory, the coefficients of high frequency are almost smaller than that of the low frequency. Moreover, the sustained calculation to the ratio-distortion-slope leads to a lower efficiency. Therefore we improve EBCOT algorithm to solve the problems. The proposed algorithm consisted of the following procedures:

Step 1: Wavelet de-noising based on Bayesian estimation.

Because of quantum noise in CT images, we consider first to eliminate the noise effect. The methods based on Bayesian estimation are a powerful tool in wavelet de-noising (Lefkimmiatis, et al., 2009). We estimate non-noise wavelet coefficients based on Bayesian theory of Maximum a Posteriori (MAP) (Masuzaki et al., 2002).

The non-noise coefficients are estimated using (3).

$$\hat{d} = sign(d_j^i) \times \left(\left| d_j^i \right| - \frac{\sqrt{2} \times \sigma_n^2}{\sigma_s^2} \right) \qquad (3)$$

where d_j denotes the high frequency coefficients of level j, $\sigma_n = median(|d_i|) / 0.6745$ is the noise variance for three high frequency sub-bands. $\sigma_s^2 = \max(0, \sigma_d^2 - \sigma_n^2)$ is the estimated variance of non-noise coefficients for each sub-band .

Step 2: Select appropriate bit plane numbers for each sub-band and then EBCOT coding and MQ coding.

The bit plane numbers are chosen respectively equaling to the maximum value of wavelet coefficients of each sub-band.

Step 3: Adaptively truncate the bit stream.

The bit stream in the proposed algorithm is truncated adaptively based on the weight value of sub-band and the coding channel numbers.

At each level there are four (LL, HL, LH, and HH) sub-bands after wavelet decomposition. As we know, the signal energy focuses on low-frequency LL sub-band. The energy of LL sub-band is higher than the HL and LH sub-band, and the energy of HL and LH sub-band is higher than HH sub-band. The importance of wavelet sub-band can be measured based on the weight value of sub-bands.

On the other hand, channel numbers each code block contained has an approximate proportional relation with length of code word. The more the channel coding number contained, the better influence is on the image quality (Zhi Tao et al. 2010).

Therefore, we decided the truncation rule of the bit stream as shown in Fig. 1.

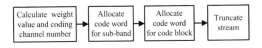

Figure 1. Adaptive truncation coding.

- **Calculation:** Calculate the weight value ω_k of the kth sub-band using (4):

$$\omega_k = (\dot{\omega}_k / \sum_{k=0}^{3j+1} \dot{\omega}_k) \times (3j+1) \tag{4}$$

Here $\dot{\omega}_k$ is the weighted coefficients of the kth wavelet sub-band (Adams, 2008), j is decomposition level.

The total code channel number $TnumPs$ is obtained using (5).

$$TnumPs = \sum numPs_i \tag{5}$$

where $numPs_i$ is the channel number of the ith code block; $numPs_i = numBPs_i \times 3 - 2$, $numBPs_i$ is the valid bit plane number of the ith code block.

- **Allocate code word for a sub-band:** The code word of each sub-band is calculated in

$$Rsub_k = R \max \times numSubPs_k / TnumPs \times \omega_k \tag{6}$$

Here R max is the code word which is related to the set compression ratio. $numSubPs_k$ is the channel number of the kth sub-band.

- **Allocate code word for a code block:** We allocate code word averagely according to the coding channel number in sub-band calculated using (7).

$$Rcb_k^i = Rsub_k \times numCbPs_k^i / numSubPs_k \tag{7}$$

Here $Rsub_k$ is the code bytes number of the ith code block in the kth sub-band, $numCbPs_k^i$ is coding channel number for the ith a code block in the jth sub-band. $numSubPs_k$ is the channel number of the kth sub-band.

- **Obtain the truncated bit stream:** When the encoding bit stream is approximate in agreement with the allocated code word length, the bit steam is truncated.

Step 4: MQ decoding and EBCOT decoding, then reconstruct the image by the inverse integer wavelet transform.

4 SIMULATION RESULTS

To verify the validity of the proposed algorithm, we implemented simulation experiments on CT images. CT images are taken from different body parts and conformed to DICOM Standard, provided by Radiology Department of First Affiliated Hospital of Zhengzhou University. The experimental environment is MATLAB R2010a. Platform is Windows XP, CPU is 2.50 GHz, and RAM is 2.00 GB.

Both quantitative and visual evaluations are presented with implementation of new algorithm in CT images.

Using the above quantitative standard, calculate PSNR and MSSIM between the original images and the reconstructed images respectively by using the proposed algorithm and the EBCOT.

The obtained results are given in Table 2. Both PSNR and MSSIM are the average value calculated from the results of 10 pair of CT images.

Table.2 shows that PSNR, MSSIM and the stability of the proposed method are better than that of EBCOT.

Table 2. Contrast of the PSNR and MSSIM for two algorithms.

Setting	PSNR		MSSIM	
CR	Proposed method	EBCOT	Proposed method	EBCOT
4	49.8095	43.6614	0.9813	0.9806
8	44.0112	39.0362	0.9805	0.9700
10	43.9409	37.8217	0.9805	0.9713
15	43.7357	35.6107	0.9774	0.9572
20	42.2780	34.8734	0.9751	0.9443
30	41.0724	34.6861	0.9678	0.9413

Fig. 2 shows the comparison of the processing effect between the proposed algorithm and the EBCOT. Fig. 2(a)–(c) respectively shows an original CT image, the reconstructed image by the proposed algorithm and the reconstructed image by EBCOT. Here CR is set to 30. In Fig.2(b), PSNR = 41.04 dB, MSSIM = 0.9678. In Fig.2(c), PSNR = 34.69 dB, MSSIM = 0.9413.

As can be seen in Fig. 2(b) and (c), the visual quality of the processed image using the proposed method is improved clearly in comparison with EBCOT. The results of PSNR and MSSIM also show that image quality has been improved effectively in the reconstructed images using the proposed method.

The comparison of the execute time between two methods is shown in Fig. 3. The execution time of the proposed algorithm is shown using the solid line while the EBCOT shown by the dotted line.

(a) Original image (b) Proposed method (c) EBCOT

Figure 2. Comparison of proposed method and EBCOT.

Fig. 3 shows that the solid line always are in the below in dotted line.

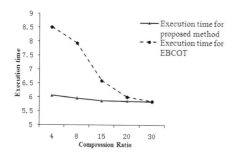

Figure 3. Contrast of execute time for two algorithms.

Therefore, the time consumption of the proposed algorithm is superior to EBCOT, especially in the low compression ratio, the performance is more obvious.

5 CONCLUSION

In this paper, we have proposed a new method to compress CT image. Through the experimental results and visual evaluations, we have demonstrated the proposed algorithm not only yields significantly superior image quality and a higher PSNR and MSSIM than the conventional methods, but also lower the computation complexity. The PSNR and the MSSIM increase by 17% and 2% respectively compared to the EBCOT algorithm with the same compression ratio, and the execute time reduces by 30%.

REFERENCES

Adams M. D. 2008. 04.20. Jasper/ JPEG2000 software reference manual version 1.900.1. http://www. ece. uvic. ca/-mdadam / jasper.

David.W et al. 2006. Perceptually lossless medical image coding. *IEEE Transaction on medical imaging* 25(3):335–344.

Ge Wang & Hengyong Yu. 2008. An outlook on X-ray CT research and development. *Medical Physics* 35(3): 1051–1064.

Lefkimmiatis, S. et al. 2009. Bayesian inference on multi-scale models for Poisson intensity estimation: applications to photon-limited image de-noising. *Image Processing, IEEE Transactions* 18(8):1724–1741.

Lei Xu. 2009. *Research of rate-control algorithms for still image compression.* Xi'an: Xi'an University of Electronic Science and Technology.

Masuzaki.T, Tsutsui.H & Izumi T. 2002. Adaptive rate control for JPEG2000 image coding in embedded systems. *Image Processing* 3: III–77.

Ping Li & Huiqin Jiang. Near-lossless compression for medical image in PACS. *Journal of Image and Graphics*, 2013, 18(6):483–490.

Ping Li & Huiqin Jiang. Near lossless ROI compression for medical images. *Optical and Precision Engineering* 2013, 21(3):759–766.

Rabbani et al. An overview of JPEG 2000 still image compression standard. *Signal Processing: Image Communication*, 2002, 17(1):3–48.

Ting Zhang & Ji'an Yang. The comparison of SPIHT and SPECK. *Journal of Guangxi University for Nationalities (Natural Science Edition)*, 2006, 12(1):99–103.

Wang Zhou et al. Image quality assessment: from error visibility to structural similarity. *Image Processing, IEEE Transactions*, 2004, 13(4):600–612.

Zandi, A et al. CREW: compression with reversible embedded wavelets. *Data Compression Conference*, 1995,2(12):28–30.

Zhi Tao et al. A lifting wavelet domain audio watermarking algorithm based on the statistical characteristics of sub-band coefficients. *Archives of Acoustics*, 2010, 35(4):481–491.

Information Technology – Wan et al. (Eds)
© 2015 Taylor & Francis Group, London, ISBN 978-1-138-02785-5

An acquisition algorithm of multipath BOC signals based on rake receiver

Yajuan Zhang, Tianqi Zhang, Wangjun Wu & Yu Liu
Key Laboratory of Signal and Information Processing, Chongqing University of Posts and Telecommunications
Chongqing, China

ABSTRACT: For the problem of ambiguous acquisition of the multipath binary offset carrier (BOC) signals, an algorithm which is based on rake receiver is proposed. By the processing of rake receiver with carrier phase estimation and the maximal ratio combining (MRC) combination method, the multipath BOC signals can be regarded as a single path BOC signal, which is captured by the method of relevant reconstruction with the compensation of frequency offset. Theoretical analysis and computer simulations shown that, under the same conditions, this algorithm has the advantage that high accuracy and easy to implement compared with other algorithms, and the higher modulation order of BOC signal is, the better performance get.

1 INTRODUCTION

In satellite navigation and positioning system, multipath fading is an inevitable factor, but multipath diversity reception technology is one of the effective measures to against it, especially the rake receiver which still has better performance in a spread spectrum system. Due to its unique split spectrum characteristic and multipath interference, which will greatly reduce the accuracy and effectiveness of acquisition for BOC signals. Therefore, it will be a good idea that the rake receiver is applied to the capture of multipath BOC signals.

Rake receiver is widely used in the direct spread spectrum system, such as Xiantao Cheng & Yongliang Guan (2011), Bin Jiang & Jianliang Xu (2013), Zhuoying Ye & Guotong Geng (2013), Choudhury S H, Bhuiyan T A, & Al-Rasheed A, et al (2010). The theory about rake receiver and its improved technology are described in many references, but there are a few relevant literatures about multipath BOC modulation signals, such as Xiaowen Sun & Shufang Zhang, et al (2010), X W Sun, S F Zhang, Q Hu, et al (2012). The study on rake receiver base band implementations case and phase difference estimation method of multipath components are proposed by Xiangdong Jia & Haiyang Fu, et al (2011), which can keep pace with all the received signals to make up for the darkness of the general rake receiver. Therefore, this paper considers using that method to diversify and combine multipath BOC signals. Due to multipath BOC signals turn into single path signal after using the rake receiver with phase estimation, so it can be used

for all the single path acquisition methods. People have studied a variety of acquisition methods for the single path BOC signal, such as auto-correlation side-peak cancellation technique (ASPeCT) which is proposed by Julien O, & Macabian C, et al (2007), filtered correlation, new orthogonal quadratic code correlation (OQCC) which is proposed by Shi, Chen & Tianqi Zhang, et al (2013), and correlation reconstruction which is proposed by Li Yang & Chengsheng Pan, et al (2009), but it did not consider the Doppler shift. Therefore, this paper proposes a new improved reconstruction method to capture multipath BOC signals.

In this paper, both the principle of rake receiver and the mathematical theory course of phase estimation method, which is proposed by Ke Chen & Qingming Gui, et al (2012), and the theory of correlation reconstruction with frequency offset compensation will be detailed presented for multipath BOC signals. Then this paper will take different modulation order of sinusoidal BOC signals as research background. Because of using a rake receiver with phase estimation, the energy of the signal is enhanced and the influence of the sub-peaks are almost entirely eliminated in the capture process.

2 THE APPLICATION OF RAKE RECEIVER

2.1 *The principle of rake receiver*

Rake receiver can effectively use the multipath components of signals, namely eliminating multipath interference and collecting multipath signal energy, and can reduce the probability of bite error rate (BER) of the system. It is mainly constituted by a group of the correlators using the same template with the different time delay corresponding to different multipath components. The received signals first pass through a bank of correlators, and then these multipath components are combined. When in the single user communication system and without intersymbol interference (ISI), MRC has the best performance. However, the general rake receiver only can achieve synchronization with the strongest signal, which will affect the performance of the receiver, and even lead to it lose effectiveness. Based on the above analysis, we will use the rake receiver with carrier phase estimation and the MRC method to solve the multipath interference. Fig.1 shows the principle block diagram of the rake receiver in this paper.

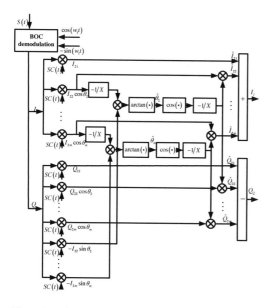

Figure 1. The principle block diagram of rake receiver.

2.2 Rake receiver with carrier phase estimation

The multipath BOC signal model is defined as:

$$S(t) = s_1(t) + \sum_{i=2}^{m} \alpha_i s_1(t - t_{di}) \qquad (1)$$

where m is the number of multipath, the ith amplitude is α_i, the ith path delay is t_{di}, $s_1(t)$ is the first path signal and the desired signal. After coherent demodulation of local carrier, in phase and quadrature signals are expressed as:

$$\begin{cases} I = I_1 + \sum_{i=2}^{m} I_i \cos\theta_i + \sum_{i=2}^{m} Q_i \sin\theta_i \\ Q = Q_1 + \sum_{i=2}^{m} Q_i \cos\theta_i - \sum_{i=2}^{m} I_i \sin\theta_i \end{cases} \qquad (2)$$

where $I_1 = d(t) SC(t)$, $Q_1 = I_1$; $d(t)$ is navigation data; $SC(t)$ is new code obtained by combining pseudo-code and sub carrier, called combination code in the paper.

This article assumes that carrier phase of the first path signal has not changed, so regards the second path signal as example to analysis. According to Fig.1, the output of the second path signal through correlators is expressed as:

$$\begin{cases} N_{I2} = I_{22} \cos\theta_2 + SC(t)\left[I_1 + \sum_{i=3}^{m} I_i \cos\theta_i + \sum_{i=2}^{m} Q_i \sin\theta_i \right] \\ N_{Q2} = -I_{22} \sin\theta_2 + SC(t)\left[Q_1 + \sum_{i=2}^{m} Q_i \cos\theta_i - \sum_{i=3}^{m} I_i \sin\theta_i \right] \end{cases} \qquad (3)$$

After passing through the low pass filter, the second path signal is changed to

$$\begin{cases} N_{I2} = I_{22} \cos\theta_2 \\ N_{Q2} = -I_{22} \sin\theta_2 \end{cases} \qquad (4)$$

So the carrier phase estimated value of the second path can be obtained from Fig.1.

$$\hat{\theta}_2 = \arctan\left(I_{22} \sin\theta_2 / I_{22} \cos\theta_2 \right) \qquad (5)$$

Similarly, the carrier phase estimation value of other paths is

$$\hat{\theta}_i = \arctan\left(I_{ii} \sin\theta_i / I_{ii} \cos\theta_i \right), (2 \le i \le m) \qquad (6)$$

Then the estimated value of multipath components of the in phase signal can be expressed as:

$$\hat{I}_{2i} = I_{2i} \cos\theta_i / \cos\hat{\theta}_i, (2 \le i \le m) \qquad (7)$$

For the estimated value of multipath components of the quadrature signal can also be obtained in the similar way, this article will not repeat them. Finally, the output of the in phase and quadrature by the maximum ratio combining are

$$I_I = Q_Q = s_1(t) + \sum_{i=2}^{m} \alpha_i s_1(t) \cos\theta_i \left(1/\cos\hat{\theta}_i\right) \approx s_1(t) + \sum_{i=2}^{m} \alpha_i s_1(t) \quad (8)$$

3 ACQUISITION OF MULTIPATH BOC SIGNALS

From the previous section, we obtain the new BOC signal which is no longer interference by multipath. For ease of analysis, we assume that the synchronization delay is accurate when the signals through rake receiver. So it can be seen as the single path BOC signal, just the energy of the single is enhanced.

3.1 The theory of correlation reconstruction

This paper uses correlation reconstruction method based on fractal theory. First, search for the Doppler frequency based on the theory that complex exponential multiplying in the time domain is equivalent to offset in the frequency domain. So make fast flourier transform (FFT) for the combined BOC signal. Then make the compensated signal correlate with the local BOC signal and pseudo code sequence by sliding them. Lastly, reconstruct the result of sliding correlation. For the BOC signal whose modulation order is N, the main peak and the first two side-peaks can be expressed by the following mathematical expression.

$$x_{BOC-N}(\tau) = tri_0(\tau/l) - \frac{N-1}{N} tri_{-1/N}(\tau/l) -$$
$$- \frac{N-1}{N} tri_{1/N}(\tau/l) \quad (9)$$

where $tri_\alpha(x/y)$ is a triangular function, whose independent variable is x, center is α, the bottom width is y, and the peak height is one. The cross-correlation function of pseudo noise (PN) sequence and the BOC signal whose modulation order is N, for two peaks around the center can be expressed as:

$$x_{BOC/PN-N}(\tau) = \frac{1}{N} tri_{-1/N}(\tau/l) - \frac{1}{N} tri_{1/N}(\tau/l) \quad (10)$$

Then the new correlation function can be got by adding equation 9 and equation 10.

$$|x| = |x_{BOC-N}(\tau)| + |x_{BOC/PN-N}(\tau)|$$
$$= tri_0(\tau/l) + tri_{-1/N}(\tau/l) + tri_{1/N}(\tau/l) \quad (11)$$

Correlation function is equivalent to the three triangles of Fig.2 (a). Reconstruction can be seen as

folding simply. The basic schematic of reconstruction is sketched in Fig.2, the specific steps are as follows.

First, define the position of C, D and E which are the three maximum value, it is theoretically speaking that they have the same size and can form a straight line, which is called the first broken line. By folding the first broken line, we can get Fig.2 (b).

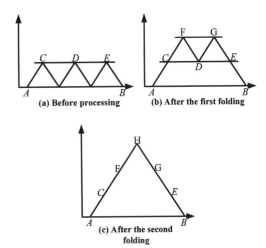

(a) Before processing   (b) After the first folding

(c) After the second folding

Figure 2. The theory of reconstruction.

Assuming the coordinates of C is (x_C, y_C) and the coordinates of any point W between C and E before folding is (x_W, y_W). After folding the coordinates of W' according to formula (12) is determined (x'_W, y'_W).

$$\begin{cases} x'_W = x_W \\ y'_W = 2 \times y_C - y_W \end{cases} \quad (12)$$

Then we can get the second broken line, according to the two new maximum points F and G. After the second folding, we obtain the Fig.2(c). Similarly, we can define any point between F and G, and finally we can get a definite peak point H. From Fig.2 (a) (b) (c) can be seen that the new correlation function without any side-peaks after two folding.

4 ALGORITHM PROCESSES

The algorithm block diagram of acquisition of multipath BOC signals represented as Fig.3.

a. Put multipath BOC signals through the rake receiver with carrier phase estimation to get the approximate single path signal.
b. Make down conversion of the approximate single path BOC signal, and sample two cycle signals,

which are expressed as r and its length is $2L$; Generate local PN sequence and BOC signal, and their length are $2L$.

c. Take FFT for r to obtain r', and then make rotate left and right k times respectively to constitute $2k+1$ frequency offset compensation sequence together with r'; Take FFT for local BOC sequence and take conjugated to get FFT(local BOC)*.

d. The frequency offset compensation sequences are multiplied by FFT (local BOC)* and make inverse fast flourier transform (IFFT) operation to get the correlation results, from which choose the maximum value that is expressed as c_k, $2k+1$ c_k constituted c.

e. Take frequency offset compensation sequence r'_k, which corresponds to the max peak of c, and make IFFT operation for r'_k to obtain r_k, which is the received signal with Doppler compensation in frequency domain.

f. Then make the local PN and local BOC signal sliding correlate with r_k L times and take modulo operation, and add them to get a new correlation function $|x|$; Then the $|x|$ is reconstructed by the previously described theory, and obtain the sequence k by finding the peak and folding operation.

g. Select the maximum value from k, which will be compared with the threshold, if the maximum value exceeds the threshold value, it is considered that manage to acquisition; if the maximum value is less than the threshold, reselect two cycles the received signals, and repeat step (b) ~ (f) until the peak exceeds the threshold value.

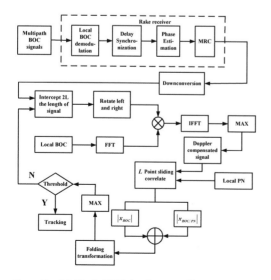

Figure 3.   Multipath BOC signal capture diagram.

5   SIMULATION AND ANALYSIS

Experiment I: In order to verify the performance of the rake receiver with carrier phase estimation in this paper, using five paths to simulate the multipath environment. The pseudo code rate of BOC (1, 1) is $f_c = 1.023$ MHz and sub carrier rate is $f_s = 1.023$ MHz. In this experiment, Simulation comparison the error performance of improved Rake receiver and general Rake receiver by the MRC. The simulation results are shown in Fig.4.

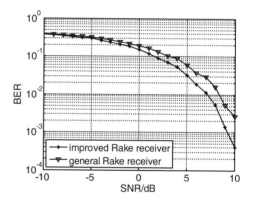

Figure 4.   The comparison of BER.

When at the same SNR, the BER of improved Rake receiver is lower than general Rake receiver.

Experiment II: To verify the effectiveness of the principle of folding, we choose BOC (1, 1) and BOC (5, 2), whose output envelope after folding are shown in Fig.5.

From Fig.5 we can see that there is no zero point, and high ratio of primary and secondary peak, so it can eliminate ambiguity and improve the accuracy of capture.

Experiment III: In this experiment, modulation signal is BOC (2,1), the Doppler frequency offset after down con- version is $f_d = 1.28$kHz, the length of pseudo code is $L_{PN} = 1023$, the pseudo code offset is $\phi = 250$, the sampling frequency is 10.23MHz. As we know that sub carrier frequency is $f_s = 2.046$MHz, the number of sampling points is $L = 20460$. So we get the frequency offset compensation precision is $f_{prec} = 250$Hz, the offset of pseudo code sampling point is $\phi' = \phi \times L/L_{PN} = 5000$. Reconstruction sequence is shown in Fig. 6, where the signal noise ration (SNR) is -10dB, and the figure only takes 200 sampling points around the reconstruction correlation peak.

As shown in Fig.6, the peak is in the 5000 sampling points, which is consistent with the simulation conditions.

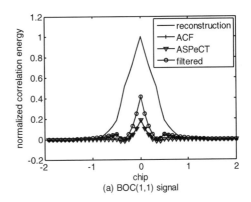

(a) BOC(1,1) signal

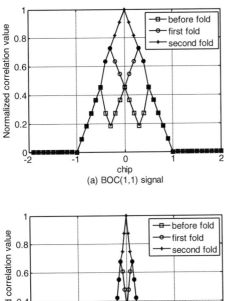

(b) BOC(5,2) signal

Figure 5. The reconstruction process of BOC (1, 1) and BOC (5, 2).

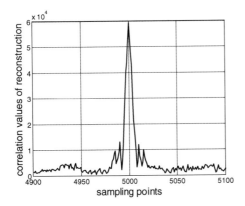

sampling points

Figure 6. The result of slid correlate.

Experiment IV: To further illustrate the advantages and effectiveness of the algorithm, make comparison with other algorithms, such as the auto-correlation function (ACF), (ASPeCT), and the filtered correlation under the same conditions. The normalized comparison of four kinds of algorithms is shown in Fig.7.

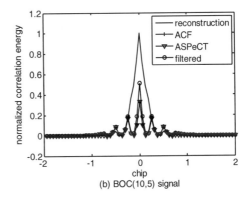

(a) BOC(1,1) signal

(b) BOC(10,5) signal

Figure 7. The comparison of normalized correlation energy.

As shown in Fig.7, the proposed reconstruction correlation algorithm has not side peak and also has good performance when the modulation order is larger than two. The higher modulation order of BOC signal is, the smaller primary peak width is.

Experiment V: The result of ratio of primary and secondary peak is shown in Fig.8. In addition,

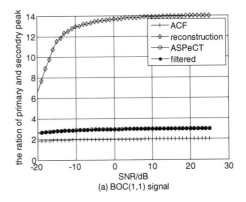

(a) BOC(1,1) signal

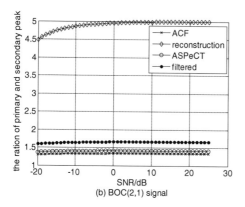

(b) BOC(2,1) signal

Figure 8. The comparison of ration of main and side peak.

in order to make the conditions more consistent, take the square root of correlation results of ASPeCT and filtered correlation.

As it can be easily seen, for both the BOC (1, 1) and BOC (2, 1) signals, the proposed unambiguous acquisition algorithm outperforms other algorithms, and the simulation results perfectly match the theoretical results.

6 CONCLUSION

For the problem of ambiguous acquisition of multipath BOC signals, this paper researches the multipath capture algorithm based on the rake receiver with phase estimation in detail. The algorithm does not only achieve the use of multipath by rake receiver, but also increase the probability of detection and reduce the probability of false alarm. The simulation results shown that, this algorithm compared with an ACF acquisition algorithm, 9.5dB~16dB has been improved in the ration of primary and secondary peak, and the precision of capture is good and easy to implement.

ACKNOWLEDTGMENT

This work is supported by the National Natural Science Foundation of China (No. 61371164, 61071196, 61102131), the Project of Key Laboratory of Signal and Information Processing of Chongqing (No. CSTC2009CA2003), the Chongqing Distinguished Youth Foundation (No. CSTC2011jjjq40002), the Natural Science Foundation of Chongqing (No.CSTC2012JJA40008) and the Research Project of Chongqing Educational Commission (KJ120525, KJ130524).

REFERENCES

Xiantao Cheng & Yongliang Guan. (2012). Pre/Post-Rake diversity combining for UWB communications in the presence of pulse overlap. *IEEE Transactions on Wireless communications*, 11(2):481–487.

Bin Jiang & Jianliang Xu. (2013). Performance simulation research on RAKE receiver in spread spectrum communication system. *Electronic Design Engineering*, 21(7), 160–162.

Zhuoying Ye & Guotong Geng. (2013). Efficient synchronization scheme for RAKE receiver at WCDMA downlink. *Journal of Beijing Information Science and Technology University*, 28(5), 43–46.

Choudhury S H, Bhuiyan T A, Al-Rasheed A, et al. (2010). A new approach in the modeling of MC-CDMA RAKE receiver and its performance evaluation under the effect of multipath fading channel. *International Conference on Computational Intelligence and Communication Networks (CICN)*, 529–533.

Xiaowen Sun & Shufang Zhang, et al. (2010). GNSS multipath signal acquisition method based on Rake architecture. *Chinese Journal of Scientific Instrument*, 31(12), 2854–2860.

X W Sun, S F Zhang, Q Hu, et al. (2012). Modified Rake architecture for GNSS receiver in a multipath environment. *Acta Astronautica*. 79(10):1–11.

Xiangdong Jia & Haiyang Fu, et al. (2011). An implementation scheme for rake receiver in base band and phase difference estimation of multipath signals. *Journal of Beijing University of Posts and Telecommunications*, 34(4), 97–100.

Julien O, & Macabian C, et al. (2007). ASPeCT: Unambiguous sine-BOC (n, n) acquisition/tracking technique for navigation applications signal. *IEEE Transactions on Aerospace and Electronic Systems*, 43(1), 150–162.

Shi, Chen & Tianqi Zhang, et al. (2013). Novel unambiguous acquisition algorithm of BOC (n, n) signal. *Application Research of Computers*, 2013, 30(3), 672–675.

Li Yang & Chengsheng Pan, et al. (2009). A new method for acquisition of BOC modulation signal based on refactoring theory to cross-correlation. *Journal of Astronautics*, 30(4), 1675–1679.

Ke Chen & Qingming Gui, et al. (2012). A multicorrelator-based GPS multipath estimation method. *Journal of Astronautics*, 33(9), 1241–1247.

Information Technology – Wan et al. (Eds)
© 2015 Taylor & Francis Group, London, ISBN 978-1-138-02785-5

An effective evaluation metric of LPI radar waveform

Hailin Li, Fei Wang, Jun Chen & Jianjiang Zhou
College of Electronic and Information Engineering, Nanjing University of Aeronautics and Astronautics, Nanjing, China

ABSTRACT: This paper presents an effective metric to evaluate different kinds of low probability of interception (LPI) waveforms. Based on the common view that white Gaussian noise is the best LPI waveform, the method introduced in this paper first use turning point method in time domain to test if a time series is white Gaussian noise. Then, if judged as a white Gaussian noise by the turning point method, we calculate a novel measure called "the statistical average signal to noise ratio (SA-SNR)" to evaluate LPI waveforms. When the SNR of the contaminated LPI radar waveform is less than or equal to SA-SNR, these data would not almost provide any information about LPI radar waveform. Simulations show that the proposed SA-SNR is effective to evaluate LPI radar signals.

KEYWORDS: Radar waveform, LPI, Evaluation

1 INTRODUCTION

Low probability of interception (LPI) radar waveforms against passive radar detection system has been developed for several decades [3]. Many evaluation metrics, such as time-bandwidth product, the peak to average power ratio and parameters agile characteristics, have been discussed and applied to compare different LPI waveforms [5] [6]. After analyzing and integrating those classical evaluation metrics of LPI radar waveforms, Fancey presented a complex metric to measure LPI performance [2].

Though LPI radar waveforms against passive radar detection system have been developed for several decades, a simple and effective metric to evaluate LPI radar waveforms is still nonexistent. The common view is that the white Gaussian noise is the best LPI waveform, and noise radar is regarded as the best LPI radar. Many methods to test white Gaussian noise have been studied in the time series domain, such as turning point test, different-sign test and rank-test. Among those methods, turning points test is the most useful to test if a time series is white Gaussian noise [1][4].

In most cases, interceptor wants to get so accurate parameters of radar signal that a great many time-frequency analysis techniques are used to show as many characteristics of signal as possible. Modern advanced radar can control the emitter power, then after propagation in a long distance, radar signal power may attenuate to such an extent that any time-frequency analysis method can never offer valuable signal information and then the intercepted data would be regarded as white noise. This paper bases this idea to try to find out the statistical average signal to noise ratio (SA-SNR) under which a certain contaminated radar waveform would be regarded as white Gaussian noise and interceptor could almost get nothing from the data. Obviously, to evaluate different LPI waveforms, the SA-SNR is more effective than those time-frequency analysis methods when the signal model is unknown.

Section 2 introduces turning point test method. Section 3 presents our estimation processes of SA-SNR. Simulations in section 4 show that our idea is effective and simple to evaluate different LPI radar waveforms.

2 TURNING POINT TEST METHOD

Let $\{X_t\}$ be a random event of independent identical distribution, and X_t, X_{t+1} and X_{t+2} are equally likely to occur in any of six possible orders as figure 1 shows.

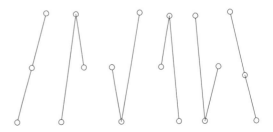

Figure 1. Four of the six are turning points.

From figure 1, if N represents the number of random events, the mean and the variance of the number of turning points would be $(N-2)2/3$ and $8N/45$ respectively. And let T be the number of turning points with N experiments, then this time series would be regarded as white Gaussian noise with 95% confidence if the following is satisfied:

$$\left| T - \frac{2N}{3} \right| \leq 1.96 \sqrt{\frac{8N}{45}} \tag{1}$$

3 ESTIMATING STATISTICAL AVERAGE SIGNAL TO NOISE RATIO

To illustrate our idea of estimating SA-SNR, this paper would show our estimation process which seems to be similar to the hypothesis test process.

First, define that:

$$H_1 : x(n) = \upsilon(n), \quad 1 \leq n \leq N \tag{2}$$

$$H_0 : x(n) = s(n) + \upsilon(n), \quad 1 \leq n \leq N \tag{3}$$

where H_1 means that data is judged as noise and H_0 represents that data is not white Gaussian noise. $s(n)$ denotes sampled zero mean signal, and $\upsilon(n)$ is zero mean white Gaussian noise whose variance is unknown.

For each experiment of hypothesis test of (2) and (3), our purpose is to estimate the current maximum SNR of $s(n)$ and $\upsilon(n)$ when $x(n)$ belongs to H_1. Figure 2 shows the detailed procedures of each experiment to test $x(n)$.

The detailed procedures to test $x(n)$ in figure 2 are described as follows:

1 Let $M = N/3$, where $N/3$ is an experience value. Set initial signal to noise ratio $SNR_I = 0dB$, and set $-0.5dB$ as step interval.
2 With certain M, randomly pick up M data from $x(n)$ whose signal to noise ratio is SNR_I and construct a new time series $x(i_1), x(i_2), \cdots, x(i_M)$, where $i_1 < i_2 < \cdots < i_M$.
3 Compute the number of turning points of $x(i_1), x(i_2), \cdots, x(i_M)$ as illustrated in section II.
4 With 95% confidence range, according to (1), using Normal Test (Z-Test) to determine if $x(i_1), x(i_2), \cdots, x(i_M)$ is white Gaussian noise. If the output of Z-Test is 1, we think that $x(i_1), x(i_2), \cdots, x(i_M)$ would be looked as white Gaussian noise. If not, let $SNR_I = SNR_I - 0.5$ and repeat step 2 to step 4.
5 Let $M = M + 1$, repeat 2–5 until $M = N$.
6 Take current SNR_I when all of $x(i_1), x(i_2), \cdots, x(i_M)$, $N/3 \leq M \leq N$, have passed the Z test. That is, the output of multiplication is equal to 1.

With L Monte Carlo experiments, we define SA-SNR as:

$$SA-SNR \triangleq \frac{1}{L} \sum_{l=1}^{L} SNR_l \tag{4}$$

From (4), it is obvious that the SA-SNR is the estimated average SNR under which any part of $x(n)$ would be largely judged as white Gaussian noise by turning point method.

Figure 2 and (4) show that the SA-SNR is independent of signal power. Since that, if the background noise to signal is known, with certain attenuation of the transmission, we could draw out how far the signal could be transmitted when the data are judged as white Gaussian noise. To estimate the distance, we need at first to know the power of background noise, signal power and the SA-SNR of the signal. And the key is to regard the background noise plus signal as a new waveform and to estimate the SA-SNR of the received signal. Now, let $s(n)$ be the new waveform, that is:

$$s(n) = s_p(n) + \varepsilon(n), \quad 1 \leq n \leq N \tag{5}$$

where $s_p(n)$ is the signal and $\varepsilon(n)$ is background noise.

Assume that the SA-SNR of $s_p(n)$ is η_p and ε_p is the SNR of $s(n)$. There is:

$$\eta_p = 10 \log \frac{P_{s_p}}{P_{N_1}} \tag{6}$$

$$\varepsilon_p = 10 \log \frac{P_{s_p}}{P_{N_2}} \tag{7}$$

where, P_{s_p} is the power of $s_p(n)$, P_{N_1} is noise power and P_{N_2} is the power of $\varepsilon(n)$.

Let SA-SNR of $s(n)$ be η. When η_p and ε_p are known and η_p is less than ε_p, from (6) and (7), it is easy to have that:

$$\eta = 10 \log \frac{10^{\eta_p/10}(10^{\varepsilon_p/10}+1)}{10^{\varepsilon_p/10} - 10^{\eta_p/10}} \tag{8}$$

Obviously, equation (8) is also independent of signal power.

4 SIMULATIONS

First, we would simulate SA-SNR of several different radar waveforms. Table 1 shows estimation results of the SA-SNR of LFM waveform and some polynomial phased waveforms, including P1, P2, P3, P4 and FRANK. From table 1, all values of SA-SNR of polynomial phased waveforms are larger than that of LFM. It means that LFM is relatively easier to be detected than polynomial phased waveforms, which is same to the result of Fancey [2].

From table 1, we also find that the values of SA-SNR of polynomial phased waveforms are similar. Considering the step interval of our experiments is $-0.5dB$, we can only provide the conclusion that P1,

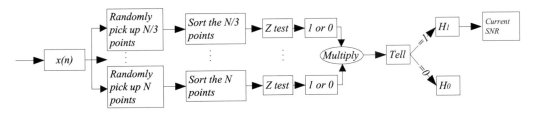

Figure 2. The procedures to estimate the current maximum SNR for each experiment.

Table 1. SA-SNR of several different radar waveforms.

Type	Bandwidth (MHz)	Pulse width (us)	Sampling frequency (MHz)	Code numbers	SA-SNR (dB)
P3	100	1.438	500	144	-11.20
P4	100	1.438	500	144	-11.22
Frank	100	1.438	500	144	-11.32
P1	100	1.438	500	144	-11.44
P2	100	1.438	500	144	-11.50
LFM	100	1.438	500	---	-14.72

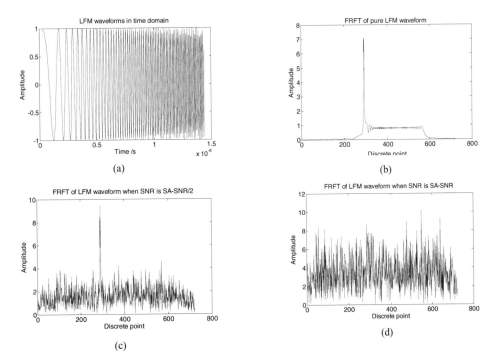

Figure 3.   (a) LFM signal, (b) FRFT of pure LFM, (c) FRFT of noisy LFM and SNR is SA-SNR/2 (-7.36dB) (d) FRFT of noisy LFM and SNR is SA-SNR (-14.72dB). (FRFT is fraction Fourier transform).

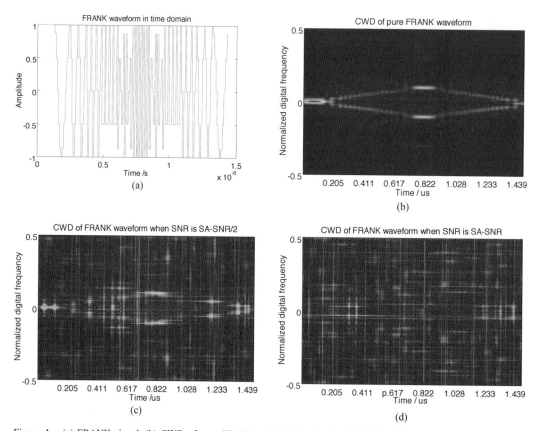

Figure 4. (a) FRANK signal, (b) CWD of pure FRANK, (c) CWD of noisy FRANK and SNR is SA-SNR/2 (-5.66dB) (d) CWD of noisy FRANK and SNR is SA-SNR (-11.32dB). CWD is Choi-Williams distribution.

Table 2. Difference of simulated and deduced SA-SNR of different waveforms.

	LFM (SA-SNR/2)	LFM (SA-SNR/3)	LFM (SA-SNR/4)	LFM
Simulated SA-SNR	-5.90	-7.90	-9.42	-14.72
Deduced SA-SNR	-5.75	-8.12	-9.14	—

Table 3. Difference of simulated and deduced SA-SNR of different waveforms.

	Frank (SA-SNR/2)	Frank (SA-SNR/3)	Frank (SA-SNR/4)	Frank
Simulated SA-SNR	-3.56	-4.84	-6.42	-11.32
Deduced SA-SNR	-3.24	-5.19	-6.01	—

P2, P3, P4 and FRANK have the similar anti-detection performance because their absolute differences of SA-SNR are less than $0.5dB$.

Then, take LFM and FRANK waveforms as examples, Figures 3 and 4 show the effect of SA-SNR where (a) is a waveform, (b) is time-frequency analysis of pure waveform, (c) and (d) are time-frequency analysis of contaminated waveforms whose SNR are equal to SA-SNR/2 and SA-SNR, respectively.

Since table 1 shows SA-SNR of several different waveforms, it is possible to deduce SA-SNR with when $s(n)$ is some contaminated data expressed in . For example, let 'LFM (SA-SNR/T)' represents contaminated LFM waveform $s(n)$ whose SNR ε_p is equal to SA-SNR/T.

Then, with it is easy to deduce SA-SNR of 'LFM (SA-SNR/T)' when SA-SNR of LFM in table 1 is known.

Table 2 and 3 show deduced SA-SNR with (8) and simulated SA-SNR for some different data. Since the estimation error of deduced and simulated SA-SNR is small, table 2 and 3 illustrate that our estimation procedure of SA-SNR is simple and effective.

5 CONCLUSION

Electronic warfare in recent years leads to a great number of applications of LPI radar waveforms. However, none of the effective metric independent on waveform parameters and time-frequency tools has been presented until now. The purpose of this paper is to provide an effective simple metric named SA-SNR which is uncorrelated with concrete waveform and any time-frequency tool, to evaluate LPI radar waveforms.

ACKNOWLEDGMENT

The authors want to give thanks to the National Natural Science Foundation of China (Grant No. 61371170), the Fundamental Research Funds for the Central Universities (Grant No. NJ20140010), Funding of Jiangsu Innovation Program for Graduate Education (CXLX13_154), the Priority Academic Program Development of Jiangsu Higher Education Institutions (PADA) and Key Laboratory of Radar Imaging and Microwave Photonics (Nanjing Univ. Aeronaut. Astronaut.), Ministry of Education, Nanjing University of Aeronautics and Astronautics, Nanjing, 210016, China.

REFERENCES

[1] Box, George EP, Gwilym M. 2011. Jenkins, and Gregory C. Reinsel. *Time series analysis: forecasting and control.* Vol. 734. Wiley.
[2] Fancey, C., and C. M. Alabaster. 2010. The metrication of low probability of intercept waveforms. *Waveform Diversity and Design Conference (WDD), 2010 International. IEEE.*
[3] Pace, Philip E. 2009. *Detecting and classifying low probability of intercept radar.* Artech House Publishers.
[4] Scharf, Louis L. 1991. *Statistical signal processing.* Vol. 98. Reading: Addison-Wesley.
[5] Schleher, D. C. 2006. LPI radar: fact or fiction. *Aerospace and Electronic Systems Magazine, IEEE* 21.5: 3–6.
[6] Zhi-ming, C. H. E. N. 2006. Stealth Waveform for Airborn Radar. *Modern Radar* 9: 1–7.

Information Technology – Wan et al. (Eds)
© *2015 Taylor & Francis Group, London, ISBN 978-1-138-02785-5*

An efficient computer-aided hyperspectral imaging system and its biomedical applications

Y. Zhang, Q. Li & H. Liu
School of Information Science and Technology, East China Normal University, Shanghai, China

Y. Wang & J. Zhu
New Institute, Gouda, NetherlandsInstitutes for Advanced Interdisciplinary Research, East China Normal University, Shanghai, China

ABSTRACT: Hyperspectral imaging has been dramatically developed and used for biomedical analysis which can help doctors or researchers with diagnosis of various diseases. This paper presents a fully automatic system for hyperspectral image capture and preprocessing. The system is composed of two parts, the hardware building and the software programming. In order to get more useful information, users can adjust parameters, such as exposure time, operational state of the Charge-coupled Device (CCD) camera, etc. The related parameters can be modified through the software before image capture conveniently. The system allow users to do preprocessing for the acquired images, and the influences of external environment on images can be eliminated. To evaluate the performance of the new system, various images acquired under different parameters settings are showed in this paper. One of the most proper CCD parameters settings is presented in the discussion section to reduce time consumption.

1 INTRODUCTION

Molecular hyperspectral imaging (MHI) technology has been widely used for the scientific research of remote sensing (Wang&Huang&Liu 2012), and the application of hyperspectral image analysis for biomedicine has been greatly developed till today. In the past decades, researchers have developed various multispectral or hyperspectral imaging systems for capturing the biochemical image of various biological tissues. For instance, Mehrube Mehrubeoglu(2012) introduced a kind of hyperspectral imaging system relies on the optical properties of materials which absorb or scatter light at different wavelengths. Then, Yongyu Li, Jiajia Shan, et al(2011) introduced a system to assess the beef-marbling grade using hyperspectral imaging technology, which was developed to collect hyperspectral images in the spectral region of 400–1100 nm. In addition, Benxue M A, Wendong XIAO, et al(2011) proposed an identification method to detect the slight bruises and bruises area of fruits based on hyperspectral imaging technology. From the above, it can be seen that a kind of appropriative microscopic hyperspectral image capture system for biological information has a bright prospect.

In this paper, the proposed system is based on acousto-optic tunable filter (AOTF) and Charge-coupled Device (CCD). And the system is equipped with a high performance Nikon DS-Qi1 CCD for acquiring images of high quality. Both single band and continuous multiband spectral images can be captured by the system automatically. And the related parameters can be modified through the computer software interface. For the acquired data, the system allow users to do image preprocessing to make the images clearer. The appropriative software was developed under the visual studio 2010 environment on our own efforts.

2 METHODS

The whole system is developed mainly based on molecular hyperspectral imaging technology and visual c++ programming technology. This system can capture images in contiguous bands with adjustable band number and step-length of wavelength. The software allows users to preview the scenes of biological samples in real-time in the preview window. Therefore, users can adjust the parameters for acquiring more applicable images. The capture process is time-saving due to the high performance of CCD (Bodegom 2010). Time consumption for capturing per image can be conditioned to millisecond.

Except for the image capture function, this system can also do image preprocessing based on radiometric correction algorithm. The preprocessing mothod can eliminates the influences of external environment on the acquired data.

2.1 Hardware configuration

This system consists of a optical microscope, an AOTF, a CCD camera and a computer. The system hardware configuration is showed in figure 1. The microscope is equipped with a manually adjustable illuminant, which provides a way to set light conditions. AOTF is served as a tunable optical band pass filter. It has a designed wavelength ranges from 550 to 1000nm and its scanning wavelength can be changed in a very high speed (minimum interval is 20ns). The system is equipped with a high performance Nikon DS-Qi1 CCD. By combining low-noise electronics and a high-quantum efficiency detector, this CCD can capture a wide dynamic range of intensities while maintaining quantitative linearity. A 1394 video capture card is attached to the CCD, and the image data transmission is accomplished through IEEE 1394 serial port (Saad 2003). This CCD has refrigeration function, so it can be operated continuously in a stable way. According to the designed parameters of CCD, the resolution of acquired images can be chosen as 320*256, 640*512 or 1280*1024 and the data depth can be set as 8bits, 16 bits, 32bits or 48 bits for per pixel.

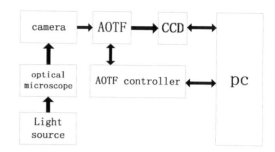

Figure 2. Structure diagram of the system.

2.2 Software function description

We have used visual c++ to develop this dialog based molecular hyperspectral image capture software. The spectral image data is acquired under the control over AOTF and CCD of the software platform. One key technology of the software development is to realize connection between visual studio environment and the hardware equipment.

The software has a main interface which contains a menu to execute operations like image live preview, image capture in single or contiguous bands. The menu is composed of seven optional items, Capture (capture mode selection), Preview (image live preview before capture), File (open and do image preprocessing), Data Conversion (image format transformation), Spectrum and Band Operation (show spectrum and band information) and Help. The main menu of user interface is showed in figure 3.

Figure 3. Main menu of user interface.

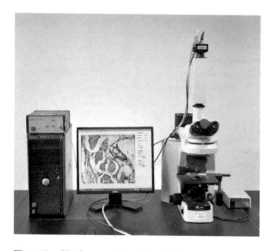

Figure 1. Hardware configuration of the system.

When the system starts, light transmitted by the illuminant firstly pass the slide which is placed on the microscope stage. Then the transmission light is collected by the microscope and is filtered by the AOTF. At last, it is captured by CCD and saved as data. The image data will get further processing by the software. During the data acquisition process, both CCD and AOTF are controlled by computer under the program execution. The system structure diagram is showed in figure 2.

Users can set the operational state of the CCD camera as live mode. Then, parameters like exposure time and gain value can be modified to get more proper displaying picture according to the live preview. When the preview picture satisfies the requirements, it is allowed to choose capture mode, for single or contiguous bands according to the main menu. For the contiguous bands capture, users can set parameters like the band number, start frequency and wavelength step in accordance with the designed performance of AOTF. All the relevant parameters are modified through corresponding edit boxes on the right end of the interface.

It is worth mentioning that images of each single band can be previewed before saved under the continuous multiband capture mode. In the mean time, progress of the capture work is showed by a progress bar. The image capture interface is showed in figure 4.

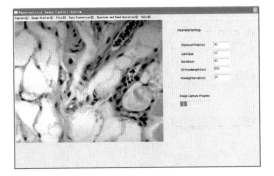

Figure 4.   Image capture interface.

In addition to the effective image capture function, this system can also do image preprocessing after capture. Due to the non-uniform of the light source of the system, there are some spots on the blank image due to the dust in the light path. For these reasons, image radiation correction is necessary.

In our work, the image preprocessing is based on a radiometric correction algorithm[11–12](Yana 2012,Koeni&Moo 2011). It eliminates the influences of different parts in the optical path of the system. The correction coefficient is calculated and demonstrated on the blank sampling image. D (i, j, n) represents the digital value (DN value) of the nth band on the blank image. The spectral mean value B (i, j) of a pixel can be expressed as (1).

$$B (i, j) = \Sigma[D (i, j, n)] / 80 \qquad (1)$$

And the correction coefficient K (i, j, n) can be expressed as (2).

$$K (i, j, n) = B (i, j) / D (i, j, n) \qquad (2)$$

The corrected blank sampling image is obtained by multiplying the K (i, j, n) and D (i, j, n), and D' (i, j, n) represents the DN value of the nth band on it, so as (3).

$$D' (i, j, n) = D (i, j, n) * K (i, j, n) \qquad (3)$$

We have imported this method into our software as one of system functions. After image capture, users should click the menu item which is named File to open the hyperspectral image data. The image will be showed according to the band number. The display interface is showed in figure 5.

In this display interface, we can do preprocessing for the selected band. The preprocessed results are showed in section III.

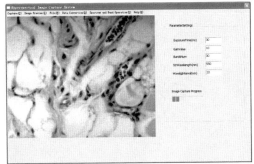

Figure 5.   Image display interface.

3   RESULTS

3.1   *Contiguous bands images and spectrum curve*

In this paper, we choose the human skin samples to get molecular hyperspectral images. By setting the data format of CCD, each pixel of the acquired image contains 16 bits data which is composed of gray-scale value and spectral information. The spectral information is useful for biological analysis. In our experiment, we have captured 80 bands images of the skin sample. Parts of the multiband images are showed in Figure 6.

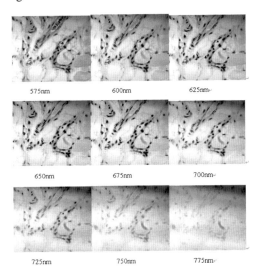

Figure 6.   One acquired series of multi-bands images.

The band number is 80 and the CCD exposure time is set as 30ms. The system has showed its high efficiency with the result that the total time consumption is merely about 5 seconds. Therefore, this system can capture hundreds of bands for motionless or moving samples in a very short time.

From figure 6, we can see that there are obvious differences among different bands, and the pixels of the same coordinates in different single-band vary in relation to spectral changes. Fundamentally, the variation occurred because biological tissues would generate different energy under spectrum of different wavelength (Yun&Qionglin&Shenping 2013). Then, spectral information of different bands can be expressed in the form of a curve. The spectrum curve of the skin samples is generated by ENVI and is showed in figure 7.

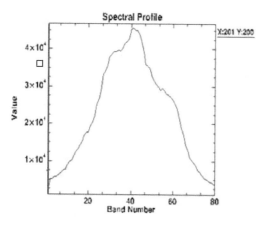

Figure 7. Typical spectra of multi-bands images in Figure 4.

From figure 7, it can be seen that one single band has a different spectrum signature from the others. Consequently, various bands of images provide a wealth of spectrum information which is potentially useful for further image analysis.

3.2 Image preprocessing

Type primary headings in capital letters roman (Heading 1 tag) and secondary and tertiary headings in lower case italics (Headings 2 and 3 tags). Headings are set flush against the left margin. The tag will give two blank lines (26 pt) above and one (13 pt) beneath the primary headings, 1½ blank lines (20 pt) above and a ½ blank line (6 pt) beneath the secondary headings and one blank line (13 pt) above the tertiary headings. Headings are not indented and neither are the first lines of text following the heading indented. If a primary heading is directly followed by

a secondary heading, only a ½ blank line should be set between the two headings. In the Word programme this has to be done manually as follows: Place the cursor on the primary heading, select Paragraph in the Format menu, and change the setting for spacing after, from 13 pt to 0 pt. In the same way the setting in the secondary heading for spacing before should be changed from 20 pt to 7 pt.

We have used our preprocessing method to realize the removal of the irrelevant spots in the original images. Figure 8 shows the results of the comparison between the original image and the processed image.

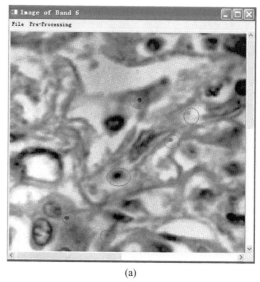

(a)

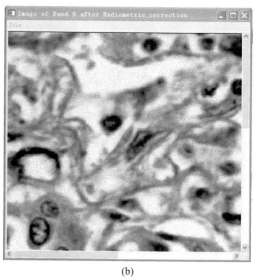

(b)

Figure 8.   (a) Original image and (b) Preprocessed image.

From figure 8, we can clearly see that there are irrelevant spots in the original image parts of which are marked by red circle in the original image. These spots were probably due to the dust in the light path. By comparison, it can be seen that the preprocessed image is clearer. The spots in the blank image and the original image is removed. It means that the influences of external environment are eliminated.

4 DISCUSSIONS

For this computer-aided hyperspectral image capture system, one of its most typical features is that parameters of CCD and AOTF can be adjusted conveniently according to the preview picture. As the response time of AOTF is conditioned to microsecond, the main time consumption is contributed by exposure time of the CCD camera.

Exposure time is directly related to the brightness of acquired images. Although increasing gain value of CCD can also enhance the image brightness, the noise is enhanced at the same time. Therefore, as a general rule, the gain value is set as minimum when we use this system. To decrease the exposure time consumption, the illuminant installed under the microscope stage should be adjusted to the strongest manually. Figure 9 shows a group of single band image of different exposure time settings.

According to figure 9, we can see that the image brightness increases rapidly due to the high sensitivity of the CCD camera. By comparison, the proper setting of exposure time is 30ms. Therefore, the system can capture hundreds of bands in a short time, and it can satisfy the requirement for capturing objective in movement.

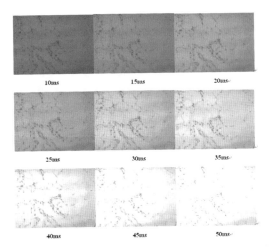

Figure 9.   Results of different exposure time.

5 CONCLUSION

In this paper, we proposed an efficient system for hyperspectral biomedicine image capture. The system can capture single band or continuous multi-bands spectral images automatically through the software interface. The results show that multiband molecular hyperspectral images of skin tissue sections can be acquired in a very short time with high quality. In addition, the system can also do image preprocessing to eliminate the influences of external environment after capture. And then, in order to get further efficiency increase, we provide a kind of proper parameters settings for CCD to decrease the system time consumption.

The next step for this work will include automatic image analysis, by the extended function of software, such as image segmentation and features extraction. In the future, this work will allow researchers to get more biomedicine information than the traditional optical methods.

ACKNOWLEDGMENT

This work is supported in part by the National Natural Science Foundation of China (Grant No. 61177011, 61377107), the Project supported by the Shanghai Committee of Science and Technology, China (Grant No. 11JC1403800), and the Project supported by the State Key Development Program for Basic Research of China (Grant No. 2011CB932903).

REFERENCES

Yiting Wang, Shiqi Huang, Daizhi Liu 2012. Research Advance on Band Selection-based Dimension Reduction of Hyperspectral Remote Sensing Images. RSETE 2012 2nd International Conference on Digital Object Identifier:1–4.

Mehrubeoglu, M.2012. Spectral Characterization of a Hyperspectral Imaging System Using Optical Standards. Imaging Systems and Techniques (IST), 2012 IEEE International Conference on Digital Object Identifier:126–129.

Yongyu Li , Jiajia Shan, Yankun Peng, Xiaodong Gao 2011. Nondestructive Assessment of Beef-marbling Grade using Hyperspectral Imaging Technology. New Technology of Agricultural Engineering (ICAE), 2011 International Conference on Digital Object Identifier: 779–783.

Benxue Ma, Wendong Xiao, Nianwei QU, Weixin WANG, Lili WANG, Jie WU 2011. Detection of Fruits Slight Bruises Based on Hyperspectral Imaging Technology. Electrical and Control Engineering (ICECE), 2011 International Conference on Digital Object Identifier:4445–4448.

Widenhorn, R., Justin C.Dunlap and Erik Bodegom 2010. Exposure Time Dependence of Dark Current in CCD

Imagers. Electron Devices, IEEE Transactions on Volume:57, Issue3:581–587.

Ashraf Saad 2003. AN IEEE 1394 – FIREWIRE – BASED EMBEDDED VIDEO SYSTEM FOR SURVEILLANCE APPLICATIONS. Advanced Video and Signal Based Surveillance, 2003. Proceedings. IEEE Conference on Digital Object Identifie): 213–218.

Stanley B.Lippman 2013. C++ Primer, 4th ED. Beijing:Posts and Telecom Press.

Meyers.Sl.2011. Effective C++ - 3rd ED. Beijing:Electronic Industry Press.

MingRi Technology 2012. Mastering C++ Programming, 2nd ED. Beijing: Tsinghua University Press.

Hongying, Liu, et al 2011. Radiometric correction of hyperspectral imaging data in spacial dimension and spectral dimension. Mechanic Automation and Control Engineering (MACE), 2011 Second International Conference on IEEE.

Guan, Yana, et al 2012. Pathological leucocyte segmentation algorithm based on hyperspectral imaging technique. Optical Engineering 51.5.

Andrew Koenig, Barbara E. Moo, 2011. Accelerated C++:Practical Programming by Example. Beijing: China Electric Power Press.

LING Yun , PENG Qionglin, XIAO Shenping,2013. The Spectrum Analysis Optimization for Band-limited Discrete Spectrum Continuous Signal Based on PSO. Intelligent System Design and Engineering Applications (ISDEA), 2013 Third International Conference on Digital Object Identifier, Page(s): 1541–1544.

Feldmann, I., Eisert, P., Kauff, P.2003. EXTENSION OF EPIPOLAR IMAGE ANALYSIS TO CIRCULAR CAMERA MOVEMENTS. Image Processing, 2003. ICIP 2003. Proceedings. 2003 International Conference on Digital Object Identifier:vol.2, Page(s): III-697–700.

Information Technology – Wan et al. (Eds)
© 2015 Taylor & Francis Group, London, ISBN 978-1-138-02785-5

An ultra-low and adjustable high-pass corner frequency variable gain amplifier using T-type pseudo-resistor

Yuliang Qin, Weilin Xu, Gongli Xiao, Dongyu Deng, Fadi Gui & Xuqiong Li
School of Information and Communication, Guilin University of Electronic Technology, Guilin, China

ABSTRACT: An ultra low and adjustable high-pass corner frequency variable gain amplifier (VGA) is presented in this paper. The proposed VGA is composed of operational trans-conductance amplifier (OTA), capacitive feedback circuit and a T-type pseudo-resistor. It has advantages of low-power, low input referred noise and smaller total harmonic distortion (THD). The T-type pseudo-resistor is applied to the additional DC biasing circuit of capacitive feedback VGA in order to form a RC feedback loop with function of high-pass filter. This method can meet the requirement of very large resistance for capacitive feedback VGA with ultra-low high-pass corner frequency. The high-pass corner frequency of the proposed VGA can be adjusted by the T-type pseudo-resistor, which is controlled by a tunable voltage. Simulation results under SMIC 0.18 μm 1P6M Salicide CMOS process show that the adjustable range of the high-pass corner frequency is 8.5 mHz to 400 mHz and the gain range of VGA is 10 dB to 49 dB. The input referred noise and THD are 2.6 μVrms and 0.23% respectively. The total current consumption is 14 uA under a 1.8 V power supply. Its characteristic makes it suitable for biomedical applications.

KEYWORDS: variable gain amplifier; capacitive feedback; high-pass corner frequency; T-type pseudo-resistor

1 INTRODUCTION

Variable gain amplifier (VGA) is an important building block for biomedical communication system. It can amplify the variable amplitude of input signal by corresponding gain stage which provides stable signal for analog-to-digital converter (ADC). The amplitudes of common physiological signals, such as ECG, EMG and EEG, are in order of tens of uV to tens of mV and the frequencies span from tens of mHz to a few kHz. The VGA should have the features of low input referred noise, reconfigurable bandwidth, low THD, and variable gain to meet requirements of different biomedical applications[1].In order to ensure that the low cut-off frequency of VGA lower than the lowest frequency of all useful input signals, an appropriate ultra-low high-pass corner frequency must be designed.

VGA can be realized either by close loop or open loop configuration. Close loop topology VGA is more commonly used for its accurate gain control and better linearity [2]. The close loop topology VGA can be divided to the resistor feedback configuration and the capacitor feedback configuration. Its gain can be realized by the ratio of resistor or capacitor. The resistor feedback configuration usually consumes large current. Capacitive feedback is widely adopted to achieve low power consumption for biomedical application. A DC bias circuit must be added for capacitive

feedback VGA to import the common-mode voltage for OTA. A RC feedback is formed by the feedback capacitor and the equivalent resistor of the DC biasing circuit. A low high-pass corner frequency can be achieved by this RC feedback circuit. Usually, the additional DC biasing circuit is handled by two serially connected PMOS transistors which call pseudo-resistor [3]. This kind of pseudo-resistor has too large storage charge and its value of equivalent resistance is not large enough. Thus it is needed to improve the resistance of pseudo-resistor. For examples, tunable pseudo-resistor [4] and balanced tunable pseudo-resistor [1] are proposed by using different structures, which can achieve larger equivalent resistance values and solve the problem of too large storage charges. Although they can achieve bigger equivalent resistance values, it is still not enough for biomedical application. Furthermore, it exhibits asymmetric and nonlinear resistances while the voltage across the tunable pseudo-resistor varies, and the structure of balanced tunable pseudo-resistor is very complex. Analysis shows that it has great challenges to achieve the ultra-low high-pass corner frequency of VGA for biomedical applications.

In this paper, a new T-type pseudo-resistor is proposed and a capacitor feedback VGA with gain rage of 10 dB to 49 dB is designed. This VGA has ultra-low high-pass corner frequency, and its high-pass corner frequency is adjustable. The additional

DC biasing circuit is handled by the T-type pseudo-resistor with the characteristic of huge and adjustable equivalent resistance. A RC feedback loop is formed by the equivalent resistor and the feedback capacitor of VGA. An ultra low high-pass corner frequency can be achieved by this RC feedback loop and the high-pass corner frequency is adjusted by the T-type pseudo-resistor. The adjustable frequency range of VGA is 8.5 mHz to 400 mHz. The paper is organized as follows: Section 2 describes the system architecture of the VGA and the T-type pseudo-resistor design and optimization. Details of theoretical analysis and simulation results are presented in Sections 3. Section 4 concludes the paper.

2 DESIGN OF VGA

2.1 Structure of variable gain amplifier

The structure of VGA using new T-type pseudo-resistor is shown in Figure 1. The gain of VGA is given by:

$$G = \log\left(\frac{R_T C_1}{R_T C_2 + 1/jw}\right) \quad (dB) \tag{1}$$

where, R_T is the huge equivalent resistance of the T-type pseudo-resistor. The DC gain of VGA is zero due to the huge resistance values of RT. Thus the gain of VGA is also given by

$$G \approx \log(\frac{C_1}{C_2}) \quad (dB) \tag{2}$$

Obviously, the gain of VGA can be controlled by the ratio of C_1/C_2. In this design, the value of C_2 is fixed. 14 different values of capacitor C_1 are designed to achieve 14 different gain stages. The achieved gain range is from 10 dB to 49 dB and the gain step is 3 dB.

2.2 T-type pseudo-resistor

As the Figure 1 (a) shows, the T-type pseudo-resistor is formed by three PMOS transistors which connected by T network. In this circuit, V_{cm} is the common-mode voltage of OTA and V_{turn} is the control voltage of the T-type pseudo-resistor. The DC biasing is handled by the T-type pseudo-resistor. The function of high-pass filter can be achieved by the DC biasing and the feedback capacitor of the VGA. The high-pass corner frequency is given by

$$f_h = \frac{1}{2\pi R_T C_2} \tag{3}$$

where, C_2 is set to 100 fF in view of area consumption. R_T is the equivalent resistance values of the T-type pseudo-resistor. Obviously, an ultra high-pass corner frequency can be achieved when the value of R_T is huge enough. The T-type pseudo-resistor can be equivalent to ∏-type network [5], thus the equivalent resistance of ∏-type network are given by

$$R_{12} = \frac{R_1 R_2 + R_2 R_3 + R_3 R_1}{R_3} \tag{4}$$

$$R_{23} = \frac{R_1 R_2 + R_2 R_3 + R_3 R_1}{R_1} \tag{5}$$

$$R_{31} = \frac{R_1 R_2 + R_2 R_3 + R_3 R_1}{R_2} \tag{6}$$

where, R_1, R_2, R_3 are the equivalent resistances of PM1, PM2, PM3 respectively. The aspect ratio of PM1, PM2 and PM3 are equal to 0.23 um/10 um, which aim to achieve huge equivalent resistance. The substrates of them are connected to ground. R_{12}, R_{23}, R_{31} are the equivalent resistance of ∏-type network respectively. As Figure 1 (a) shows, R_1, R_2 are almost the same and their equivalent resistance values are very large. Base on formula (5) and (6), the formula (7) can be developed.

$$R_{23} = R_{31} = R_2 + 2R_3 \tag{7}$$

where the R_3 can be achieved an appropriate values by modulated the value of Vturn. Based formula (4) and formula (7), the formula (8) is achieved.

$$R_{12} = \frac{R_1 R_2}{R_3} + 2R_2 \tag{8}$$

If R_3 was small enough, it would be achieve

$$R_{12} \gg R_{23} + R_{31} \tag{9}$$

Thus the improving DC biasing circuit by used new T-type pseudo-resistor can achieve ultra big equivalent resistance. An ultra low and adjustable high-pass corner frequency can be achieved by used new T-type pseudo-resistor.

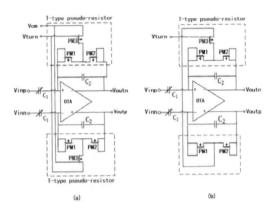

Figure 1. Structure of VGA. (a): The T-type pseudo-resistor; (b): The optimized T-type pseudo-resistor.

2.3 Optimization of T-type pseudo-resistor

Two PMOS transistors of PM3 in Figure 1(a) may exist asymmetry factor due to process variation. Then this imbalance will change the common-mode input voltage and it will be converted to differential input signals for fully differential OTA. That will eventually affect the normal operation of VGA. In this design, the T-type pseudo-resistor is optimized to solve the problem of process variation. The optimized circuit is shown in Figure 1 (b), where the two T-type pseudo-resistors connect together and share the same PMOS transistor of PM3. Based on pseudo-resistor theoretical analysis [3], PM1, PM2 and PM3 are connected to the same point and the voltage of this point is equal to the common-mode voltage Vcm of the OTA. Therefore, the voltage of the source and drain for transistor PM3 are respectively fixed to Vcm. Furthermore, the two T-type pseudo-resistors are restricted by the PMOS transistor of PM3. Thus the two DC biasing circuits are balanced by the optimization T-type pseudo-resistor and the CMRR is improved by this method. Based theoretical analysis, an ultra low and adjustable high-pass corner frequency can be achieved by the VGA using T-type pseudo-resistor, and two DC biasing circuits of fully differential OTA are balanced by the optimization T-type pseudo-resistor. They make huge contribution to biomedical chip process the ultra low frequency.

3 POST-SIMULATION RESULTS AND ANALYSIS

In this design, a T-type pseudo-resistor is adopted to achieve an ultra low and adjustable high-pass corner frequency VGA in SMIC 0.18 μm 1P6M Salicide CMOS process. The layout of the VGA is showed in Figure 2. For the requirement of small area consumption in biomedical application, the area of VGA is 0.425 mm * 0.372 mm, and the gain control of VGA is 4-bit.

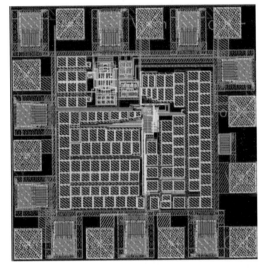

Figure 2. The layout of VGA circuit.

Generally, the amplitude rang of the biomedical signal is 0.01 mV to 100 mV, after magnified by the preamplifier and attenuated by the low filter, the amplitude range is up to 1 mV to 250 mV. According to formula (2), the gain range from 10 dB to 49 dB is achieved by controlling the values of C_1 in this design. In order to obtain enough gain control precision, the gain step of VGA is set to 3 dB. The simulation results are shown in Figure 3. As the Figure 3 shows, the high-pass corner frequency of 0.016 Hz is achieved in this design. The low-pass corner frequency is 3.6 kHz. Commonly, the frequency distribution of ECG, EEG, EOG is 0.1 Hz to 400 Hz. Thus the proposed VGA satisfies the requirements of biomedical applications.

The simulation results of the adjustable high-pass corner frequency with the gain of 31 dB are shown in Figure 4. The adjustable high-pass corner frequency range of the other gain levels are almost the same. According to formula (3) and (9), the high-pass

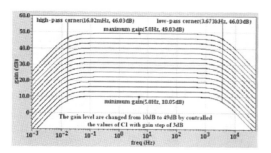

Figure 3. The simulation result of the gain range, high-pass corner frequency, and low-pass corner frequency.

corner frequency with adjustable range from 8.5 mHz to 400 mHz is achieved by controlling the voltage values of V_{turn} in this design. The voltage range of V_{turn} is 550 mV to 750 mV. The simulation results show that the low cut-off frequency of useful biomedical signals is included in the adjustable high-pass corner frequency. Thus an accurate low cut-off frequency can be adjusted by different voltage of V_{turn}.

The noise of VGA has a direct effect on back-end circuit. Therefore, this VGA should have low noise feature. By increasing aspect ratio of input MOSFET in OTA, this design has capability to reduce flicker noise in low frequency. VGA equivalent input noise is showed in Figure 5. As the simulation indicated, this VGA can obtain a very low equivalent input noise in the bandwidth of application. For instance, the ECG frequency is approximated 1Hz. When VGA is working in 10dB, the input noise is 55.53uVrms in 1Hz. In 49dB, the input noise is 0.6uVrms in 1Hz. Hence, this VGA has a well performance in terms of equivalent input noise.

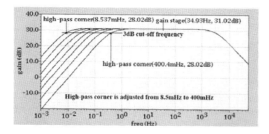

Figure 4. The simulation results of the adjustable high-pass corner frequency.

The detail VGA gain step transient response obtained from post-layout simulation are shown in Figure 6. Its gain range is 10 to 49 dB with the gain step of 3 dB. From the Figure, the output amplitude change with variable of control signal. The function of VGA is implemented.

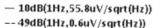

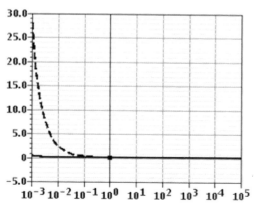

Figure 5. The equivalent input noise of T-type pseudo-resistor VGA.

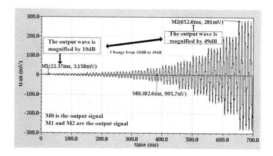

Figure 6. Transient response of the VGA at different C1 with gain step of 3 dB.

The performance comparison of the proposed VGA and other VGAs reported is summarized in Table I. Compared with other reported works list in Table I, the VGA has improvements that it has low input noise, small THD, and ultra-low and adjustable high-pass corner frequency. It meets all requirements of specific design for biomedical application.

4 CONCLUSION

We have demonstrated an adjustable ultra low high-pass corner frequency VGA using a new T-type pseudo-resistor for biomedical application in SMIC 0.18 μm 1P6M Salicide CMOS process. The gain range of 10 dB to 49 dB is achieved and the maximum overall gain is 39 dB. The adjustable range of ultra low high-pass corner frequency of this VGA is

8.5 mHz to 400 mHz. The current dissipation is less than 14 uA from a 1.8 V supply voltage. The proposed VGA is well-suited for biomedical application.

Table 1.   Performance summary and comparison.

	[6]	[7]	**This design**
Technology	0.18 µm CMOS	0.18 µm CMOS	0.18 µm CMOS
Supply voltage	1.8 V	1 V	1.8 V
DC Current	NC	10 µA	14 µA
Gain range	54 dB ~ 67 dB	8 dB ~ 58 dB	10 dB ~ 49 dB
Gain step	NC	NC	3dB
High-pass corner frequency	20 Hz ~ 200 Hz	15 mHz	8.5 mHz ~ 400 mHz
Low-pass corner frequency	7 kHz	100 Hz ~ 2 kHz	3.6 kHz
Input noise	11.93 µVrms	3.35 µVrms @ 49 dB 370 µVrms @ 10 dB	2.7 µVrms @ 49 dB 239.7 µVrms @ 10 dB
THD	NC	0.65%	0.23%

ACKNOWLEDGMENT

This work was supported by the Natural Science Foundation of China (61264001, 61161003, 61166004), Guangxi Natural Science Foundation (2013GXNSFAA019333) and Scientific Research Foundation of GUET (UF12001Y).

REFERENCES

Zou X.D., Xu X.Y., Yao L.B. & Lian (2009). A 1-V 450-nW Fully Integrated Programmable Biomedical Sensor Interface Chip. IEEE J. of Solid-State Circuits. 44, 1067–1077.

Zou X.D., Liu L., Tan Y.S., Minkyu J. & Seng Y.K. (2012). Integrated Circuits Design for Neural Recording Sensor Interface. IEEE J. of Asia Pacific Circuits and Systems Conference. 51, 17–20.

Mohseni P. & Najafi K. (2004). A Fully Integrated Neural Recording Amplifier with DC Input Stabilization. IEEE J. of Engineering in Medicine and Biology Society. 51, 832–837.

Chae M., Liu W., Chen T., Kim J., Sivaprakasam M. & Yuce M. (2008). A 128-channel 6 mW Wireless Neural Recording IC with On-the-Fly Spike Sorting and UWB Transmitter. IEEE J. of ISSCC Digital Tech. Papers. 146–603.

Paymond A.D. & Pen-Min L. (2001). Linear Circuit Analysis (Second Edition), Oxford University press, New York: Oxford.

Seung-In N., Susie K., Taehoon K. & Hyongmin L. (2013). A 33-Channel Neural Recording System with a Liquid-Crystal Polymer MEA. IEEE J. of Biomedical Circuits and Systems Conference. 13–16.

Liu X. & Zheng Y.J. (2012). An Ultra-Low Power ECG Acquisition and Monitoring ASIC System for WBAN Applications. 2, 60–70.

Information Technology – Wan et al. (Eds)
© 2015 Taylor & Francis Group, London, ISBN 978-1-138-02785-5

AugECE – A framework for audio signal extraction, classification, and enhancement

Hongfei Wang

China Ship Development and Design Center, Wuhan, China

ABSTRACT: In this report, we present a framework, AugECE, which accomplishes the task of audio signal extraction, classification, and enhancement. Specifically, AugECE first extracts the targeted and interested signals from a mixture of audio signals using independent component analysis, and then classifies them by k-nearest neighbor for their corresponding labels. This classification procedure is conducted in both time and frequency domain, of which the latter aims to address the computation and memory issues from real world application. At the last stage, wavelet-based hidden Markov models (HMM) and principal component analysis (PCA) are further employed in AugECE to enhance the quality of the extracted signals by reducing background noise and/or residual signals originated from other sources as well. Experiments are performed to compare the performance of wavelet-based HMM with wavelet-based PCA. The results are presented with analysis and conclusion.

1 INTRODUCTION

Extracting interested or pre-targeted signals from mixtures of audio has long been a hot research topic. Further processing and analyzing the extracted audio signals is therefore worth of serious attention and study.

Classification aims to identify to which of a set of categories an extracted signal belongs. It is the process in which a signal is fully recognized, differentiated, and understood. Classification of the extracted signals lays the basis for computer music (Gunawan & Sen, 2009, Chafe & Jaffe, 1986) and various biometric authentication (Mehra et al., 2010). Particularly, as opposed to manually identifying each of the sources, an automatic approach should be developed and thereby applied when then number of sources is large.

Enhancing the quality of audio signals can be achieved by filtering out the noise, which typically accounts for the main factor that deteriorates the quality.

For real world applications, the processing of audio signals is often expected to be completed within a reasonable and acceptable time duration. Most classification models require historical data so that the queried source can be compared with that of making decision of class assignment. The amount of data thus stored can be significant in many scenarios, and consumes much of the memory. This is especially crucial for the application in embedded systems where there is stringent requirement of power and storage.

To address the above issues, we present a framework called **AugECE** (**Au**dio Signal **E**xtraction, **C**lassification, and **E**nhancement) that accomplishes the extraction, classification, and enhancement of audio signals. Specifically, AugECE first extracts the signals of interests from a mixture of audio signals via independent component analysis (ICA); AugECE then classifies them by k-nearest neighbor (kNN) to match their corresponding labels; last, wavelet-based hidden Markov models (HMM) and principal component analysis (PCA) are further employed to enhance the quality of the extracted signals by reducing background noise and/or residual signals originated from other sources as well. A high-level schematic diagram is illustrated in Fig. 1.

At this point, it would be tempting to perform the enhancement prior to the classification, given that the classification procedure would benefit from more *pure* signals, and as a consequence, the classification accuracy would increase considerably. However, quality enhancement usually costs over 10× or even more than 100× of CPU time than generic classification procedure does. Such expense would be wasted if some signals were later classified and then deemed unwanted or little interested. Plus, the classification results can be satisfactory before enhancement. Therefore, we must resist this temptation and leave the enhancement to the last stage.

The remainder of this paper is organized as follows. Section 2 describes in details each method applied in AugECE. Section 3 presents experiments and results, with analysis and discussion. Section 4 summarizes the work.

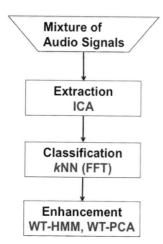

Figure 1. The high-level schematic diagram for AugECE. The methods applied in each stage are displayed in red. Note the classification procedure is conducted in both time and frequency domain, of which the latter aims to address the computation and memory issues from real world application.

2 METHODS

2.1 ICA for signal extraction

Assume there are M source signals and N mixtures of them. The source signal is represented as s, where the random variable S_m represents the mth source. Similarly, X_n represents the nth mixture. Ignoring multipath effect and noise, the output of each signal source is

$$X_n = \sum_{m=1}^{N} A_{nm} S_m,$$

or equivalently, $X = AS$.

AugECE extracts S by estimating A. Some assumptions have been made about signal sources. For example, different people have different voice pattern, from common sense. It is thereby reasonable to assume their voices are independent. ICA provides a statistical technique for revealing independent sources given mixed observations (Roberts & Everson, 2001, Lee et al., 2000).

2.2 kNN and FFT for classification

Identifying an extracted signal for its label is a classification problem. Among the many existing classification algorithms, k-nearest neighbor (kNN) is selected to fit into this framework. Unlike other learning algorithms, kNN (k-nearest neighbor) (Cover & Hart, 1967) does not require a separate training process. It searches all the original sources and finds k instances that have the nearest distance from the

extracted source. These k instances are the k-nearest neighbors of the current extracted signal in this distance-based approach. Using their corresponding values from the response vector Y, kNN applies majority voting to determine the final identity. kNN is simple but can be very accurate in many cases (Wang, 2012).

One major concern rises from real-world application is the volume of the memory required to store all the historical data. Despite numerous approaches to implement kNN, we realized that the problem can be re-formulated in a sequence query problem, i.e., given a query sequence, find the nearest k sequences from the database. This problem has been addressed in several research works such as in (Agrawal, 1993). The main idea stems from the observation that if the sequences are mapped into frequency domain by certain transform, such as discrete Fourier transform, only the first few frequencies are strong and thereby can be used for indexing/ representing the entire sequence in its original time domain, as illustrated in Fig. 2.

By Fourier transform and analysis in frequency domain, not only huge memory can be saved for only storing the first few significant frequencies, but also the computation can be completed in a reduced dimension, with the expectation of a much faster speed. Consequently, we adopted the idea in (Agrawal, 1993) and performed classification of the extracted signals in both time and frequency domains.

2.3 Wavelet-based models for enhancement

Signal in real world is always companied by noise. Hence de-nosing the extracted signal obtained from ICA is expected to enhance the overall quality. Here it is assumed that the extracted signals are added with the Gaussian white noise, which holds true for practice. Some existing methods are based on the spectral analysis, followed by low-pass or band-pass filtering. Although these methods are easy to implement, it fails to filter out the effective band noise. Moreover, they often face the dilemma of choosing bandwidth over high resolution. On the contrary, wavelet transform projects original signal into time and frequency domain simultaneously with multiple resolution. Thus wavelet-based methods become more and more popular in signal de-nosing, including wavelet modulus maxima de-noising, related de-noising and thresholding methods. They all exploit the signal property in wavelet domain. For example, useful signals have relatively small number of large wavelet coefficients, while noise has large number of small wavelet coefficients. Thresholding methods are thereby useful to separate the two. While the choice of threshold seems trivial, considering the locality property of the wavelet-transformation, statistical methods could be applied. Furthermore, since noise is with small amplitude and useful signals with large amplitude, it is natural to consider the observed wavelet coefficients are from mixture distribution. In (Baraniuk, 1998), the wavelet-based hidden Markov model (WT-HMM) is proposed with the assumption that each individual coefficient is modeled by a mixture Gaussian distribution of a hidden state variable.

Despite its elegance in statistical analysis in wavelet domain, one disadvantage of wavelet-based HMM is that only one signal could be de-noised at one time. Different parameters for different signals are trained separately, even if they were generated from the same source. Since signals in wavelet domain have a lot of properties suitable for machine learning approaches, AugECE further employs Wavelet-based principle component analysis (WT-PCA) from paper (Aminghafari, 2006), which analyzed the covariance matrix of noise in wavelet domain.

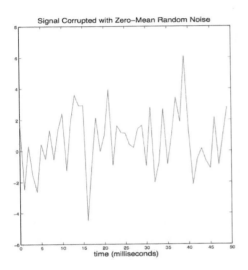 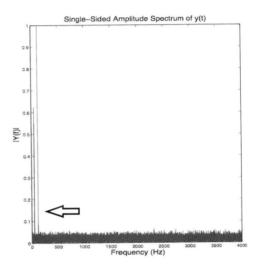

Figure 2. A sequence in time domain (left, with 50 data points) and frequency domain (right) after taking Fourier transform. The characteristic of this sequence in frequency domain can be captured by two low frequencies, as shown by the two spikes (pointed out by the arrow), which suggests a parsimonious yet still precise representation in frequency domain.

3 EXPERIMENTS

We downloaded a number of famous speeches from the Internet as audio signal sources. We also recorded plain speeches from several colleagues. All these signals were then segmented into small periods, which can be randomly mixed to produce the artificial mixtures as the input to the designed framework shown in Fig. 1. AugECE employs ICA for individual signal extraction. The qualities for extracted speech sources are generally acceptable; they are loud and clear even with random noise added. However, occasionally, we do hear noise and additional speeches, usually in very low volume, from other sources. Hence, the ICA algorithm adopted from (Hyvärinen, 1999) (namely, FastICA) may be less effective compared to alternative ICA algorithms, such as the one in (Lee et al., 2000). Nevertheless, through experiments, FastICA is found to be able to converge fast and it is thereby recommended for real-time use.

For classification, the experiments were set into two scenarios: (1) the original signals corresponding to the extracted signals are presented in the database, and (2) the original signals are absent. The classification accuracy for (1) is always 100% as we experimented. However, the accuracy for (2) decreases to approximately 75%. An absolute number for this accuracy is difficult to estimate or obtain, as it depends on the variety of the signals, the number of segmentations, the length of source periods, the actual frequencies of each sources, the amount of noise added, etc.

To address the issue of storing all the signals in memory and for fast identification, besides the above analysis in time domain, AugECE transforms all signals into frequency domain via fast Fourier transform

(FFT). The saving from indexing and representing all the signals in frequency domain is significant for over 99.9%. However, the accuracy decreases for both experiment settings. Moreover, since sources have different frequencies among themselves, it is hard to determine a uniform threshold as a cut-off frequency. Varying the threshold frequency for different sources not only increases the complexity of algorithms, but also produces the signals from the same length to different ones, which makes the problem more complicated.

For the enhancement part, Fig. 3 shows an extracted signal plotted before and after being de-noised. The extracted signal was added with randomly generated noise from a normal distribution and its amplitude is regulated to 10% of the signal on average. Both WT-PCA and WT-HMM were used for de-noising, whose results are similar. The shape and trends of the two signals (before and after enhancement) are highly similar, except that the de-noised one is more compact and thin, as a result of removing the added noise. The SNR of the extracted signal to the original source signal increased after de-noising. However, we found the auditory effect does not improve much. It is occasionally not as smooth as even before the enhancement. Therefore, it is possible and very likely that some portion of the signal is also filtered out along with the added noise.

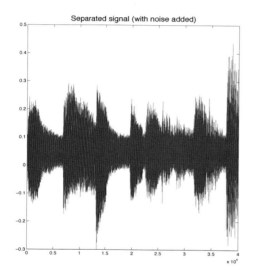 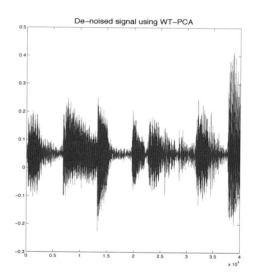

Figure 3. Plots of an extracted audio signal prior to enhancement (left) and after enhancement (right). Note in this paper *extracted signals* and *separated signals* are used interchangeably, since *separated* also means *apart from the audio mixture*.

Contrary to our intuition, the SNR does not improve monotonically with more the increasing of the HMM length, as shown in the last column of Table 1.

Given that WT-HMM has similar results with WT-PCA for this data set, we further compare the two enhancement methods. This time the standardized data set in Matlab called is used. *ex4mwden* contains two 1024*4 matrices, with every column being a signal. Both WT-HMM and WT-PCA are evaluated on *ex4mwden*, with the resutls presented in Figs. 4. and 5, respectively.

It is difficult to tell if the auditory effect of the signals de-noised by WT-HMM is better than WT-PCA, though both of them did improve the SNR. Thus we resort to the quantitive evaluation results. Three out of the four cases (except signal 3) demonstrate that WT-HMM slightly outperforms WT-PCA by log-based SNRs, though the difference is not significant.

Consequently, considering the fact that WT-PCA is able to run multiple signals simultaneously, AugECE will first invoke WT-PCA for enhancement; in case the results are not acceptable, fine-tuning with WT-HMM is thereafter applied.

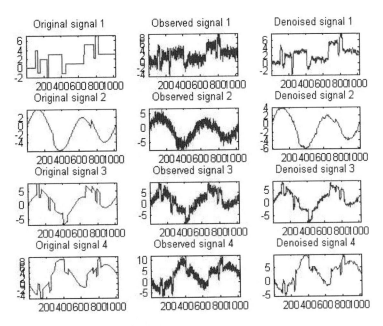

Figure 4.   Results for running WT-HMM on *ex4mwden*.

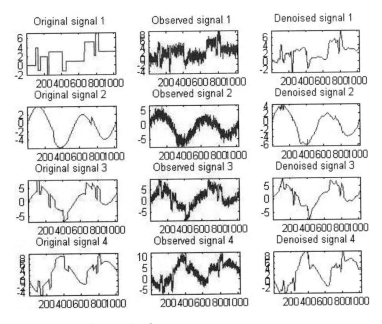

Figure 5.   Results for running WT-PCA on *ex4mwden*.

Table 1. This table shows the results of using WT-HMM for de-noising the same signal, by taking a different length (4096, 8192, 16384, 32768) at each time. The runtime of WT-HMM is much longer than WT-PCA. Unlike WT-PCA, WT-HMM is unable to perform analysis for multiple signals simultaneously. Note in this paper *extracted signals* and *separated signals* are used interchangeably.

Length	Before enhancement	After enhancement	SNR
4096			-0.013
8192			2.7001
16384			0.2035
32768			7.8761

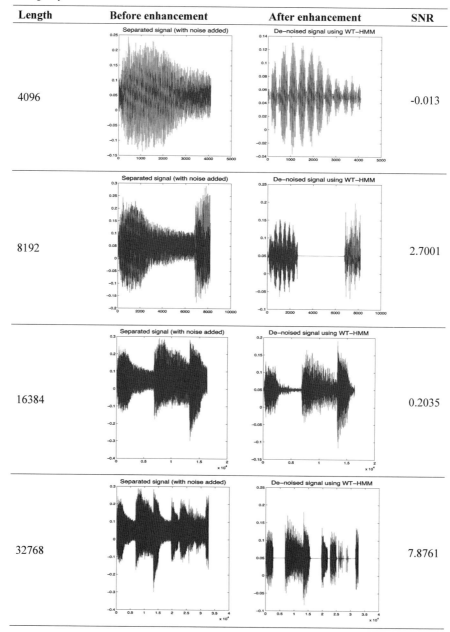

4 CONCLUSIONS

We implemented a framework called AugECE that employs various methods for the purpose of audio signal extraction, classification and enhancement. The framework is validated through analysis and experiments. The results prove that AugECE is efficient and effective for audio signal extraction and subsequent analysis and enhancement.

REFERENCES

Gunawan, D. & Sen, D. Music source separation synthesis using multiple input spectrogram inversion. *In IEEE International Workshop on Multimedia Signal Processing (MMSP 09)*, 2009, pp.1–5.

Chafe, C. & Jaffe, D. Source separation and note identification in polyphonic music. *In IEEE International Conference on Acoustics, Speech, and Signal Processing (ICASSP 86)*, 1986, pp. 1289–1292.

Mehra, A., Kumawat, M., Ranjan, R., Pandey, B., Ranjan, S., Shukla, A. & Tiwari, R. Expert system for speaker identification using lip features with PCA. *In IEEE 2nd International Workshop on Intelligent Systems and Applications (ISA)*, 2010, pp. 1–4.

Roberts, S. & Everson, R. 2001 *Independent Component Analysis: Principles and Practice.* Cambridge University Press.

Lee, T-W., Lewicki, M. S. & Sejnowski, T. J. ICA mixture models for unsupervised classification of non-Gaussian sources and automatic context switching in blind signal separation. *IEEE Transactions on Pattern Analysis and Machine Intelligence*, 2000, 22(10): 1078–1089.

Lee, T. W., Girolami, M., Bell, A. J. & Sejnowski, T. J. A unifying information-theoretic framework for independent component analysis. *Computers & Mathematics with Applications*, 2000. 39(11): 1–21.

Hyvärinen, A. 1999. Fast and robust fixed-point algorithms for independent component analysis. *IEEE Trans. on Neural Networks.*

Cover, T. & Hart, P. Nearest neighbor pattern classification. *IEEE Transactions on Information Theory*, 1967, IT-11: 21–27.

Wang, H. et al. Test-data volume optimization for diagnosis. *In 49th ACM/EDAC/IEEE Design Automation Conference (DAC)*, 2012.

Agrawal, R., Faloutsos, C., & Swami, A. N. Efficient similarity search in sequence databases. *In Proceedings of the 4th International Conference on Foundations of Data Organization and Algorithms*, 1993, pp. 69–84.

Baraniuk, R. G., Crouse, M. S., & Nowak, R. D. 1998. Wavelet-based statistical signal processing using hidden Markov models. *IEEE Transactions on Signal Processing*, 1998, 46(4): 886–902.

Aminghafari, M., Cheze, N. & Poggi, J.-M. 2006. Multivariate de-noising using wavelets and principal component analysis. *Computational Statistics & Data Analysis*, 2006, 50:2381–2398.

Information Technology – Wan et al. (Eds)
© 2015 Taylor & Francis Group, London, ISBN 978-1-138-02785-5

Automatic facial image standardization based on active appearance model

Qiaoliang Li, Zhengxi Cheng, Suwen Qi, Huisheng Zhang, Xinyu Liu, Yongchun Deng,
Menglu Yi, Qing Yuan, Tianfu Wang & Siping Chen
Department of Biomedical Engineering, School of Medicine, Shenzhen University, china

ABSTRACT: Image standardization can benefit the facial image in many applications, such as face recognition, virtual reality, pose rectification and so on. However, as far as we know, there is no automatic facial image standardization method proposed until now. In this paper, we have developed an automatic facial image standardization algorithm based on Active Appearance Model (AAM). The right and left eyes' coordinates are accurately extracted by AAM, and used to compute the displacement, by which an affine transformation model is estimated to translate the original image into a standard image. The face in the standard image is located at a typical coordinate with standard size. Experimental results demonstrate that the proposed method could provide an accurate and reliable way for facial image standardization.

1 INTRODUCTION

In the face recognition and the virtual reality application field, the quality of a face image was usually affected by the size and the location of the face. A standard image could be processed more effectively. Automatic facial image standardization is a process that detects the location of the face and transforms the original images into the object standard images automatically by using affine transformation [5].

In 1998, T.Cootes and C. Taylor [1] first put forward the Active Appearance Model algorithm and then had an extensive survey on this topic. The Active Appearance Model (AAM) has been one of the most popular models for face image detection [8]. It has been extensively applied in many computer vision tasks, such as facial image processing, medical image analysis [6], object appearance modeling [7], etc. Although the AAM has been proposed for several years, there is little open-source implementation on it. P. JIA [3] opened his algorithm in the Vision Open Statistical Models (VOSM). In this paper, we make the original image standardization through affine transformation which includes translation, scale and rotation transformation.

There are mainly three parts of this algorithm. First, we take advantage of the AAM algorithm to obtain the information of feature points. Second, these feature points of the eyes will be used to build an affine model. Third, the image was normalized automatically according to the affine model that we have built. The experimental results indicate that our proposed method could provide an accurate and reliable way for facial image standardization.

2 METHOD

The flow diagram of our proposed method is organized as follows:

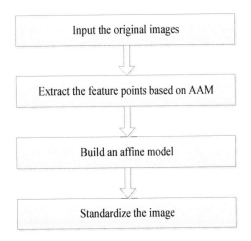

Figure 1. Flow diagram of our proposed method.

2.1 Extract the feature points based on AAM

The Active Appearance Model (AAM) is a local search method which combines the full shape model and texture variation learned from a training set [1]. It tries to match both shape and holistic texture that represented the object simultaneously. The objective of basic AAM is to minimize the mean square error

(MSE) of the difference between the real image I_r and the image built from the model parameter I_m:

$$I(c) = I_r - I_m, \qquad (1.1)$$

$$E(c) = I(c)^T I(c), \qquad (1.2)$$

The above problem is a classical optimization problem; we can use numeric differentiation method to simply cope with it in an iterative way. Our aim is to find the most suitable δc; Equation (1.3) goes to minimum for each iteration.

$$E(c + \delta c) = I(c + \delta c)^T I(c + \delta c), \qquad (1.3)$$

With each iteration, c is updated to $c + \delta c$. The iterations will proceed continuously until $E(c + \delta c)$ and $E(c)$ are of negligible difference. So, from the Equation (1.3), we can see this problem is changed to calculate the renewed value δc in terms of the current c in each iteration. First, 1-order Taylor expansion is carried out, and then the partial derivative on (1.3) is also unfolded in order to compute δc.

$$
\begin{aligned}
\frac{\partial E(c+\delta c)}{\partial \delta c} &\\
= \frac{\partial I(c+\delta c)^T I(c+\delta c)}{\partial \delta c} &, \qquad (1.4)\\
= 0 + 2I(c)^T \frac{\partial I(c)}{\partial c} + 2\delta c^T (\frac{\partial I(c)}{\partial c})^T \frac{\partial I(c)}{\partial c}
\end{aligned}
$$

Then let the above equation (1.4) equal to zero gives:

$$\delta c = -HI(c), \qquad (1.5)$$

$$H = ((\frac{\partial I(c)}{\partial c})^T \frac{\partial I(c)}{\partial c})^{-1} (\frac{\partial I(c)}{\partial c})^T, \qquad (1.6)$$

How to choose the pre-computed H in (1.6) is necessary for AAM fitting algorithm. Displace each parameter from the known optimal value on training images, $\frac{\partial I(c)}{\partial c}$ could be estimated by numeric differentiation, so H also could be computed as well.

In order to operate the VOSM, it is very necessary to configure the operating environment. The OpenCV and the BOOST libraries must be set for Visual Studio 2010 correctly. The AAM tries to match both shape and holistic texture that represented the object simultaneously. In this method, we mainly use the shape information and mark the feature points. Hence,

we modify the procedure by neglecting the texture information and mark several key points. Four points was marked for the eyes, one point for the nose, and two for the mouth.

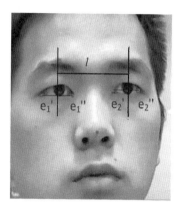

Figure 2. The marked feature points.

2.2 Build an affine model

We have obtained the two eyes' coordinates. Each eye was labeled with two points, where, $e_1'(x', y')$, $e_1''(x_1'', y_1'')$ for the left eye; and $e_2'(x_2', y_2')$, $e_2''(x_2'', y_2'')$, for the right eye. Thus, we can calculate the center point which regarded as the location of the eyes. Assuming the two eyes' center point coordinate is $e_1(x_1, y_1)$, $e_2(x_2, y_2)$, so the $e_1(x_1, y_1) = ((x_1'+x_1'')/2, (y_1'+y_1'')/2)$, $e_2(x_2, y_2) = ((x_2'+x_2'')/2, (y_2'+y_2'')/2)$. The distance ($l$) of two eyes is

$$l = \sqrt{(x_1 - x_2)^2 + (y_1 - y_2)^2}, \qquad (2.1)$$

where l stands for the quantity of pixels in an image. A standard eye width (std_len) was set manually at the beginning of our procedure. Then, the scale ratio (s) could be expressed as

$$s = \frac{std\_len}{l}, \qquad (2.2)$$

The deflection angle θ is the angle of two eyes on horizon could be constructed as

$$\theta = 360° + \frac{\arctan \frac{(y_2 - y_1)}{(x_2 - x_1)} * 180°}{\pi}, \qquad (2.3)$$

The coordinate of the center point which used to make rotation transformation is $c((x_1 + x_2)/2, (y_1 + y_2)/2)$.

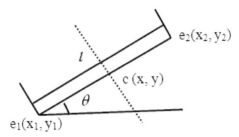

Figure 3. Geometry used to build an affine model from the coordinates of the key points.

2.3 Standardize the image

Affine Transformation is a fundamental transformation in the affine geometry. In the Two-dimensional Euclidean space of the affine transformation could be expressed:

$$\begin{bmatrix} x' \\ y' \end{bmatrix} = \begin{bmatrix} a_1 & a_2 \\ a_4 & a_5 \end{bmatrix} \begin{bmatrix} x \\ y \end{bmatrix} + \begin{bmatrix} a_0 \\ a_3 \end{bmatrix}, \quad (3.1)$$

where (x, y), (x', y') and $(a_0, a_3)^T$ are two points in the x-y plane, and is a translation vector respectively. $\begin{bmatrix} a_1 & a_2 \\ a_4 & a_5 \end{bmatrix}$ is a matrix that composed of translation, rotation, and scale transformation. Parameter $a_i (i = 0, 1, 2, 3, 4, 5)$ is the real number [4].

We have obtained deflection angle (θ), scale ratio (s), center point (c) according to the equation (2.1) ~ (2.3), then use the built model to make the image affine transformation automatically.

3 RESULTS AND ANALYSIS

We choose four images in which the face has some excursion as the original images, as shown in the Fig 4. The size of the original image is random. Fig 5 shows the process of the affine transformation. Fig 6 shows the results of the standardization. The object image has an identical size that the width is 301 pixels and the height is 367 pixels (We adjust the size 3.16*2.0 centimeters for display conveniently here). The eyes are located in the identical coordinate each image.

We record the scale ratio and the deflection angle in table.1. It shows the scale ratio and the deflection angle of object image compared with original image (The value of the deflection angle smaller than 360° means the orientation of the rotation is clockwise; otherwise is anticlockwise).

As far as I know, there are no related papers on standardizing facial image automatically. Some software on image processing such as Photoshop [12] can easily make the translation, rotation, scale

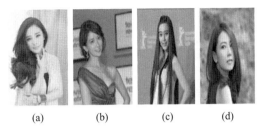

(a)　　　　(b)　　　　(c)　　　　(d)

Figure 4. The original images (a), (b), (c) and (d).

Figure 5. The process of the affine transformation is composed by the four images. The original image (a) was used to show this process which includes translation, rotation, scale and the three all transformation respectively. The forth is the last result of the standardization.

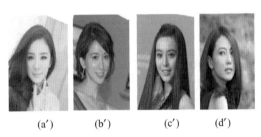

(a′)　　　　(b′)　　　　(c′)　　　　(d′)

Figure 6. The result of the standardization. The four images have the same size and the eyes are located in the identical coordinate.

Table 1. Scale ratio and deflection angle.

Original Image	Original Size (pixels)	Scale Ratio	Deflection Angle (°)
(a)	299*396	1.476	342.031
(b)	700*1050	0.692	339.955
(c)	853*1280	0.737	353.660
(d)	1500*2258	0.267	353.875

transformation, but all these are operated manually. Our proposed method is the first to solve this problem. For an image, we detect the location of the face and make the affine transformation so that every different image could have uniform standards [13]. It can improve efficiency in image processing field, especially batch processing.

4 CONCLUSION

In this paper, the location of the original facial image could be detected steadily and accurately based on AAM algorithm. We build an affine model of the facial image using the location information. These original images are standardized automatically based on this model. The experimental results show that our proposed method works reliably. In the following work, we would concentrate on building a more accurate affine model which will be applied to multiple faces in one image [9] [10] [11].

ACKNOWLEDGMENT

This work was supported by the Project of the National Science Foundation of China under Grants 81000637, 81271651, 61031003, and 61401285 and Guangdong Innovation Training Platform Outstanding Youth of China (2012LYM_0114).

REFERENCES

T. Cootes and C. Taylor. Active Appearance Models. Pattern Analysis and Machine Intelligence, 2001.6, pp.681~686.

T. Cootes and C. Taylor. Statistical models of appearance for computer vision. Technical report, Imaging Science and Biomedical Engineering, University of Manchester, March 8, 2004.

P.JIA, 2D Statistical Models, Technical Report of Vision Open Working Group, 2st Edition, October 21, 2010.

James D. Hamilton, Jing Cynthia Wu, Identification and estimation of Gaussian affine term structure models, Journal of Econometrics 168 (2012) 315–331.

Feeman, T. G., & Marrero, O. Affine transformations, polynomials and proportionality. The American Mathematical Monthly, 108(10), (2001) 972–975.

R. Beichel, H. Bischof, F. Leberl, M. Sonka. Robust active appearance models and their application to medical image analysis, IEEE Transactions on Medical Imaging 24 (9) (2005) 1151–1169.

E. Jones, S. Soatto. Layered active appearance models, in: Proc. 10th Int. Conf. on Computer Vision, vol. 2, Beijing, China, 2005, pp. 1097–1102.

T. Cootes, D. Cooper, C. Tylor, J. Graham. A trainable method of parametric shape description, in: Proc. 2nd British Machine Vision Conference, Glasgow, UK, September 1991, pp. 54–61.

Xiaoming Liu. Video-based face model fitting using Adaptive Active Appearance Model, Image and Vision Computing Volume 28, Issue 7, July 2010, Pages 1162–1172.

A. M. Bronstein, M. M. Bronstein, and R. Kimmel. Three-dimensional face recognition, Int. J. Compute. Vis., vol. 64, no. 1, Aug. 2005 pp. 5–30.

S. Berreti, A. Del Bimbo, P. Pala, and F. J. S. Mata. Geodesic distances for 3D-3D and 2D-3D face recognition, in Proc. IEEE Int. Conf. Multimed. Expo, 2007, pp. 1515–1518.

Evening, Martin. Adobe Photoshop CS3 for photographers: a professional image editor's guide to the creative use of Photoshop for the Macintosh and PC. Taylor & Francis, 2007.

Marissa N. Lassere. Imaging: the need for standardization, Best Practice & Research Clinical Rheumatology, Volume 22, Issue 6, December 2008, Pages 1001–1018.

Information Technology – Wan et al. (Eds)
© 2015 Taylor & Francis Group, London, ISBN 978-1-138-02785-5

Change detection method design and analysis based on Geographical National Conditions Monitoring

Jianping Pan & Chunhua Yang
Applications Centre, Chongqing Institute of Surveying and Mapping, NASMG, Chongqing, China

Li Li
Chongqing Institute of Land Surveying and Planning, Chongqing, China

ABSTRACT: Based on the development needs of the key project Geographical National Conditions Monitoring project in China, we analyzed lots of change detection algorithm prototypes in domestic and foreign, and then probed vector method based on pixels' gray value and hierarchical cross correlation analysis method. The vector method based on pixels' gray value was designed based on the change vector analysis method which can extract much more information, hierarchical cross correlation analysis method was designed based on the Expectation Maximization detection method which has higher precision. The test result shows that, the accuracy of, change information is not affected by the accuracy of classification using the 2 designed methods, and hierarchical cross correlation analysis method costs less time, demands less memory. At last, we analyzed the constraint conditions from image preprocessing for these 2 methods.

KEYWORDS: Hierarchical cross correlation analysis method; vector method based on pixels' gray value; Geographical National Conditions Monitoring.

1 INTRODUCTION

China is vast in territory, complicated in geographical conditions, with the rapid development of the economy, the land cover changes frequently, we need to try to insure that the basic geographical information is always the latest, so that the needs of government, enterprise and social sectors can be satisfied.

First Survey of National Geographical Conditions is ongoing in China, it will lay the database foundation for carrying out normalized geographical condition monitoring in the future. Geographical conditions monitoring can not only provide geographic space-time database, but also the development tendency and evolvement rules of geographic features are included. In this case, change detection turns out to be the key task of Geographical conditions monitoring. So comprehensive utilization of various technologies in data acquisition and processing is needed to find out, identify, extract and update the geographic features.

2 ANALYSIS ON EXISTING CHANGE DETECTION METHODS

Currently, many countries have launched special work on dynamic monitoring of the geographical environment, and implemented a number of significant action plans. They provide us technical experiences which can be used as reference. Many scholars around the world have studied a lot of change detection methods and theoretical models from different angles, scholars generally believe that change detection is a complex and comprehensive process, the most optimal methods cannot be found among the existing detection method in actual production applications, no single method can fit all applications. Due to the complexity and difficulty of the change detection problem, automation is at low stage, no groundbreaking research has been found throughout the world.

In the existing research, change detection methods can be divided into 4 categories as a vector to vector data, image to vector data, image to image data and multi-source data according to the study object.

Funding Projects: 1. Funded by Key Laboratory of Geo-informatics of State Bureau of Surveying and Mapping,2012: Theory improving and production application research on change detection method of change vector and cross analysis; 2.scientific research and technical innovation of Chongqing Institute of Surveying and Mapping, NASMG, 2012: Research on automatically extract change vector of remote sensing image with different resolution; 3.science and technology plan projects of Chongqing land and resources bureau in 2010: oriented land use classification system of high resolution remote sensing image by change detection technology research.

In geographical condition monitoring, land cover information is extracted from images, so this paper will analyze change detection methods on image to image data.

Change detection methods of image to image data can be divided into 3 levels, according to its focus and research objectives, they are change judgment, determination of change type and determination of change tracking. The level of change judgment has the most abundant research resources, and it's also the basic requirement in geographical condition monitoring, it can guide the operating of change information interpretation according to the judgment, thus improve the production efficiency.

Through in-depth analysis of existing change detection methods, this paper selected those methods which have more possibility to be used for reference based on the needs of change detection technology of land cover classification in geographic condition monitoring (as shown in Table 1).

Table 1. Analysis table of change detection methods.

Method Name	Application features	Advantage	Shortcoming
Difference value method	Calculate difference value by pixel	Straightforward and easy to implement	High threshold selection requirements
Change vector analysis	The overall change amplitude of each pixel changes is calculated by Euclidean distance which decide the endpoint of n-dimensional space	Can handle multi-bands, and can obtain detailed information on change detection	Difficult to determine the trajectory of land cover change
Principal Component Analysis	Assuming multi-temporal data is highly relevant and change information will highlight in new components	Reduce data redundancy among the bands, and emphasize the difference information component	Difficult to interpret and label change results, and the changed regions are decided by threshold
EM (Expectation Maximization) detection method	Using the expectation-maximization algorithm to estimate the prior joint probability of the two phase images based on classification method	Its accuracy is higher than other methods	Asked to estimate the probability of a prior joint class
Mixed change detection	Supervised classification and create a binary mask, the mask select change informations from land cover image	Category contain no change cells, reducing the classification error	Classification threshold selection is required, difficult to identify the trajectory
Visual interpretation	Put the same band of difference into multi-temporal to different color channels, and identify change area by the color combination, and then do interactive heads-up digitizing based on overlap data	Prior knowledge is useful, two or three data can be analyzed simultaneously, and texture, shape, size and pattern all can be considered	Could not provide detail change information, the results depend on the analyst's interpreting skills, time-consuming and difficult to update results

3 CHANGE DETECTION METHOD DESIGN

According to Table 1, this paper designed two algorithms based on the needs of land cover classification in geographic condition monitoring, as follows:

3.1 *Vector method based on pixels' gray value*

This method was extended from the algorithm prototype of the change vector analysis method. An algorithm designed to determine the upper and lower range of the optimal segmentation threshold by the histogram curvature of change intensity map, and then to do traversal calculations within the scope of the upper and lower threshold value use the improved Otsu method, then to get the optimal segmentation threshold when the interclass variance reaches the maximum value, at last, to divide the change

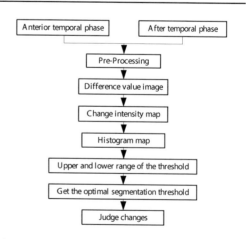

Figure 1. Algorithm design sketch for vector method based on pixels' gray value.

intensity map by the optimal threshold to judge changes. Design ideas indicated in Figure 1:

3.2 *Hierarchical cross correlation analysis method*

This method was designed based on the algorithm prototype of the EM detection method, in consideration of the advantages of the cross-correlation analysis method, object-oriented classification, unsupervised change detection method and mixed change detection method. This method does cluster analysis to the image data of anterior temporal phase first of all, so that all operations latter use the clustering figure spot as detection unit. The second step is to do an overlay analysis with anterior and after temporal phase image data based on same figure spots, including calculating the maximum statistical gray value and the average gray value of each figure spot in each band for the 2 images, and calculate the probability distribution of mean deviation. The third step is to set thresholds in different ways at different levels based on the results of step 2, and then extract change figure spots at different levels. Finally overlay all change figure spots to get change results. Design ideas indicated in Figure 2:

4 EXPERIMENTS AND RESULT ANALYSIS

To realize these 2 new methods, we did program development in the IDL language, and took 2 scenes of Quick Bird image around the university city of Chongqing to verify the effectiveness of the designed methods.

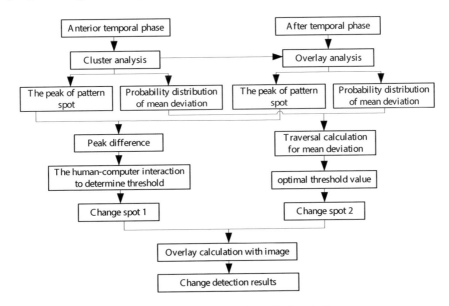

Figure 2. Algorithm design sketch for hierarchical cross correlation analysis method.

a. Anterior temporal phase

b. After temporal phase

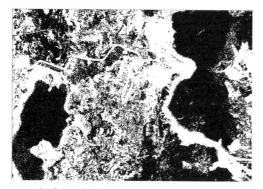

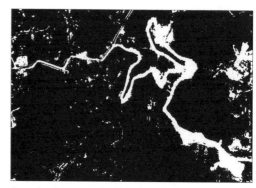

c. result of post classification comparison method

d. result of vector method based on pixels' gray value

Figure 3.   Comparison of post classification method and vector method based on pixels' gray value.

Firstly, we did contrast analysis with the post classification, comparison method and vector method based on pixels' gray value, the post classification, comparison method realized by auto-classification, and vector method based on pixels' gray value realized by our program. The first contrast analysis results are shown in Figures 3:

From figure 3 we can see that:

1 Result in d has significantly changed area but c is not, misclassification in c is much more.
2 The accuracy of the vector method based on pixels' gray value is not affected by classification, but the accuracy of the post classification, comparison method depends on classification accuracy, especially vegetation and road.

a. Anterior temporal phase

b. After temporal phase

c. result of vector method based on pixels' gray value

d. result of hierarchical cross correlation analysis method

Figure 4.   Comparison of vector method based on pixels' gray value and hierarchical cross correlation analysis method.

Table 2. Confusion matrix and accuracy result of vector method based on pixels' gray value.

Confusion matrix			Accuracy		
Realistic class Classification result	Did not change	change	Miss rate (%)	false detection rate (%)	Overall accuracy (%)
Did not change	625	90	3.48	6.97	89.55
change	45	532			

Table 3. Confusion Matrix and accuracy result of hierarchical cross correlation analysis method.

Confusion matrix			Accuracy		
Realistic class Classification result	Did not change	change	Miss rate (%)	false detection rate (%)	Overall accuracy (%)
Did not change	552	59	8.52	4.91	87.84
change	87	503			

So in the second contrast analysis, we selected vector method based on pixels' gray value which got better result to do a comparison with hierarchical cross correlation analysis method.

From the comparison result, hierarchical cross correlation analysis method can be more efficiently by removing vision disparity which may caused by radiation from objects, so the extracted information would be more pure. In addition, our experiment shows that, hierarchical cross correlation analysis method does better than vector method based on pixels' gray value in the running time, the demand for memory and other aspects, so we speculate that it has better applicability of large data.

5 CONSTRAINT CONDITIONS FROM IMAGE PREPROCESSING

This paper designed 2 change detection methods based ideal state, it is difficult to avoid inapplicability in the practical application of information extraction. National Administration of Surveying, Mapping and Geo-information is carrying out the First Survey of National Geographical Conditions, in order to improve the applicability of these 2 methods to satisfy the future demands in normalized geographical condition monitoring, we must pay attention to these constraints during preliminary work:

1 *Image option:* to avoid wrong information caused by terrain, try to choose images with smaller inclination if possible; try to keep away from image covered by cloud, ice and snow to avoid unnecessary non-change information.
2 *Preprocessing:* enhance effective information for images which have weak information; try to improve the registration accuracy to pixel level; do resemble to the two temporal phase image data in order that they have the same resolution.

REFERENCES

[1] YuShuang Bai, HongWan Chen. The necessity to carry out geographical condition monitoring in Surveying and Mapping departments. http://fazhan.sbsm.gov.cn/article/rdjj/dlgqjc/201111/20111100093730.shtml. March 12, 2011.
[2] DeRen Li, HaiGang Sui, Jie Shan. Probe into the technical support of geographical conditions monitoring. Geomatics and Information Science of Wuhan University. 2012, 5: 505–512.
[3] Qiming Zhou. Summary of change detection for multi-temporal remote sensing image. World international photogrammetry and remote sensing dynamic themes on geographic information, 2011, 4(2):28–33.
[4] Volker Walter. Object-based classification of remote sensing data for change detection. ISPRS Journal of Photogrammetry & Remote Sensing, 2004, vol 58: 225–238.
[5] D.Lu, P.,ausel, E.Brondízio, etc. Change Detection Techniques. International Journal of Remote Sensing, 2004, vol 25(12):2365–2407.
[6] Richard J. Radke, Srinivas Andra, Omar Al-Kofahi, etc. Image Change Detection Algorithms: A Systematic Survey. Image Processing, 2005, vol 14(3):294–307.
[7] Wei Hu, Jie Ma, HaiTao Li, etc. Design and implementation of multi-temporal image change detection system. Computer and Digital Engineering, 2011, 3: 51–55.
[8] GuoHua Xu, BaoMing Zhang, Xu Li. Change Detection based on improved Otsu method Surveying and Mapping, 2012.
[9] Turagy Celik. Change Detection in Satellite Images Using a Genetic Algorithm Approach. IEEE Geoscience and Remote Sensing Letters, 2010, 7(2):386–390
[10] Ya-Qiu Jin,Dafang Wang. Automatic Detection of Terrain Surface Changes After Wenchuan Earthquake, May 2008, From ALOS SAR Images Using 2EM-MRF Method. IEEE Geoscience and Remote Sensing Letters, 2009, 6(2):344–348.
[11] Sicong Liu, Peijun Du, Shaojie Chen. Change detection for multi-resolution remote sensing images using decision-making level fusion. Journal of Remote Sensing, 2011 Geomatics and Information Science of Wuhan University, 2011.

Information Technology – Wan et al. (Eds)
© *2015 Taylor & Francis Group, London, ISBN 978-1-138-02785-5*

Chaotic attractor of the image

Wanbo Yu
School of Information, Dalian University, Dalian, Liaoning, China.

Shuo Yu
Management Department, Dalian University of Finance and Economics, Dalian, Liaoning, China.

ABSTRACT: The chaotic attractor is discovered as a new feature for image recognition in this paper. The chaotic characteristic of a dynamic system is composed of a wavelet function and an image function. Analyze the chaotic characteristic by calculating the Lyapunov exponent and drawing the bifurcation diagram. Then it is found that the number of attractor points can be changed by adjusting the parameter of wavelet function, but the shape of the attractor barely changed when the parameter changed slightly. The experimental result shows that using the chaotic attractor to distinguish different human faces is measurable.

1   INTRODUCTION

It is necessary and meaningful to study the chaotic characteristics of the image function. The applications of chaotic research in image recognition are discussed in literature [1–3]. Up until now, there are no such applicative examples that are taking an image as function to construct dynamic system. In curve iteration, when one curve is adjusted to a standard one in unit area, the other is a random generated one, the probability of chaos is above 10% [4]. We call the standard curve in the unit area as auxiliary surface (auxiliary function). In literature[5], it turns out that when sine functions or wavelet functions are taken as auxiliary surface, the other surface is generated randomly, chaos can be brought more easily. Based on these works, this paper researches chaotic characteristics of the image. Section 2 shows that dynamic system of Gaussian function and image can bring on chaos. In section 3, construct three dynamic systems by Gaussian function and three different groups of human face images, it is found that the shape of attractors are close to each other when human faces are similar.

Attractor in this paper is approximate attractor without specification.

2   DYNAMIC SYSTEM OF CURVE AND IMAGE

2.1   *Image function and sine function*

Intercept Figure 1a from Lena. Cut off the parts that can hardly bring on chaos, keep line 110 to 195, row 110 to 175. Then use function Imresize in Matlab to resize it to 256*256. Figure 1b is the three dimensional graph displaying of Figure 1a.

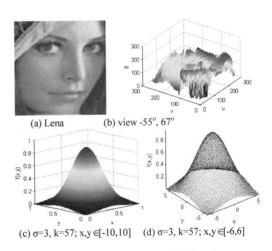

(a) Lena   (b) view -55°, 67°

(c) σ=3, k=57; x,y ∈[-10,10]   (d) σ=3, k=57; x,y ∈[-6,6]

Figure 1.   Image Function and Gaussian Function.

Two dimensional Gaussian function is (1).

$$f(x,y) = \frac{1}{2\pi\sigma^2} e^{-\frac{1}{2\sigma^2}(x^2+y^2)} \tag{1}$$

For better application, adjust height of (1) and then get (2).

$$f(x, y) = \frac{k}{2\pi\sigma^2} e^{-\frac{1}{2\sigma^2}(x^2+y^2)} \qquad (2)$$

Let σ=3,k=57, x,y∈ [-10,10], graph of (2) is Figure 1c. Adjust x,y to [-6,6], graph of (2) is Figure 1d.

The dynamic system (3) is consisted of Gaussian function $f(x,y)$ and image function $g(x,y)$. Since the range of Gaussian function is [0,1], let domain and range of image function be [1,256] for convenience.

$$\begin{cases} f(x, y) = \dfrac{k}{2\pi\sigma^2} e^{-\frac{1}{2\sigma^2}(x^2+y^2)} \\ g(x, y) \end{cases} \qquad (3)$$

2.2 Iterative method and image attractor

First, translate image to gray image, then adjust size to 256*256.

The iterative method is following: Give the initial value of x and y, then get the value of $f(x,y)$ in (3). Enlarge the value 256 times, let u be the rounding value just got; Round x and y after they are enlarged 256 times, and then to calculate the gray value of image $g(x,y)$ (use function Interp2 in Matlab). Let v be the rounding value of the gray value, then get the new number pairs (u,v). Let x=(u/256-0.5)*12, y=(v/256-0.5)*12, get the new (x,y). Let (x,y) be into $f(x,y)$, (u,v) be into $g(x,y)$ without reducing Repeat.

In fact, the operation above is the basic operation of dynamic systems. For convenience, image function is not unitized with wavelet function. It would be simpler to implement and describe if the image function are translated to one that range and domain both is [0,1].

Figure 2a-c are attractors drawn by this iterative method. Initial value x=0.2539, y=0.3516; k is 57, σ in Figure 2a is 3, in Figure 2b is 3.1623.

It can be seen that chaos appears from the graphs. To confirm chaos, 2.3 plot the bifurcation diagram and Lyapunov exponential diagram.

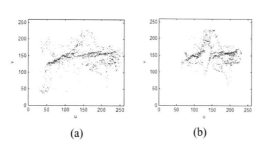

| (a) | (b) |

Figure 2. Graphs of the image function and Gaussian function attractors.

2.3 Bifurcation diagram and Lyapunov exponential diagram

Let σ=3, make k change from 35 to 55. The bifurcation diagrams of u and v are Figure 3a and Figure 3b. Make k change from 35 to 55, σ change from $\sqrt{8}$ to $\sqrt{10}$. The three dimensional bifurcation diagrams of u and v are Figure 3c and Figure 3d. Figure 4a is the Lyapunov exponential diagram when k changes from 35 to 55 and σ=$\sqrt{8}$. σ=$\sqrt{10}$ in Figure 4b.

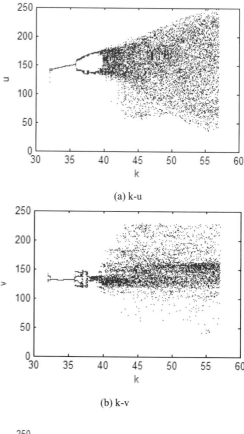

(a) k-u

(b) k-v

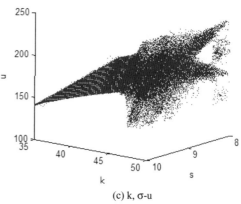

(c) k, σ-u

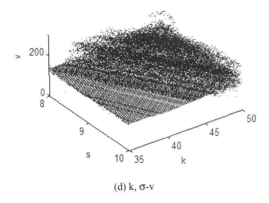

(d) k, σ-v

Figure 3. The bifurcation diagram of image function and sine function.

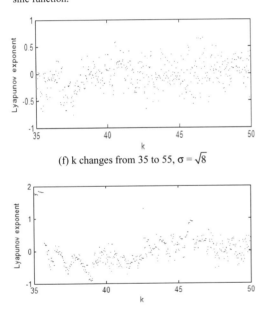

(f) k changes from 35 to 55, $\sigma = \sqrt{8}$

(b) k changes from 35 to 55, $\sigma = \sqrt{10}$.

Figure 4. Lyapunov exponential diagram.

3 IMAGE CHAOS ATTRACTOR

3.1 *Image chaos attractor of human face*

Process 3 groups of human faces of yalefaces(No.5, 6,7), each group has 11 images. The image and attractor are shown in Figure 5.

The concrete procedure is following: Cut off the edge of each image, keep line 65 to 200, row 50 to 150. Use function Imresize in Matlab to resize the image to 256*256. Turn image function to a continuous function by Iterp2 in Matlab. Initial value x=0.4, y=0.9, iterate 300 times of each image. Draw the

sequence points (approximate attractor). In Gaussian function, k=57, σ=3.

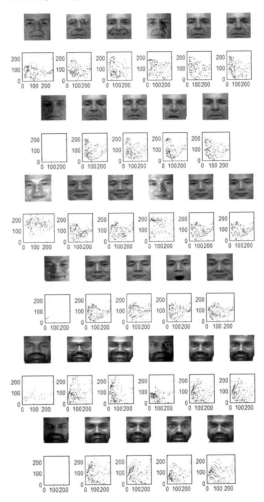

Figure 5. 3 groups of human face images and image attractors.

Calculate the Lyapunov exponent of 33 images in Figure 5:

0.1780 0.4977 0.3321 0.2071 0.2778 0.5119
-0.8116 0.2000 0.3437 0.2335 0.2289
0.4089 0.3288 0.2294 0.5175 0.2171 0.2171
0.1537 0.1883 0.1461 0.1939 0.1040
0.1828 0.1575 0.3232 0.0623 0.2416 0.2416
-0.7115 0.1468 0.3597 0.4032 0.1544

Since eigenvalue can be 0 sometimes, an infinite negative can be get in the calculation of logarithm. This happens is not because of the low chaotic degree, but the calculation method and the speciality of the image. The method of Lyapunov exponent calculation needs to be improved.

3.2 Attractors of different images

Research shows that attractors of different images differ greatly. In Figure 6 for example, the 11 images are different from each other, so are the attractors.

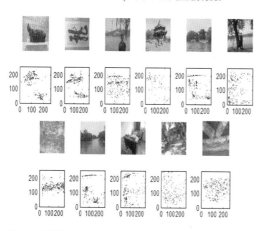

Figure 6. Different images and their attractors.

Calculate the Lyapunov exponent of 11 images in Figure 6:

0.6078 0.3167 1.2237 0.5086 0.1527 0.9825
1.4737 -0.2768 0.4818 2.3152 1.2919

In fact, Lyapunov exponent is not the best way to describe chaos, especially image chaos. A new judging method or description of chaos should be given in further work.

3.3 Effects on attractors when curves or images change

The attractor can be changed by changing the image. For example, use Matlab function Imadjust to adjust gray value of 3 images in Figure 7(the 7th image of each group in Figure 5): imadjust(G,[0 0.6],[0.1 1]); imadjust(G,[0 0.8],[0 1]); imadjust(G,[0 0.6],[0.1 1]); The function of imadjust(G,[0 0.6],[0.1 1]) is to adjust the gray value in [0,255*0.6] of G to [0.1*255, 255] by interpolation. The attractors are shown in Figure 7.

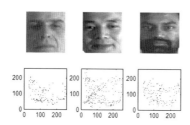

Figure 7. Images and attractors after adjusting gray value.

If no chaos attractor appears in the iterate process of dynamic system, there may be the following situations: the image is too smooth; large area of the image is white or black; the sequence falls in the cycle. Dynamic system falls in the cycle is the difficulty in structuring chaotic iterative sequence. Using Gaussian function surface can largely solve this difficulty.

4 CONCLUSION AND FURTHER WORK

This paper works with chaotic characteristics of the image and auxiliary function iteration. Researches show that attractors can be generated when parameters meet some condition in a dynamic system of image and Gaussian function. Test and verify the chaotic phenomena by plotting the bifurcation diagram and Lyapunov exponential diagram. Then do more researches on yalefaces images. It is found that image attractor can be used in image recognition as a feature of the image. More work should be done further: do projection (Radon transform for example) in multiple angles of attractor set, then images can be classified, tracked or detected by the projection. In addition, the chaotic sequence structured by an image can be used in image encryption, image coding. Significantly, gave the mathematic condition which the function (include image) can fall into chaos.

REFERENCES

Huang, H & He, H T 2011. Super-Resolution Method for Face Recognition Using Nonlinear Mappings on Coherent Features. *IEEE Trans. Neural Networks*, 22:121–130.

Mohammadzade, H & Hatzinakos D 2013 Iterative Closest Normal Point for 3D Face Recognition. , 35:381–397.

Li, R & Tian, T P & Divide, S S 2012 Conquer and Coordinate: Globally Coordinated Switching Linear Dynamical System. *IEEE Trans. Pattern Analysis and Machine Intelligence*, 34:654–669.

Yu, W B & Zhou, Y 2013 Chaos analysis of the rational Bézier biquadratic surface in the unit area. *Acta Phys. Sin.* 62:220509.

Yu, W B & Zhao, B 2014 A new chaotic attractor graphics drawing method based on the curved iteration. *Acta Phys. Sin.* 63:120502.

Yu, W B 2014 Chaotic characteristics of three-dimensional function determined by cross-section geometric shape. *Acta Phys. Sin.* 63:120501.

Information Technology – Wan et al. (Eds)
© 2015 Taylor & Francis Group, London, ISBN 978-1-138-02785-5

Chinese handwritten writer classification based on affine sparse matrix factorization

Jun Tan
School of Mathematics and Computer Science Sun Yat-Sen University, Guang Zhou, PR China.

Wei-Shi Zheng
School of Information and Science and Technology Sun Yat-Sen University, Guang Zhou, PR China.

ABSTRACT: In this article, we propose a new approach for writer classification of Chinese handwriting. In this method, we deal with writer classification of Chinese handwriting using Chinese character structure features (CSFs) and affine sparse matrix factorization (ASMF). To extract the features embedded in Chinese handwriting characters, special structures have been explored according to the trait of Chinese handwriting, where 20 features were extracted from the structures; these features constitute patterns of writer handwriting. We also propose an improved sparse matrix factorization that integrates affine transformation, namely ASMF, for automatically classifying writers from Chinese handwriting. Here, the affine transformation is vitally important because it can address the poor-alignment problem of handwriting image. This algorithm learns much faster than traditional popular learning algorithms. Experimental results demonstrate CSF/ASMF method can achieve better performance than other traditional schemes for writer classification.

1 INTRODUCTION

Writer classification is one of the important methods in the biometric individual identification; it has been widely used in the fields of forensic, bank check, historic document analysis, archaeology, and identify personality (Impedovo & Pirlo 2007), some approaches have been developed (He, Tang, & You 2005) (Bulacu, Schomaker, & Vuurpijl 2003). According to the various input methods, writer classification is usually classified into online and off-line.

The stroke shapes and structures of Chinese words are quite various from those of other languages such as English, so it is still challenge to identify Chinese handwriting (He, Tang, & You 2005). The approaches proposed for English handwriting writer classifications are not suitable for the Chinese handwriting (Schlapbach & Bunke 2004) (Schomaker & Vuurpijl 2000). This article proposes Chinese structure feature (CSF) as algorithm of feature extraction and combine CSF with affine sparse matrix factorization (ASMF) as a new scheme for writer classification.

Handwriting features except CSF (Tan, Lai, Wang, & Feng 2011) such as textures, edges, contours, and character shapes have been researched recently. Several researchers (Zhu, Tan, & Wang 2000) (Chen, Yan, Deng, & Yuan 2008) took handwritings as an images consist of special texture, and so regarded writer identification as the texture identification. Among them, Zhu et al. (2000) extracted the texture features by 2-D Gabor filtering, at the same time, Chen et al. (Chen, Yan, Deng, & Yuan 2008) used the Fourier transform. Xu et al. (2011) proposed an approach named *Gabor surface feature* (GSF), a histogram-based feature, which is extracted from edges of the scanned handwritten images.

Once discriminant features have been extracted, they are submitted to a classifier whose task is to identify writer that they represent. The widely used classifiers at least include weighted Euclidean distance (WED) classifier (He, Tang, & You 2005) (Zhu, Tan, & Wang 2000) (Chen, Yan, Deng, & Yuan 2008), Bayesian model, back propagation (BP) neural networks (Bulacu, Schomaker, & Vuurpijl 2003), likelihood ranking (Zhu, Zheng, Doermann, & Jaeger 2009), and *support vector machine* (SVM) (Imdad, Bres, Eglin, Emptoz, & Rivero-Moreno 2007). Bulacu et al. (2003) adopted neural networks and Bayesian classifiers as the classifiers. Imdad et al. (2007) used steered Hermite features to identify writer from a written document, and the algorithm was trained and tested on SVM. The traditional algorithms for the issues such as BP need many iterative steps to calculate the optimal values of the input weights and the output weights, so their speeds are very slow in general.

To overcome this limitation, the nonnegative matrix factorization (NMF) has been successfully applied to font identification (Lee & Jung 2005) and classification of gene expression data (Liu, Yuan, Wu, Ye, Ji, & Chen 2009). In traditional NMF (Lee & Jung 2005) (Lee, Seung, et al. 1999), component and coefficient are constrained to be nonnegative in accordance to reality evidences, but there is experimental result that shows nonnegativity does not usually generate sparsity. Zheng et al. (Zheng, Li, Lai, & Liao 2007) proposed a general framework named constrained sparse matrix factorization (CSMF), in which the nonnegativity is substituted by other constrains. With the affine transform, ASMF (Tan, Xie, Zheng, & Lai 2012) approach can solve the poor-alignment problem caused by internal diversity of handwriting image segmentation, so the affine transform is very important in ASMF method.

This article proposes a writer classification method based on the ASMF (Tan, Xie, Zheng, & Lai 2012) technique that automatically form a set of handwritings images. Feature vector of the handwriting image are extracted automatically using CSF (Tan, Lai, Wang, & Feng 2010). In a testing phase, the test sample is compared with the prototype templates that consist of training samples. The handwritings images are finally classified to different writers. By integrating the affine transform, we solved the poor-alignment problem caused by internal diversity of handwriting images segmentation, which make sparse transformation work more effective in writer classification.

The rest of the article is organized as follows. First, we reviewed algorithms of CSF feature extraction in Section 2. The method of ASMF is briefly explained in Section 3. We proposed relative scheme in Section 4 and analyzed the experimental results in Section 5. Finally, the conclusion is given in Section 6.

2 CHINESE STRUCTURE FEATURES

Features are directly extracted from each single Chinese character. Because the stroke shape and structure of Chinese character are more various from those of other languages such as Arabic, where the handwriting characteristics are embedded, we propose the use of stroke shape and structure for writer classification.

By several experiments, we discover that the difference in handwriting characteristics lies in the two structures (Tan, Lai, Wang, & Feng 2011). They are a special quadrilateral, which we call TBLR quadrilateral, and the bounding rectangle, as shown in Fig. 1(a) and (b), respectively.

The following nine CSF are obtained from the bounding rectangle as shown in Table 1.

$f1$: Width to height ratio of the bounding rectangle A; $f2$, $f3$: relative horizontal and vertical position of the gravity centroid; $f4$, $f5$: relative horizontal and vertical gravity centroid; $f6$, $f7$: distance between the gravity centroid G1($x1$, $y1$) and the geometric centroid G2($x2$, $y2$); $f8$: ratio of the foreground pixel numbers to the area of the bounding rectangle; and f9: stroke width property, where Pt is the binary pixel after refining the preprocessed image A. Given a structuring element $B = \{C, D\}$ consisting of two elements C and D, the refining operation keeps repeating the hit-or-miss operation, until convergence, that is, the change stops.

(a) Bounding rectangle   (b) TBLR quadrilateral

Figure 1. Two special structures of Chinese handwriting character. (a) Bounding rectangle. (b) TBLR quadrilateral.

Table 1. CSF features from the bounding rectangle.

ith	Eqs.	Comments
1	A_w / A_h	A_w and A_h are the width and height of A.
2	$\dfrac{\sum_{i=1}^{A_w} i \times P_x(i)}{\sum_{i=1}^{A_w} P_x(i)}$	Foreground pixel number i-th vertical
3	$\dfrac{\sum_{j=1}^{A_h} j \times P_y(j)}{\sum_{j=1}^{A_h} P_y(j)}$	Foreground pixel number j-th horizontal
4	$f2 / A_w$	$f2$ is 2th CSF feature
5	$f3 / A_h$	$f3$ is 3th CSF feature
6	$\| G1 - G2 \|$	Gravity center $G_1(x_1, y_1)$
7	$(y_2 - y_1) / (x_2 - x1)$	Geometric center $G_2(x_2, y_2)$
8	$\dfrac{\sum_{i=1}^{A_w} \sum_{j=1}^{A_h} \times P(i,j)}{A_w \times A_h}$	Foreground pixel number $P(i,j)$
9	$\dfrac{\sum_{i=1}^{A_w} \sum_{j=1}^{A_h} \times P(i,j)}{\sum_{i=1}^{A_w} \sum_{j=1}^{A_h} \times P_t(i,j)}$	Binary pixel after refining $P_t(i,j)$

$$A \circledast B = (A \ominus C) - (A \oplus \hat{D}) \tag{1}$$

Similarly, from the TBLR quadrilateral, we can obtain the following seven CSF features as shown in Table 2.

Table 2.   CSF features from the TBLR quadrilateral

ith	Eqs.	Comments
10	S_{up}/S	S_{up} is the area of the top half part.
11	S_{left}/S	S_{left} is the area of left half part.
12	$\cos(a,b)$	a and b are the direction vectors of diagonal
13	P_{inner}/P_{total}	Foreground pixel number P_{inner}
14	P_{inner}/S_{TBLR}	Area of the TBLR quadrilateral S_{TBLR}
15	P_{left}/P_{total}	Foreground pixel number of left half part P_{left}
16	P_{top}/P_{total}	Foreground pixel number of left half part P_{top}

$f\,10$: Ratio of the area of the top half part S_{up} to the area of the whole quadrilateral S; $f\,11$: ratio of the area of the left half part S_{left} to S; $f\,12$: value of cosine of the angle of the two diagonal lines, where a and b are the direction vectors of the two diagonal lines, respectively. The $f\,10$, $f\,11$, and $f\,12$ measure the global spatial structure of the character; $f\,13$: ratio of foreground pixel numbers P_{inner} within the TBLR quadrilateral to the total foreground pixel number P_{total}, it measures the global degree of stroke aggregation; $f\,14$: ratio of the P_{inner} to the area of the TBLR quadrilateral S_{TBLR}; $f\,15$: the ratio of foreground pixel number of the left half part P_{left} with the TBLR quadrilateral to P_{total}; $f\,16$: ratio of foreground pixel numbers of the top half part P_{top} with the TBLR quadrilateral to P_{total}.

Apart from the above sixteen features, we obtain another four CSF features as follows:

$f\,17$: Number of connected component. The feature measures the joined-up writing habit; $f\,18$: number of hole within the character; $f\,19$: number of stroke segment. It can be obtained by deleting all crossing point of a character, and the number is the total segment number; $f\,20$: ratio of the longest stroke segment to the second longest stroke segment, where the stroke segments are obtained same as that of $f\,19$.

3   AFFINE SPARSE MATRIX FACTORIZATION

NMF (Lee, Seung, et al. 1999), which is one of the representative spare decomposition algorithms,

seeks to decompose a nonnegative data matrix $V \in \mathbb{R}^{n \times N}$ into the product of two nonnegative lower rank matrices, termed basis matrix $W \in \mathbb{R}^{n \times l}$ and coefficient matrix $H \in \mathbb{R}^{l \times N}$, respectively, so that V is approximately equal to W times H as follows:

$$V_{i\mu} \approx (WH)_{i\mu} = \sum_{a=1}^{l} W_{ia} H_{a\mu} \tag{2}$$

After decomposition, the obtained W and H are both sparse. Because of the nonnegative constraint, NMF meets many physical requirements in real applications. However, the nonnegative constraint, which may affect the sparseness, intuitionally, we in this article introduce CSMF (Zheng, Li, Lai, & Liao 2007), which is an improved form of NMF. CSMF guarantees the sparseness by bringing in penalized function but releasing nonnegative constraint. In our method, V and W correspond to handwriting feature set and writer set (basis), respectively, and H is the decomposition coefficient with respect to W. CSMF can be described as follows:

$$\min_{W,H} G(W,H) = \|V - WH\|_F^2 + g_1(W) + g_2(W) \tag{3}$$

$$s.t. W \in D_1, H \in D_2$$

where D_1 and D_2 are value regions and g_1 and g_2 are the penalized functions defined as follows with respect to W and H, respectively.

$$g_1(W) = \alpha \sum_{j1=1} \sum_{j2=1, j2 \neq j1} \sum_{c=1}^{n} |W_{j1}(c)\| W_{j2}(c)|$$

$$+ \beta \sum_{j=1}^{l} \sum_{c=1}^{n} |W_j(c)| \tag{4}$$

$$g_2(W) = \lambda \sum_{k=1}^{N} \sum_{j=1}^{l} |h_k(j)| \tag{5}$$

where α, β, and λ are nonnegative weighing coefficients. Affine transform is denoted as f. By combining affine transform with CSMF, we describe the proposed ASMF as follows:

$$\min_{W,H} G(W,H) = \|A - WH\|_F^2 + g_1(W) + g_2(W) \tag{6}$$

$$= \|f(V) - WH\|_F^2 + g_1(W) + g_2(W)$$

$$s.t. W \in D_1, H \in D_2$$

167

where

$$A_{i\mu} = f(V_{i\mu}) \approx (WH)_{i\mu} = \sum_{a=1}^{r} W_{ia} H_{a\mu} \qquad (7)$$

and the affine transform f on a Chinese handwriting feature set V is detailed as follow:

$$\begin{aligned}A_{i\mu} &= f(V_{i\mu}) = AV_i + b \\ &= A_s A_t A_u A_\theta V_i + b\end{aligned} \qquad (8)$$

where A_s, A_t, A_u, A_θ, and b denote the scaling, stretching, skew, rotation, and translation transformations, respectively. The affine transformation helps to improve the performance of sparse factorization on weak-alignment character images. Experiment shows that ASMF get more legible results with adjusting reconstructed handwriting into normal structure.

4 OUR SCHEME

Some of the results in this article were first presented in the work of Tan et al. (2011). In this article, we present more technique details and effectiveness of CSF/ASMF approach. Fig. 2 demonstrates the flowchart of the proposed approach. There are three main steps for Chinese handwritten writer classification. The first step is handwritten image preprocessing, which removes noises and normalizes the images into the same size. The second step is feature extraction, for which we propose CSF. The last step is to apply ASMF learning method to classify different writers.

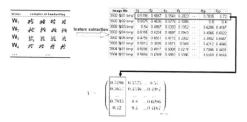

Figure 2. The flowchart of the proposed approach.

For example, the entire process of CSF/ASMF-based handwritten writer recognition is as follows:

- Step 1: The appropriate training strategy based on the selected training set. We randomly selected from a handwritten database the training set $TrainSet = Tr_i, i = 1, ..., N_1$, where N is the total of training samples, and the remaining samples as the test set $TestSet = Te_i, i = 1, ..., N_2, N = N_1 + N_2$.

- Step 2: Image preprocessing for training set, through the noise removal and standardization.
- Step 3: CSF feature extraction method to extract the optimal recognition feature vector. A total of 20 features were extracted from structures of Bounding Box and TBLR quadrilateral.
- Step 4: ASMF training phase. Using **Algorithm ASMF** and Eqs. (3), (6), and (8), initialize W, H randomly, set the input parameters α, β, λ and error estimation $\in$ to compute iteratively. We got the output matrix W and H until $\| G(W, H) \| \le \varepsilon$. W is sparse matrix, each row of W represents a writer, $H = \{h_1, ..., h_n\}$ where h_1–h_n are projection coefficients. When the model of training has been trained, the training process is complete.
- Step 5: ASMF testing phase. Testing the model parameters obtained from the training model W for each test sample v. Then, v is projected onto corresponding dictionary W.

$$h = [h_1, h_2, ..., h_r]^T = W^+ v \qquad (9)$$

where W^+ is the pseudo-inverse matrix of W, and h is projection coefficient.

Among coefficients h, there are prominent values corresponding to writers. Since each coefficient h_i corresponds to a learned writer model, we can determine the writer of test samples by using following operation:

$$h_{\text{opt}} = \arg\max_{1 \le i \le n} \qquad (10)$$

where h_{opt} essentially provides the writer coding of the referred model. Since all reference writers have been coded in a library W, we can recognize v through looking up the reference library using the code generated from $\{h_i\}$.

5 EXPERIMENTS AND ANALYSIS

5.1 Handwritten database

To test the performance of the proposed method in the writer classification, we experimented two Chinese handwritten databases: SYSU (Tan, Lai, Wang, & Feng 2011) and KAIST Hanja1 (Kim & Kim 2003). Among them, SYSU database was generated and collected by us as follows: 150 volunteers were asked to sign his (or her) name and one of the others names twice, and write Chinese number from "1/yi" to "10/shi" five times. The KAIST Hanja1 database contains 783 frequently used Chinese characters, where each character consisting of 200 samples written by

200 writers, respectively. Fig. 3 shows some samples of the databases.

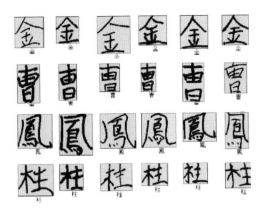

Figure 3.   Examples of the HanjaDB1 database.

5.2   *Comparative results*

In experiment, because the features may have large differences in value, to avoid large values submerge the contributions of the small value of features, all samples were normalized between 0 and 1 before sending to the learning algorithms as input.

We compared the proposed method with three feature extraction methods by Zhu et al. (2000), Chen et al. (2008), and Xu et al. (2011), which use different features (Zhu, Tan, & Wang 2000, Chen, Yan, Deng, & Yuan 2008, Xu, Ding, Peng, & Li 2011) and classifiers. These methods used Gabor feature (Zhu, Tan, & Wang 2000), grid microstructure feature (GMSF) (Xu, Ding, Peng, & Li 2011), and Fourier feature and mathematical expectations (Chen, Yan, Deng, & Yuan 2008). Each of the compared features was well adjusted/trained to generate the best results. Fig. 4 shows the performance of different feature.

Furthermore, learning method is also an important problem. We also compared the proposed method with four well-known unsupervised classification methods: Bayesian NN (Chang, Wang, & Suen 1993), MLPS (Al-Shoshan 2006), RBFS (Armand, Blumenstein, & Muthukkumarasamy 2006), and SVM (Srihari, Xu, & Kalera 2004). Both the recognition accuracy of writer classification and average time cost were reported and compared. First, we used learning methods for testing, and the average training time and test time was calculated. The cost time of the experiments is shown in Table 3. From the table, we can see that Bayesian NN (Chang, Wang, & Suen 1993), MLPS (Al-Shoshan 2006), and RBFS (Armand, Blumenstein, & Muthukkumarasamy 2006) training times are relatively much more than the ASMF training time. The average total time of ASMF, SVM, Bayesian NN (Chang, Wang, & Suen 1993),

MLPS (Al-Shoshan 2006), and RBFS (Armand, Blumenstein, & Muthukkumarasamy 2006) were 5.4, 15.3, 12.2, 11.8, and 7.7 seconds, respectively. Methods based on neural network are slower than the ASMF and SVM methods. Therefore, the method of ASMF has the highest speed.

Finally, we compared our scheme with Bayesian NN (Chang, Wang, & Suen 1993), MLPS (Al-Shoshan 2006), RBFS (Armand, Blumenstein, & Muthukkumarasamy 2006), and SVM (Srihari, Xu, & Kalera 2004) approaches for the classification accuracy. In comparison with Chinese handwritten database Hanja and SYSU, our method adopted CSF as feature. Table 4 gives the classification accuracies of the four methods on different writer numbers such as 200, 100, 50, and 20. The classification accuracies of the algorithms on the handwritten database are shown in Table 4. From the table, we can see the accuracies corresponding to different number of writers. The lowest accuracy of method (Chen, Yan, Deng, & Yuan 2008) is 43.5% when number of writers is 200. Less writers can improve the accuracy; our method achieved highest accuracy of 92.1%, it had the similar accuracy as that of SVM method (Srihari, Xu, & Kalera 2004). It is obvious that our method is more effective for identifying writer in Chinese handwriting.

Table 3.   CSF features from the TBLR quadrilateral.

Method	Average training time	Average test time	Average total time
ASMF	12.32	0.457	5.4
Bayesian NN	34.81	1.58	15.3
MLPS	26.6	0.972	12.2
RBFS	23.52	0.861	11.8
SVM	16.09	0.782	7.7

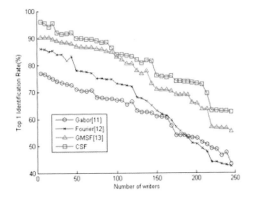

Figure 4.   Performance of different feature method.

169

Table 4. Classification accuracies of different methods.

Writers (n)	200		100		50		20	
Method	SYSU	Hanja	SYSU	Hanja	SYSU	Hanja	SYSU	Hanja
NN (Chang, Wang, & Suen 1993)	–	43.5	53.4	47.2	64.8	57.2	68.3	61.5
MLPS (Al-Shoshan 2006)	–	49.7	63.5	52.7	71.2	59.3	71.9	64.9
RBFS (Armand, Blumenstein, & Muthukkumarasamy 2006)	–	52.2	62.4	57.6	72.3	67.5	73.4	68.4
SVM (Srihari, Xu, & Kalera 2004)	–	59.3	67.9	62.8	76.7	72.8	80.2	78.8
Ours	–	64.5	71.6	68.4	82.3	75.6	92.1	84.3

6 CONCLUSION

In this article, we propose an ASMF method used in Chinese handwritten writer classification. CSF uses original handwriting images and tries to find out the writing structure of the writer. ASMF is a classifier for writer-used matrix factorization. From the experimental results, we can compare the performance of our method and some other methods for Chinese writer classification. The recognition accuracy of our method using the CSF/ASMF seems better than the other methods for Chinese writer classification. Compared with traditional learning algorithm, ASMF has faster speed, better generalization performances. The effectiveness of CSF/ASMF for Chinese handwriting writer classification is proved by the experiments.

It is expectable that our approach can be used for multilingual handwriting including western handwritings and Arabic number.

ACKNOWLEDGMENT

This work was supported by the NSFC (11471012), Guangdong Provincial Government of China through the "Computational Science Innovative Research Team" program, and Guangdong Province Key Laboratory of Computational Science at the Sun Yat-sen University.

REFERENCES

Al-Shoshan, A. (2006). Handwritten signature verification using image invariants and dynamic features. In *Computer Graphics, Imaging and Visualisation, 2006 International Conference on*, pp. 173–176. IEEE.

Armand, S., M. Blumenstein, & V. Muthukkumarasamy (2006). Off-line signature verification based on the modified direction feature. *In Pattern Recognition, 2006. ICPR 2006. 18th International Conference on*, Volume 4, pp. 509–512. IEEE.

Bulacu, M., L. Schomaker, & L. Vuurpijl (2003). Writer identification using edge-based directional features. *writer 1*, 1.

Chang, H., J. Wang, & H. Suen (1993). Dynamic handwritten Chinese signature verification. In *Document Analysis and Recognition, 1993., Proceedings of the Second International Conference on*, pp. 258–261. IEEE.

Chen, Q., Y. Yan, W. Deng, & F. Yuan (2008). Handwriting identification based on constructing texture. In *Intelligent Networks and Intelligent Systems, 2008. ICINIS'08. First International Conference on*, pp. 523–526. IEEE.

He, Z., Y. Tang, & X. You (2005). A contourlet-based method for writer identification. In *Systems, Man and Cybernetics, 2005 IEEE International Conference on*, Volume 1, pp. 364–368. IEEE.

Imdad, A., S. Bres, V. Eglin, H. Emptoz, & C. Rivero-Moreno (2007). Writer identification using steered hermite features and SVM. *In Document Analysis and Recognition, 2007. ICDAR 2007. Ninth International Conference on*, Volume 2, pp. 839–843. IEEE.

Impedovo, S. & G. Pirlo (2007). Verification of handwritten signatures: an overview. *In Image Analysis and Processing, 2007. ICIAP 2007. 14th International Conference on*, pp. 191–196. IEEE.

Kim, I. & J. Kim (2003). Statistical character structure modeling and its application to handwritten Chinese character recognition. *Pattern Analysis and Machine Intelligence, IEEE Transactions on* 25(11), 1422–1436.

Lee, C. W. & K. Jung (2005). NMF-based approach to font classification of printed English alphabets for document image understanding. *In Modeling Decisions for Artificial Intelligence*, pp. 354–364. Springer.

Lee, D., H. Seung, et al. (1999). Learning the parts of objects by non-negative matrix factorization. *Nature* 401(6755), 788–791.

Liu, W., K. Yuan, J. Wu, D. Ye, Z. Ji, & S. Chen (2009). Combining generalized NMF and discriminative mixture models for classification of gene expression data. *International Journal of Pattern Recognition and Artificial Intelligence* 22(8), 1587.

Schlapbach, A. & H. Bunke (2004). Off-line handwriting identification using hmm based recognizers. In *Pattern Recognition, 2004. ICPR 2004. Proceedings of the 17th International Conference on*, Volume 2, pp. 654–658. IEEE.

Schomaker, L. & L. Vuurpijl (2000). Forensic writer identification: A benchmark data set and a comparison of two systems [internal report for the {Netherlands Forensic Institute}].

Srihari, S., A. Xu, & M. Kalera (2004). Learning strategies and classification methods for off-line signature verification. In *Frontiers in Handwriting Recognition, 2004. IWFHR-9 2004. Ninth International Workshop on,* pp. 161–166. IEEE.

Tan, J., J. Lai, C. Wang, & M. Feng (2010). Off-line Chinese handwriting identification based on stroke shape and structure. In *Information Engineering and Computer Science (ICIECS), 2010 2nd International Conference on,* pp. 1–4. IEEE.

Tan, J., J. Lai, C. Wang, & M. Feng (2011). A stroke shape and structure based approach for off-line chinese handwriting identification. *International Journal of Intelligent Systems and Applications (IJISA) 3*(2), 1.

Tan, J., X. Xie, W.-S. Zheng, & J.-H. Lai (2012). Radical extraction using affine sparse matrix factorization for printed Chinese characters recognition. *International Journal of Pattern Recognition and Artificial Intelligence 26*(03).

Xu, L., X. Ding, L. Peng, & X. Li (2011). An improved method based on weighted grid micro-structure feature for textindependent writer recognition. In *Document Analysis and Recognition (ICDAR), 2011 International Conference on,* pp. 638–642. IEEE.

Zheng, W., S. Li, J. Lai, & S. Liao (2007). On constrained sparse matrix factorization. In *Computer Vision, 2007. ICCV 2007. IEEE 11th International Conference on,* pp. 1–8. IEEE.

Zhu, G., Y. Zheng, D. Doermann, & S. Jaeger (2009). Signature detection and matching for document image retrieval. *Pattern Analysis and Machine Intelligence, IEEE Transactions on 31*(11), 2015–2031.

Zhu, Y., T. Tan, & Y. Wang (2000). Biometric personal identification based on handwriting. In *Pattern Recognition, 2000. Proceedings. 15th International Conference on,* Volume 2, pp. 797–800. IEEE.

Information Technology – Wan et al. (Eds)
© *2015 Taylor & Francis Group, London, ISBN 978-1-138-02785-5*

Construction of a cloud computing platform for medical logistics service

Fang-zhong Qi & Jun-feng Wang
College of Economics and Management, Zhejiang University of Technology, 288 Liuhe Road, Hangzhou, China

ABSTRACT: Medical logistics enterprises can't meet the rapid growth of data processing by self-built logistics service platform, for the large amount of information in the process of logistics operation, as well as strong dynamic, high timeliness, fast reaction and wide varieties. It is imperative to establish a logistics service platform based on cloud computing. This research analyzed the necessity of building a medical logistics service platform based on cloud computing, and then demonstrated its feasibility from the current practice of theoretical research. We presented a cloud computing service platform architecture, taking the service as the core and focusing on medical logistics process management and process performance, according to service level agreements (SLA) and Quality of Service (QoS) properties of Service Oriented Architecture (SOA). Then it was briefly proved in the application of a dentist cloud computing platform.

KEYWORDS: cloud computing; medical logistics service platform; business process management; process performance

1 INTRODUCTION

The advent of the big data era leads people to the "third industrial revolution" [1]. The people's information application has constantly changed by the emerging application mode like social media, collaborative creation and virtual service, which is under the support of mobile computing, cloud computing, Internet of things, and a series of new technologies.

The particularity of medical logistics information mainly displays in the following three aspects when compared to other areas of information: firstly, due to it is a broader range activity of the medical logistics, information sources are distributed in a large range with many information source points and huge informative [2]. Secondly, the medicine, logistics information has strong, dynamic, the value of information attenuated quickly, this demands the information with high timeliness. In a large system, it specially emphasized on timeliness, the information collection and processing should be dealt with quickly. Finally, medical logistics information has wide varieties, not only the internal system has different kinds of information, but also collects information of other categories such as a production system, marketing system and consumption system, because the medicine logistics system are closely related to these systems [3]. All these particularities increased the difficulty of medical logistics information classification, research and screening.

The modern operation of medical logistics is based on information technology, in the process of its informatization, traditional medical logistics service platform needs a lot of manpower and material resources, and with the relatively independent software between principals, data sharing is unavailable. It cannot meet the rapid growth of enterprise information system and diversified service requirements.

Cloud computing got a wide range of applications when it is put forward, for its rapid deployment of resources or services, expand and use, according to demand, pay by the number of usages and provide services via the Internet [4,5]. Cloud computing offers all kinds of resources in a paradigmatic form (infrastructure, platform, software, etc.) to the relevant customer to meet their needs. Cloud computing providers mainly provide three levels of Service: (1) Software as a Service (SaaS), providing a complete application as a service, such as customer relationship management (CRM) software 6]; (2) Platform as a Service (PaaS), providing a platform which can be embedded into other application software as a service, such as Google App Engine (GAE) [7]; (3) Infrastructure as a Service (IaaS), providing an environment of deployment, operation and management of the virtual machines and storage as a service [8].

Based on modern medicine logistics and the characteristics of cloud computing, applying cloud computing to medicine logistics information

This research was supported by the Natural Science Foundation of China (No. 71371169), the National Social Science Foundation of China (No. 12BJY116), and the Natural Science Foundation of Zhejiang province, China (No. Y6110796).

management to establish a medical logistics service cloud platform will be an important development direction of medicine logistics and cloud computing providers. The application of cloud computing to solve the large amount of information problem in the process of medical logistics operation, such as variety, strong dynamic, high timeliness and fast reaction rate, which will bring great changes to the development of the whole industry.

2 APPLICATION AND RESEARCH STATUS OF MEDICAL LOGISTICS SERVICE CLOUD COMPUTING PLATFORM

Medical logistics closely associated with cloud computing, the establishment of medical logistics service cloud platform is an example of the use of cloud computing on the application dimension. Many experts and scholars had come to this challenging new area in both theory and practical application, but barely attempt of applying cloud computing technology into the field of medical logistics service.

2.1 Application practice

IBM has set up a wisdom logistics cloud platform in Ningbo, China, providing an end-to-end service for enterprise's supply chain, from the material supply plan to purchasing, production and sales. The logistics operation management system could help enterprises or logistics companies, from the dimension of logistics informationization and optimized to build transport order that dominated by price.

Angela et al. analyzed the recognition and acceptance of cloud computing by IT specialists, they found the IT companies in Hong Kong are not willing to accept the cloud computing, unless the cloud computing can eliminate the uncertainty of its own, such as security and standardization [9]. Osvaldo and Fernando depicted the main feature of cloud computing in detail, and analyzed the cognitive and implementation status of cloud computing among Portuguese companies [10].

2.2 Research status

Many scholars' research on medical logistics service cloud platform mainly stays on the design of the framework, such as put forward the fourth party logistics service platform based on SOA architecture, studied its core function using the JAVAEE technology under complete platform implementation [11]. Using the ontology technology, the establishment of the resource management framework based on the SLA, bringing the opportunity for SLA management

of cloud computing platforms and resources optimization [12].

Another hot topic is the security issues of medical logistics service cloud platform, such as the overall research, medical logistics service cloud platform, security framework model in a cloud computing environment. In the proposed logistics service platform which is based on the basic framework model, obtained the fusion of IaaS, PaaS, SaaS cloud computing service platform.

Li et al. studied the service selection problems of packaging case and logistics service virtualization, using the particle swarm algorithm, under the constraints of service quality [13]. Adamskj proposed a wide spectrum in the study of hierarchical intelligent integrated logistics system platform, using Hiils platform solutions to solve the existing problems in the urban logistics [14]. The platform multi-level choice of logistics system integrated management, coordination, adaptation, optimization, scheduling, monitoring and intelligent monitoring, direct control layer is devoted to the dynamic, random, multi-criteria decision making problem of intelligent logistics.

As can be seen from the review above, research on medical logistics service cloud platform is still in its infancy, mainly concentrated in the framework model building and safety technology to solve the problem. The study of medical logistics service platform construction is very little, the purpose of this paper is to build a medical logistics service based on a cloud computing platform. We proposed on an overall architecture of medical logistics service cloud platform after referencing a variety of frame model (see Fig. 1), and then emphatically illustrated it from the business process management (BPM) and process performance, as well as the key technical support required.

3 BUSINESS PROCESS MANAGEMENT AND PROCESS PERFORMANCE MANAGEMENT

From the overall architecture diagram of medical logistics service cloud platform (Fig. 1), we played down the three cloud computing service model (SaaS, PaaS, IaaS) partition in the architecture, and took the entire cloud computing services as a whole. In real applications, the software itself, platform and infrastructure are inseparable, they must be a whole service for the customer, otherwise cannot achieve satisfactory results.

The ideal state of the medical logistics service cloud platform construction is that the medical logistics service demander input service requirements at his own terminal and then easily obtain the required services, needn't to worry about how the process works, who provide the services scheme, and operate

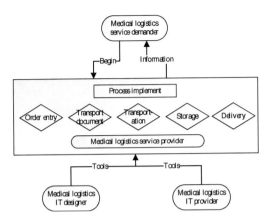

Figure 1. Overall architecture of medical logistics service cloud platform

by which person. There is no doubt that the process is very complex, how the participants involved can efficiently cooperate with each other, these need to have a clear and effective process. Thus, an important problem is effective planning and management of the whole platform operation process, when building a cloud computing medical logistics service platform.

3.1 Medical logistics management system platform design based on business process management

Business process management (BPM), originated in the field of the industrial engineering work process analysis technology (PAT), originally used to analyze manufacturing process improvement

only, later expanded to the application scope of improving white-collar workers work [15]. Different from the traditional development mode, BPM is process-oriented rather than code-oriented, providing a set of visualization, flexible development tools, and can be closely connected with the real business. What's more, it integrated SCM, CRM and ERP, realized the connectivity from customers to suppliers, completely integrated enterprise internal and external trade.

Applying the BPM method, combined with SOA based open standard interface and loose coupling way, to construct the combination of the implementation process. The medical logistics system based on BPM framework is shown in Fig. 2, we take the warehouse management system of medical logistics services cloud computing platform as an example. The subsystem of the overall architecture is divided into four layers (system platform layer, BMP platform layer, processing layer and application layer).

System platform layer mainly supports to complete the system resources, as well as the interactive integration of the medical enterprise resources of the original system and warehousing logistics management system. Data, realized the unified management to various data sources, when maintained access to the system data correctly. The BPM platform layer provides software platform supports to the application of the BPM implementation process. Processing layer create, manage, and maintain all business processes in the system. The application layer provides functional modules according to the specific requirements, such as basic information maintenance, equipment procurement, leasing, maintenance, inventory management, financial management, and so on.

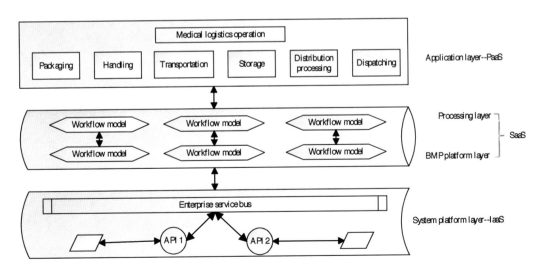

Figure 2. Architecture of a medical logistics management system based on BPM

3.2 Process performance management of medical logistics service cloud computing platform

Process performance management focus more on customer needs, which is different from the traditional performance management, it deploys resources from the process to manage the enterprise, making the enterprise a great change from the basic business activities to the value concept [16], specific displays in the following four aspects:

- The process becomes the basis of the established assessment index system;
- Take customer satisfaction as the basic goal, and pass the no difference terminal customers information to the process of every link and post through the process, then each node has its own direct customers (internal and external customer);
- The goals of the process rely on the coordination of each section of the process;
- The direct purpose is to improve the process efficiency.

From the characteristics of process performance, we can see the fundamental is service, which can be measured by the QoS attributes of SOA. QoS generally refers to the bandwidth, delay, jitter and loss rate, is a network traffic parameter, presented as an objective value. But the QoS in the area of SOA is not just limited to the network layer and transport layer parameters, also related to the host performance, security, transactions and cost.

Of course, QoS attributes have different metrics from different point of view, we study the process performance management here, therefore, how to use these indicators to manage is the key point. In this paper, the application can be divided into four processes: discovery, measurement, monitoring and management. Thio et al. set up two models to monitor the service performance attributes and load situation, automatic measurement of dynamic QoS properties, respectively from the viewpoint of users and service providers [17].

The QoS management range of SOA is very wide, including service composition, monitoring, measurement. Daniel et al. put forward a negotiated QoS architecture in the SOA [18]. In the SOA application environment, the service provider and the requester may have different requirements according to the service performance, reliability, timeliness and safety. And the ideal result of process management and process performance is processed intelligence (or Smart-flow) of medical logistics service cloud computing platform. Research in this field is still less, and has broad space for development.

3.3 Service component architecture

Service Component Architecture for SOA realized the separation of business components and transmission protocol, and could handle the integration of all kinds of platform component [19]. In the SCA, one of the most important concepts is service, its connotation essentially is independent from the specific implementation technology of a service oriented architecture.

It is tight coupling between the existing components and transmission protocol, and the biggest difference between the SCA and the traditional business component is that it implements two functions: the separation of components and transmission protocol; and the separation of interface and the real language. The goal is to create an integrated service component runtime environment, the medical logistics information platform based on cloud computing using SCA, can integrate logistics functions as SCA services, or to reconstruct new services, helping software provider provides a quick software upgrades.

Web service made it easier to implement, which is an online application service formed by the completion of the enterprise deployment of specific functions, and other companies or application software can access and use the online service through the Internet [20]. It needs a set of protocols to implement the creating of distributed applications, like XSD, XML, SOAP, WSDL and UDDI.

4 APPLICATION OF THE CLOUD COMPUTING PLATFORM

We applied the medical logistics cloud computing platform structure to a large dental enterprise information system designing. The company has a lot of chain dental clinics in many provinces of China, its operation involving coordinating and dealing with the medical instrument, equipment, material, medicine, preparation and medical waste logistics distributed across the country.

In order to realize the unified management of the resources and services, the company purchasing all of the equipment and medicines and then distributed them to each chain clinic. However, the clinic and related items supplier distribution around brought a huge burden for the enterprise's logistics operation, sporadic demand caused logistics resource waste, and sudden emergency orders can't respond in time, which reduced the service level.

In the new dentist cloud computing platform, the clinic release order and required delivery time according to their demand, suppliers organize production and shipment according to the processed order of the system, and the third party logistics enterprise matches the corresponding task based on the published service needs, then the relevant logistics process completed. It's obvious to find that the dentist cloud computing platform has obvious advantages in aspects of optimization of resource utilization, improving the response speed and efficiency.

5 CONCLUSION

The construction of medical logistics service platform based on cloud computing is a new field, it relies on the large-scale processing capability, standard operation process, flexible portfolio, precise control, intelligent decision support and deep information sharing of cloud computing, satisfy the demand of large information capacity, dynamic link each activity, high timeliness, and complex requirements in the medical industry. From the example of the dentist cloud computing platform, the medical logistics service cloud computing platform which strengthens service concept and focus on process management and process performance, is a big exploration of cloud computing applications and medicine logistics operation mode and the development trend.

REFERENCES

[1] D. Du, A. Li, L. Zhang, "Survey on the Applications of Big Data in Chinese Real Estate Enterprise," Procedia Computer Science, Vol. 30, pp. 24–33, 2014.

[2] S. D. Lapierre, A. B. Ruiz, "Scheduling logistic activities to improve hospital supply systems," Computers & Operations Research, Vol. 34(3), pp. 624–641, March 2007.

[3] D. Prajogo, J. Olhager, "Supply chain integration and performance: The effects of long-term relationships, information technology and sharing, and logistics integration," International Journal of Production Economics, Vol. 135(1), pp. 514–522, January 2012.

[4] M.G. Avram, "Advantages and Challenges of Adopting Cloud Computing from an Enterprise Perspective," Procedia Technology, Vol. 12, pp. 529–534, 2014.

[5] M. Yigit, V. C. Gungor, S. Baktir, "Cloud Computing for Smart Grid applications," Computer Networks, Vol. 70, pp. 312–329, September 2014.

[6] M. Cusumano, "Cloud computing and SaaS as new computing platforms," Communications of the ACM, Vol. 53 (4), pp. 27–29, 2012.

[7] E. Ciurana, "Developing with Google App Engine," Berkeley, CA: USA, 2009.

[8] R. Buyya, C. Yeo, S. Venugopal, J. Broberg, I. Brandic, "Cloud computing and emerging IT platforms: vision, hype, and reality for delivering computing as the 5th utility," Future Generation Computer Systems, Vol Vol. 25(6), pp. 599–616, 2009.

[9] A. Lin, N. Chen, "Choud computing as an innovation: Percepetion, attitude, and adoption," International Joural of Information Management, Vol. 32, pp. 533–540, 2012.

[10] O. Ferreira, F. Moreira, "Cloud Computing Implementation Level in Portuguese Companies," Procedia Technology, Vol. 5, pp. 491–499, 2012.

[11] J. J. Jung, "Service chain-based business alliance formation in service-oriented architecture," Expert Systems with Applications, Vol. 38(3), pp. 2206–2211, March 2011.

[12] A. G. García, I. B. Espert, V. H. García, "SLA-driven dynamic cloud resource management," Future Generation Computer Systems, Vol. 31, pp. 1–11, February 2014.

[13] W. Li, Y. Zhong, X. Wang, Y. Cao, "Resource virtulization and service selection in cloud logistics," Journal of Network and Computer Applications, In Press.

[14] A. Adamski, A. Turnau, "Simulation Support Tool for Real-Time Dispatching Control in Public Transport," Transportation Research, Vol. 32, pp. 73–87, 1998.

[15] H. Bae, S. Lee, I. Moon, "Planning of business process execution in Business Process Management environments," Information Sciences, Vol. 268, pp. 357–369, June 2014.

[16] A. Sujova, R. Rajnoha, M. Merková, "Business Process Performance Management Principles Used in Slovak Enterprises," Procedia-Social and Behavioral Sciences, Vol. 109, pp. 276–280, January 2014.

[17] N. Thio, S. Karunasekera, "Automatic Measurement of a QoS Metric for Web Service Recommendation," Proceedings of the 2005 Australian Software Engineering Conference, pp. 202–211, 2005.

[18] D. A. Menasccea, H. Ruana, H. Gomaa, "QoS management in service-oriented architectures," Performance Evaluation, Vol. 64, pp. 646–663, 2007.

[19] R. Mirandola, P. Potena, E. Riccobene, P. Scandurra, "A reliability model for Service Component Architectures," Journal of Systems and Software, Vol. 89, pp. 109–127, March 2014.

[20] M. Oriol, J. Marco, X. Franch, "Quality models for web services: A systematic mapping," Information and Software Technology, Vol. 56(10), pp. 1167–1182, October 2014.

Information Technology – Wan et al. (Eds)
© 2015 Taylor & Francis Group, London, ISBN 978-1-138-02785-5

Construction of product name dynamic management system for different classification criteria

Huali Cai, Fang Wu, Jian Kang & Yawei Jiang
China National Institute of Standardization

ABSTRACT: Currently, there is a wide variety of product classified catalogs, and a majority of these classified catalogs are not connected with each other, resulting commodity-related information sharing among various departments difficult to carry out. This article aims to analyze the current situation and characteristics of existing product classification; design the function of product name management system; develop a data structure to cover all the classified catalogs, develop a management system to store various trade names information; use expert resources to carry out synonyms mapping of different product classification; achieve a high degree of correlation between each category; design and develop the system.

1   OVERVIEW

Unified product name is the basic prerequisite for national import and export transaction, an important guarantee for orderly development of social economy, as well as the key to achieving inter-industry data sharing basis. Currently, there are some popular and widely used product catalogs or standard specifications in the world as well as China's import and export. However, in many of Chinese government departments, industry associations and enterprises, many different product classified catalogs are used, resulting in inconsistent granularity of commodity type, and different scope of application. Besides, products with the same name may be not in the same catalog, and bear different meanings. This phenomenon makes it difficult to achieve data sharing between different industries. For example, the data of complaints from consumers of different industries are incomparable, and different results may show up when consumers search for product in different e-commerce platforms.

This article aims to analyze existing commodity classification, set a unified data structure, establish a database and database management system to store various product catalogs. It also uses expert resources to carry out synonyms mapping of the classification of different commodities, and to achieve a high degree of correlation between different categories, so that different names under different classification and the names above and below it will show up when retrieving any name.

2   MAJOR COMMODITY CLASSIFICATION AT HOME AND ABROAD

2.1   Common international classification

1   United Nations *Central Product Classification* (CPC)[1] is an international standard of commodity classification developed by the United Nations Statistical Office. It is a complete commodity classification covering both goods and services. It aims to classify goods and services, which are the production result of any economy.

2   *Harmonized Commodity Description and Coding System*(H.S)[2] is an all-purpose international trade product classified catalog made by World Customs Organization on the basis of *Customs Co-operation Council Nomenclature* (CCCN) and UN *Standard International Trade Classification*(SITC), and in reference to the classified catalogs of tariff, statistics, transport and others of major countries.

3   *United Nations Standard Products and Services Code* (UNSPSC) [3], developed by United Nations Development Programme in 1998, is a common classification of products and services used specifically for B2B e-commerce and government procurement. In May 2003, UNDP officially commissioned the USA's Uniform Code Council (UCC) to provide real-time maintenance and management of UNSPSC. UNSPSC is different from CPC and HS. First of all, the main classification framework of UNSPSC is to classify

according to the common functions and main purposes of a commodity, which facilitates international retrieve and inquiry of products and services. Second, UNSPSC is a real-time maintenance management, and users can put forward requirements of products and services classification at any time. Third, UNSPSC uses GSMP (Global Standards Management Process) to achieve interaction between user and maintenance organizations.

2.2 Common chinese classification

1 China's National Standards

Two national standards GB/T 7635.1-2002 have been formulated in order to establish a national product classification and coding standards system that is suitable for different departments and consistent with internationally accepted product catalogs, and to meet the requirements of China's economic management and domestic and foreign trade. National central product classification and codes –Part 1: transportable product[4] and National central product classification and code–Part 2: non-transportable product[5].

2 Classification of Government Departments

According to the demands of socio-economic development, some Chinese governments have formulated a number of product catalogs or specifications, such as *Product Classified Catalogs in Statistics* by National Bureau of Statistics. This catalog meets the demands of national economic accounting and statistics, fully reflects physical production volume and business volume of China's products and services, and provides basis for national macro-economic management and adjustment of economic policy. The catalog can be divided into 5 layers, involving detailed trade names. AQSIQ also formulates *Key Industrial Products Catalog* in order to carry out supervision and inspection work. This catalog can be divided into 3 layers, containing fewer products.

3 Industry Classification

Many industries have also developed their own product catalogs or systems based on requirements of industrial development, e.g. medical devices catalog and machinery and equipment industry commodity classification. Industrial catalog is always specific, containing virtually all products of the industry.

4 Enterprise Classification

Many enterprises, especially e-commerce enterprises such as Jingdong Mall and Ali Baba, have established classification system or catalog related to their products based on requirements of industrial

development. Commodity classification system is essential to their further development.

2.3 Characteristics of different commodities classification

1 Form a system within specific business scope

The classified catalog of each product can provide service to the industry of the drafter of the catalog. The drafter would take into account every specific issues of the industry to make the catalog as specific as possible to get a complete system.

2 Classify from broader range to smaller range

All the product classified catalogs are made from broader range to smaller range, such as "consumer goods – furniture – spring mattress". There are generally 3–5 layers.

3 Name and additional information

Apart from product name, number and other basic information, there is always additional information, such as product description, its number in other catalogs, etc.

3 DESIGN AND DEVELOPMENT OF PRODUCT NAMES MANAGEMENT SYSTEM

3.1 Functional design

3.1.1 Collection of product name information

Collection of product name information can be divided into two steps: (1) enter the basic information of the catalog, e.g. catalog number, catalog name, classification levels, etc.; (2) enter the basic information of the product, e.g. name, corresponding catalog number, upper-level number and additional information. It can be done by bulk import.

3.1.2 Relationship mapping between product names of different standards

The different classified catalogs stored in database shall be associated, with a detailed catalog as the main line to carry out synonyms mapping between products in other catalogs and this catalog. Note down number of synonyms.

3.1.3 Maintenance of product name synonyms

After matching the existing catalogs, the industry experts will provide product name synonyms, which are generally the useful expressions of ordinary consumers, instead of formal expressions. The storage of such expressions is very important to product name information retrieval.

3.2 Design of product name database

Two data tables are designed according to the above functional analysis, see Tables 1 and 2.

Table 1. ClassifyStd (table of product classified catalogs).

No.	Column name	Description
1	StdID	Self-add ID
2	StdNO	Classification catalog number
3	StdName	Category Name
4	ClassifyLevel	Number of classification layer
5	ClassifyStd_backup	Spare field

Table 2. Product classify (maintenance table of product classification).

No.	Column name	Description
1	ProductID	Self-add ID
2	ProductName	Product name
3	StdCode	Classification Catalog Number
4	ProductLevel	Number of classification layer
5	StdParentCode	Parent code
6	SynonymsCodes	Synonyms number collection (separated by semicolon)
7	SynonymsNames	Synonyms collection (separated by semicolon)
8	AdditionalInformation1	Additional Information 1
9	AdditionalInformation2	Additional Information 2
10	AdditionalInformation3	Additional Information 3

3.3 Design of product name management system

Product name management system is designed according to the above functional requirements. The system can update product name database at any time, see Figs. 1 and 2 for maintenance interfaces of classified catalog and product name.

Figure 1. Maintenance interfaces of classified catalog.

Figure 2. Maintenance interfaces of product name.

ACKNOWLEDGMENTS

This work was funded by the Chinese Quality Inspection Commonweal Projects under Grant No. 2012104018-16, the National Social Science Foundation of China under Grant No. 11AZD096, the basic research funds from Chinese government under Grant No. XXX and the National Key Technology R&D Program of the Ministry of Science and Technology under Grant No. 2013BAK04B01.

REFERENCES

[1] UN CPC Classification Introduction. Chinese Commerce, 1993.2
[2] People's Republic of China Customs Import and Export Tariff. Economic Daily Press, 2014.1
[3] Shanghai Xunluo Electronic Technology Co., Ltd., http://www.signalcn.com/News/News_265.Html [EB/OL] 2014.8.1.
[4] GB/T 7635.1-2002 National Major Product Categories and Codes Part 1: Transportable Products Standards Press of China.
[5] GB / T 7635.2-2002. National Major Product Categories and Codes Part 2: Non-transportable Products[S] China Standard Press

Information Technology – Wan et al. (Eds)
© 2015 Taylor & Francis Group, London, ISBN 978-1-138-02785-5

Data transformation network performance analysis

Yanping Chen
Department of Comupter and Information Enginer
Harbin University of commerce, Harbin, China

Yulong Gao
Communication Research Center
Harbin Institute of Technology, Harbin, China

ABSTRACT: As the application of dynamic routing, cognitive computing and network coding, etc., routs not only store and forward but also computing traffic data, which induces new traffic aggregation and change traffic model. Modeling and evaluating the network service processes are the foundation for protocols and algorithms designing. In this paper, the data transformation process is defined to characterize the change of traffic model; based on the definition and taken the coding network as example, cross traffic model under the aggregation of traffic coding is obtained, using this model together with the left service curve theorem, the service provided to current traffic and its end-to-end performance can be acquired. Taken the delay as example, traffic simulation results satisfy the limitation of delay bounds of numerical results, which illustrates that the model presented in this paper characterizes the network well.

1 INTRODUCTION

Future wireless communication will focus on wideband multi-media communication, its best-possible obstacle is imbalance and scarcity of frequency source. Cognitive radio provides a technology to resolve above contradiction, which can exploit the limited frequency source effactually by spectrum assign to adjust some parameters such as modulation type, central frequency(Ciucu et al., 2006). Its key technologies include spectrum sensing, spectrum assign, channel state estimation and so on.

In the traditional communication network, the route just stores and forwards data, thus the data is not alter and is described by commodity flow in the research vividly. As the exploration of cognitive network, autonomic network, etc., the routes not only stores and forwards data, but also have the function of computing and cognition, therefore the data is likely altered in the path in the communication network. The data is changed initiatively to increase the network throughput or enhance the performance, which is different from the passive change arose by the error or losses. At present, network coding is the active technique to increase the network throughput. New data transformation is explored as new network. In this paper, data active or passive transformations are all called data transformation. Analyzing and modeling the network

with data transformation is the foundation of network performance evaluation, network design and etc.

Network calculus is emerging network performance analysis theory after queuing theory. Its theory foundation is (min, +) or (max, +) which has systems theory structure and realize the scaling of queue measure. The determined network calculus can analyze the network performance in the worst case, and recently the stochastic network calculus can perform the statistic analysis for traffic and apply to kinds of networks performance models widely, such as IntServ and DiffServ, and the wireless sensor network. Network calculus has many open problems to tackle accompanied with new network techniques. Constructing the performance model for the network with data transformation presented in the paper is the one. Jing Xie and Ciucu et al. (2006) consider the network error in performance model, Fidler & Schmitt (2006) first present the data scaling, and use it to model the data transformation network. Ciucu et al. (2011) extend the results based on determined network caluslus in Fidler & Schmitt (2006) to the ones based on stochastic network calculus. However, the scaling element (scaling function) is a specific function which cannot cover the general case.

Network information flow is presented in 2000 by Alhswede et al., and it becomes the theory base of network coding. It not only characterize the lossless

Supported by national science foundation of China(NSFC)(61301101)

traffic but also traffic with data transformation. In this paper, we introduce the concept to the network performance model and explore the traffic model based on information flow. The data in traffic information flow can superimpose and aggregate, which is different from the counting process that is used in the conventional traffic model. To characterize this type of traffic from the perspective of the functional analysis, we use the vector in the probability space to describe the traffic arrival process, and the traffic change in the transmission is the projection of the original traffic random vector in the subspace. Based on the above theory, the data transformation is defined; using the data transformation definition, the traffic arrival model after data transformation is obtained.

The remainder of the paper is organized as follows: in Section 1 we take the network coding as example to analyze the detail process when the network node takes the calculus on traffic, then the data transformation is defined and the traffic arrival curve after data transformation is presented; In Section 2, the left service curve under the aggregation way of network coding is presented; In Section 3, the simulation and numeral example is dune to prove the presented models. Finally, brief conclusions are given in Section 4.

2 NETWORK PERFORMANCE MODEL WITH DATA TRANSFORMATION

2.1 Network sceno: Network coding

We take network coding as an example to explain the basic idea of data transformation using a very simple example consisting of three wireless nodes as shown in Fig. 1(a) (Sengupta et al., 2010). In the figure, node 1 wants to send packets to node 3, while node 3 wants to send packets to node 1. Fig. 2 shows the queue when nodes just store and forward data. In Fig. 2, to simplify, data from different traffic flow in different queues. The process of modeling the network is as follows:

1 The arrival and depart process are denoted by $A = (A(t))_{t \geq 0}$ and $D = (D(t))_{t \geq 0}$, here $A(t)$ and $D(t)$ present the cumulate bits of traffic arrival at and leave the node at time t. Obviously, A and D are not decreasing and left continuous, and $\forall t \geq 0$, $D(t) \leq A(t)$. The traffic arrival model is $P(A(s,t) - \alpha(t-s) > x) \leq \varepsilon(x)$, and here $A(s,t) = A(t) - A(s)$.

2 If the path $p(1,2,3)$ provides the service S to traffic A satisfy $P(D(t) < A \otimes S(t)) \leq \varepsilon(x)$, then using the service and arrival model, the performance factor of delay W and backlog B can be obtained:

$$P(W(t) > d(x)) \leq \varepsilon(x) \tag{1}$$

$$P(B(t) > G \oslash S(0) + \sigma) \leq \varepsilon(\sigma) \tag{2}$$

Here

$$d(x) = \inf\{d : S(s+d) \geq G(s) + x\} \tag{3}$$

and G is the traffic arrival envelope.

Different from the store-forward pattern, network coding takes advantage of the broadcast of wireless channels, and make the node have the calculus function except the store-forward (Wu et al., 2009). The queue of network with network coding in Fig.1 is shown in Fig.3, the flow of traffics A and B take part in the network coding in node 2, then the node 2 broadcast the coded data, and nodes 1 and 3 receive the data. As the sender of traffic A and B, nodes 1 and 3 caches the original data, then the decoded data can be obtained in nodes 1 and 3. From Fig. 3 and the analysis above, we can find that the data change in the transmission, therefore the arrival model of the cumulate bits in time t cannot characterize the type of traffic.

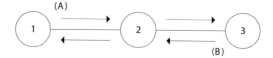

Figure 1. The topology of the example

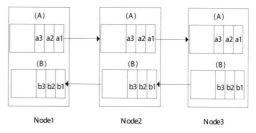

Figure 2. Queue in the node with store-fowared.

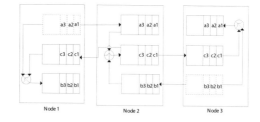

Figure 3. Queue in the node with network coding.

Except network coding, P2P content distribution system, media streaming applications with some transcoding, and dynamic routing, etc, all can change the data in the transmission. The change leads to that the original arrival model is powerless to the data transformation.

2.2 Performance model of network with data transformation

To get the general formula of data transformation, we introduce the probability space $L(\Omega, F, \mathbb{P})$. Assume it is square integrable, thus it is Hilbert space. The random variable of arrival process $A \in L(\Omega, F, \mathbb{P})$. In this paper, the calculus of traffic is abstracted as projection operator in probability space.$L(\Omega, F, \mathbb{P})$ Traffic flow is transformed to information flow under the projection operator. The information flow no more satisfies the model of traffic flow. Next, we define the data transformation to characterize the calculus process, and based on it, present the model of information flow which is the result of data transformation from traffic flow.

Definition1 (data transformation): Random variable $X, P \in L(\Omega, F, \mathbb{P})$, X denotes the traffic arrival process, and the data transformation P denotes the projection operator of X in subspace $L(\Omega, F, \mathbb{P})$, and satisfies $\| PX \| \leq \| X \|$, here $\| \cdot \|$ denotes norm of probability $L(\Omega, F, \mathbb{P})$.

When x is discrete, the norm of X is$\| X \| = \sum_{i=0}^{t} \xi_i$; When x is continuous, the norm of X is $\| X \| = \int_a^b x(t) dt$.

To increase the throughput of network is the main purpose of adding calculus function to network node, then data transformation is a contraction operator, thus $\| PX \| \leq \| X \|$. The original traffic flow model and the data transformation both determine the information flow model, and specifying the projection operator is the crucial point of data transformation.

Theorem 1 (arrival curve of information flow):Given the arrival curve of traffic flow $P(A(t,s) - f(x) > \sigma) \leq \varepsilon(\sigma)$, and the data transformation P, then arrival curve of the information flow satisfies $P(A'(t,s) - pf(x) > \sigma) \leq \varphi(\sigma)$, here $\varphi(\sigma) = \frac{E[e^{-\theta A'(t,s)}]}{e^{\theta(pf(x)+\sigma)}}$.

Proof: Traffic arrival process $A(t,s)$ under the projection p is transformed to $A'(t,s)$, and the envelope is transformed to $pf(x)$. Set $\theta > 0$, there is

$$P(A'(t,s) > pf(x) + \sigma) = P(e^{A'(t,s)} > e^{pf(x)+\sigma})$$

According to Markov inequality, we have

$$P(A'(t,s) > pf(x) > \sigma) \leq \frac{E[e^{-\theta A'(t,s)}]}{e^{\theta(pf(x)+\sigma)}}$$

Here, statistics parameters of $A'(t,s)$ can be obtained through online measurement.

3 PERFORMANCE ANALYSIS OF DIFFERENT TRAFFIC IN NETWORK CODING

3.1 The network model

Network coding is the typical data transformation. Now, we apply the theories presented above to the network shown in Fig. 4. The node 0 sends traffic a_i and a_j to nodes 5 and 6. In the paths from node 0 to nodes 5 and 6, there is a cross sub-path if traffic a_i and a_j pass through the path $p_1 = (0,1,3,4,5)$ and $p_2 = (0,2,3,4,6)$, respectively. The idea of network coding is to take advantage of the feature in the transmission, code the two traffic data in the cross sub-path to increase the throughput. In Fig. 4, network coding happens in node 3. To simply the analysis, assume that there is one coding opportunity in a node, and all data participate in the coding if the traffic does. Next, we use the theories of data transformation to explore the end-to-end delay of flow a_i through the path $p_1 = (0,1,3,4,5)$ and analyze the impact of network coding on delay.

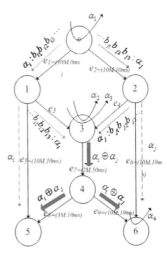

Figure 4. The topology of network with network coding.

The EBB (Exponentially Bounded Burstiness) traffic model is used, and it satisfies

$$P(A(t,s) > \rho(t-s) + \sigma) \leq Me^{-\theta\sigma}$$

Here the envelope $f(x) = f(t-s) = \rho(t-s)$, p is the mean rate, and the error function $\varepsilon(\sigma) = Me^{-\theta\sigma}$.

Network coding maybe takes in same traffic flow or different traffic flows, Fig. 4 follows the later. When different traffic flows take part in network coding, coding is also a type of aggregation of traffic, the aggregation makes the amount of data decrease, and also lead to the left service provided to cross traffic changed. Next, we introduce the detail of data transformation and present the left service curve under the aggregation of network coding.

3.2 Traffic transformation

The rate of traffic flows participating in coding is not same probably. According to the network coding analysis in Fig. 3, the rate of information flow that the result of both traffic a_i and a_j coding is the faster one in a_i and a_j. Assume traffic a_i and a_j satisfies EBB model, and respectively is

$$P(A_i(t,s) > \rho_i(t-s) + \sigma) \le M_i e^{-\theta\sigma}$$

$$P(A_j(t,s) > \rho_j(t-s) + \sigma) \le M_j e^{-\theta\sigma}$$

Here $\rho_i < \rho_j$. Then, the projection p of data transformation of traffic a_i is defined as follows:

$$p = \{\max(\rho_x)/\rho_i\}$$

where x denotes all traffic flows participate in coding. According to theory 1, the transformation of traffic flow a_i is

$$P(A_i'(t,s) > p\rho_i(t-s) + \sigma_i) \le M_i' e^{-\theta_i\sigma_i}$$

Where $M_i = e^{-\theta\rho_i} E[e^{-\theta A(t,s)}]$ and $M_i' = e^{-\theta\rho_i} E[e^{-\theta A(t,s)}]$.

3.3 Left service curve under the aggragation of network coding

To obtain left service curve, traffic in node should be classified, the traffic that participate in coding or not is distinguished.

We use the coding structure in Jiang & Emstad (2005) to present a coding opportunity. For a node i, a coding opportunity at node i is completely specified by a structure S that consists of elements of the form $s = (e_1 e_2, v)$, where e_1 is the incoming link of the packet, e_2 is the outgoing link of the packet, and $v=(c,n)$, depending on whether the packet was received as coded(c) or native(n).

For the necessary of the following performance analysis, we define several variables.

Let $x_i^t(S)$ denote the traffic associated with coding structure S at node i during time t, let $z_i^t(p), e_x e_y \in p$ be the portion of the traffic on path p that is transmitted as native from node i, and the set of traffic flows that goes out as native is defined as

$F_{cnr}(t) = \{F_j \mid F_j \in x_i^t(s), s = (e_x e_y, n) \in S\}$

The set of traffic flows that participates in coding as native-received flows is defined as

$F_{ccr}(t) = \{F_j \mid F_j \in x_i^t(s), s = (e_x e_y, c) \in S\}$

The set of traffic flows that participates in coding as coded-received flows is defined as

$F_{cr}(t) = F_{cnr}(t) \cup F_{ccr}(t)$ $F_{native} = \{F_j \mid F_j \in z_i^t(p)\}$

Thus, the set of flows associated with coding structure S at node i is

The coding structure S can be obtained by the method introduced in Jiang & Emstad (2005).

According left service curve theory, the service provided to cross traffic flows subtracts from the capacity in node gives the service provided to the current flow, here all flows except current flow in node is called cross traffic. The cross traffic is dynamic because of network coding.

(1). if current flow participate in coding, that is $a_i \in F_{cr}$, then the envelope of cross traffic in node i is

$$\alpha_{cross} = \sum\nolimits_{\alpha_{nj} \in F_{native}} \alpha_{nj}$$

(2). if current flow does not participate in coding, that is $a_i \in F$, then the envelope of cross traffic in node i is

$$\alpha_{cross} = \sum\nolimits_{\alpha_{nj} \in (F_{native} - \alpha_i)} \alpha_{nj} - (\odot_{\alpha_{cj} \in F_{cr}} \alpha_{cj})$$

Here $\odot$ denotes the coding operator. To obtain the unified formula, we define the index function

$$I_c = \begin{cases} 0, \alpha_i \in F_{cr} \\ 1, \alpha_i \notin F_{cr} \end{cases}$$

The a_{cross} is

$$\alpha_{cross} = \sum\nolimits_{\alpha_{nj} \in (F_{native} - I_c \alpha_i)} \alpha_{nj} - I_c(\odot_{\alpha_{cj} \in F_{cr}} \alpha_{cj})$$

According to Jiang & Emstad (2005), error function of cross traffic flows is $\varepsilon(\sigma) = \varepsilon_1 \oplus \cdots \oplus \varepsilon_N(\sigma)$. Using the left service theory (Ciucu et al., 2006), the envelope of service provided to current flow is $\beta_{v_i}^0 = \beta_{v_i} - \alpha_{cross}$. And the error function of the service curve is $\varepsilon(\sigma)$.

3.4 The service model of network with coding

The service curve captures the characteristics with which data is forwarded by the node. It abstracts from the specifics and idiosyncrasies of the link layer. The feature of coding determines that its service curve is a specified one. In network coding, as the different arrival rate and independent send time of traffic flows, traffic flows are asynchronous. Therefore, to synchronize the traffic flows is needed in node, which means that the node stores and forward data, also code data. A typical example of a service curve from traditional traffic control in a packet-switched network can characterize the feature, which is rate-latency service curve:

$$\beta_{R,T}(t) = R[t-T]^+ = \begin{cases} R(t-T) & \text{if } t > T \\ 0 & \text{otherwise} \end{cases}$$

Latency T describes the delay caused by the coding in the node.

Assume the schedule scheme is DPS (Dynamic Priority Scheduling), according to the lemma 3 in Jiang & Emstad (2005), respectively, the delay D and backlog B is

$$P(D(t) > x) \le g(x)$$

$$P(B(t) > y) \le h(y)$$

H e r e , $g(x) = f_i \otimes \hat{f}'(x)$, $h(y) = f_i \otimes \hat{f}'((R - \hat{r})y)$, $\hat{f}'(x) = \hat{f}(x + RT)$ .And the current traffic flow arrival curve satisfies $A_i \sim \langle f_i, \alpha_i \rangle$ the cross traffic flows arrival curve is $A \sim \langle \hat{f}, \hat{r} \rangle$.

4 NUMERICAL RESULTS

We take the network in Fig. 4 as example, use the performance derivation process presented in Section 2 to do numerical analysis, and compare it with the results from the simulation to verify theory presented, here the simulation is conducted using NS. First, we measure the delay of traffic from nodes 0 to 5 in shortest path and coding opportunity aware route respectively, and analyse the affect of coding on delay by comparison. We use the following simulation settings. Exponential on-off sources are used for traffic a_i, the average on/off time is 100 ms, the transmission rate during on period is 1 mbps; The CBR sources are used for traffic a_j, the rate is 2 mbps. The coding operator is XOR, the simulation time is 300s, a_i sends data at 3-4s randomly.

In the simulation, the end-to-end delay and its cumulate distribution is obtained by analyzing the traffic trace file using trace function in NS. Figs. 5 and 6 show that delay distribution in shortest path and coding opportunity aware route respectively. In shortest path route, p1=(0,1,5) is the path of traffic ai, and in coding opportunity aware rout, path p2=(0,1,3,4,5) is deliberately chosen for a_j to take advantage the coding opportunity. The delay increases as the bottle link 3–4 in the path p2. This is shown in Fig. 5 and 6 that the range of delay is (0.0216s-0.0224s) in path p1 and (0.282s–0.287s)in path p2. Obviously, the delay is larger in p2. It illustrates that although the throughput can be increased in coding opportunity aware route, the route algorithm does not provide the Qos guarantee for traffic, on the contrary the performance of traffic may reduce. Therefore, the performance of traffic should be considered in route algorithm design, and the performance model should be constructed to evaluate the performance of traffic.

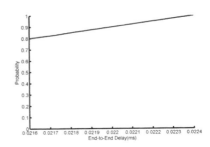

Figure 5. Delay distribution in shortest path route.

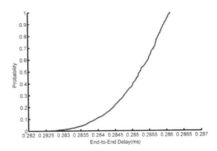

Figure 6. Delay distribution in coding opportunity aware route.

We compare theory and simulation result of end-to-end delay to prove the performance models presented. On-Off exponential traffic belonging to EBB model is used for current and cross traffic and coding opportunity aware route is chosen.

Given the current traffic model (4.1), cross traffic model (4.2) and the service model (6), according to definition of delay bound (7), we can have

$$g(\sigma) = f_i \otimes \hat{f}'(\sigma + RT)$$

Here $\hat{f}'$ denotes the error function of cross traffic and $\hat{f} = M_j e^{-\theta(\sigma + RT)}$, then $g(\sigma) = M' e^{-\theta\sigma}$, where $M' = M_i + M_j e^{-\theta RT}$. According to (2), the delay bound can be obtained

$$d = \frac{\log(M'/\varepsilon)}{\theta(C - \rho_j)}$$

It follows from theorem X with σ replaced by $d(C - \rho_j)$ that

$$P(D(t) > d) \le M' e^{-\theta d(C - \rho_j)} \tag{4}$$

Solving for the delay bound in (4) with the right hand side set equal to ε, we get

$$d(\sigma) = \frac{\sigma}{C - \rho_j}$$

In Fig. 7 we show the delay bounds obtained by (10) and by simulation under the various utilization levels of network node. The current traffic delay measured by simulation is in range of delay bound obtained by theory, which proves the performance present in the paper.

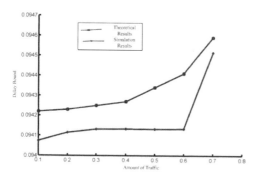

Figure 7. Theoretical and Simulation Results on End-to-End Delay Bound.

5 CONCLUSION

We have extended the state-of-the-art of the stochastic network calculus by deriving the general traffic curve of data transformation. The network coding is taken as an example of data transformation, the cross traffic envelope, traffic transformation and service curve in the coding aggregation is presented. Using the network arrival curves and service curve, we calculated statistical end-to-end delay bounds and compare it with the simulation results. In various utilization levels of network node, the delays measured in the simulation are all in range of delay obtained by calculation, which prove the performance model for data transformation network presented in the paper.

REFERENCES

Ciucu F. et al. Scaling properties of statistical end-to-end bounds in the network calculus. *IEEE Transactions on Information Theory*, 2006, 52(6), 2300–2312

Ciucu F., Schmitt J. & Wang H.. On expressing networks with flow transformations in convolution-form. *Proceedings of IEEE INFOCOM*, 2011, 1979–1987.

Fidler M & Schmitt J B(2006) On the way to a distributed systems calculus: an end-to-end network calculus with data scaling . *Performance Evaluation Review,* 34 (1): 287–298.

Jiang Y. & Emstad P. J.. Analysis of stochastic service guarantees in communication networks: a traffic model. *Proceedings of 19th International Teletraffic Congress,* 2005.

Jing X. & Jiang Y. M.. An analysis on error servers for stochastic network calculus. *Proceedings of IEEE 33rd Local Computer Networks Conference,* 2008, pp. 184–191.

Sengupta S, Rayanchu S & Banerjee S. Network coding-aware routing in wireless networks. *IEEE/ACM Transactions on Networking,* 2010, 18(4), 1158–1170.

Wu K et al(2009). Performance modeling of stochastic networks with network coding. *Workshop on Network Coding, Theory, and Applications, Lausanne, Switzerland.*

Information Technology – Wan et al. (Eds)
© 2015 Taylor & Francis Group, London, ISBN 978-1-138-02785-5

Design and implementation of knowledge management system for research teams based on content based image retrieval

Jing Liu & Fang Meng
Information Engineering School, Communication University of China, Beijing, China

Yuhang Yao
China Art Science and Technology Institute, Beijing, China

ABSTRACT: With the rapid development of science and technology, the form of research teams has great changes, which makes the management of research teams more important. This paper designs and implements a knowledge management system for research teams based on content based image retrieval. In general retrieval algorithms, the scale invariant feature transform (SIFT) algorithm becomes popular due to its robustness for scale variation, rotation, illumination, etc. An improved SIFT algorithm is introduced in this paper in order to reduce the processing time without destroy the superiority of the original SIFT algorithm. The management system, implemented in this paper, can not only provide a better way to retrieval the explicit knowledge, but also a good environment for team members sharing and exchanging the tacit knowledge.

1   INSTRUCTION

With the development of information technology and knowledge economy, a new kind of thought and way of management comes into being, which is called knowledge management [3]. The aim of knowledge management is to build a knowledge system, which can have a cyclic utilization of knowledge from the feedback of information and knowledge. The feedback can be obtained through several methods such as creating, sharing, integration, recording, and so on. Knowledge can be divided into two categories in knowledge management theory: explicit knowledge [2] and tacit knowledge [7]. Explicit knowledge is knowledge that has been articulated, codified, and stored in certain media [2]. It can be readily transmitted to others [2]; tacit knowledge, as opposed to formal, codified or explicit knowledge, is the kind of knowledge that is difficult to transfer to another person by means of writing it down or verbalizing it, such as subjective understanding and experience [7]. An impeccable knowledge management system can make knowledge especially tacit knowledge have a great sharing. The ability of innovation in research teams will be promoted with the help of a perfect knowledge management system.

Research teams are formal groups, made up of researchers in order to achieve a certain scientific research goal. The form of research teams has great changes with the rapid development of science and technology. Nowadays, research teams are made up

by researchers in different disciplines, come from different institutions and even different countries. Then communication between team members cannot be timely and accurate. Besides, one research project is often initiated by a director and other members choose to join or quit according to their abilities, interests and schedule. In this way, the liquidity of team members is tremendous. Some valuable knowledge is often taken away when a member drop out, especially the tacit knowledge which has potential to be created more valuable knowledge. At present, colleges and universities have become important groups to undertake research projects. These projects always tend to be broken up into many sub-topics which are in charge of by students. Then, the related tacit knowledge, which is very important to the projects ongoing or in the future, will disappear when students graduate from school. An efficient management system will give a great support for the teams to do research.

An efficient and accurate retrieval method is essential to a knowledge management system because of there are a variety of explicit knowledge. Therefore, the content based image retrieval is introduced in this paper to improve the retrieval effect. Content based image retrieval method can extract features through image itself, such as color and texture. Therefore, content based image retrieval method can avoid the subjectivity that comes from the text based image retrieval method. In addition, content based image retrieval method will bring great convenience to a new member, when he is not familiar with the query images or even does not know what he wants to retrieve. For example, a new member, who is unfamiliar with the research content of the team, will understand the related subjects as soon as possible through the way of CBIR. At the same time, image features can also be associated with some other information such as the information of relevant researcher, results or related research content with the images. Thus, the new member can understand what should do, get in touch with related researchers and integrate into the team as quickly as possible. An improved retrieval method based on SIFT [4][5] and saliency map detection algorithm is introduced in this paper, which can computational efficiency relative to the original algorithm.

This paper is organized as follows: firstly, it introduces the importance of the theory of knowledge management and content based image retrieval method to the management of research teams. Secondl, the detail of the design scheme of the management of explicit knowledge and tacit knowledge is elaborated. And the retrieval algorithm is emphatically introduced. Third, the two main modules of the system, content based image retrieval module and tacit knowledge management module, are shown. Finally, the whole paper is summarized.

2 SYSTEM STRUCTURE

This chapter mainly introduces the overall design scheme of management of the two kinds of the knowledge, and the detail of the improved SIFT algorithm.

2.1 System framework

Different management strategies are introduced for the tacit knowledge and explicit knowledge according to their special characteristics. For explicit knowledge, the system can make the members find knowledge quickly. For tacit knowledge, the system provides an open environment for communicating between members. The overall system model is shown in Fig. 1.

2.2 Quick access to explicit knowledge

The management of explicit knowledge is focused on how to make team members find the knowledge they want quickly and accurately. Then the theory of knowledge map [6] is introduced in this paper. Knowledge map is the navigation system of knowledge, which can provide a high efficient search engine to users. Not only the knowledge itself but

also the related information such as the researchers or papers will be found through the logical relationship of the knowledge map. Then the member who is finding knowledge can also turn to be the knowledge source to others, which can realize the sharing and exchanging of knowledge. In addition, knowledge map provides fuzzy retrieval function, which allows users easily locate the required knowledge and knowledge source when they cannot express what they want to find. Content based image retrieval method is a good way to implement this function.

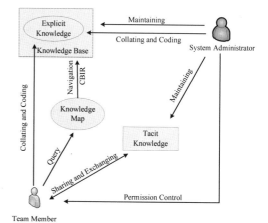

Figure 1.   The overall system model.

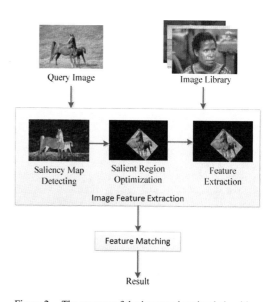

Figure 2.   The process of the improved retrieval algorithm.

2.3 CBIR method based on improved SIFT algorithm

SIFT algorithm has become popular in feature extracting and matching due to its high accuracy and good robustness. An improved SIFT algorithm is implemented in this paper which can improve the computational efficiency without destroying the superiority of the original algorithm or even increased. The process of the improved algorithm is shown in Fig. 2.

2.3.1 Saliency map detecting

a. Algorithm selection: FT (frequency-tuned) algorithm [1] is selected to detect saliency map in this paper due to its computational efficiency and well-defined borders (as shown in Fig. 3) which can be beneficial to the next step, saliency region optimization. FT algorithm exploits features of luminance and color of the images to detect saliency map which is simple to implement and computationally efficient. And it can also output full resolution saliency maps with well-defined boundaries of salient objects [1].

(1) Original 1   (2) Original 2   (3) Original 3

(4) Saliency map 1   (5) Saliency map 2   (6) Saliency map 3

Figure 3.   Saliency maps detected by FT algorithm.

b. Process of salient map detection: There are three steps in the process of salient map detecting: First, smooth the input image with Gaussian filter. Second, calculate the mean value of pixel color of the input image. Third, the Euclidean distance is found between the lab pixel vector in the Gaussian filtered image and the average Lab vector

c. Salient region optimization: due to the irregular shape of the salient regions, a lot of noise will be produced if the salient regions are used for feature detection directly. So it should be optimized before feature detecting.

First, the saliency map (Fig. 4(2)) is binarized, as shown in Fig. 4(3). Second, the largest one of the contour of the highlighted areas is chosen (Fig. 4(4)). Third, the minimum bounding rectangle (Fig.. 4(5)) is generated according to the points of the largest contour. Finally, the points outside the minimum bounding rectangle will be set to black.

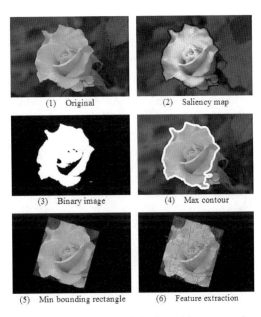

(1)   Original   (2)   Saliency map

(3)   Binary image   (4)   Max contour

(5)   Min bounding rectangle   (6)   Feature extraction

Figure 4.   Salient region optimization and feature extraction.

Different calculation methods are used to judge whether the point is outside the minimum bounding rectangle. It is assumed that the four apexes of the rectangle ordering by abscissa size of the four apexes are $p_1(x_1, y_1)$, $p_2(x_2, y_2)$ $p_3(x_3, y_3)$ and $p_4(x_4, y_4)$. Then the points inside the rectangle should satisfy the following conditions:

The first condition: if $x_1=x_2$, then $x_2 <x<x_3$ and min $(y_2, y_3) <y<$max (y_2, y_3);

The second condition: if $\dfrac{y_2 - y_1}{x_2 - x_1} \leq 0$, then

$$\frac{y_2 - y_1}{x_2 - x_1} < \frac{y - y_1}{x - x_1} < \frac{y_3 - y_1}{x_3 - x_1} \text{ and } \frac{y_3 - y_4}{x_3 - x_4} < \frac{y - y_4}{x - x_4} < \frac{y_2 - y_4}{x_2 - x_4}$$

The third condition: if $\dfrac{y_2 - y_1}{x_2 - x_1} \geq 0$, then

$$\frac{y_3 - y_1}{x_3 - x_1} < \frac{y - y_1}{x - x_1} < \frac{y_2 - y_1}{x_2 - x_1} \text{ and } \frac{y_2 - y_4}{x_2 - x_4} < \frac{y - y_4}{x - x_4} < \frac{y_3 - y_4}{x_3 - x_4}$$

After all the steps, the areas used to extract features can be reduced significantly in order to achieve the purpose of reducing processing time.

2.3.2 Feature extraction

First, SIFT algorithm gets the extreme value points by comparing of the adjacent pixels in the scale space. Second, every extreme value point will be given direction and size though gradient formula. Finally, each feature transform to a 128-dimension feature vector. In the improved retrieval algorithm of this paper, features just can be extracted from the regular salient region (Fig. 4(6)).

2.3.3 Feature matching

The Euclidean distance is chosen to judge whether the feature vector is similar or not.

2.4 Sharing of tacit knowledge

The management of tacit knowledge mainly includes recognition, development and utilization, retain and innovation [7]. As opposed to the explicit knowledge, tacit knowledge is difficult to be expressed and summed up by the sign language, because it is always related to personal experience and inspiration. Coupled with the liquidity and randomness of the members, making tacit knowledge is less likely to be retained. Therefore, taking into account of the above points, tacit knowledge mining is the focus of this part. The system, in this paper, has provided an open environment for members to sharing exchanging tacit knowledge, which can create more valuable knowledge.

3 IMPLEMENTATION OF THE MAIN FUNCTIONAL MODULES

This chapter mainly introduces the two modules according to the system design scheme: the content based image retrieval module of the repository and the discussion board module.

3.1 Experimental Platform

The management system, based on Web technology and B/S architecture programming implementation, is performed on a computer with a 2.0 GHz Intel Xeon processor and 16GB of RAM. Server-side programming language uses PHP, client technology using HTML, CSS and JavaScript. And MySQL is chosen as the database of the system.

3.2 The content based image retrieval of the repository module

Repository is a collection of all the explicit knowledge of the research teams, including various related documentation, project research results, experimental material, etc. Fig. 6 shows the result of the content based image retrieval method.

Figure 5. Result of the content based image retrieval metod.

As shown in Fig. 5, one query image not only can retrieval the similar images, but also retrieval the knowledge source.

3.3 The discussion board module

The discussion board can be used to publish the learning experience or put forward problems in the study or even in our life.

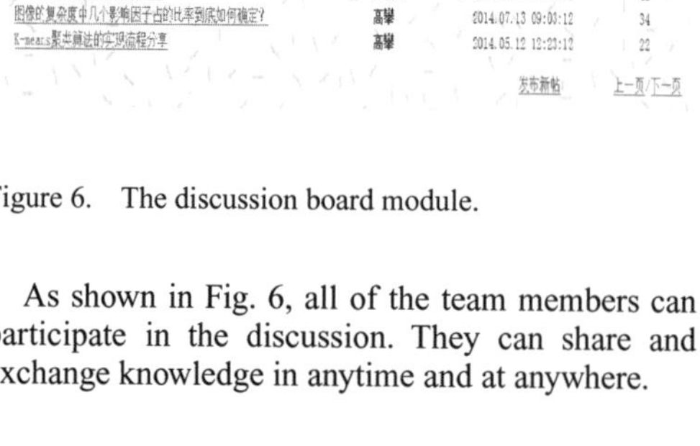

Figure 6. The discussion board module.

As shown in Fig. 6, all of the team members can participate in the discussion. They can share and exchange knowledge in anytime and at anywhere.

4 CONCLUSION

The practical test result shows that the management system implemented in this paper can not only help team members find the explicit knowledge they want conveniently, but also provides a harmonious environment for team members sharing and exchanging tacit knowledge.

This paper just introduces the technology of content based image retrieval. Considering that there are many other kind of resources such as audio, video and 3D resources, the future research will concentrate on the retrieval method for these resources. If the content based retrieval method for these resources can be implemented, the application field for the management system will be more extensive.

Avoid excessive notes and designations.

ACKNOWLEDGMENTS

This work is supported by the Key Project of the National Twelfth-Five Year Research Program of China. (2012BAH01F00).

REFERENCES

[1] Achanta, R., Hemami, S., Estrada, F., and Susstrunk, S., "Frequency-tuned salient region detection, IEEE Conference on Computer Vision and Pattern Recognition, CVPR 2009, pp. 1597–1604, June 2009.

[2] Explicit knowledge .http://en.wikipedia.org/wiki/Explicit_knowledge.

[3] Knowledge management.

[4] Lowe, D. G., "Distinctive image features from scale-invariant keypoints. *International Journal of Computer Vision,*" 2004, 60(2): 91–110.

[5] Lowe, D. G., "Object recognition from local scale-invariant features," *The Proceedings of the Seventh IEEE International Conference on Computer Vision,* vol. 2, 1999, pp. 1150–1157.

[6] Ling He, Jian Huang, Youguo Pi, "The research on the management system for research teams", *Science and Technology Management Research*, 2010, 30(4): 187–190. http://en.wikipedia.org/wiki/Knowledge_management_system

[7] Tacit knowledge .http://en.wikipedia.org/wiki/Tacit_knowledge

[8] Wen, Z., Gao, J., Luo, R., and Wu, H., "Image Retrieval Based on Saliency Attention." Foundations of Intelligent Systems, Springer Berlin Heidelberg, 2014, pp. 177–188.

Information Technology – Wan et al. (Eds)
© 2015 Taylor & Francis Group, London, ISBN 978-1-138-02785-5

Dynamic simulation analysis of engine valve train based on ADAMS

Hongmei Xu & Xinbo Ma
College of Engineering, Huazhong Agricultural University, Wuhan, China

Jinhua Yang
Mechatronics Engineering School, Harbin Institute of Technology, Harbin, China

ABSTRACT: As an important part of internal combustion engine, structure design of valve train greatly affects the mechanical and kinematic properties of internal combustion engine directly. It is of great significance to simulate the kinematics of valve train. Taking the diesel engine of 4108 as the research object, dynamic simulation analysis of valve train was carried out based on the working mechanism and dynamic calculation method. The software of ADAMS / View was adopted to establish the multi-body dynamic simulation model, and various parameters of valve train were achieved. At different speeds of cam, the valve lift, valve velocity, valve acceleration, contact force between cam and tappet, dynamic characteristics of valve spring and the impact between valve and valve seat were subsequently analyzed in depth. The results show that the vales don't fly off or bounce, the seated speed of valve and the shearing stress of valve spring is allowable. The research results can provide references for optimizing the structure of gas distribution mechanism.

1 INTRODUCTION

Valve train is one of the most important components of internal combustion engine. Design of valve train affects its economic performance, vibration, noise and working reliability directly. The main function of valve train is to make fresh and combustible mixture enter into the cylinder timely, thus the exhaust gas can be removed in accordance with the requirements of valve timing. Valve train mainly contains camshaft, valve tappet, rocker arm, push rod and valve spring.

The camshaft drives the valve through the driving mechanism, and makes it open and close according to the predetermined motions. This kind of valve timing system is widely used in internal combustion engine [1]. In order to make the valve train system meet the above requirements and achieve excellent movement characteristics, it is necessary to simulate and analyze the related parameters of the valve train.

Taking the diesel engine of 4108 as the research object, the software of ADAMS / View was adopted to establish the dynamic simulation model of the valve train, and the kinematics and dynamics of the mechanism were simulated and analyzed [2,3]. The results can provide references for the optimal design of gas distribution mechanism.

2 MODEL

2.1 Mechanical system

According to the arrangement of camshaft, gas distribution mechanism can be divided into three categories: down-setting camshaft, centrally-mounted camshaft and overhang camshaft. At present, down-setting camshaft is the most widely used in the internal combustion engine of low velocity. The way it works is that the crankshaft drives the camshaft by a transmitting belt or gears. The cam transfers the driving force to the swing arm via tappet and pushrod, and makes it rotate around the rocker arm. When the rocker rotates to a certain position, it will contact with the valve and run the valve. The advantage of gas distribution mechanism with down-setting camshaft is that the motion and transmission of the system is simpler and more reliable because the camshaft is near from crankshaft [4]. In this work, the diesel engine of 4108 was investigated, which utilizes the down-setting camshaft. The gas distribution mechanism mainly contains a camshaft, tappet, pushrod, adjusting bolt, rocker arm, valve spring, valve spring seat and so on. The related design parameters are shown in Table 1 According to the design parameters, the three-dimensional model was established as Figure. 1.

Table 1. Margin settings for A4 size paper and letter size paper. Parameters of 4108 type valve cam.

Parameter	Value
Datum circle radius of cam (r_0)	18 mm
Lift of buffer segment (h_0)	0.3 mm
Lift of basic segment (h)	7.5 mm
Wrap angle during the buffer (α_0)	16°
Half wrap angle of basic segment(α_B)	59°
Camshaft Speed (n)	1300 r/min
Tappet speed of end of buffer segment (v_{oT})	0.167 m/s
Rocker ratio (k)	1.667
Valve clearance	0.2 mm

Figure 1. The general assembled drawing.

2.2 *Model preprocessing*

Mechanical systems are typically composed of many moving components. Constraint relations exist between each contact member, which is known as kinematic pair or hinge. In order to truly simulate the movement status of system, it is necessary to draw out the corresponding motion pair based on the actual situation and define the kinematical pair between the members. Table 2 shows the constraint relationship between the members of the gas distribution mechanism.

Table 2. Constraint of simulation model.

Number	Constrained object	Constraint
1	Cam and ground	Revolute Joint
2	Cam and tappet	Cam couple
3	Tappet and pin	Fixed Joint
4	Pin and pusher	Revolute Joint
5	Pusher and regulating bolt	Spring modeling impact[1]
6	Rocker and regulating bolt	Fixed Joint
7	Rocker and bearing	Revolute Joint
8	Bearing and ground	Fixed Joint
9	Rocker and valve	Spring modeling impact
10	Valve and ground	Translational Joint
11	Valve and valve seat	Spring modeling impact
12	Pusher and ground	Translational Joint

3 RESULTS

3.1 *Valve lift*

The valve movement pattern determines the gas exchange process of engine, especially the valve lift curve, which reflects the relationships between opening degree of valve and time directly. Figure 2 shows the valve lift curve of system at six different speeds of cam.

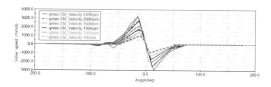

Figure 2. Valve velocity curves at six different speeds.

As shown in Figure 2, the valve lift at the rated speed is 13.9mm. Additionally, lift curve is continuous, and that is to say there is no valve bounce. The valve lift curve coincides with each other at six different speeds. It means the speed has little effect on valve lift.

3.2 *Valve velocity and valve acceleration*

Figure 3 shows the valve velocity curves at six different speeds. According to the results, the maximum and maximum seating velocity of valve was extract, and the results are shown in Table 3.

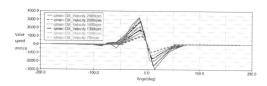

Figure 3. Valve velocity curves at six different speeds.

The maximum speed was 1.852m/s and the maximum seating velocity was 0.193m/s at the rated speed, which comply with the design requirements. The maximum velocity and maximum valve seating

Table 3. Valve velocity analysis table under 6 kinds of speed.

	700 r/min	1000 r/min	1300 r/min	1600 r/min	2000 r/min	2400 r/min
Maximum (m/s)	0.99	1.38	1.85	2.24	2.81	3.33
Maximum seating (m/s)	0.13	0.13	0.19	0.22	0.29	0.34

velocity increase with the cam speed, but the valve seating velocity is less than 0.6m/s even at the speed of 2400r/min. In order to express the change of acceleration clearly, the speeds of 700~1300r/min are defined as low speed, while the speeds of 1600~2400r/min are defined as high speed. The curves of acceleration are shown in the Figure 4 and Figure 5.

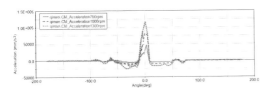

Figure 4. Valve acceleration curves at the low speed.

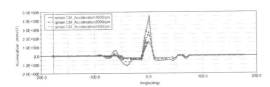

Figure 5. Valve acceleration curves at the high speed.

Based on the above analysis, the results are as follows:

1 The maximum acceleration of valve was 1176 m/s2 at the rated speed. It can reduce the acceleration impact of valve by utilizing the cam ramp with constant acceleration and speed.
2 The acceleration of valve increases with engine speed. The acceleration impact arising from the opening and closing of valve increases accordingly.
3 As one of the most important indexes, the greater the width of negative acceleration section is, the longer the opening time of valve is and the higher the time area of cam is. In addition to this, the speed affects the width of negative acceleration section significantly. The width of negative acceleration section is smaller at the low speed. With the increase of speed, the width of negative acceleration section will be gradually broadened. It

indicates that engine can obtain higher time area of cam at the high speed, and the time area is lower at the low speed.

3.3 Contact force between cam and tappet

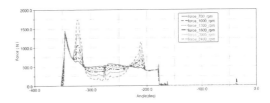

Figure 6. Contact force between cam and tappet.

Compare the valve lift curve with the contact force, the conclusions can be drawn as follows:

1 The movement pattern determined by the cam profile is transferred to the valve accurately. No high distortion occurs, which can avoid the impacts of parts against each other effectively.
2 The contact force is continuous and has no mutations when the cam moves over its top. There was no phenomenon of flying off.
3 The curves of contact force were similar to each other at six different speeds. The maximum contact stress between cam and tappet almost occurred when the relative opening of valve reaches the maximum.
4 There is a contact force at the end of the rising and falling cam ramp, but the value is less than that at work.

3.4 Dynamic characteristics of valve spring

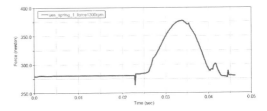

Figure 7. Valve spring force curve.

Figure 7 and Figure 8 show the force and shear stress of valve spring in a single working circle respectively. In the initial position, the force of valve spring is 164 N, which is not equal to the preload of 771N because of the gravity of valve. The maximum force of valve spring is 1006 N. The shear stress curve of spring is similar to that of valve lift. It suggests that

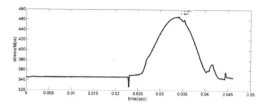

Figure 8.   Shear stress of spring.

the movement pattern of cam has been transferred to the valve. According to the calculation, the maximum and minimum shear stress is 465 MPa and 326 MPa respectively. Because the allowable stress is 198.54N and the working ultimate load is 830MPa, the valve spring of the mechanism can meet the working requirements.

4   CONCLUSIONS

Taking the diesel engine of 4108 as the research object, the software of ADAMS / View was adopted to establish the dynamic simulation model of the valve train. At six different speeds, the valve lift, valve velocity, valve acceleration, contact force between cam and tappet, dynamic characteristics of valve spring, impact between valve and valve seat were subsequently analyzed in depth. The results show that the acceleration of valve increases with engine speed. The acceleration impact arising from the opening and closing of valve increases accordingly. Additionally, the contact force is continuous and has no mutations when the cam moves over its top. In a word, the vales don't fly off or bounce, the seated speed of valve and the shearing stress of valve spring is allowable. The results can provide references for the optimal design of gas distribution mechanism.

ACKNOWLEDGMENT

This research was supported by Specialized Research Fund for the Doctoral Program of Higher Education of China "Research on the prediction and optimization mechanism of the acoustic performance for vehicle-use sound packages based on the hybrid method" under grant Nos. 20110146120008, National Innovation Experiment Program for University Students" Prediction and experimental study on the acoustic performance of vehicle-use sound packages based on FE-SEA hybrid model " under grant Nos. 1210504013, and Student's Science & Technology Innovation Fund of Huazhong Agricultural University under grant Nos. 11028.

REFERENCES

Hao, Y. G. 2006. Dynamics of the gas valve based on the multi body system. Master Thesis, Hangzhou: Zhejiang University.

Chen, T. 2009. Dynamics characteristics of combustion engine. Master Thesis, Wuhan: Huazhong University of Science and Technology.

Niu, W. B. 2008. Multi body dynamics modeling and analysis of valve-train. Master Thesis, Tianjin: Tianjin University.

Cemal, B., Mehmet, P.2004. A New Approach to Simulation of Hydraulic Valve lifter of a Car Engine, Turkiah J.eng. env.Sci: 1–11.

Engelhard, T., Ulbrich, H., Remmelgas, J. 2005.Nonlinear Dynamics of Timing Gear Transmissions Including Valve Trains, VDI-Tagung: International Conference on Gears 200, Munchen, VDI-Bericht 1904.

Martin, F. 2007.Thomas ENGELHARDT and Heinz ULBRICH, Contacts within Valve Train Simulations: a Comparison of Models, Journal of System Design and Dynamics, 1(3):513–523

Wei, G.D. 2010.Study of Effects of Valve Train Optimization for Engine Performance. Master Thesis, Tianjin: Tianjin University.

Zhang, X.R, Zhu, C.C, Wu, J.Y.2008. System Dynamic Analysis of Engine Valve-train, Journal of Chongqing University, 3:294–298.

Information Technology – Wan et al. (Eds)
© *2015 Taylor & Francis Group, London, ISBN 978-1-138-02785-5*

Electronic counter measures against reconnaissance satellites

Xiongjun Fu, Ting Li, Shengqi Qian, Jiayun Chang & Min Xie
School of Information and Electronics, Beijing Institute of Technology, Beijing, P.R.China

Jin He
Shanghai Aerospace Electronic Communication Institute, Shanghai, P.R.China

ABSTRACT: Electronic warfare against Radar reconnaissance satellites and electronic reconnaissance satellites becomes urgent nowadays with the development of space-based radar and electronic reconnaissance technology. By the analysis on the targets to be jammed in detail, including spaceborne imaging surveillance radar, spaceborne Ground moving target indication (GMTI) warning radar, spaceborne air moving-target indication (AMTI) warning radar and spaceborne electronic reconnaissance facilities, the weaknesses of them are revealed, the feasibility of jamming against these facilities is studied, and the operational schemes are proposed.

1 INTRODUCTION

The technology of spaceborne imaging surveillance radar and spaceborne electronic reconnaissance equipment has got great achievement in recent decades on the microwave band and millimeter wave band. Spaceborne synthetic aperture radar (SAR) satellites can make long-term mapping for global or some local areas with high imaging resolution.

Owing to the serious atmospheric attenuation effect of electromagnetic wave with the wavelength close to 2.7 us or 4.3 us, it is difficult for spaceborne infrared warning system to detect the ballistic missiles flying inside the atmosphere. Thereby, the spaceborne warning radar can provide accurate range and velocity of ballistic missiles and thus the warning time could be longer.

The countermeasure methodology against space-based radars and reconnaissance equipments discussed in this paper are as follows:

1 Countermeasure against imaging surveillance radars.
2 Countermeasure against ground moving target indication (GMTI) warning radars.
3 Countermeasure against air moving target indication (AMTI) warning radars.
4 Countermeasure against electronic reconnaissance equipments.

There are some advantages for the jammers against the spaceborne radars:

1 As for the warning radar, the signal to noise ratio (SNR) is inversely proportional to the quartic of the range from the radar to a target; As for the synthetic aperture radar (SAR), the SNR is inversely proportional to the cubic power of the range; while as for the receiver of jammers, the SNR is just inversely proportional to the quadratic power of the range. Therefore, it is more advantageous to the interferers as the range becomes longer.
2 The spaceborne warning radar obeys the law of two-body motion, which makes its orbit easy to be detected.

There are some advantages for the jammers against the spaceborne electronic reconnaissance equipments:

1 Just as the spaceborne warning radar, the spaceborne electronic reconnaissance equipments obey the law of two-body motion, which makes its orbit easy to be detected.
2 Ground-based jamming equipments could have high enough transmitting power to mask the adjacent emitters. Therefore, the electronic reconnaissance equipments could not sense the presence of the adjacent emitters to be protected.

According to the analysis above, it is feasible to jam against spaceborne radar and electronic reconnaissance equipment.

2 ANALYSIS ON THE JAMMING OBJECTS

2.1 Spaceborne imaging reconnaissance radars

As a kind of high resolution, high SNR, all-time and all-weather imaging radar, SAR has been deployed in spaceborne platform for reconnaissance to the Earth owing to its capacities of remote sensing or ground/foliage penetrating.

The state-of-art spaceborne imaging reconnaissance radars have some characteristics such as:

1 The operating frequency band becomes wider. For isntance, SIR-C/X-SAR covers L, C, and X bands [1].
2 Single-polarization mode is replaced by full-polarization mode to some extent [1].
3 The incidence angle of incidence could be mechanically or electronically scanned instead of fixed, so that the width of surveying and mapping can be adjusted. SIR-C/X-SAR adopts phased array antenna for instance [1].
4 The bandwidth increases so that the slant-range resolution increases.
5 Multiple operation modes including stripmap SAR, spotlight SAR, scan SAR and hybrid SAR are integrated, such as ASAR on the ENVISAT.
6 Netted SAR systems spring up, such as TerraSAR-1.

In all, multi-band, multi-polarization, multi-angle, high-resolution, and multi-mode are the distinctive features of the new generation of space-based SARs, and netted SARs and distributed SARs are the key issues in recent years.

2.2 *Spaceborne warning radars*

Spaceborne warning radars can be used to detect satellites, space debris, ballistic missiles and strategic bombers. The warning time can be longer than ground-based, sea-based or aerial facilities due to the absence of national boundaries in space. Spaceborne warning radars can be divided into spaceborne GMTI warning radar and spaceborne AMTI warning radar.

The spaceborne GMTI warning radar is the one that has orbital motion, and has the capacities of high-resolution imaging and ground moving target indication. It is a good replacer of the aerial warning radar [2].

The spaceborne AMTI warning radar is the one that has orbital motion, and has the capacity of air moving target indication. It is good at detecting ballistic missiles and high-altitude bombers. However, it is difficult to be put into practice compared with the space-based GMTI warning radar. The major reasons include much lower radar cross section (RCS), higher data rate requirement and higher cover area compared with GMTI application [3].

2.3 *Spaceborne electronic reconnaissance equipments*

Spaceborne electronic reconnaissance equipments make electromagnetic reconnaissance, intercept the signals and then acquire the parameters of emitters

on passive mode. The parameters such as operational frequency, pulse width, bandwidth, pulse repetition frequency, direction of arrival (DOA), and polarization are useful to analyze the deployment state of emitters. Low Earth orbits in the range from 300 km to 500 km or the range from 1000 km to 1400 km are common selection. Synchronous orbits are also good candidates.

Precise positioning for the emitters is the core technology of the spaceborne electronic reconnaissance equipments. The three coordinates of an emitter's location could be resolved by data processing based on the parameters extraction of the emitter's signals. The reconnaissance modes include location by angle measurement from single station (i.e., using single satellite), DOA location from single, double or multiple stations, time difference of arrival (TDOA) location from double or multiple stations, etc.

Location by angle measurement from single station is the most basic approach; it cannot measure the distance but the direction (azimuth and elevation) of an emitter.

In DOA location approach from single station, the station resolves the location of a radiation source by receiving and processing the signals at different times, etc., at different observation angles. Since considerable position changing is required for a reconnaissance platform itself, it will take a long time to complete the task.

As for DOA location approach from double or multiple stations, the location of a radiation source is the intersection of the bearing lines, which is resolved by two or more stations receiving the radiation signals at the same time. The precision of this approach is higher than that of DOA location approach from single station. The location system is fairly complex since cooperation between or among satellites is required.

As for time difference of arrival (TDOA) location approach from double or multiple stations, two or more adjacent satellites receive the signals from emitters and then forward them to the ground stations, at which the TDOA and the difference of Doppler frequency are measured. Therefore, the real location of the emitters could be determined according to the geometries. This approach has the highest measurement precision compared with other approaches.

3 JAMMING METHODS

3.1 *Jamming against spaceborne SARs*

There are some disadvantages for the jammers against the spaceborne SARs:

1 Very high jamming power is required owing to the facts that the procedure of SAR imaging is a

two-dimensional matched filtering, and the jamming signal usually feed into the antenna of SAR from the sidelobe.

2　As for noise jammers, higher transmitting power is necessary since the SNR of SARs is inversely proportional not to the quartic but to the cubic of the range from the radar to a target.

3　As for deception jammers against phased array SARs, owing to the effect of aperture fill time, false-target deception jamming could not works well only if the jamming signal feeds into the antenna of SARs from the mainlobe or the adjacent sidelobes. Therefore, multiple jammers are required in order to protect wide area.

Fig. 1 shows the common SAR jamming methods, including passive and active jamming.

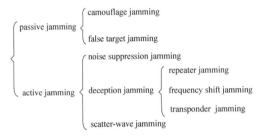

Figure 1.　Jamming methods.

In spite of these unfavorable factors, jamming against SARs is feasible due to the facts below:

1　The space-based SAR obeys the law of two-body motion, which makes its orbit easy to be detected.

2　Many SARs transmit the same waveform in different pulses.

3　The beam direction of SARs should meet the requirement of tasks, which is beneficial for jammers. For instance, when operating on stripmap mode, the angle between the beam direction and the direction of flight should maintain a fixed angle. As to spotlight mode, the antenna beam is always pointing to the observation area.

The essence of passive jamming is to add jamming channels on SARs' signal transceiver path. For instance, the chaff echo from the illumination of SARs directly or from the scattered return wave of the ground can be an effective jamming owing to its good coherence with the transmitting signal of SARs. The major materials of passive jamming against SARs are many kinds of metal reflectors and radio frequency (RF) absorbers, which can be used to increase or decrease the RCS of ground or military equipments and make the SARs get a false radar image or make the image be distorted.

The main effects of various passive jamming methods are as follows:

1　Mix the spurious with the genuine.

2　Damage the true distribution of ground's RCS.

3　Damage the calibration to SARs especially to interferometric SARs.

4　Make the receiver of SARs saturated.

5　Yield false image texture to increase the difficulty of SAR image interpretation [4].

Active jamming is the core approach against the SARs, including noise suppression jamming, deception jamming and scatter-wave jamming.

Noise suppression jamming can reduce the SNR of SAR echoes by transmitting high power random noise. The resolution of SAR images will be degraded to some extent and result in failure in serious cases. Coherent noise jamming with the operation frequency band equals to the SAR can obtain the gain of pulse compression.

Deception jamming can achieve signal processing gains on range dimension and azimuth dimension by digital radio frequency memory (DRFM) approach, therefore the requirement of transmitting power is reduced amazingly. Noisy Doppler frequency and time delay are modulated on the received signal to produce false targets. Repeater jamming, frequency shift jamming and transponder jamming all belong to deception jamming.

Transponder jamming is the most effective approach in the meantime is not easy to put into practice. A transponder jamming system against spaceborne SARs includes six components: signal interceptor, modulation information generator, jamming signal generator, transmitter, parameter measurement sub-system, and false target database as shown in Fig. 2.

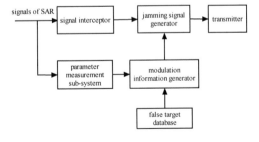

Figure 2　Diagram of a transponder jammer against spaceborne SARs.

Scatter-wave jamming [5] does not transmit jamming signal directly to the radar, but intercepts the radar transmitting signal and generates coherent pulse train, then emits the pulse train into the area to be protected. Thus, SAR imaging and target recognition many be destroyed since the real image is blurred by a false image. This kind of jamming has good capability of self-protection and is easy to be realized.

A single jamming approach may not work well due to the fact that spaceborne SARs are developing toward to multiple frequency bands, full polarizations, multi-resolutions, multiple perspectives and multiple modes. A comprehensive solutions combined with many kinds of jamming approaches against SARs should be proposed, such as passive/active jamming, noise-suppression/deception jamming, direct-wave/scatter-wave jamming, true target screening/false target jamming, etc.

In order to implement jamming or even netted jamming against space-based SARs more effectively, it is necessary to identify the operation mode of SARs by electronic support measure (ESM) and electronic intelligence (ELINT).

The possible operation modes of spaceborne SARs include: side-looking stripmap SARs, forward or backward squint side-looking stripmap SARs, Scan SARs, multi-look mapping SARs, spotlight SARs, ground moving target indication (GMTI) SARs, interferometric SARs (InSAR) mode (single pass or multi-pass).

Pulse description word (PCW) generation and parameter analysis on statistical properties of PCW are essential for SARs operation mode estimation, so the SAR antenna's pattern estimation and the recorded data of interceptor system parameters are likewise. An approach of SARs operation mode recognition is presented here:

1 The curve of azimuth angle vs. time could be drawn in the light of interception equation, according to the sidelobe power received within a period of time and the ranges between the SAR satellite and the interception receiver in the corresponding time. If the variation of the curve obeys the law of an actual stripmap SAR, i.e., the SAR antenna is not rotated as for a mechanical antenna or not electronic-scanned as for a phased array antenna, then we could draw a conclusion that the spaceborne SAR is operating on stripmap mode. It is the same to other mode.

2 Space-based SARs are prone to remain the same ground resolution and azimuth resolution in different sub-swath when operating in Scan SAR mode. To this end, narrower bandwidth of transmitted signals is adopted to reduce the slant range resolution as the incident angle increasing; in the mean time, more pulses are increased. Therefore, if the bandwidth of an emitter's signal intercepted by a jammer is periodically changed, it is evident that the radar is working as a scan SAR.

3.2 Jamming against spaceborne warning radars

It is conducive to the long-time reconnaissance for ground-based reconnaissance receiver to acquire precise radar parameters that spaceborne warning radars, including GMTI warning radars and AMTI warning radars that work for longer time continuously.

The main jamming methods against spaceborne warning radars are noise suppression and deception. The noise jamming against warning radars is easier than against SARs. Coherent noise jamming, also called smart noise jamming, should be adopted to save power. High fidelity false target jamming is the main approach of deception jamming. The most dazzling feature of deception jammers is that the characteristics of inner-pulse phase are preserved owing to the technique of DRFM.

The Doppler frequency to be modulated should be consistent with rationally predesigned dimensions of the false target and the law of its motion. The RCS, the velocity and the dimensions of a false target should fit the actual situation. For example, as to AMTI warning radars, the RCS, the velocity and the dimensions can be designed as 0.1 m2, 5 Mach and 0.5m × 4m, respectively.

In major scenario, the jamming power feeds into the spaceborne warning radars from the sidelobes of the antenna. A novel sidelobe jamming solution is presented here. Multiple jammers are located on different regions, and each of the jammers works or shuts down according to a pseudo-random sequence all by itself. Different jamming signal with irregular transmitting power feeds into the radar's antenna from different sidelobe. The number of operating jammers is irregular at any time. As a result of this jamming, the anti-jamming adaptive beamforming of the radar does not work well.

3.3 Jamming against spaceborne electronic reconnaissance equipments

The best countermeasure against spaceborne electronic reconnaissance equipments is to keep radio silence if the orbital parameters of the equipments are known. More jamming approaches are required since it is difficult to acquire orbital parameters of all the equipments and some satellites have the ability of orbital transfer.

High-power noise suppression spot jamming is the basic approach to reduce the probability of intercept for emitters to be protected. In general, the transmitting power of jammers should be at least 20 dB higher than the highest sidelobe of an emitter to be protected.

Active deception jamming is a significant approach. Let a large number of jammers deployed around the radar to be protected radiate confused electromagnetic waveforms, which parameters such as carrier frequency, pulse repetition interval (PRI) and so on are close to the transmitting signal of our radar. Thus as for the electronic reconnaissance equipments the signal sorting becomes more difficult. Unmanned aerial vehicle (UAV) can be adopted in practice. As for electronic reconnaissance equipments which locate emitters by angle measurement from single station, active deception jamming will bring about location error and an elliptical ambiguity zone which size is relative to the orbital altitude of the electronic reconnaissance equipment and the range between the equipment and the radar [6]. As for electronic reconnaissance equipments which locate emitters by DOA from single, station, active deception jamming will bring about false location. Assume there n emitters including the radar to be protected and the jammers, n^2 cross-points will appear while $n(n-1)$ cross-points are false.

A technology called low probability of intercept (LPI) radar is a comprehensive method against spaceborne electronic reconnaissance equipment. LPI radars can avoid been intercepted by jammers when detecting the targets of interest. There are some solutions to acquire LPI characteristics such as ultra-low side-lobe antenna, omnidirectional antenna with multi-beam receiving and long-time coherent integration processing, wideband frequency agility, frequency diversity, coherent integration, RF power management, low peak power and LPI radar signals.

In general, LPI radar signals have large time-bandwidth product and complex forms and parameters, such as non-linear FM signals, pseudo-random multi-phase coded sequences, and hybrid modulated signals. In addition, the technology of orthogonal waveform random agility utilized among pulses or pulse groups can reduce the probability of being intercepted. The radar receivers could be safety if bistatic system or multi-transmit multi-receive distributed coherent system is adopted.

4 CONCLUSIONS

This paper makes a comprehensive discussion and presents some solutions on the jamming approaches against the spaceborne radar and the electronic reconnaissance equipments. It is beneficial to acquire more prior orbital information of the targets and update it frequently. Large antenna is helpful to increase the gain of antenna and the receiver's sensitivity. High speed sampling and high precision parameter measurement with low computational overhead are significant for coherent jamming. The technology of low probability of intercept (LPI) radar is a potential approach against electronic reconnaissance.

ACKNOWLEDGMENT

This work was supported by National Natural Science Foundation of China under Grant No. 61271373, and by Shanghai Aerospace Science and Technology Innovation Foundation under Grant No. SAST201240.

REFERENCES

[1] Ian G. Gumming, Frank H. Wong, "Ditigal Processing of Sythetic Aperture Radar Data-Algorithms and Implementation" Boston | London: Artech House, 2005.
[2] J. Cao, B. R. Wang, Y. Wang, G. Y. Zhang, "Development trend of space based radars", Electronic Engineer, 2005, 31(5):1–5.
[3] K. Xie, L. K. Lin, W. An, Y. Y. Zhou, "Analysis of system concept of space-based radar system", Modern Radar, 2006, 28(9): 1–5..
[4] G. Y. Zhang, "Electronic Counter-measures against synthetic aperture radars", Modern Radar, 2008, 30(7): 1–9.
[5] D. H. Hu, Y. R. Wu, "The scatter-wave jamming to SAR", ACTA Electronica Sinica, 30(12): 1882–1884.
[6] X. Jiao, Y. G. Chen, "On study of countermeasures against electronic reconnaissance satellites", Electronic Information Warfare Technology, 2004, 19(1): 34–37.

Information Technology – Wan et al. (Eds)
© *2015 Taylor & Francis Group, London, ISBN 978-1-138-02785-5*

Electrostatic potential properties along the ionic pathway of ClC-ec1 protein

T. Yu, X. Guo, X. Ke & J.P. Sang
Department of Physics, Jianghan University, Wuhan, China

ABSTRACT: The ClC-type protein family is widely distributed throughout the biological world, and functionally important for metabolism of the organism. Biological function of this type protein closely relates to the ionic transport rate and the situation of the channel. In this paper, we analyzed the electrostatic potential properties of ClC-ec1 protein, which belongs to *Escherichia coli*. According to comparison between the wild-type structure and the protonated E148 structure, we found that there two functionally important factors for ionic transport: the width of the channel and the electrostatic potential distribution along the channel. The narrowest part of the channel determines the ionic transport quantity. Protonation of the gating amino acid E148 can produce positive potential values along most parts of the channel. These positive potential is beneficial for Cl permeation through the channel. This work provides a guide rule to estimate the importance of amino acids in the channel and how these amino acids influences the ionic transport in ClC proteins.

1 INTRODUCTION

ClC-type chloride channel proteins are widely distributed throughout the biological cell membrane, and functionally divided into two types of categories: passive Cl channels and secondary active Cl/H$^+$ transporters (Accardi & Picollo 2010, Jentsch 2008, Miller 2006). In the passive Cl channels, the selective chloride ions flow down its electrochemical gradient. The transporters, which are also known as Cl/H$^+$ exchangers or antiporters, couple a fixed stoichiometric exchange of chloride ions and protons in the opposite direction (Accardi & Miller, 2004). Although they have different functions, similar molecular architectures existed among them (Dutzler, 2007, Feng et al., 2010). In the human genome, there are nine ClC members which are strictly divided functional properties. Among them, four ClC proteins (ClC-1, ClC-2, ClC-Ka, and ClC-Kb) are divided as ion channels and the other five ClCs (ClC 3-7) are considered as transporters (Jentsch et al., 2005, Achcroft et al., 2009, Jentsch, 2008). These proteins play critical functional roles such as, regulating the membrane action potential, muscle excitability, renal intravascular transport, cell flexibility, and so on (Accardi & Picollo, 2010, Miller, 2006, Jentsch, 2008). Several genetic diseases in humans are directly associated with functionally defection of ClC mutants (Chen, 2005, Lourdel et al., 2012).

Among these channels and transporters, a prokaryotic transporter ClC-ec1 protein which belongs to *Escherichia coli* is the most representative one because its high-resolution crystal structure is obtained by using X-ray diffraction (Dutzler et al., 2002, 2003).

Since this protein is readily over expressed, purified, and reconstituted (Maduke et al., 1999, Accardi et al., 2004), it provides a more favorable molecular framework for experimental and theoretical research. The systematic experimental research indicates that ClC-ec1 protein transports its substrate ions with an obligatory stoichiometry of 2 Cl for 1 H$^+$, and they move in opposite directions (Accardi & Miller, 2004). Based on the wild-type crystal structure of ClC-ec1 protein, there have a narrow 15 Å selective filter region inside each subunit with two Cl binding sites: the central site S_{cen} and the internal site S_{int} near the intracellular entrance. Moreover the putative gating glutamate residue E148 occupies the third binding site S_{ext} near the extracellular entrance of the selectivity filter. The side chain state of glutamate E148 is functionally important for the gating mechanism of ClC-ec1. According to structural properties of ClC proteins, many theoretical and computational algorithms were developed, and valuable molecular details about the ion transit pathway and the ion transport mechanism were obtained (Han et al., 2014, Ko & Jo, 2010, Nguitragool & Miller, 2007, Kuang et al., 2008, Jayaram et al., 2008, Miller & Nguitragool, 2009). Since chloride ion has a negative charge, it's obvious that the charging distribution of ClC protein directly affects the movement of Cl when permeating across the channel. Differences of the electrostatic potential along the channel make Cl have different staying time of different binding site, and this phenomenon can be reflected by measuring binding energy of Cl (Picollo et al., 2009). Since chloride ions bind to S_{int} is much weaker than the other two binding sites (S_{int} and S_{ext}), it reflects that the electrostatic potential distribution among these binding sites is distinguished. Therefore, the relationship between Cl binding properties and electrostatic potential distribution of ClC protein should be investigated systematically.

In this paper we analyzed the electrostatic potential properties of ClC-ec1 channel protein based on the all-atom molecular dynamics (MD) simulation. Magnitude of the calculated electrostatic potential reflects the influence on Cl functional transport. Moreover, we attempt to find some potential relationships between our calculated results and the ionic transport mechanism of ClC proteins.

2 MATERIALS AND METHODS

2.1 MD simulation for the ClC-ec1 protein

The original crystal structure of wild-type ClC-ec1 protein and its mutants can be downloaded from Protein Data Bank (PDB). It is a huge molecule which contains 5 amino acid chains. Chain A and chain B are the functional part across the cell membrane. Fig. 1 shows the wild-type structure of ClC-ecl. It resembles a homodimer and each chain contains an independent pole. There are two binding sites (central binding site: S_{cen} and internal binding site: S_{int}) which are occupied by chloride ions in each chain. If the electronegative residue E148 is protonated by a proton or mutates to the neutral amino acid E148A, the third binding site S_{ext} of chloride ion will appear nearby the extracellular side. Then, the ClC-ec1 protein obtained above was embedded in an POPE membrane with size of 120 Å × 120 Å, and a 10 Å water layer was added on both top and bottom of membrane by using soft VMD (Humphrey et al., 1996). At last 57 Cl and 47 Na$^+$ ions were mingled in water to make the solution concentration at 300 mM and the total charge of the system was zero. The atomic charges were obtained from the CHARMM27 force field parameter (Mackerell et al., 1998). To make the system in the steady state with low energy, the energy minimization was carried out for 5000 steps under temperature 298.15 K by using software NAMD (Phillips et al., 2005).

2.2 Calculating the electrostatic potential

The function of ClC proteins is mainly expressed as Cl$^-$ transport which is closely related to the electrostatic distribution along the pathway. Therefore, how much role in the function of ClC proteins is played by certain chain can be investigated through calculating the electrostatic potential of chains A and B. The electrostatic potential Φ can be obtained by solving linear Poisson–Boltzmann equation (Davis & McCammon, 1990, Honig & Nicholls, 1995).

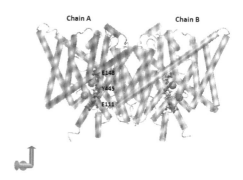

Figure 1. Topological structure of ClC-ec1 protein. The backbone of protein is represented by cylinder. Chain A and B are represented by color blue and red, respectively. Three functional crucial residues E111, E148, and Y445 are denoted by the bead-bond model. Two chloride ions (green spheres) are located at the internal and central binding sites in each chain.

$$-\nabla \cdot (\varepsilon(r)\nabla\Phi(r)) + \bar{\kappa}^2(r)\Phi(r) = 4\pi \sum_{i=1}^{N} q_i \delta(r - r_i), \quad (1)$$

where q_i and r_i are the charge and coordinate of the ith atom, respectively, N is the number of atoms in the whole system, $\varepsilon(r)$ denotes the position-dependent dielectric constant. The Poisson–Boltzmann equation was solved by soft APBS 0.5.1, in which a finite element method with a discrete grid was used (Holst & Saied, 1993). In this process, the automatically configured sequential method was used. The APBS parameter settings were taken as follows. The number of grid points was 161 × 161 × 129. The size of the coarse grid region was 130 Å × 130 Å × 110 Å and that of the focusing fine grid region was 60 Å × 60 Å × 60 Å. The dielectric constant ε is taken with different value in two regions: $\varepsilon = 4$ for ClC-ec1 channel protein and membrane, and $\varepsilon = 78$ for the surrounding solvent. The focusing procedure was repeated for eight different locations of system relative to the center of the 3D-grids (shifted half of the fine grid spacing in each direction), and the resulting values were averaged.

3 RESULTS AND DISCUSSIONS

3.1 Channel properties of ClC-ec1 protein

There are two identical pathways in chains A and B of ClC-ec1 protein respectively, and each channel is functionally independent. In order to describe its geometrical properties, we measured the dimension of this ionic pathway by using soft HOLE (Smart et al., 1996). Fig. 2a shows the diameter of the Cl$^-$ ion pathway along the Z coordinate when the channel is in the open state. We can see that the channel is broad for two sides and narrow in the central section. The narrowest part is from Z = 1 to 3 Å with diameter about 3.5 Å. This narrowest section is just the position of side-chain for residue E148. When H⁺ goes to neighborhood of E148, it will neutralize the side-chain of this glutamate residue. Then this neutral side-chain is mobile, and moves to another place. So the channel is open at the same time, Cl ions can cross in or out from it. For the wild-type structure, however, the side chain of E148 occupies the external binding site and the diameter of this region is almost zero. Fig. 2b is the 3 dimensional representation of the pathway with the open state. We can see that this curving pathway goes across the membrane.

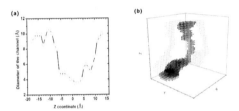

Figure 2. The geometrical properties of Cl⁻ transport pathway. (a) The diameter of the Cl- ion pathway along the Z coordinate when the channel is in the open state. (b) Three dimensional view of the pathway. Points represent the accessible positions of Cl-.

The channel width directly affects the ionic transport rate, so a wider space can conduct more Cl under the same condition such as: electrostatic potential, pH value, concentration gradient and so on. Obviously, the narrowest part of the channel determines the ionic transport quantity. In the ClC-ec1 protein structure, the channel space around E148 is the narrowest part and only allows one Cl pass at a time. This phenomenon demonstrates the importance of glutamate E148 works as the gating amino acid. If the negative side chain E148 mutates to a neutral and smaller one such as A148 (E148A), the width of channel and transport rate will be increased (Jayaram et al., 2008). Mutant not only changes the width of the channel, but also the potential are changed in the neighborhood, so it is necessary to investigate the electrostatic potential distribution caused by the changing of charge.

3.2 Electrostatic potential distribution in different situations

When negatively charged Cl moves into the channel, it will interact with the around amino acids. These interactions between each other can be represented by the electrostatic potential. Fig. 3 plots distribution of

the average electrostatic potential along the channel direction (Z coordinate) with different state situation. In the wild-type structure, negatively charged glutamate E148 occupies the external binding site firmly and makes the channel in closing state. If no Cl is in the channel, the electrostatic potential is negative number around E148 and positive in the other place (see Fig. 3a). When Cl travels into the channel, its negative charge makes the electrostatic potential decrease sharply. It can be seen from Fig. 3a that the potential reach to the minimum value if the two chloride ions occur in the binding sites S_{cen} and S_{ext} at the same time.

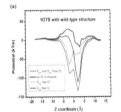 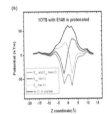

Figure 3. The electrostatic potential distribution of ClC-ec1 channel protein along the channel direction. (a) The electrostatic potential data were obtained in the wild-type structure. (b) The electrostatic potential data were obtained in the protonated E148 structure.

When the channel of the ClC-ec1 protein is in open state, how will the electrostatic potential change along the channel direction? Fig. 3b shows a different potential distribution of ClC-ec1 protein with the side chain of glutamate E148 is protonated by a proton. Protonation of this side chain leads to its charge becoming neutral and the potential along the most parts of the channel has positive values. These positive potential is beneficial for the Cl permeation through the channel. If the structural information is determined, the ionic transport is also affected by other minor factors such as: the concentration gradient and difference pH values between two sides of the membrane. Relative to the intrinsic structure features, these factors can be manipulated easily. Systematically research and analysis have been accomplished during the past 20 years.

4 CONCLUSIONS

In this paper we have analyzed the electrostatic potential properties of ClC-ec1 channel protein based on the linear Poisson–Boltzmann equation. Based on the comparison between the wild-type structure and the structure with protonated E148, we found that there are two functional important factors for ionic transport: the width of the channel and the electrostatic potential distribution. The channel width directly affects the ionic transport rate, and the narrowest part of the channel determines the ionic transport quantity. Protonation of the gating amino acid E148 can make the potential along the most parts of the channel are positive values. These positive potential is beneficial for the Cl permeation through the channel. It provides a guide rule to estimate how chloride ions have influences on the function of ClC proteins.

ACKNOWLEDGMENT

This work was supported by the National Natural Science Foundation of China (Grant No. 11304123 and 31270761), the China Scholarship Council Foundation, and Scientific Research Foundation of Jianghan University (Grant No. 2013016).

REFERENCES

Accardi, A., Kolmakova-Partensky, L., Williams, C. & Miller C.. Ionic currents mediated by a prokaryotic homologue of CLC Cl⁻ channels. *J. Gen. Physiol.*, 2004, 123: 109–119.

Accardi, A. & Miller, C. 2004. Secondary active transporter mediated by a prokaryotic homologue of ClC Cl⁻ channels. *Nature*, 427: 80–807.

Accardi, A. & Picollo, A. CLC channels and transporters: Proteins with borderline personalities. *Biochim. Biophys. Acta*, 2010, 1798: 1457–1464.

Achcroft, F., Gadsby, D. & Miller, C. Introduction. The blurred boundary between chnnels and transporters. *Phil. Trans. R. Soc. B*, 2009, 364: 145–147.

Chen, T.Y. 2005. Structure and function of clc channels. *Annu. Rev. Physiol.* 67: 809–839.

Davis, M.E. & McCammon, J.A. Electrostatics in biomolecular structure and dynamics. *Chem. Rev.*, 1990, 94: 7684–7692.

Dutzler, R. A structural perspective on ClC channel and transporter function. *FEBS Lett.*, 2007, 581: 2839–2844.

Dutzler, R., Campbell, E.B., Cadene, M., Chait, B.T. & MacKinnon, R. X-ray structure of a ClC chloride channel at 3.0 Å reveals the molecular basis of anion selectivity. *Nature*, 2002, 415: 287–294.

Dutzler, R., Campbell, E.B. & MacKinnon, R. Gating the Selectivity filter in ClC chloride channels. *Science*, 2003, 300: 108–112.

Feng, L., Campbell, E.B., Hsiung, Y. & MacKinnon, R. Structure of a eukaryotic CLC transporter defines an intermediate state in the transport cycle. *Science*, 2010, 330: 635–641.

Han, W., Cheng, R.C., Maduke, M.C. & Tajkhorshid, E. Water access points and hydration pathways in CLC H⁺/Cl⁻ transporters. *Proc. Natl. Acad. Sci. U. S. A.*, 2014, 111: 1819–1824.

Holst, M. & Saied, F. Multigrid solution of the Poisson-Boltzmann equation. *J. Comput. Chem.*, 1993, 14: 105–113.

Honig, B. & Nicholls, A. Classical electrostatics in biology and chemistry. *Science*, 1995, 268: 1144–1149.

Humphrey, W., Dalke, A. & Schulten, K. VMD-Visual molecular dynamics. *J. Mol. Graphics*, 1996, 14: 33–38.

Jayaram, H., Accardi, A., Wu, F., Williams, C. & Miller, C. Ion permeation through a Cl⁻-selectivie channel designed from a CLC Cl⁻/H⁺ exchanger. *Proc. Natl. Acad. Sci. U. S. A.*, 2008, 105: 11194–11199.

Jentsch, T.J. CLC chloride channels and transporters: from genes to protein structure, pathology, and physiology. *Crit. Rev. Biochem. Mol. Biol.*, 2008 43: 3–36.

Jentsch, T.J., Neagoe, I. & Scheel, O. CLC chloride channels and transporters. *Curr. Opin. Neurobiol.*, 2005, 15: 319–325.

Ko, Y.J. & Jo, W.H. Secondary water pore formation for proton transport in a ClC exchanger revealed by an atomistic molecular-dynamics simulation. *Biophys. J.*, 2010, 98: 2163–2169.

Kuang, Z., Liu, A. & Beck, T.L. 2008. TransPath: a computational method for locating ion transit pathways through membrane proteins. *Proteins,71:* 1349–1359.

Lourdel, S., Grand, T., Burgos, J., González, W., Sepúlveda, F.V. & Teulon, J. ClC-5 mutations associated with Dent's disease: a major role of the dimer interface. *Eur. J. Physiol.*, 2012, 463: 247–256.

Maduke, M., Pheasant, D.J. & Miller, C. High-level expression, functional reconstitution, and quaternary structure of a prokaryotic ClC-type chloride channel. *J. Gen. Physiol.*, 1999, 114: 713–722.

Mackerell, Jr.A.D., Bashford, D., Bellett, M., Dunbrack, R.L., Evanseck, Jr.J.D., Field, M.J. & Fischer, S. All-atom empirical potential for molecular modeling and dynamics studies of proteins. *J. Phys. Chem. B*, 1998, 102: 3586–3616.

Miller, C. ClC chloride channels viewed through a transporter lens. *Nature*, 2006, 440: 484–489.

Miller, C. & Nguitragool, W. A provisional mechanism for Cl⁻/H⁺ exchange in CLC transport proteins. *Phil. Trans. R. Soc. B*, 2009, 364: 175–180.

Nguitragool, W. & Miller, C. CLC Cl⁻/H⁺ transporters constrained by covalent cross-linking. *Proc. Natl. Acad. Sci. U. S. A.*, 2007, 104: 20659–20665.

Phillips, J.C., Braun, R., Wang, W., Gumbart, J., Tajkhorshid, E., Villa, E., Chipot, C., Skeel, R.D., Kale, L. & Schulten, K. Scalable molecular dynamics with NAMD. *J. Comput. Chem.*, 2005, 26: 1781–1802.

Picollo, A., Malvezzi, M., Houtman, J.C.D., & Accardi, A Basis of substrate binding and conservation of selectivity in the CLC family of channels and transporters. *Nat. Struct. Mol. Biol.*,2009, 16: 1294–1301.

Smart, O.S., Neduvelil, J.G., Wang, X., Wallace, B.A. & Sansom, M.S.P. HOLE: a program for the analysis of the pore dimensions of ion channel structural models. *J. Mol. Graphics*, 1996, 14: 354–360.

Information Technology – Wan et al. (Eds)
© 2015 Taylor & Francis Group, London, ISBN 978-1-138-02785-5

Expression of HIF prolyl hydroxylases in the ovary during the follicular development of adult sprague-dawley rats

Fan Wang, Liyun Chen, Zonghao Tang, Zhenghong Zhang & Zhengchao Wang
Provincial Key Laboratory for Developmental Biology and Neurosciences, College of Life Sciences, Fujian Normal University, Fuzhou, P. R. China

ABSTRACT: Our previous studies have demonstrated that hypoxia-inducible factor (HIF)-1alpha plays an important role in the ovarian functions, which is regulated by HIF prolyl hydroxylase (PHD)-mediated degradation. For further understanding the regulation of HIF-1alpha pathway and the functions of PHDs in the ovary, the present study first developed the model of ovarian follicle development of adult female rats through vaginal smears and ovarian histology, and then investigated the expression and localization of three different kinds of PHD isoenzymes, PHD1, PHD2, and PHD3 in each ovary through immunohistochemistry. The present results showed the model was developed successfully, and all of these three PHD enzymes were expressed in the ovary and changed in a stage-specific manner. Further observation found that PHD2 was mainly expression isoenzyme during the regulation of HIF-1alpha-related functions in the ovary. Together, our present study indicated PHDs may be directly involved in the regulation of ovarian physiological functions through HIF-1alpha signaling, such as the follicular development in the ovary.

KEYWORD: hypoxia-inducible factor-1alpha; HIF prolyl hydroxylase; follicular development; ovary

In mammals, a follicle is the basic unit of ovary, which consists of an oocyte and surrounding granulosa cells and theca cells. Thus, follicular development includes many complex changes in the oocyte and its surrounding cells (Lu and Zhu 2008), which is an orderly, continuous process. Under normal circumstances, most of the follicles will undergo atresia after growth, recruitment, selection, but only one dominant follicle can mature and discharge eggs (Jin and Liu 2003). Meanwhile, the ovarian follicular growth and development was controlled by many factors. They not only include the regulation of hypothalamic-pituitary-ovarian axis (HPOA) and the factors secreted by the ovary itself in an autocrine or paracrine manner, but also include the role of other organizations, which produce the substances that can impact follicular development and exogenous substances or environmental factors, which affect the normal follicular development and ovulation. Therefore, ovarian follicular development models are usually constructed and contributed to investigate the effects of these factors on the regulation of protein functions in the ovary.

Follicular growth is characterized by angiogenesis, which is accompanied with the decreased supply of oxygen, since the pressure of oxygen in follicular fluid is negatively related to the size of the follicle and gradually reduce the follicle during the follicular growth process (Basini 2004). Hypoxia-inducible factor-1 (HIF-1) is a major regulator of cell oxygen balance and widely present in the human and mammalian body, which is involved in anoxic responses. HIF-1 may directly participate in the physiological regulation of ovarian functions in mammals, since this transcription factor specifically expressed in ovarian cells (Wu 2012a, b). HIF-1 is a heterodimer composed of alpha and beta subunits (Wu 2013b). HIF-1beta is stably expressed in all cells, and its changes are not sensitive to oxygen tension, but varying the oxygen tension is required for the changes of HIF-1 heterodimer mediated hypoxia-inducible transcription (Nakayama 2012). HIF-1alpha is the functional subunit accumulated in the absence of oxygen. Hypoxia-stable HIF-1alpha subsequently transferred from the cytoplasm to the nucleus, where the dimerization with HIF-1β occurs and then a transcriptional activity of HIF-1 forms. HIF-1 regulates the transcription of target genes by combining with hypoxia response element (HRE) on its target genes enhancer and promoter regions (Brahimi-Horn and Pouyssegur 2009). However, HIF prolyl hydroxylases (PHDs) make an oxygen atom add into Pro402 or Pro564 at the ODD region to form a prolyl residue in HIF-1alpha through oxygen molecular as a substrate in the conditions of normal oxygen partial pressure, Fe^{2+} ion, and alpha-ketoglutarate. Hydroxylated HIF-1alpha proteins rapidly bind to von Hippel-Lindau tumor suppressor protein (pVHL) for degradation via the ubiquitin-proteasome pathway rapidly. Therefore,

PHDs are a rate-limiting enzyme during the HIF-1 degradation process (Li and Zhang 2007), so it is very important to regulate HIF-1alpha signaling *in vivo*.

Our previous studies have demonstrated that HIF-1alpha plays an important role in the periodical changes of ovarian functions. Therefore, PHD expression and localization in the ovary will help us learn more about ovarian physiological functions. The present study used immunohistochemistry (IHC) to examine the expression and cellular localization of PHDs and then analyzed the regulatory roles of ovarian follicular development during Sprague-Dawley (SD) rat estrous cycle through the semiquantitative analysis of PHDs. IHC detected the target protein expression through the chemical reaction of the labeled antibody based on the principle of specific binding of antigen and antibody. So the present study will contribute to further illuminate the mechanism of HIF-1alpha in regulating ovarian physiological function.

1 MATERIALS OF METHODS

1.1 *Animals*

Mature male and female SD rats were purchased from Wushi Experimental Animal Supply Co. Ltd. (Fuzhou, China). The animals were maintained under a 14-h light/10-h dark schedule with continuous supply of chow and water. The experimental protocol was approved in accordance with the Guide for the Care and Use of Laboratory Animals prepared by the Institutional Animal Care and Use Committee, Fujian Normal University.

1.2 *Experimental design*

Adult rats were monitored during normal estrous cycles. Vaginal cytology was assessed each day throughout the experiment as a measure of ovarian function and to determine reproductive senescence. Vaginal smears were stained with 1% toluidine blue and read under photomicroscope. Only those displaying at least 3 consecutive 4-day estrous cycles were considered regularly cycling; regular cycles included proestrus (P), estrus (E), metestrus (M), and diestrus (D). The day of vaginal cornification was designated as the day of estrus, and the next day was designated as metestrus.

The ovaries of the adult rat model were obtained in the morning at approximately 1000 h. One ovary from each rat was fixed in 4% paraformaldehyde for histological evaluation and IHC, while the other ovary was snap-frozen and used for mRNA and protein extraction.

1.3 *Histological examination*

To evaluate the ovarian development of neonatal rats, ovaries kept in 70% alcohol were dehydrated through increasing alcohol concentrations, and embedded in paraffin. The slides were stained with hematoxylin and eosin (H&E) and we observed histopathologic changes under a light microscope (Nikon Tokyo, Japan). The number of individual follicles was determined by examining five sections (the middle cross section and four other sections) per ovary. Every section was at least 10 μm away from another section. The stages of follicles (primordial, primary, secondary, and antral) were divided based on classification criteria derived mainly from our previous laboratory studies.

1.4 *Immunohistochemistry*

After fixation, the ovaries from the three groups were embedded in paraffin, and then 5-μm sections were cut and mounted on slides. The sections were then processed for immunohistochemical analysis using an antibody against PHD1, PHD2, or PHD3 (rabbit polyclonal IgG, Novus Biologicals) diluted 1:500 in PBS containing 1.5% rabbit serum. The immunoreactivity assay of specific protein was visualized by the Elite ABC kit (Bio Genex, San Ramon, CA, USA) and reaction with 0.05% 3,3'-diaminobenzidine tetrachloride (DAB; Sigma Chemical Co.) in 10 mmol/L phosphate-buffered saline (PBS) containing 0.01% H_2O_2 for 2 min. The negative control was the use of normal rabbit serum instead of primary antibody (Boster Biological Technology). Three independent observers were asked to examine the pictures and assess the intensity of staining using the following scale: −, no staining detected; +, weak; ++, moderate; +++, strong staining. All observers evaluated all slides, and observations outside the 5%–95% confidence interval of the remaining observations of the treatment group were considered to be outlying data and were excluded from analysis. Relative levels of immunostaining were evaluated and repeated at least four times. The results represented consistently observed patterns of immunohistochemical staining.

1.5 *Statistics*

Data are presented as mean±SE. The significance of differences in mean values within and between multiple groups was evaluated using an ANOVA followed by a Duncan's multiple range test. $P < 0.05$ was considered statistically significant.

2 RESULTS

2.1 *Observations of vaginal smear in adult rats*

Vaginal smear is one of the common methods used to detect the estrous cycle of adult female rats. The estrous cycle is specifically divided into four stages and each stage is characterized by the different type of cells in mucus as following. During the proestrus stage (P), there are many swelling oval nucleated epithelial cells, containing a small amount of keratinized epithelial cells and scattered leukocytes (Fig. 1A), while in the estrus stage (E), nucleated epithelial cells disappear, the keratinized epithelium from the scattered state to set block (Fig. 1B). During the metestrus stage (M), the keratinized epithelial cell continue to accumulate, nucleated epithelial cells and numerous leukocytes (Fig. 1C), while during the diestrus stage (D), main cell type is leukocyte besides a few nucleated and keratinized epithelial cells (Fig. 1D). These results are consistent with previous reports (Chateau 2007).

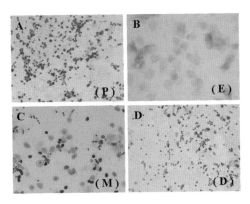

Figure 1. Vaginal mucus smear observation in adult SD rats P: proestrus day, E: estrus day, M: metestrus day, D: diestrus day.

2.2 Changes of uterine wet weight in the rats

During the estrous cycle in adult female rats, the uterus also will be changed subsequently for pregnancy. Therefore, uterine weight also acts as an important indicator to identify the estrous cycle stage of adult rat. It can be seen from Fig. 2 that uterine wet weight in the proestrus and estrus stage was

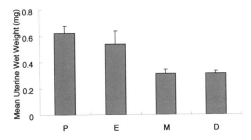

Figure 2. Changes of uterine wet weight in SD rats at different stages during the estrus cycle P: proestrus day, E: estrus day, M: metestrus day, D: diestrus day.

significantly higher than that in the metestrus and diestrus stage ($P < 0.05$) throughout the development cycle.

2.3 Observations of ovarian histology in adult SD rats

In different stages of estrus, ovarian histology is changed, which mainly reflects ovarian follicles at the different development stages in the ovaries. By HE staining method, histologic observation found that there are a lot of antral follicles at proestrus on ovarian (Fig. 3A); there are corpus luteum and some big follicles at estrus stage (Fig. 3B), there are basically corpus luteum and some degradation of follicle at metestrus stage (Fig. 3C), the ovary begins to recruit the next developed follicular cells (Fig. 3D).

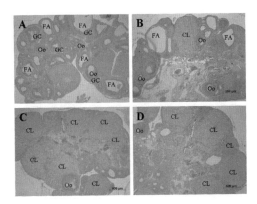

Figure 3. Representative photographs of follicular development in adult SD rats A: a proestrus day (P), B: an estrus day (E), C: a metestrus day (M), D: a diestrus day (D). All sections were stained with hematoxylin and eosin (H&E). GC = granula cell, Oo = oocyte, FA = follicle autrum, CL = corpora luteum.

2.4 Expression and cellular localization of PHD1 in the rat ovaries

The results of PHD1 protein IHC showed PHD1 expression in the ovarian cells of rats (Fig. 4), though the expression was weak. The cellular localization found that it expressed in the oocytes, granulosa cells, luteal cells, and stromal cells in a stage-specific manner by relative quantitative analysis (Table 1). These results showed PHD1 is one of PHDs expressed in the ovaries, which may participate in the regulation of ovarian functions.

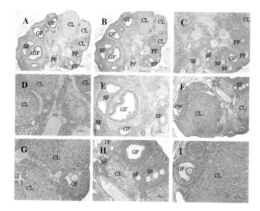

Figure 4. Expression and cellular localization of PHD1 in the rat ovaries pF: primordial follicle, PF: primary follicle, SF: secondary follicle, GF: Graafian follicle, CL: corpus luteum.

Table 1. Immunohistochemical localization of PHD1 in the ovary of the rats at the different estrous stage.

Follicular development	Staining intensity			
	P	E	M	D
Oocyte				
Primordial	+	+	+	+
Primary	+	+	+	+
Secondary	NA	NA	NA	NA
Tertiary	NA	NA	NA	NA
Granulosa cells				
Primordial	+	+	+	+
Primary	NA	NA	NA	+
Secondary	NA	NA	NA	NA
Theca cells	NA	NA	NA	NA
Theca cells				
Secondary	+	+	+	+
Tertiary	+	+	+	+
Corpora luteum	+	+	++	++
Endothelial cells	+	+	+	+

Note: +: weak, ++: moderate, +++: strong, NA: not available

2.5 Expression and cellular localization of PHD2 in the rat ovaries

PHD2 protein IHC was also done during the analysis of the possible role of PHD proteins in the ovaries. The present results indicated PHD2 expressed stronger than PHD1 (Fig. 5), which may be the main one of PHD proteins regulating HIF-1alpha signaling in the ovaries. The results of relative quantitative analysis showed that PHD2 also expressed in

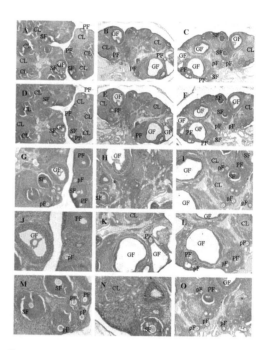

Figure 5. Expression and cellular localization of PHD2 in the rat ovaries pF: primordial follicle, PF: primary follicle, SF: secondary follicle, GF: Graafian follicle, CL: corpus luteum.

Table 2. Immunohistochemical localization of PHD2 in the ovary of the rats at the different estrous stage.

Follicular development	Staining intensity			
	P	E	M	D
Oocyte				
Primordial	++	++	++	++
Primary	++	++	++	++
Secondary	++	++	++	NA
Tertiary	++	++	NA	NA
Granulosa cells				
Primordial	+	+	+	+
Primary	+	++	+	+
Secondary	++	++	++	NA
Theca cells	+	+	NA	NA
Theca cells				
Secondary	++	++	++	NA
Tertiary	++	++	NA	NA
Corpora luteum	+++	+++	+++	+++
Endothelial cells	++	+++	++	++

Note: +: weak, ++: moderate, +++: strong, NA: not available

the oocytes, granulosa cells, luteal cells, and stromal cells in a stage-specific manner (Table 2), implying the important role in the regulation of the ovarian physiology.

2.6 Expression and cellular localization of PHD3 in the rat ovaries

For further understating the role of PHD proteins in the ovarian physiology, the present study also examined the expression pattern of PHD3 protein during the follicular development. The results showed the weak expression of PHD3 in the ovaries (Fig. 6) with the similar expression pattern in the follicles (Table 3), further indicating PHD2 is the main one regulating HIF-1alpha signaling in the ovary.

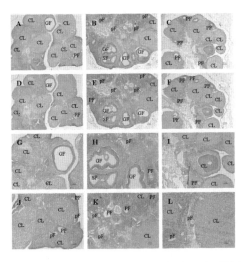

Figure 6. Expression and cellular localization of PHD3 in the rat ovaries pF: primordial follicle, PF: primary follicle, SF: secondary follicle, GF: Graafian follicle, CL: corpus luteum.

Table 3. Immunohistochemical localization of PHD3 in the ovary of the rats at the different estrous stage.

Follicular development	Staining intensity			
	P	E	M	D
Oocyte				
Primordial	+	+	+	+
Primary	+	+	+	+
Secondary	NA	NA	NA	NA
Tertiary	NA	NA	NA	NA
Granulosa cells				
Primordial	+	+	+	+
Primary	NA	NA	NA	+
Secondary	NA	NA	NA	NA
Theca cells	NA	NA	NA	NA
Theca cells				
Secondary	+	+	+	+
Tertiary	+	+	+	+
Corpora luteum	+	+	+	+
Endothelial cells	+	+	+	+

Note: +: weak, ++: moderate, +++: strong, NA: not available

3 DISCUSSION

In the present study, an ovarian follicular development model of adult female rats was constructed through vaginal smear, the observation of uterus weight changes and the histological examination of each ovary. And then the expression and localization of PHD proteins was investigated by IHC during the follicular development cycle in the ovary. The results showed PHD protein expressed in the different ovarian cells in a stage- and cell-specific manner during the follicular development, implying it may play an important role in the regulation of ovarian functions in mammals.

At present, the method of vaginal smear is always used for the identification of different estrous stage in the rodent rats, as different stage of estrus is characterized by different kinds of cells existing in the vaginal mucus. At the proestrus stage, the vaginal epithelium cells grow very slow because of the lower level of blood estrogen, therefore, some cells shed into the vaginal cavity with oval shape. While at the period of estrus, the increase of blood estrogen stimulates vaginal epithelial cells proliferation and thickening, so there are leukocytes and fewer non-keratinized epithelial cells at this time in the vagina mucus. During metestrus, the level of blood progesterone is elevated and the level of estrogen is decreased, which make epithelial thinning and glycogen synthesis reducing, thus vaginal smears mainly consist of leukocytes at this period. Anestrus is the end of the estrous cycle and the beginning of the next cycle, and most of cell types are leukocyte, nucleated epithelial cells may be occasionally seen (Rubenstein 1937). That is why vaginal smears may distribute each stage of the estrous cycle and help to obtain more accurate results during experiments (Li 2011). Therefore, the four stages of estrous cycle in mammals is a process of gradual and orderly change, accompanying with the changes of blood hormone levels and the changes of compositions of vaginal smears.

Moreover, uterus is the place for the implantation and development of embryos, so it also changes periodically during each estrous cycle. Our present results showed that the uterus weight in proestrus and estrus was significantly higher than that in metestrus and anestrus, which may be caused by the change of ovarian philology. Uterus wall becomes thickened during proestrus and estrus, as many nutrition need to be reserved. While uterine wall turns thinner and lighter during metestrus and anestrus, as no fertilized egg is implanted. Uterus has many other functions. During estrus, the cervical mucus is benefited for sperm to pass through uterus, while the rhythmic contractions of uterine muscle help the sperm target toward the fallopian tubes when mating. The strong contraction of uterine smooth muscle promotes natural childbirth. Prostaglandin is secreted by uterine, which causes corpus luteum dissolved through the uterus-ovaries cycle. Thus, the examination of uterus helps not only to precisely determine the stage of estrus cycle, but also to understand its role in the process of follicular development during the estrous cycle. And the changes of uterus size and shape are accompanied by this cycle.

In mammals, there is a series of continuous process in the mature animals during each cycle of follicular development, including recruitment, selection, maturity, ovulation/luteinization, and so on. Therefore, the stage of estrous cycle is always determined by vaginal smear in vitro for further investigating the cycle of ovarian follicular development in vivo, as the estrous cycle is actually the reflection on the developmental cycle of ovarian follicles in vivo. For example, the follicles grow rapidly and mature during proestrus, the matured follicles rupture and ovulate during estrus, the follicles become the corpus luteum and then secret progesterone during metestrus, and finally, the corpus luteum is degenerated and a second new cycle is begun. This is consistent with our present results. There are mainly secondary follicles in proestrus ovaries, preovulatory follicles, and some early corpus luteum existing in estrus ovaries, mature corpus luteum in metestrus phase, and the phenomenon of corpus luteum degradation observed in diestrus ovaries.

Recent studies have shown that maintaining ovarian function is inseparable from ovarian granulosa cells. Follicular development is closely related to granulosa cells proliferation, and follicular atresia is also closely related to granulosa cells apoptosis (Matsuda 2012). Alam et al. found that FSH stimulates granulosa cells to induce the elevated HIF-1alpha protein levels and HIF-1 activity is required for FSH-induced expression of related genes. Due to an important role of HIF-1alpha in ovarian physiological functions, the expression and localization of PHD proteins, the key regulator of HIF-1alpha degradation, was detected in the ovary from the adult female rats. The present results found that three isoenzymes of PHDs (PHD1, PHD2, and PHD3) are specific to cyclical changes during ovarian follicular development, especially its expression in ovarian granulosa cells, which is consistent to the expression of HIF-1alpha. These results indicate that it may be involved in regulating physiological function of the ovaries by HIF-1alpha signaling pathways. It was also found that the expression of PHD2 in the ovaries is significantly higher than PHD1 and PHD3, suggesting PHD2 may be a main isoenzyme of PHDs in the ovaries. Although all of these PHD families can hydroxylate proline residue in HIF-1alpha, the expression levels of these members are not the same in different tissues and cells. The in vitro studies showed that the intensity of their activity is PHD2 >> PHD3> PHD1, due to PHD2 having a unique structure of the N-terminal zinc finger, which conservatively exists in *C. elegans* EGL-9 (Gilbert 2011; Hu 2013). As Pang et al. found PHD2 may play a major role in granulosa cells from the ovaries (Pang 2011). Thus, PHD2 is a major isoenzyme and cyclically expressed in ovarian follicular development, suggesting that PHD2 may be involved in the regulation of ovarian function by controlling the levels of HIF-1alpha protein.

In summary, HIF-1alpha plays an important role in the physiological function of the ovary (Wu 2012a, b; Zhang 2011a, b and 2013), so the regulation of HIF-1alpha is critical to maintain the function and balance of HIF-1a signaling in mammal. To our knowledge, this is the first time to investigate the role of PHD proteins in the ovarian function in vivo. In the present study, the follicular development model of adult rats was used to investigate the expression and localization of PHD proteins in the ovary by immuohistochemistry. The results demonstrated that all of three PHD proteins expressed and PHD2 is the main one in the ovary. Further analysis found that PHD protein expressed in a cell- and stage-specific manner. These results indicate PHD proteins may be involved in the regulation of ovarian function by controlling the levels of HIF-1alpha protein. However, more detailed molecular mechanisms remain to be further studied.

ACKNOWLEDGMENT

This study was supported by the National Natural Science Foundation of China (31101032 and 31271255), Program for New Century Excellent Talents in University of Ministry of Education of China (NCET-120614) and Doctoral Foundation of the Ministry of Education in China (20113503120002).

REFERENCES

Basini, G., Bianco, F., Grasselli, F., Tirelli, M., Bussolati, S. & Tamanini, C. 2004. The effects of reduced oxygen tension on swine granulosa cell Regul Pept 120: 69–75.

Brahimi-Horn, M.C. & Pouyssegur, J. 2009. HIF at a glance. J Cell Sci 122:1055–1057.

Chateau, D., Geiger, J.M., Samama B. & Boehm, N. 1996. Vaginal keratinization during the estrous cycle in rats: a model for evaluating retinoid activity. Skin Pharmacol 9: 9–16.

Gilbert, I., Robert, C., Dieleman, S., Blondin, P. & Sirard, M.A. 2011. Transcriptional effect of the LH surge in bovine granulosa cells during the peri-ovulation period. Reproduction 141:193–205.

Hu, Y., Liu J. & Huang, H. 2013. Recent agents targeting HIF-1α for cancer therapy. J Cell Biochem 114: 498–509.

Jin, X. & Liu, Y.X. 2003. Follicular growth, differentiation and atresia. Chinese Sci Bull 48: 1786–1790.

Li, P.Y. & Zhang, C.G. 2007. Cellular low oxygen sensor: advance research on hypoxia inducible factor-1 prolyl hydroxylases. Prog Phys Sci 38:62–65.

Li, Y.S., Li, B., Ma, L.H., Zhu, Q. & Chen, J. 2011. The effect of copulation achievement ratio by selecting estrous female rats. Modern Med J China 13:5–7.

Lu, J. & Zhu, Y. 2008. The progress of follicular development and its regulatory factors, Medical Recapitulate 14: 507–509.

Matsuda, F., Inoue, N., Manabe, N. & Ohkura, S. 2012. Follicular growth and atresia in mammalian ovaries: regulation by survival and death of granulosa cells. J Reprod Dev 58:44–50.

Nakayama, K. 2012. Cellular signal transduction of the hypoxia response. J Biochem 146: 757–765.

Pang, X.S., Wang, Z.C., Yin, D.Z. & Zhang, Z.H. 2013. Overexpression of hypoxia-inducible factor prolyl hydoxylase attenuated by HCG-induced vascular endothelial growth factor expression in luteal cells. African J Biotech 10:8227–8235.

Rubenstein, B.B. 1937. The relation of cyclic changes in human vaginal smears to body temperature and basal metabolic rates. American J Phys 119:635–641.

Wu, Y.Q., Chen, L.Y., Huang, X.H. & Wang, Z.C. 2012a. Regulatory effects of HIF-NOS signaling pathway on NO-dependent functions in mammalian ovary. Chinese J Biochem Mol Biol 28:297–303.

Wu, Y.Q., Zhang, Z.H., Luo, Q.P. & Wang, Z.C. 2012b. Regulation of HIF-1 in the process of ovarian corpus luteum development of angiogenesis. Chinese J Cell Biol 34:1042–1048.

Zhang, J., Zhang, Z., Wu, Y., Chen, L., Luo, Q., Chen, J., Huang, X., Cheng, Y. & Wang, Z. 2012. Regulatory effect of hypoxia-inducible factor-1α on hCG-stimulated endothelin-2 expression in granulosa cells from PMSG-treated rat ovary. J Reprod Dev 58: 678–684.

Zhang, Z., Yu, D., Yin D. &Wang, Z. 2011a. Activation of PI3K/mTOR signaling pathway contributes to induction of vascular endothelial growth factor by hCG in bovine developing luteal cells. Anim Reprod Sci 125:42–48.

Zhang, Z., Yin D. & Wang, Z. 2011b. Contribution of hypoxia-inducible factor-1a to transcriptional regulation of vascular endothelial growth factor in bovine developing luteal cells. Anim Sci J 82:244–250.

Information Technology – Wan et al. (Eds)
© 2015 Taylor & Francis Group, London, ISBN 978-1-138-02785-5

Face detection and facial feature points localization in complex background

Li Ke
Institute of Biomedical and Electromagnetic Engineering Shenyang University of Technology, Shenyang, Liaoning, China

Mo Yang
Institute of Information Science and Engineering Shenyang University of Technology, Shenyang, Liaoning, China

ABSTRACT: In the current face detection and localization algorithm, there is multi-face detection error cannot be corrected in complex background, and to the small area, low-resolution facial features location difficulty, in order to solve this problem, we proposed a new weight update rules and improved adaBoost algorithm. We also present a facial feature localization algorithm based on image gray statistic, and verify whether to face each other based on facial feature localization. The experimental results indicate that the proposed method can achieve a single facial features localization accuracy of 100%, and it also effectively eliminates the phenomenon of false detection under complex and multi-face detection background. Using this method also might achieve a rapid and accurate positioning of facial features and lay a foundation for face recognition.

1 INTRODUCTION

Face detection and recognition technology is a key application technology of computer vision field which has great potential value in economic, security, social security, military, and so on. In recent years, many studies carried out and achieved great success at home and abroad. Since the 1990s, face detection achievements in this stage are mainly to improve detection accuracy and various perspectives face detection (Sung & Poggio 1998, Yang & Roth 1999). After 10 years of development, the face detection accuracy has been greatly improved, the detection rate can reach more than 90% (Rowley & Baluja 1998). However, face detection to move towards practical application, the detection speed is a key problem needed to resolve. Therefore, the detection accuracy is improved at the same time and researchers pay more and more attention to the detection speed (Sanderson & Palial 2003, Feraud & Olivier 2001). The adaBoost based method proposed by Viola gains great popularity due to better detection results in 2001, but the error detection rate is higher (Viola & Jones 2001). In order to achieve higher detection rates and lower false positive rates, in this paper, we proposed an improved cascade adaBoost algorithm model training face classifier. We also present a facial feature localization algorithm based on image gray statistic, achieving rapid positioning of facial features, and verify whether to face each other based on facial feature localization, effectively eliminate the phenomenon of false detection under

complex and multi-face detection background. Using this new weight update rules and improved adaBoost algorithm is more accurate classification results, can effectively remove the background interference. Comparative experiments on images from the FERET and the self-made image databases demonstrate that our method is more robust than traditional methods to scale variation, illumination changes, partial occlusion, and complex facial expressions. The tests also show that the proposed method improves the detection rate and achieves high precision of localization.

2 FACE DETECTION

2.1 *Haar-like rectangle features*

This article uses the Haar-like features as the key feature of the human face. Haar-like features can effectively characterize the face of some of the features and the feature extraction speed (Lienhart & Maydt 2002). Haar-like features template of face detection reflects the changes in local gray face image (see Figure 1). As can be seen from the figure, the feature selected seems to focus on the property that the region of the eyes has been often darker than the region of the nose and cheeks (see Figure 1). The first feature capitalizes on the observation that the eye region is often darker than the cheeks. The second feature compares the intensities in the eye regions to the intensity across the bridge of the nose.

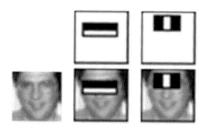

Figure 1. Haar-like features template matching in faces.

Rectangular eigenvalues refer to two or more in the image shape inside the rectangle with the same difference between the sums of all the pixel gray value. Rectangle features can be computed very rapidly using an intermediate representation for the image which we call the integral image. Integral figure just for scanning images again each eigenvalue calculation can be completed the picture (Ke & Wen 2012).

2.2 The improved adaBoost algorithm

According to the selected rectangle features, for a 20×20 detection window, the exhaustive set of rectangle features is quite larger over 18,000. Even though each feature can be computed efficiently; computing the complete set is prohibitively expensive. Therefore, using adaBoost algorithms to extract a large number of valid features of the training image rectangle features, distinguish human faces and non-face effectively. That is, each rectangle features corresponding to a classifier, these weak classifiers to form a strong classifier by linear superposition, so as to improve the classification performance of the classifier.

AdaBoost algorithm to solve the key problem is how to choose the optimal Haar-like features from a large number of features eigenvalues to construct face detection classifiers (Chen 2009). The classifier construction method is as follows:

Given example images (x_1,x_2), (x_2,x_2),... (x_n, x_n), where $y_i = 0$,1 for negative and positive examples respectively. Where n is the total number of training samples. Initialize, $D_t(i) = \dfrac{1}{2l}, D_t(i) = \dfrac{1}{2m}$, weights, for $y_i=0$, 1 respectively, where l and m are the number of negatives and positives respectively. To train the detector, a set of 2000 face and 2000 non-face training images were used by the MIT image library.

For t=1,..., T:

Normalize the weights:

$$q_t(i) = \frac{D_t(i)}{\sum_{i=1}^{n} D_t(i)} \qquad (1)$$

For each feature j, train a classifier h_t which is restricted to using a single feature. To determine the optimal classifier threshold f_θ and p_j a parity indicating the direction of the inequality sign. The error is evaluated with respect to q_t, where $\xi_t = q_i |h((x_i, f, p, \theta) - y_i)|$.

Choose the classifier h_t with the lowest error ξ_t.
Update the weights:

$$D_t(i) = D_t(i)\beta_t^{1-e_i} \qquad (2)$$

where $e_i=0$ if example x_i is classified correctly, otherwise $e_i=1$, and $\beta_t = \dfrac{\xi_t}{1-\xi_t}$.

Weak classifier cascade optimizing to form a strong classifier. The final strong classifier is:

$$H(x) = \begin{cases} 1, & \Sigma_{t=1}^{T}\alpha_t h_t \geq 0.5\Sigma_{t=1}^{T}\alpha_t \\ 0, & others \end{cases} \qquad (3)$$

Where $\alpha_t = \log\dfrac{1}{\beta_t}$.

Compared with traditional adaBoost algorithm, using new weights updating rules and improved adaBoost algorithm is more accurate classification test results, the detection rate of increase. The improved adaBoost algorithm and cascade model combining build detector, just fewer weak classifier can achieve better classification performance, not only to speed up the training speed, while achieving high detection accuracy.

3 FACIAL FEATURE LOCALIZATION

Due to single individuals face image pixels in the images of many face detection is lower, it hard for a single facial feature location, therefore, it needs to be amplified to a high resolution image (see Figure 2).

Figure 2. Enlarged images.

Based on image gray statistic of facial features location using the face core region gray value is lower than the surrounding area characteristics, first the statistics of the gray-function value X or Y direction, find out the change of the specific points, and then combining different direction of change point location, face the core region of the position can be calculated out. Using the facial features localization algorithm based on gray level statistics, locate the eyes, nose, the midpoint of the mouth of each feature point location (see Figure 3).According to the stability of facial features location, determine the position of each feature, and then use the facial statistical relationship was obtained through extensive analysis, to verify by AdaBoost algorithm whether the candidate region is the face.

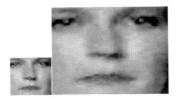

Figure 3. Facial features localization results based on gray statistics.

Compared with the complex of Hough transform algorithm (Bergen & Shvayster 1991), affected by the eyes closed to the eye as ellipse template positioning method and other eye location method, based on the gray statistics of facial features localization algorithm to locate eyes, in the actual application scenarios with high accuracy, strong practicability. The smaller the effect of nose by other facial organ, and unless there is a lot of noise interference or thick beard, in other cases can quickly and accurately locate the midpoint of the nose and mouth.

4 EXPERIMENTAL RESULTS AND ANALYSIS

4.1 Face detection and location in simple background

Using the proposed algorithm in this paper on images form FERET and self-made image database, the test results are shown in Table 1.

Table 1. The detection results of the human face.

Databases	Sum of images	Size	Correct	Accuracy
FERET	1477	384 × 256	1419	96.07%
Self-made	203	640 × 480	197	97.04%

The FERET dataset includes faces under a very wide range of conditions including: illumination, different color of skin, deflection angle and decorations (such as glasses, etc.), the image size is 384 × 256. The results of face detection is 96.07%.Self-made image database collected without any decoration, size is 640 × 480 single front face image, a total of 203 images. The result of face detection is 97.04%. The proposed method can achieve a single facial features localization accuracy of 100%.Figure 4 shows some faces detected in simple background.

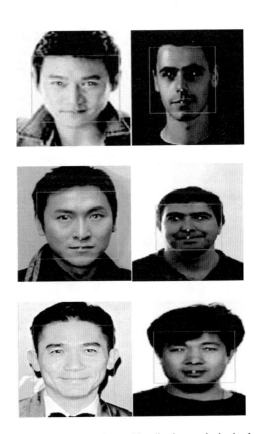

Figure 4. Face detection and localization results in simple background.

4.2 Face detection and location in complex background

Using CMU dataset and self-made image database, the complex background of the face image detection result as shown in figure 5.

The experimental results show that the improved adaBoost algorithm are successfully tested with various types of faces affected by scale variation,

Figure 5. Face detection and localization results in complex background.

illumination changes, partial occlusion, complex facial expression, have achieved high detection accuracy, strong adaptability and fast detection speed, provided the conditions for the candidate regions of facial feature analysis. The proposed facial feature localization method based on the gray image statistics to locate the eyes, nose, and mouth feature points has achieved high positioning accuracy and speed. At the same time, this method can effectively eliminate the phenomenon of false detection under a complex background of many faces. However, in terms of feature point's location, the presented method is not able to accurately locate eyes wear glasses. It will affect the level of the mouth gesture judgment for the rotated image and needs to be improved.

5 CONCLUSION

In this paper, we have demonstrated that the improved adaBoost algorithm greatly reduces the training time and requires less weak classifiers can achieve better classification performance. It also owns the benefits of strong adaptability and robustness can be effectively used in different size, different posture, different color of skin, complex and multi-face detection background. The proposed facial features, localization algorithm based on the gray image statistics to locate the eyes, nose, and mouth feature points has achieved a high precision of localization and can effectively correct face false detection.

REFERENCES

Bergen, J.R. & H.A. Shvayster (1991). Probabilistic algorithm for computing Hough Transform. Journal of Algorithms: 639–656.

Chen, X.B. (2009). Research of Face Detection Based on Skin Color and AdaBoost Algorithm, Master's Dissertation, Dalian University of Technology.

Feraud, R. & J.B. Olivier (2001). A Fast and Accurate Face Detector Based on Neural Network. IEEE Transion Pattern Analysis and Machine Intelligence 23(1).

Ke, L. & L.P. Wen (2012). Face Detection Based on Modified AdaBoost Algorithm, Opto-Electronic Engineering 39(1): 114–118.

Lienhart, R. & J. Maydt (2002). An Extended Set of Haar-like Features for Rapid object, IEEE ICIP: 900–903.

Rowley, R. & S. Baluja (1998). Neural network-based face detection, IEEE Transion Pattern Analysis and Machine Intelligence: 23–38.

Sung, K. & T. Poggio (1998). Example-based learning for view based human face detection, IEEE Trans Pattern Analysis and Machine Intelligence: 39–51.

Sanderson, C. & K.K. Palial (2003). Fast Features for Face Authentication under Illumination Detection Changes, Pattern Recognition Letters 24(14).

Viola, P. & M. Jones (2001). Rapid Object Detection Using a Boosted Cascade of Simple Features, on Computer Vision and Pattern Recognition.

Yang, M.H. & D. Roth (1999). A SNoW-Based Face Detector, Neural Information Processing Systems. MIT Press: 855–865.

Information Technology – Wan et al. (Eds)
© 2015 Taylor & Francis Group, London, ISBN 978-1-138-02785-5

GIS for healthy Georgia communities

Rui Zheng
Second Monitoring and Application Center, CEA, Xi'an, China

ABSTRACT: Data and information regarding the relationships between various measures of health and the socioeconomic status of Georgia communities are studied using ArcGIS softwares to determine if there are patterns that would be helpful in developing policies that target programs and allocate resources.

1 INTRODUCTION

Measures of health that represent the seven leading causes of years of potential life lost (YPLL) due to death before the age of 75 were chosen to explore. YPLL was chosen as a realm of study, because health issues affecting people under 75 produce a greater economic and social burden on society. In addition, these health problems can decrease through special programs and education. The leading causes of YPLL in Georgia are in this order:

- Motor-vehicle accidents
- Lung cancer
- Heart disease
- HIV/AIDS
- Suicide
- Homicide
- Stroke

Data for these measures is gathered via death records that are maintained in a standardized repository. We examined additional measures of health of special interest, including substantiated infant

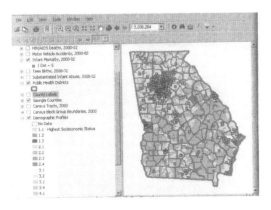

Figure 1. ArcReader interface shows the infant mortality rate in relation to demographic profiles.

maltreatment, teen births and infant mortality. All mortality and other health records are routinely geocoded to the individual's residence address so that mapping for each measure of health is a straightforword process.

2 DEMOGRAPHIC PROFILES

The demographic profiles of Georgia are the result of a segmentation analysis similar to those used by many marketing and polling agencies.

Demographic profiling is a means of reducing data into distinct categories with unique properties so that many rows of individual records are accumulated into a few categories with specific properties. Relevant measures of health can be analyzed in terms of the categories and the determination of variation in the measures of health by category can be discovered.

The profiles were developed in a statistical software package from a variable set of socioeconomoc data at the census-block group level. The set consisited of the following:

- Age distribution as a percentage of the population
- Education as a percentage of the population age 25 and older
- Occupation as a percentage of the population age 16 and older
- Family structure as a percentage of households
- Number of dependents
- Population density
- Housing features
- Median household income
- Median number of vehicles per household
- Urban versus rural population

Four major socioeconomic clusters were initially defined: prosperous, middle class, lower middle class and lower income. From these four major clusters, eighteen demographic clusters were developed. Using ArcMap software, we joined the assigned demographic profile to each block group, creating a new shapefile with a standard legend. Detailed descriptions of each of the eighteen profiles, from the prosperous blue areas through the middle-class greens and yellows to the lower-income reds and purples are available.

3 VISUALIZATION IN A PUBLISHED MAP FILE

A map was created in ArcMap displaying the demographic profiles as a base layer and each of the measures of health as point layers. Using ArcGIS Publisher software, the map documents were converted to ArcReader format.

We can examine how the various measures of health and demographic profiles are ralated spacially. We may note that any correlation present among the measures of health and demographic profiles seems to be more readily visible in the urban areas of the state, as opposed to rural areas. This could be because of the large differences in population density between the areas.

Focusing on the metropolitan Atlanta area, where population density is uniform, HIV/AIDS deaths are highest in the lower-socioeconomic areas. Homicide and suicide present an interesting juxtaposition. Homicides fall primarily within the lower socioeconomic areas, whereas suicides occur in middle-class and prosperous areas. Substantiated infant abuse displays a pattern similar to HIV and homicide, occurring mostly in the lower-socioeconomic neighborhoods.

Northern Fulton Country, which has a low rate of substantiated maltreament, is in stark contrast to south Fulton, which has a high rate of substantiated

Figure 3. Homicides (green circles) occur in lower-socioeconomic areas of Atlanta, whereas suicides (orange squares) tend to occur within middle-class and prosperous (blue) areas.

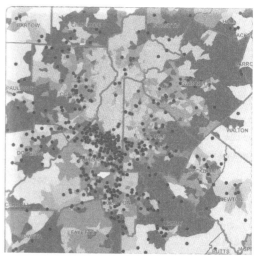

Figure 4. Cases of substantiated infant abuse (black circles) fall primarily within lower-socioeconomic (red) areas.

Figure 2. Blue squares representing HIV/AIDS deaths are more prevalent in the lower-socioeconomic (red) areas of Atlanta.

maltreatment. Northern DeKalb Country, immediately adjacent to Fulton, also has a lower rate than southern DeKalb. Recognizing the visual patterns apparent in infant maltreatment data led to further statistical analysis beyond the visualization arena. Statistical analysis demonstrates that there is an exponential correlation between the rate of infant maltreatment and the socioecomomic level of the eighteen demographic profiles.

4 POLICY DEVELOPMENT

Policy implications from population segmentation are far reaching, with two primary dimensions. One focuses on the discovery of fragile populations whose rate of poor education (high school or less within population age 25 and older) is greater than 70 percent. A high burden of poor education tends to sustain lower socioeconomic status and dependency on social services. The lower-socioecomic groups in demographic profiles 3.4 through 4.7, primarily, exhibit this property. A striking difference between segments 3.4 and the lower 4.1 through 4.7 is population density, making those segments in urban areas extremely fragile. Their low population combined with high density may concentrate disease (e.g., HIV) and dysfunctional behaviors (e.g., homicide).

The second dimension relates to the strategic marketing of health information to members of each demographic profile. Health information marketing plans for higher socioeconomic groups 1.1 through 1.3 and lower socioeconomic groups 4.1 through 4.7 will, of necessity, be very different for one reason: Groups 1.1 through 1.3 pay for their own health care, whereas groups 4.1 through 4.7 have no health care or are dependent on social services. In the latter case, marketing strategies would focus more on where social services exist and include promotional information about when and how to use those services. The least a health system can be expected to accomplish is to find fragile populations and create relevant direct marketing of health information.

Another important aid to policy makers is the allocation of health resources. Having discovered fragile populations and created strategic marketing plans, the health system must assure that relevant services exist for health consumers based on their demographic profile. Again, groups 4.1 through 4.7, which exhibit high levels of female-headed households with children younger than eighteen years old, would need different services than groups 2.2 or 3.1, which have an older population. Moreover, allocation of health-system information resouces reqires centralization and integration, so that the fiscal nature of services, consumers and providers can be managed to assure conditions of optimal health are achievable.

5 DISCUSSION

Analysts and policy makers can use the visualization tool, in conjunction with ancillary information, to develop an understanding of the various measures-of –health distributions in relation to the demographic clusters, as well as to target intervention programs in locations where they will be most beneficial to the citizens of Georgia. A more coordinated effort, prompted by this new tool, will result in more efficient public-health policies and practice.

The ArcReader documents can enhance the ability to communicate of YPLL and other measures of health in a timely and efficient manner.

ACKNOWLEDGMENT

The Division of Public Health and the Office of Health Information and Policy are gratefully acknowledged for providing the health and demographic data.

REFERENCES

Millard, F. H. 2005. *Fragile population segments in Georgia, USA*. In a USDA ForestService conference, Georgia.

Bartonek, D. & Bureg, J. 2009. *Platform for GIS tuition usable for designing of civil engineering structures at Brno University of Technology*. Proceedings of the ICA Symposium on Cartography for Central and Eastern Europe, Research Group Cartography, Vienna University of Technology, Vienna, Austria: 955–957.

Bartonek, D. & Bureg, J. & Drfib, A. & Mengik, M. 2009. *Usage of a multidisciplinary GIS platform for the design of building structures*. Proceedings of the Professional Education, 2009-FIG International Workshop Vienna, Austrian Society for Surveying and Geoinformation, Vienna Austria: 108–118.

Pundt, H. & Bringkotter-runde, K. 2000. *Visualization of spatial data for field based GIS*. Computer & Geosciences, 26 (1): 51–56.

Hitchcock, A. & Pundt, H. & Birkkotter-runde, K. & Streit, U. 1996. *Data acquision tools for geographic information system*. Proceedings of the 13th WELL-GIS Workshop on technologies for land management and data supply, Remote Sensing and GPS Research and Education, Geographical Information System International Group (GISIG), Budapest, Hungary.

Charvat, K. & Kocab, M. & Konecny, M. & Kubicek, P. 2007. *Geographic Data in Information Society*. Research Institute of Geodesy, Topography and Cartography, Zdiby, Prague, Czech Republic, in Czech, 43: 270.

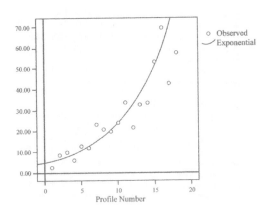

Figure 5. The rate of infant abuse has an exponential correlation with demographic profiles.

Information Technology – Wan et al. (Eds)
© 2015 Taylor & Francis Group, London, ISBN 978-1-138-02785-5

GPU based grabcut for fast object segmentation

Qiaoliang Li, Yongchun Deng, Suwen Qi, Huisheng Zhang, Zhengxi Cheng, Menglu Yi, Xinyu Liu, Jing Li, Tianfu Wang and Siping Chen
National-Regional Key Technology Engineering Laboratory for Medical Ultrasound,
Guangdong Key Laboratory for Biomedical Measurements and Ultrasound Imaging,
Department of Biomedical Engineering, School of Medicine, Shenzhen University, Shenzhen, China

ABSTRACT: Grabcut is a widely used image segmentation algorithm based on graph cut. GrabCut requires less user interaction and segmentation with high accuracy, but it is computationally complex in estimating parameters of Gaussian mixture model (GMM) based on enormous amounts of pixels. The image segmentation which establishes the graph model in pixels is time-consuming, in order to overcome this disadvantage, a GPU based Grabcut algorithm is proposed in this paper. The algorithm is implemented by CUDA framework on Nvidia GTX 580 GPU. The experimental results show that the proposed method is about 100 times faster than the standard implementation with accurate segmentation results. The proposed method has potentials in real-time video object segmentation.

1 INTRODUCTION

Image segmentation is a fundamental problem in image processing, and it remains to be a challenging problem. Recently, Graph cuts has attracted much attention and well developed [1–6].Graph-cut was first presented by Greig [7] and later Boykov [8] is renewed the graph-cuts theory to image segmentation, proposing an interactive image segmentation algorithm. Rother [1] has proposed the Grabcut algorithm which is an effective image segmentation technology based upon the Graph Cut algorithm. The Grabcut algorithm extends Graph Cut to color images and uses the color and border information in an image and to incomplete labelling and iterative estimation. These improvements greatly increase the usefulness and robustness of Graph Cut. Grabcut, whose interaction is very simple, just needing the user to drag a rectangle around the desired object. User interaction is further improved by Hui zhang et al. [6] and Hernandez et al. [7]. Chen [9] improved Grabcut performance by optimizing the Gaussian mixture model (GMM).Thus makes Grabcut able to get more accurate object extraction results, but the disadvantage of computational load is still in high level .

The contemporary graphics processor unit (GPU) has huge parallel computation power, with which the GPU has great efficiency in dealing with many data-parallel tasks [10]. In other areas, GPU has also been used, such as non-graphic applications and computer vision. The Compute Unified Device Architecture (CUDA) from Nvidia [11][12] is a good user interface for GPUs and brings us great convenience for using GPUs. We present a fast method of

Grabcut algorithm based CUDA on Nvidia GTX 580 GPU in this paper. This method has enormous potential for real-time video object segmentation.

2 METHOD

2.1 *Grabcut segmentation algorithm*

Grabcut was proposed by Rother [2] that the segmentation is produced by iterative energy minimization. The image is taken to consist of pixels Zn in RGB color space instead of the monochrome image. A full covariance GMM with K components (typically $k=5$) is defined for background pixels, and another one for the foreground pixels. We introduce an additional vector $k = (k_1,...,k_n,...,k_N)$, $k_i = \{1,2,...k\}$, $i \in [1,...,N]$ assigning, to each pixel, a unique GMM component, one component either from the background or the foreground model, according as $\beta_n = 0$ or 1.

The Gibbs energy function for segmentation is then

$$E(\beta,\text{k},\theta,\text{z}) = U(\beta,\text{k},\theta,\text{z}) + V(\beta,\text{z}) \qquad (1)$$

where θ is the gray histogram of the background and the foreground and $\theta = \{h(z,\beta),\beta=0,1\}$.

The data term U is now defined, take into consideration of the color GMM models, as

$$U(\beta,\text{k},\theta,\text{z}) = \sum_n D(\beta_n,k_n,\theta,z_n) \qquad (2)$$

where

$$D(\beta_n, k_n, \theta, z_n) = -\log p(z_n | \beta_n, k_n, \theta) - \log \pi(\beta_n, k_n) \quad (3)$$

in which $p(z_n | \beta_n, k_n, \theta)$ is a Gaussian probability distribution, and $\pi(\beta_n, k_n)$ are mixture weighting coefficients, so that (up to constant):

$$D(\beta_n, k_n, \theta, z_n) = -\log \pi(\beta_n, k_n) + \frac{1}{2}\log \det \sum(\beta_n, k_n)$$
$$+ \frac{1}{2}[z_n - \mu(\beta_n, k_n)]^T \sum(\beta_n, k_n)^{-1}[z_n - \mu(\beta_n, k_n)] \quad (4)$$

Therefore, the parameters of the model now becomes

$$\theta = \{\pi(\beta, k), \mu(\beta, k), \sum(\beta, k), \beta = 0, 1, k = 1...K\} \quad (5)$$

in which the weights π, mean μ and covariance $\sum$ of the 2K Gaussian components for the background and foreground distributions.

The smoothness term V is a regularizing prior assuming that segmented regions should be coherent in terms of color, taking into account a neighborhood C around each pixel:

$$V(\beta, z) = \gamma \sum_{(m,n) \in C} [\beta_n \neq \beta_m] \exp -\alpha \|z_m - z_n\|^2 \quad (6)$$

2.2 CUDA implementation

CUDA is a parallel computing platform and programming model that makes the use of a GPU for general purpose computing simple. CUDA has a library that it has a number of algorithms as extensions of the C language. The CPU regards a CUDA device as a multi-core co-processor when CPU and GPU are running at the same time. For the programmer, CUDA consists of a collection of threads that are scheduled for execution simultaneously on a multiprocessor. In addition, the programmer can select the number of threads to be executed.

Grabcut is iterative version of graph-cut, it takes iterative energy minimization for the segmentation, so the algorithm is computationally heavy. In order to accelerate the Grabcut, we use the parallel architecture of GPU. The proposed method is performed by built-in functions from CUDA and Open Source Computer Vision Library. Nvidia CUDA SDK 6.0 provides graph-cut built-in function that is a convenience for using GPU to accelerate the Grabcut algorithm. The algorithm framework of Grabcut on CUDA is shown in Fig.1.

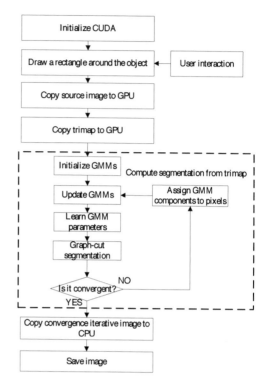

Figure 1.   The algorithm framework uses the Grabcut based on CUDA.

The algorithm needs to initialize the device of Nvidia GTX 580 GPU. The Nvidia GTX 580 graphics card has 1536 MB memory, so it has an adequate memory even if large images can reside in this memory. In the step of user interaction, users need only dragging a rectangle around the desired object, but if the color of the object and the background are very similar, the foreground and background must be marked artificially in the touchup step as is shown (Fig. 2) by which we can get accurate segmentation results. The users can choose different keyboard shortcuts to set background (CTRL+ left mouse button), foreground (SHIFT+ left mouse button), possible background (CTRL+ right mouse button) and possible foreground(SHIFT+ right mouse button).

The proposed method needs to copy images from CPU to GPU, then using Grabcut algorithm for segmentation the image. For computing segmentation from trimap, we exploit a Grabcut procedure of CUDA. The function implements the algorithm of the big framework (Fig. 1), and the algorithm guarantees convergence properties and the border matting result is obtained.

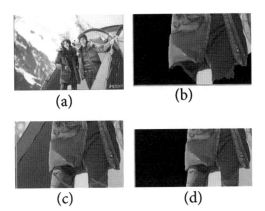

(a)   (b)

(c)   (d)

Figure 2. User editing set up. (a) Original image. (b) The result obtained by Grabcut based on CUDA only dragging a rectangle around the desired object. (c) The picture displays user setting foreground. (d) The result is obtained by (c).

3 RESULTS

3.1 *Accuracy*

In an experiment (Fig. 3) we compare the Rother's Grabcut with Grabcut based on CUDA. The result

is that the Grabcut based on CUDA algorithm at almost the same level of accuracy as Rother's Grabcut algorithm, but it is faster than the Rother's Grabcut algorithm.

3.2 *Speed*

The algorithm of Grabcut based on CUDA was tested on several standard images and Fig. 4 shows the results of image segmentation which has used our method of the algorithm on the Person image, Sponge image and the Flower image. The running times are tabulated in Table1 along with the time for Rother's sequential implementation of Grabcut. Our proposed method is surpassing 100 times faster than the previous Rother's algorithm. In general, the table shows that the detected time will be obtained more if we apply bigger images for the segmentation. But if the image (Footballers picture) has complex background, the more time is needed to iterative convergence.

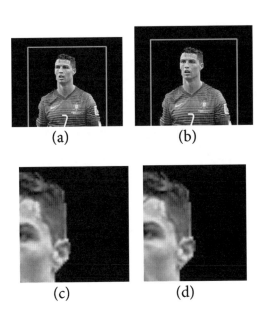

(a)   (b)

(c)   (d)

Figure 3. Comparing results for both methods. (a) and (c) The result obtained by Grabcut based on CUDA. (b) and (d) The result obtained by rother's Grabcut.

Figure 4. Binary Image Segmentation: Lovers, Horse, Flower, and Footballers images.

Table 1. Comparison of running times of CUDA method with that of Rother on different images.

Image	Size	Time(s) Rother[2]	Time(s) Proposed Method	Ratio
Lovers	940×627	39.00	0.29	134.48
Horse	1024×640	56.80	0.54	105.02
Flower	1024×768	269.00	1.59	169.18
Footballers	900×600	66.60	0.53	125.66

4 CONCLUSION

In this paper, we presented a method of grabcut on GPU using CUDA architecture. Experimental analysis shows that it can successfully speed up the Grabcut algorithm. This algorithm has potentials in real time video object segmentation.

ACKNOWLEDGMENTS

This work is supported by the National Science Foundation of China (Grant No. 81000637, 81271651, 61031003, 61401285), Guangdong Innovation Training Plan for Outstanding Youth of China (2012LYM_0114).

REFERENCES

[1] Rother C, Kolmogorov V, Blake, A, "Grabcut-interactive foreground extraction using iterated graph cuts". *ACM Transactions on Graphics*, 2004, 23(3): 309–314.

[2] Nhat V, Manjunath B. "Shap prior segmentation of multiple objects with graph cuts", *IEEE Conference on Computer Vision and Pattern Recognition*, June 2008, pp. 1–8.

[3] Freedman D, Zhang T,"Interactive graph cut based segmentation with shape priors", *IEEE Computer Society Conference on Computer Vison and Pattern Recognition*, 2005, 1: 755–762.

[4] Zhang Hui, Fritts J E,Goldman S A,et al,"Image segmentation evaluation:A survey if unsupervised methods",Computer *Vison and Image Understanding*, 2008, 110(2): 260–280.

[5] Jahangiri M. Heesch D. "Modified Grabcut for unsupervised object segmentation", *International Conference on Information Processing*, 2009, pp. 2389–2392.

[6] Hernandez A., Reyes M., "Spatio-temporal GrabCut human segmentation for face and pose recovery", *Proceedings of the IEEE Conference on Computer Vision and Pattern Recognition Workshops*, 2010, pp. 33–41.

[7] D.Greig, B. Porteous, and A. Seheult,"Exact maximum a posteriori estimation for binary images", *Journal of theRoyal Statistical Society*, 1989, 51(2):271–279.

[8] Boykov Y., Jolly M. P.,"Interactive graph cuts for optimal boundary and region segmentation of objects in n-d images", 2001,1:105–112.

[9] Chen D., Chen B., Mamic G,"Improved Grabcut segmentation via GMM optimization", *Digital Image Computing: Techniques and Applications*, 2008, pp. 39–45.

[10] Harish P.,Narayanan P.J., "Accelerating large graph algorithms on the GPU using CUDA", In Intnl.Conf. on High Performance Computing (HiPC), LNCS 4873, December 2007, pp. 197–208.

[11] Wei-Nien Chen, Dept. of Electron. Eng., Nat. Chiao-Tung Univ., Hsinchu, Hsueh-Ming Hang," H.264/AVC motion estimation implmentation on Compute Unified Device Architecture (CUDA)", *Technical report 2008 IEEE International Conference on Multimedia and Expo*, 2007, pp. 697–700.

[12] A. Corporation.Ati ctm (close to metal) guide. Technical report, AMD/ATI, 2007.

Information Technology – Wan et al. (Eds)
© 2015 Taylor & Francis Group, London, ISBN 978-1-138-02785-5

Image threshold segmentation in the colonies based on difference method of statistical curve

Sixiang Zhang, Xuefeng Wu, Wei Zhou, Pengpeng Wang & Zheng Wang
Institute of Modern Measurement and Control Technology
Hebei University of Technology, Tianjin, China

ABSTRACT: When we count the colonies with the computer image processing technology, threshold segmentation is one of the key methods with the maximum between-cluster variance method, maximum entropy methods the threshold value we can't overcome the interference of background noise in some colonies images. This paper proposes a colonies image threshold segmentation based on difference method of statistical curve, and introduces the principle and calculation method of the Image in the colonies threshold segmentation at length. The experimental results show that when we separated two typical images of colonies with this algorithm, the fault rate of the selected objects dropped into 0.17% and 0.17% respectively.

1 INTRODUCTION

In many fields such as food, pharmacy, medicine, quality and technical supervision and so on, we usually need colony counting. Today it is often realized by using computer image processing technology. This method can decrease the huge labor intensity of the traditional operator in artificial counting, and does not need too much prior knowledge of the colony recognition, Yang Fan (2011). With people using image processing techniques to count the images of the colonies, avoiding interference background noise on image processing as far as possible and determining the threshold value have become a big problem because of the colony background noise making statistical accuracy very different. So far, the researchers made a lot of researches on how to get the threshold of the image, and put forward a lot of classic methods, such as maximum between-cluster variance method, Otsu, N (1979) (also called Da-Jin method, OTSU method), maximum entropy value method, Kapur, J.N & Sahoo, P.K (1985), maximum correlation method Yen, J.C & Chang, F.J (1995), 2-mode method, Lee, S & Chung, S.A (1990), p parameters method, Doyle, W (1962) and so on. Other scholars also put forward many corresponding algorithms in different image processing fields. However, the interference of background noise remains serious with these methods. In the field of the cell counting, Dahle, J (2004) and his colleagues used the OTSU method to get the threshold value in his paper named for Automated counting

of mammalian cell colonies by means of a flatbed scanner and image processing. Correspondingly, the OTSU method which we can see in the papers" An Automated Bacterial Colony Counting System", Zhang, C.C & Chen, W.B (2008) and" An automated bacterial colony counting and classification system", Chen, W.B & Zhang, C (2009) is also used to get that value in the field of the colony counting. According to the current literatures, researchers have not the corresponding studies on the image threshold segmentation algorithm of the colonies. Therefore, this paper proposes a different method of statistical curve after a study in which we discussed the relation between the colony counting result and corresponding Threshold.

2 THE SELECTION THRESHOLD

2.1 *The grayscale distribution of the colony image*

Sometimes colonies images may exist background noise, threshold segmentation needs to eliminate the background noise, as shown in Figure 1.

The colonies image can be explained as a rugged mountain model, as shown in Figure 2. Use the gray value instead of the model's altitude. The place colonies appear is towering peaks, and the peaks are basically the same height. And the width of the peaks is corresponding to the size of the colonies. There are also some small valleys and hills are corresponding to the various background noise in the colonies gray scale image.

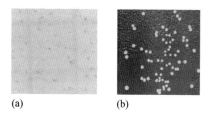

(a)   (b)

Figure 1. Colonies images of complex background. (a) Small colonies color image; (b) Large colonies color image; (c) Small colonies gray scale image; (d) Large colonies gray scale image.

To sum up, the gray scale distribution of the colonies image has the following two features:

1 All the gray values of the single colonies are in one range and changes little.
2 The gray values of Background image are choppy and there is no rule.

Based on the structural analysis of the colonies' grayscale image, the interval which is close to the top of the mountain is defined as the peak of gray interval (b, c). Using any value you choose in this interval, we separate the colonies' image with the threshold segmentation and selected the connected region, Wang, K.Q & Kangas, J.A (2003). After all these procedures, the counting result, we called Si changes little. Evidently, other valleys and hills are not in the peak of gray interval. We define the range below the peak of gray interval as a background noise interval (a, b). Because of the existence of ups and downs of the valley and hills, the counting result changes intensely. It is the upper part of the peak of gray interval we define as a colony disappeared interval (c, d). The three intervals constitute the normalization of the colonies image's gray scale interval (0~1).

2.2 Selection of connected regions

The selection of connected regions is also known as connected component labeling. The pixel p whose coordinates are (x, y), has four adjacent pixels which include two vertical and two horizontal pixels. We sign these four adjacent pixels as N4(p), as shown in Figure 3a. In Figure 3b, a pixel has two vertical, two horizontal and four diagonal neighbors. These 8 neighbors are denoted by N8 (p). If the pixel q∈ N4(p), p and q are said to be 4-adjacent, as shown in Figure 3c. Similarly, the pixels p and q are said to be 8-adjacent if q∈ N8(p), as shown in Figure 3d. We often call the pixel whose value is 1 an object. Two objects p and q are said to be 4-connected if there exists a 4-connected path between them, consisting entirely of objects, as shown in Figure 3e. Similarly,

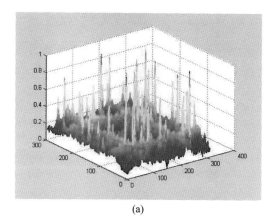

(a)

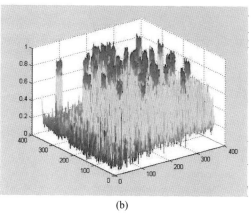

(b)

Figure 2. Mountain models of colonies image. (a) Mountain model of small colonies gray scale image; (b) Mountain model of large colonies of gray scale image.

0	1	0		1	1	1		0	0	0		0	0	q
1	p	1		1	p	1		0	p	q		0	p	0
0	1	0		1	1	1		0	0	0		0	0	0

(a)   (b)   (c)   (d)

p	1	0	0	0
0	1	1	0	0
0	0	1	0	0
0	0	1	0	0
0	0	1	1	q

(e)

p	1	0	0	0
0	0	1	0	0
0	0	1	0	0
0	0	1	0	0
0	0	1	1	q

(f)

Figure 3. 4 connections and 8 connections. (a) Four pixels adjacent to the pixel p,N4(p); (b) Eight pixels adjacent to the pixel p,N4(p); (c) Pixel p and q is 4 adjacency and 8 adjacency; (d) Pixel p and q is 8 adjacency; (e) Pixel p and q is 4 connections and 8 connections; (f) Pixel p and q is 8 connections.

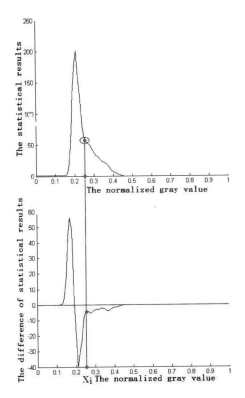

Figure 4.   Connected component labeling.

after using 8-connected labeling method, we get the counting result is 2 from the Figure 4d. The following calculation defaults 8-connected.

2.3   *Difference method of statistical curve*

Firstly, the gray value of the colonies image should be normalized. Secondly, separate the image, respectively, with the equally spaced thresholds starting from 0. Thirdly, select connected area and get the counting results. Finally, draw the statistical curve. The abscissa of the curve is equally spaced threshold, and the ordinate of it is the statistical results. We can see this curve from both the above images of the Figure 5 and the above image of the Figure 6. In mathematics, we can use derivative to

Figure 5.   Small colonies of the statistical curve difference. Above: Small colonies graph statistical curve; Below: Small colonies differential curve.

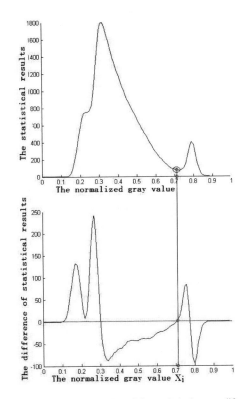

Figure 6.   Large colonies of the statistical curve difference. Above: Large colonies graph statistical curve; Below: Large colonies differential curve.

they are said to be 8-connected if there exists a 8-connected path between them, consisting entirely of objects, as shown in Figure 3f.

When we mark all the connected components in the colony counting, the number of connected components we finally mark is just the counting results. We can see that the results will be correspondingly different with different connected component labeled from Figure 4a is the original binary image data, as shown in Figure 4b, after using 4-connected labeling method, we get the counting result is 3. Similarly,

get the rate of change in a continuous function. However, the value of statistical curves are discrete points. So we use difference process instead of derivation process. We will look for a point whose difference values close to zero, and that is just the threshold we

231

Table 1. Three methods corresponding the fault rate of the selected objects.

Method	Colonies image	
	Small colonies image	large colonies image
OTSU Method	29.88%	9.87%
Maximum entropy value method	0.27%	10.46%
The proposed method	0.17%	0.017%

Table 2. Three methods corresponding the missing number of selected objects.

Method	Colonies image	
	Small colonies image	large colonies image
OTSU Method	0	0
Maximum entropy value method	2	0
The proposed method	2	0

want. As is shown in the above image of both Figures 5 and 6, the threshold point Xi is what we need(the statistical curve and the differential curve have been carried out filter processing).

The function $\phi (v)$ turns into the discrete function $\phi (v)$ when the independent variable is assigned a positive integer. The first order difference is as follows:

$$\phi (v) = \phi (v+1) - \phi (v) \tag{1}$$

3 EXPERIMENTAL RESULTS AND ANALYSIS

We separate the two different images of the colonies as shown in Figure 1 with the new algorithm this paper proposed. The size of the left image is 296×310, and the size of the right is 354×362. The experiments are performed in MatlabR2011a.Compare our method with maximum between-cluster variation method and maximum entropy value method. The results are shown in Figure 7. We can get the fault rate of the selected objects (see Table 1) and the missing number of selected objects(see table2) with all the three methods as we want.

$$F_{\mathrm{r}} = \frac{T_f}{T_p} \tag{2}$$

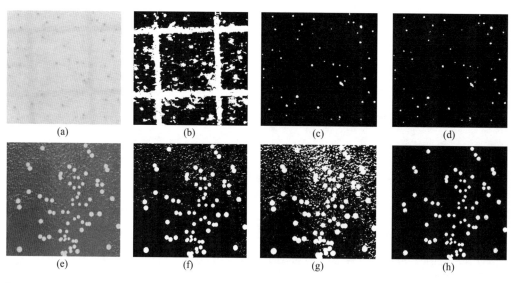

Figure 7. The result of different approaches. (a), (e) Small colonies and large colonies of gray scale image; (b), (f) Using OTSU method of segmentation results; (c), (g) Using the maximum entropy value method of segmentation results; (d), (h) Using the programs method segmentation results.

Where F_r = The fault rate of the selected objects; T_f = Total number of pixels fault rate of the selected objects; T_p = The total number of pixels of the image.

From Figure 7, it is can be seen, when we use the OTSU method the interference of background noise is very big. When we use the maximum correlation method the result is good. For a big colonies image, these two methods can't overcome the interference of background noise. However the method this paper proposes can overcome it effectively, as shown in Figure 7(d),(h).

We can see clearly from the table1 that the fault rate of the selected objects is the smallest when we use our method.

As is shown in Table 2, the missing number of selected objects are very small in all the conditions. For the images of small colonies, there is no missing selected object when we use OTSU method. Because many background noises are taken as objects. It is useless for application, as is shown in Figure 7b. We also can know that it is still hard to distinguish the missing two objects which both OTSU and our method miss even when use the artificial counting method. So actually the statistical result is not affected.

4 SUMMARY

This paper expounds a kind of threshold selection method in the colonies image counting processing. Through the analysis of the grayscale distribution of the colony image, we start our analysis of the significance of the colonies image's application and make a research on the relation between the counting result and the threshold value. We establish the statistical curve and use the principle of difference. Finally, we obtain an ideal binary image after choosing the threshold value. Experimental results show that the colonies objects can be divided from the background noise interference with the proposed method. When we separated the two classical images of colonies with this method, the fault rate of the selected objects dropped into 0.17% and 0.17% respectively. In various researches which need to count the number of objects in the image, the proposed method is also applicable.

The deficiency of this method:

a. Computing time is longer than the traditional OTSU method's, which requires further optimization of the algorithm.

b. In the suppression process of the small colonies image's background noise interference, a small part of the image was eliminated. Brought some colonies leakage.

REFERENCES

Chen, W.B. & Zhang, C. (2009). An automated bacterial colony counting and classification system. Information Systems Frontiers. 11(4), 349–368.

Dahle, J. & Kakar,, M. (2004). Automated counting of mammalian cell colonies by means of a flat bed scanner and image processing. Cytometry A. 60, 182–188.

Doyle, W. (1962). Operation useful for similarity-invariant pattern recognition. Journal Association Computer. 9(3), 259–267.

Kapur, J.N. & Sahoo, P.K. (1985). A new method for gray-level picture thresholding using the entropy of the histogram. Computer Vision, Graphics and Image Processing. 29, 273–285.

Lee, S. & Chung, S.A. (1990). A comparative performance study of several global thresholding techniques for Segmentation. Computer Vision Graphics and Image Processing,. 52, 171–190.

Otsu, N. (1979). A threshold selection method from gray-level histogram. IEEE Transactions on System Man Cybernetics. 9(1), 62–66.

Pal, N.R. & Bhandari. D. (1993). Image thresholding some new techniques. Signal Processing. 33(2), 139–158.

Pun, T. (1980). A new method for gray level picture thresholding using the entropy of the histogram. Signal Processing. 2(3), 223–237.

Sauvola, J. & Pietikäinen, M. (2000). Adaptive document image binarization. Pattern Recognition. 33(2), 225–236.

Song, F.X. & Yang, J.Y. (2006). Face representation based on the multiple-class maximum scatter difference. Acta Automatica Sinica. 32(3), 378–385.

Wang, K.Q. & Kangas, J.A. (2003). Character location in scene images from digital camera. Pattern Recognition. 36, 2287–2299.

Yang Fan. (2011). Automatic colony analyzer rapid microbial detection applications. Food Safety Guide. 12, 28–29.

Yen, J.C. & Chang, F.J. (1995). A new criterion for automatic multilevel thresholding. IEEE Trans. on Image Processing. 4(3), 370–378.

Zhang, C.C. & Chen, W.B. (2008). An Automated Bacterial Colony Counting System. IEEE International Conference on Sensor Networks, Ubiquitous and Trustworthy Computing. 233–240.

Information Technology – Wan et al. (Eds)
© 2015 Taylor & Francis Group, London, ISBN 978-1-138-02785-5

Investigation for the dates of polysomnography from obstructive sleep apnea-hypopnea

Yajun Hou
College of Science, Shenyang Ligong Universitiy, Shenyang, Liaoning, China

Feng Gao
College of Economics and Management, Shenyang Ligong Universitiy, Shenyang, China

Tingting Zhao
College of Information and Engineering, Shenyang Institute of Engineering, Shenyang, China

ABSTRACT: Laboratory polysomnography (PSG) is considered the "gold standard" for diagnosing and assessing the benefit of a treatment intervention for obstructive sleep apnea-hypopnea (OSAH). In clinical practice, doctors usually focus on the reports and results from PSG and may ignore special information of date analysis. This paper discussed the criteria of diagnosing OSAH and analyzed the characters of dates such as SaO_2 and AHI, which may help doctors to judge efficiently and accurately the severity and characters of OSAH and make more efficient treatment.

KEYWORDS: Polysomnography; obstructive sleep apnea-hypopnea; apnea-hypopnea index

1 INTRODUCTION

Obstructive sleep apnea-hypopnea is characterized by repetitive episodes of upper airway obstruction during sleep that disturb sleep architecture and induce intermittent hypoxia [1-5]. OSAH causes daytime hypersomnia and drowsiness, memory loss, and neurocognitive impairment and has been associated with hypertension, cardiovascular diseases, stroke, and diabetes. Laboratory polysomnography (PSG) is considered the "gold standard" for diagnosing and assessing the benefit of a treatment intervention for OSAH, which describes the severity of sleep apnea according to the apnea-hypopnea index (AHI). AHI is a numerical measure that accounts for the number of pauses in breathing per hour of sleep.

The sleep study called a polysomnography is typically used to diagnose sleep apnea. A lot of information is collected, and part of the study consists of tracking your breathing patterns through the night. This is accomplished with a sensor that sits in the nostril as well as two belts that stretch across the chest and stomach. In addition, a sensor called an oximeter measures the oxygen by shining a light through the fingertip. All of these information are analyzed to determine how many times you breathe shallowly or stop breathing altogether during the night. Any partial obstruction of the airway is called a hypopnea and a complete cessation in breathing is called apnea.

In order to count in the AHI, these pauses in breathing must last for 10 seconds and be associated with a decrease in the oxygen levels in the blood or an awakening called an arousal.

In clinical application, doctors usually focus on the reports and results from PSG and ignore the other characters of date analysis. This paper discusses the criteria to diagnose OSAH and analyzed the characters of dates such as SaO_2 and AHI, which may help doctors to judge efficiently and accurately the severity and characters of OSAH and make more efficient treatment.

2 METHOD

The dates in this paper come from the optical laboratory of university of Kentucky. Fifteen (15) adults participated in this study with signed consent forms approved by the University of Kentucky Institutional Review Board. The study was conducted at the University of Kentucky Good Samaritan Hospital Sleep Disorders Center. Subjects underwent continuously overnight PSG tests while they were sleeping in the sleep laboratory. The subjects were asked to lie supine on the sleeper beds and a video camera was used to remotely monitor the patient activities in the sleep room. Standard PSG sensors were installed on the bodies for simultaneous monitoring of multiple physiological parameters including nasal airflow and SaO_2.

Table1 reports the characteristics of 15 subjects and results from PSG tests. The sequence of the subjects in this table was based on the value of AHI in the order from lowest to highest.

Table 1. The characteristics of subjects and dates from PSG tests.

NO.	Age (year)	AHI	ESS	BMI (kg/m2)	LSaO2	SEI
01	58	0.5	16	24.4	98	75.9
02	33	0.9	15	44	97	85
03	57	4.1	21	48	94	74.8
04	30	12.0	18	29	88	90.0
05	43	14.3	13	25	93	86.7
06	64	14.7	11	27	84	84.9
07	61	15.8	3	47	71	46.9
08	46	21.7	16	42	93	78.4
09	56	23.5	13	34	91	58
10	51	24.9	16	33	73	66.7
11	54	33.7	10	38	83	77.5
12	42	58.9	19	56	87	74
13	61	61.7	16	32	85	76.1
14	54	68	19	34	81	53.4
15	42	74.3	12	63	80	90.9

$LSaO_2$: The Lowest of Oxygen Saturation
SEI: Sleep Efficiency Index.
ESS: Epworth Sleepiness Scale Score
BMI: Body Mass Index.

The criterion of diagnosing OSAH is currently the value of AHI that can only describe the frequency of apnea and hypopnea happing, but can't tell the time of apnea and hypopnea lasting. We defined v as the measurement of the time of apnea and hypopnea lasting that was computed as the fraction of the total time of apnea and hypopnea lasting over the total sleep time overnight for one subject. According to the results of PSG tests, V was first calculated and then the correlation between v and AHI was investigated.

In order to discuss the characters of oxygen saturation changing, the mean value (Mean)and standard deviation (SD)of SaO_2 overnight were computed for each subject and SIT_{94}(the ratio of the time of SaO_2 being smaller than 94% over total sleep time) was also computed. At last, the correlations between Mean, SD and SIT_{94} and AHI were investigated.

3 RESULT

Table 2. The results of V 、 Mean、 SD and SIT_{94}. for 15 subjects.

Table 2. The results of V '',Mean、 SD and SIT_{94}.

NO.	AHI	V	Mean	SD	SIT_{94}
01	0.5	0.009	97.0	1.5749	0
02	0.9	0.020	98.2	1.0848	0.001
03	4.1	0.025	98.6	2.2688	0.029
04	12.0	0.074	97.9	2.7863	0.033
05	14.3	0.025	98.6	1.4935	0.009
06	14.7	0.02	95.3	2.3937	0.05
07	15.8	0.023	95.4	2.4296	0.139
08	21.7	0.007	98.2	1.0536	0.003
09	23.5	0.106	97.1	2.1218	0.023
10	24.9	0.053	93.9	1.9483	0.428
11	33.7	0.091	98.4	1.5821	0.057
12	58.9	0.194	97.3	3.6273	0.08
13	61.7	0.194	92.5	4.032	0.34
14	68	0.302	95.7	3.366	0.229
15	74.3		94.1	3.0591	0.326

Figure 1 shows the correlations between V、 Mean 、 SD and SIT_{94} and AHI.

4 DISCUSSION

Sleep apnea is by far the most common sleep disorder encountered in clinical practice. Studies show sleep apnea can increase the risk of high blood pressure and stroke, because sleep apnea can cause drops in blood oxygen levels and induce Hypoxia. AHI is currently considered as the 'golden standard' for diagnosing OSAH and is used to classify the severity of sleep apnea. The problem is that AHI is just associated with the frequency of apnea and hypopnea and is not associated with the time of apnea and hypopnea lasting. It is not complete to describe the severity of OSAH with AHI. In fact, hypoxia for a long time will result in severe impairment to the healthy, The longest time of hypoxia, the greater the risk of impairment. V is associated with the time of sleep apnea lasting, which is significantly correlated with AHI (see Figure 1). For subjects No. 8, 9 and 10, their values of AHI are very closed, but the values of V are very different (see

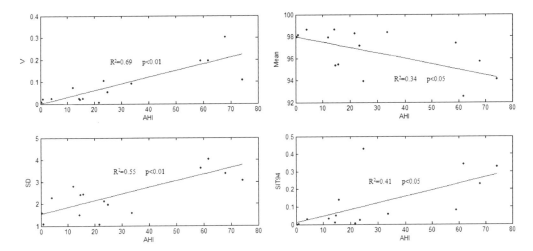

Figure 1. The correlations between V、Mean、SD and SIT_{94} and AHI.

Table 2). However, V is not associated with the frequency of intermittent arousals, so we recommend that both AHI and V should be considered together as the criteria of diagnosing OSAH.

Mean、SD and SIT_{94} describe the change of SaO_2 from different aspects, they are all significantly correlated with AHI (see Figure 1). Among these correlations, SD is most significant, this means SaO_2 will fluctuate largely during sleep apnea for the subjects with high AH; Mean is less significant with AHI. For subjects No. 11 and 12, the values of AHI are very different, but Means are almost equal (see Table 2), this means SaO_2 is not only associated with AHI but also by other factors from each subject.

In addition, the time of hypoxia should be noticed. For subjects No. 8 and 9, Means are very closed, but the time of hypoxia for No. 9 is significantly longer. Furthermore, there are significant correlations between Mean, SD and SIT_{94} and $LSaO_2$, this is natural results. $LSaO_2$、Mean、SD 和 SIT_{94} can describe completely the change of SaO_2.

We investigated the correlations between V, Mean 、SD 和 SIT_{94} and character parameters of subjects such as Age, BMI, ESS and EP, no significant differences were found among them.

REFERENCES

[1] Khayat R, Patt B, Hayes D, Jr. Obstructive sleep apnea: the new cardiovascular disease. Part I: obstructive sleep apnea and the pathogenesis of vascular disease. Heart Fail Rev 2008.

[2] Somers VK, White DP, Amin R, et al. Sleep apnea and cardiovascular disease - An American Heart Association/American College of Cardiology Foundation Scientific Statement from the American Heart Association Council for High Blood Pressure Research Professional Education Committee, Council on Clinical Cardiology, Stroke Council, and Council on Cardiovascular Nursing. J. Am. Coll. Cardiol. 2008;52(8):686–717.

[3] Schwartz AR, Patil SP, Laffan AM, Polotsky V, Schneider H, Smith PL. Obesity and obstructive sleep apnea: pathogenic mechanisms and therapeutic approaches. Proc. Am. Thorac. Soc. 2008;5(2):185–92.

[4] McGinley BM, Schwartz AR, Schneider H, Kirkness JP, Smith PL, Patil SP. Upper airway neuromuscular compensation during sleep is defective in obstructive sleep apnea. Journal of Applied Physiology 2008;105(1):197–205.

[5] Reichmuth KJ, Dopp JM, Barczi SR, et al. Impaired Vascular Regulation in Patients with Obstructive Sleep Apnea. American Journal of Respiratory and Critical Care Medicine 2009;180(11):1143–50.

[6] Guilleminault C, Connolly SJ, Winkle RA. Cardiac arrhythmia and conduction disturbances during sleep in 400 patients with sleep apnea syndrome. Am J Cardiol 1983;52(5):490–4.18

[7] Partinen M, Guilleminault C. Daytime sleepiness and vascular morbidity at seven-year follow-up in obstructive sleep apnea patients. Chest 1990;97(1):27–32.

[8] Palomaki H. Snoring and the risk of ischemic brain infarction. Stroke 1991;22(8):1021–5.

[9] Guilleminault C, Suzuki M. Sleep-related hemodynamics and hypertension with partial or complete upper airway obstruction during sleep. Sleep 1992;15(6 Suppl):S20–4.

[10] Reichmuth KJ, Austin D, Skatrud JB, Young T. Association of sleep apnea and type II diabetes: a population-based study. Am J Respir Crit Care Med 2005;172(12):1590–5.

[11] Rosen CL, Auckley D, Benca R, et al. A Multisite Randomized Trial of Portable Sleep Studies and Positive Airway Pressure Autotitration Versus Laboratory-Based Polysomnography for the Diagnosis and Treatment of Obstructive Sleep Apnea: The HomePAP Study. Sleep 2012;35(6):757–67.12.

Information Technology – Wan et al. (Eds)
© 2015 Taylor & Francis Group, London, ISBN 978-1-138-02785-5

Iris region extraction based on a new combined method

Xiangyan Meng, Yumiao Ren & Haixian Pan
College of Electronic Information Engineering Xi'an Technological University
Xi'an, China

ABSTRACT: The final performance of iris recognition system is partly dependent on the accuracy of iris segmentation (iris localization or iris edge detection). The iris edge was automatically extracted using a new method based on texture feature and watershed transform. Tamura texture was chosen for pre-treatment. To avoid over-segmentation problem, a watershed transform with a marking function and minima imposition were proposed. From the result, we can see that the method we used can improve the speed of the image segmentation. It was applied on 185 iris images with 92.67% of the test targets successfully segmented. The algorithm proposed in this paper proved the combination method is an effective method for segmentation of iris region.

1 INTRODUCTION

The human iris is a complex textured entity. Iridology, also known as iris diagnosis, is a subject which is the study of determining the body organ's health through the inspection of iris texture. The iris is an internal organ of the eye, which is located behind the cornea and in front of the lens. Publications related to segmenting the iris region constitute a significant fraction of the published work in iris biometrics. Iris segmentation (iris localization or iris edge detection), which aims to isolate the actual area of the eye image, is an important stage of iris recognition. Iris segmentation algorithms that assume circular boundaries for the iris region continue to appear in some conferences. Some iris segmentation methods which were focused on removing the assumption of circular boundaries were proposed. [Z. He,2009, S. Shah,2009] Some researchers [R.D. Labati,2009; E.P. Wibowo,2009; J. Zuo, 2010; Ning Wang, 2012), 8334] have considered various approaches to segmenting the iris with boundaries not constrained to be circles. A number of researchers prefer to use methods based on Hough-transform to segment the iris edge from image, they are Ma [L. Ma, 2004], Huang [J. Huang, 2004], Lim [S. Lim, 2001], and Yuan [X. Yuan, 2005] respectively. Iris image contains abundant types of textures, most of them are visible with naked eyes and their structural features are quite obvious. In this section, we present how to efficiently represent the iris textures and measure the features of different kinds of textures. About the texture methods, Wildes [R. Holonec, 2006] proposed an iris texture analysis method, Boles [A. Discant, 2006] proposed an iris recognition method based on the detection of

zero-cross points. These are three most recognized iris recognition methods and the Daugman's is the most applied one [J. Daugman, 2002]. Tamura texture was selected to present the plaque characteristic. In the most cases, only the first three Tamura's features are used for the CBIR. These features capture the high-level perceptual attributes of a texture well and are useful for image browsing. However, they are not very effective for finer texture discrimination [V. Castelli, 2002]. For this reason, we proposed a combined method. The object region was pre-segmented by tamura texture method, and the watershed transform was used to segment the boundary of iris. The combination of watershed segmentation and texture feature can resolve the weaknesses.

2 METHODS

In this paper, we study the iris extraction pre-treatment method based on computer vision. Currently, there is no much literature about the detection of iris based on texture feature.

2.1 *Characteristic value calculation method*

Six texture features were selected, including coarseness, contrast, degree of direction, linearity, regularity and roughness. These six attributes are vector reflects the iris texture feature. In general, high-level visual features of three kinds of attributes can be used to capture the texture, commonly used in the content based image retrieval, and better texture feature selection.

(1) Coarseness

1 It relaZtes to distances of notable spatial variations of gray levels, that is, implicitly, to the size of the primitive elements forming the texture. The proposed computational procedure accounts for differences between the average signals for the no At each pixel (x,y), compute six averages for the windows of size $2^k \times 2^k$, $k=0,1,...,5$, around the pixel.

2 At each pixel, compute absolute differences $E_k(x,y)$ between the pairs of non-overlapping averages in the horizontal and vertical directions.

3 At each pixel, find the value of that maximizes the difference $E_k(x,y)$ in either direction and set the best size $S_{best}(x,y)=2^k$.

4 Compute the coarseness feature F_{crs} by averaging $S_{best}(x,y)$ over the entire image.

Instead of the average of $S_{best}(x,y)$, an improved coarseness feature to deal with textures having multiple coarseness properties is a histogram characterizing the whole distribution of the best sizes over the image.

(2) Contrast

Contrast measures how grey levels q; $q = 0, 1, ..., q_{max}$, vary in the image g and to what extent their distribution is biased to black or white. The second-order and normalized fourth-order central moments of the grey level histogram (empirical probability distribution), that is, the variance, σ^2, and kurtosis, α_4, are used to define the contrast:

$$F_{con} = \frac{\sigma}{\partial_4^n} \tag{1}$$

$$\partial_4 = \frac{\mu_4}{\sigma^4}; \sigma^2 = \sum_{q=0}^{q_{max}} (q-m)^2 \Pr(q|g); \mu_4 = \sum_{q=0}^{q_{max}} (q-m)^2 \Pr(q|g)$$

where m is the mean grey level, i.e. the first order moment of the grey level probability distribution. The value $n=0.25$ is recommended as the best for discriminating the textures.

(3) Direction degree

Degree of directionality is measured using the frequency distribution of oriented local edges against their directional angles. The edge strength $e(x,y)$ and the directional angle $a(x,y)$ are computed using the *Sobel* edge detector approximating the pixel-wise x- and y-derivatives of the image:

$$e(x,y) = 0.5(|\Delta_x(x,y)| + |\Delta_y(x,y)|)$$
$$a(x,y) = \tan^{-1}(\Delta_y(x,y) / \Delta_x(x,y))$$

$\Delta_x(x,y)$ and $\Delta_y(x,y)$ are the horizontal and vertical grey level differences between the neighbouring pixels, respectively. The differences are measured using the following 3×3 moving window operators:

$$\begin{matrix} -1 & 0 & 1 & 1 & 1 & 1 \\ -1 & 0 & 1 & 0 & 0 & 0 \\ -1 & 0 & 1 & -1 & -1 & -1 \end{matrix}$$

A histogram $H_{dir}(a)$ of quantized direction values a is constructed by counting numbers of the edge pixels with the corresponding directional angles and the edge strength greater than a predefined threshold. The histogram is relatively uniform for images without strong orientation and exhibits peaks for highly directional images. The degree of directionality relates to the sharpness of the peaks:

$$F_{dir} = 1 - rn_{peaks} \sum_{p=1}^{n_{peaks}} \sum_{a \in w_p} (a - a_p)^2 H_{dir}(a) \tag{2}$$

where n_p is the number of peaks, a_p is the position of the *pth* peak, w_p is the range of the angles attributed to the *pth* peak (that is, the range between valleys around the peak), r denotes a normalizing factor related to quantizing levels of the angles a, and a is the quantized directional angle (cyclically in modulo 180°).

2.1 Marker-controlled watershed method

Marker-controlled watershed segmentation algorithm was selected for region segmentation. The classical edge detection operator is sensitive to the noise when detecting the edge of goal. So we chose mathematical morphology operator to compute the gradient magnitude.

The gradient magnitude was calculated as following:

$$G[f] = (f \oplus M) - (f \ominus M) \tag{3}$$

where, $G[f]$ is the gradient magnitude; f is original image; M is structural element; $\oplus$ is dilation operation; $\ominus$ is erosion operation.

The gradient magnitude was chose as segmentation function. To avoid the over-segmentation, foreground objects and background objects of image should be labeled before watershed segmentation. Morphological techniques called "opening-by-reconstruction" and "closing-by-reconstruction" were used to mark the foreground objects and background objects. Set up the original image (f) as a mask, g as the marker,

$$g = f \ominus M \tag{4}$$

240

where, let $h_1 = f$, then iterative operation of opening-by-reconstruction is defined as following:

$$R_O: \quad h_{k+1} = (h_k \oplus M) \cap f \tag{5}$$

where, R_O is opening-by-reconstruction image, and iteration stops when $= h_k$, let $j_1 = R_O$, then iterative operation of closing-by-reconstruction is defined as following :

$$R_C: \quad j_{k+1} = (j_k \oplus M) \cap R_O \tag{6}$$

where, R_C is closed reconstructed image, and iteration stops when $j_{k+1} = j_k$.

After segmentation, the inner and outer edges were obtained by 4-connected component labeling algorithm. The aim of iris segmentation is to get the inner and outer boundaries of iris. Target segmentation accuracy of object region depends on the quality of segmentation. The actual object region was extracted by manual method. The object region was erased out by graphics processing software, and then the number of pixels in foreground and background were obtained. The region would be compared to the segmented region by image processing for accuracy evaluation. We have chosen the ultimate measurement accuracy (*UMA*) and misclassified error(*ME*)[Zhang Y J,1992; Lee S U,1990]for valuation of iris region segmentation. And ultimate measurement accuracy was used to evaluate the segmentation result of inner and outer boundary.

$$UMA_f = (1 - \frac{|R_f - S_f|}{R_f}) \times 100\% \tag{7}$$

3 RESULTS AND DISCUSSIONS

185 iris images were captured for testing our method. The segmentation results were obtained in Fig.1 (For example, segmentation of inner boundary from a volunteer).

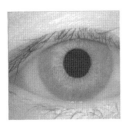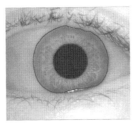

Figure 1. The segmentation result of combination method we proposed.

3.1 *Comparison of two methods*

In order to see the effect of our method, the traditional method such as Hough method was selected.

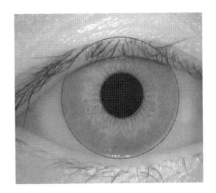

Figure 2. Segmentation result of Hough method.

The segmentation method by Hough transform (seen from Figure 2) considers iris contour as circles. The drawback of this method was that the outer boundary segmented including some non-iris regions. After comparing the two methods, we can see that the method we proposed has the good segmentation result (with no non-iris regions).

3.2 *Performance valuation*

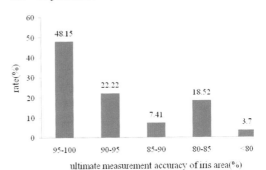

Figure 3. Evaluation of iris area.

After the calculation, the results showed that the ultimate measurement accuracy of iris area was 92.67 ± 6.72%. The results proved that the improved combined method was efficient, accurate and reliable.

3.3 *Algorithm processing time*

The current implementation runs on a Pentium 4 PC workstation, running Matlab 7.5.0. All images have good quality without heavily occluded iris, blurred or unfocused conditions. But few of them are images of people wearing glasses. The effect and processing speed of iris location are presented in Table 1.

Table 1.   Processing speed (same size of iris image).

Processing speed	Average time for our algorithm (Milli-seconds)	Average time for Hough method (Milli-seconds)
Pupil segmentation	62.03	35.3
Sclera segmentation	67.45	45.66

From Table 1, we can see that the computing time was different among the two methods. Our algorithm ran a lot of time comparing with the Hough method. But the accuracy of it is higher than the traditional method.

4   CONCLUSIONS

The eyelids and eyelashes normally occlude the upper and lower parts of the iris region. So when measuring the iris edge, some errors may creep into the segmentation results. By numerical experiments presented in this paper, it can not only detect the target objects correctly, but also is more robust to noise than traditional methods of iris location.

In this paper, we briefly explain the marker-based watershed and texture feature. The results showed the accuracy of segmentation is high.

REFERENCES

A. Discant, S. Demea, C. Rus.2006. Extraction of Relevant Information for Iris Clasification. *18th International EURASIP BIOSIGNAL2006*
E.P. Wibowo & W.S. Maulana. 2009. Real-Time Iris Recognition System Using a Proposed Method. *International Conference on Signal Processing Systems*, 15–17, 98–102.
J. Daugman.2002.Gabor wavelets and statistical pattern recognition. The Handbook of Brain Theory and Neural Networks, 2nd ed., MIT Press, 2002
J. Huang, L. Ma, Y. Wang & T. Tan.2004. Iris model based on local orientation description. *Proc. Of Asian Conf. on Computer Vision*, 954–959.

J. Zuo & N.A. Schmid.2010.On a Methodology for Robust Segmentation of Nonideal Iris Images. *IEEE Transactions on Systems, Man, and Cybernetics, Part B: Cybernetics*, 40 (3): 703–718.
Lee S U, Chung S Y, Park R H. 1990. A comparative performance study of several global thresholding techniques for segmentation. *Computer Vision, Graphics, and Image Processing*, 52(2),171–190.
L. Ma, T. Tan & Y. Wang.2004. Efficient iris recognition by characterizing key local variations, *IEEE Transactions on Image Processing*, 13(6):739–750.
Ning Wang, Qiong Li, Ahmed A. Abd El-Latif.2012. An accurate iris location method for low quality iris images. *Fourth International Conference on Digital Image Processing (ICDIP 2012)*, 8334
R.D. Labati, V. Piuri & F. Scotti. 2009. Neural-based iterative approach for iris detection in iris recognition systems. *IEEE Symposium on Computational Intelligence for Security and Defense Applications (CISDA 2009)*, 1–6.
R.D. Labati, V. Piuri & F. Scotti. 2009. Agent-based image iris segmentation and multiple views boundary refining. *IEEE 3rdInternational Conference on Biometrics: Theory, Applications,and Systems (BTAS 09)*, 1–7.
R. Holonec, S. Demea.2006. A Computerized Method ofIris Image Analysis for Evaluating Prognostic Factor in Some Medical Pathological Groups. *IEEE-TTTC Conferinta nternationala de Automatica, Testare, Calitate si Robotica AQTR 2006 ClujNapoca, Romania*, 25–28 mai2006
S. Lim, K. Lee, O. Byenon & T. Kim.2001. Efficient iris recognition through improvement of feature vector and classier. *ETRI Journal* , 23(2):61–70.
S. Shah & A. Ross. 2009.Iris Segmentation Using Geodesic Active Contours. *IEEE Transactions onInformation Forensics and Security*, 4 (4), Part 2, 824–836.
V. Castelli & L. D. Bergman (Eds.). *Image Databases: Search and Retrieval of Digital Imagery.* Wiley: New York, 2002.
X. Yuan and P. Shi. 2005. Iris feature extraction using 2-D phase congruency, Proc. of the 3rd *International Conf. on Information Technology Application*, Sydney, 33, 437–441.
Zhang Y J, Gergrands J J. 1992.Segmentation evaluation using ultimate measurement accuracy. *SPIE*, 1657,449–460 .
Z. He, T. Tan, Z. Sun & X. Qiu.2009. Toward Accurate and Fast Iris Segmentation for Iris Biometrics. *IEEE Transactions on Pattern Analysis and Machine Intelligence*, 31 (9): 1670–1684.

Information Technology – Wan et al. (Eds)
© *2015 Taylor & Francis Group, London, ISBN 978-1-138-02785-5*

Modeling of the bioelectrical impedance of cell suspension by state space method

S.A. Akulov & A.A. Fedotov
Samara State Aerospace University, Samara, Russian Federation

ABSTRACT: This paper introduces a new method of modeling bioelectrical impedance of cell suspension, allowing getting electrical equivalent circuit which may be used for the estimation of cell suspension composition. The method is based on pulse testing of suspension sample, defining the frequency characteristics of bioimpedance, finding transfer function and representation of state-space model as an electrical equivalents. Results of bioelectrical impedance modeling for cell suspension are presented. If vital cell population increases, alteration of impedance components in low frequencies is registered. These results prove the possibility of using the model parameters for the estimation of the cell suspensions vitality. The process of estimation (i.e. affection on suspension) is proceeding during the time of pulse action so this makes possible to estimate the vitality rapidly.

1 INTRODUCTION

Modeling of the bioelectrical impedance by creating the equivalent current model at different frequency ranges is widely used for tissue impedance spectroscopy, impedance tomography and for structure analysis of human body and multicomponent environments (Grimnes & Martinsen, 2000). During the electrical stimulation process, it is convenient to consider the biological tissues, placed between the electrodes, as an equivalent current model with the properties similar to a biological tissue. This makes possible to model innervation processes, and also allows using this equivalent circuit as a load equivalent for eletrostimulators' testing. Analysis of bioelectrical impedance is also used for quick estimation of blood biophysical characteristics, e.g. hematocrit level may be obtained after analysis of these parameters (Kalakutskiy & Akulov, 2009).

Since the equivalent electrical current model should have similar frequency properties to bioelectrical impedance, the frequency response method is used for mathematical modeling. Range of frequencies, where bioelectrical impedance measurement is being taken, is based on the frequency band of model application. For observed cases the upper threshold frequency is not exceeding 100 kHz. Active or capacitive component of biological tissues impedance prevails at this range of frequencies. Active component of impedance describes the current transmission in electrolytes (plasma, lymph, extracellular liquids etc.) and relates to charges transfer. Capacitive component of impedance is caused by polarization currents and charge separation (Gimsa, 2012).

The main feature is that components of bioelectrical impedance appear to be non-linear when analyzed in frequency domain. Consequently, for the purposes of modeling bioelectrical impedance as an equivalent electric circuit, the frequency characteristic of bioelectric impedance could be approximated by frequency characteristic of equivalent electric circuit impedance, represented as combination of active and capacitive elements. Pole-zero methods and method of frequency responses may be employed (Nise, 2011).

In this paper, the method on the basis of representation bioelectrical impedance in operating transform is offered. Transfer function of impedance can be analyzed as state space equations, describing electrical characteristics of tissues that permit to obtain the invariant model of bioelectrical impedance in form of equivalent electrical current model.

2 THEORY

Proposed method of bioelectrical impedance modeling is based on the functional identification of biological objects. Functional identification is realized by means of methods based on determination of the transfer function. Different kinds of impacts may be applied to the object under test, such as sine wave, unit-step, noise, etc. (Pintelon & Schoukens, 2001). Processing of responses to the test impacts allows defining the bioelectrical impedance frequency characteristic of the biological object and its transfer function.

State-space method (Durbin & Koopman, 2001) is used to transform the transfer function of bioelectrical impedance to the system of first order differential equations. Understanding of these equations gives a way towards structural modeling, which describes the real processes in biological tissue. As one of the test impacts is applied to the system, electric current may be considered as one of the testing means and voltage recorded from the object may be taken as a response.

The Laplace-domain impedance, which determines the relation of voltage and current, i.e. transfer function of the object, can be found as Karson's transform from transient function $z(t)$ of the object:

$$Z(s) = s \cdot \int_0^\infty z(t) \cdot \exp(-st) dt \qquad (1)$$

If $s = j\omega$, frequency characteristic of the bioelectrical impedance can be determined.

The description of electric properties of multi component tissues will be defined by the differential equation of the high order received from Eq. (1) that can be reduced to system of the differential equations of the first degree by using state-space model.

The state-space models are common representations of dynamic models. They allow presenting the linear differential equation of the high order as system of the first order differential equations concerning the additional variables – state variables which make it possible to characterize a condition of system.

We suppose that denominator's roots are different, and the equation for operator impedance can be described as:

$$Z(s) = d_0 + \sum_{i=1}^n f_i / (s - a_i) \qquad (2)$$

where $d_0 = \lim Z(s)$, a_i, f_i – constants.
From (2):

$$U(s) = d_0 \cdot I(s) + \sum_{i=1}^n \frac{I(s) f_i}{(s - a_i)} \qquad (3)$$

where I – a vector of entrance influences (current), U – a vector of output reactions (voltage).
A state variable is:

$$q_i(s) = \frac{I(s)}{s + a_i} \qquad (4)$$

Then Eqs. (3) and (4) can be submitted as state-space model or system of the first order differential equations:

$$\begin{cases} Q' = AQ + i(t) \\ U = FQ + d_0 i(t) \end{cases} \qquad (5)$$

where

$$A = \begin{bmatrix} a_1 & \cdots & \cdots & 0 \\ 0 & a_2 & \cdots & 0 \\ \cdots & \cdots & \cdots & \cdots \\ 0 & \cdots & \cdots & a_n \end{bmatrix}, \quad Q = \begin{bmatrix} q_1 \\ \vdots \\ q_n \end{bmatrix}, \quad F = [f_1 \cdots f_n]$$

Q – state-vector, A, B, F – matrix of constant coefficients.

The model described by the equations (5) shown in Figure.

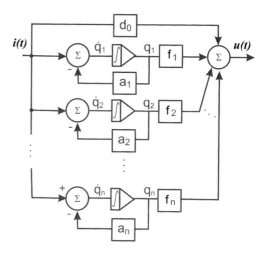

Figure 1. State-space model of the bioelectrical impedance of suspension.

In case of research of a bioelectric impedance of multi component tissues the model can be submitted as the electrical equivalent circuit containing a consecutive circuit (Fig. 2), which consists of active resistance $R0$ and n parallel cells R_iC_i, with Laplace-domain impedance:

$$Z(s) = R_0 + \sum_{i=1}^n \frac{R_i}{1 + sR_iC_i} \qquad (6)$$

where $R_0 = d_0$, $C_i = 1/f_i$, $R_i = f_i/a_i$

It could be observed that the state variables here have physical sense of an electric charge.

Identification of the bioelectrical impedance model (6) includes the following steps (Fig. 3). The unit-step

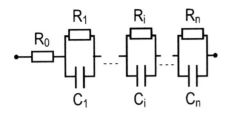

Figure 2. Equivalent electrical current model.

current is passed through a suspension by test current generator. Transient function $z(t)$ is measured as a response. Frequency characteristic $Z(j\omega)$ is calculated from (1). Then it is necessary to define the transfer function $Z(s)$ (for example, by using methods Bode, Levy (Eykhoff, 1981)), afterwards to define polynomial representation of transfer function and by equating constant coefficients receive values of parameters of model R_0, R_1, C_1.

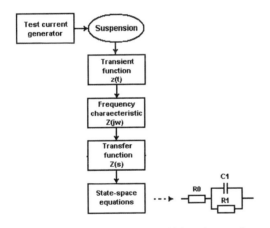

Figure 3. Identification of the bioimpedance of a suspension.

Interpretation of the model parameters gives the information on the structure of researched suspension.

3 RESULTS

Proposed method was used for modeling bioelectrical impedance of fibroblast suspension. The electrodes of measurement system are dropped on a fixed depth into a measurement sample with investigated suspension. Testing current from impulse generator flows through the suspension and cause voltage drop, registered by amplifier. From the output of the amplifier impulse voltage response is received by microcontroller for calculating the frequency characteristic of cell suspension impedance. System performed

measurements every 30 min up to 4 hwith a total amount of 9 measurements. Obtained frequency characteristics for fibroblast suspension in various time points are shown on Figs. 4 and 5.

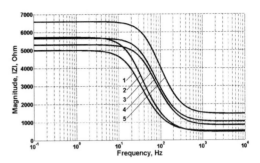

Figure 4. Frequency characteristics of absolute value of impedance for fibroblast suspension: 1 – zero time, 2 – 30 minutes later, 3–60 min later, 4–90 min later, 5–120 min later.

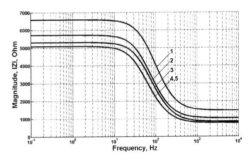

Figure 5. Frequency characteristics of absolute value of impedance for fibroblast suspension: 1–120 minutes later., 2–150 minutes later, 3–180 minutes later, 4–210 minutes later, 5–240 minutes later.

The fibroblast suspension, which is a heterogeneous structure, has the frequency characteristics of bioelectric impedance of the complex form. At low frequencies (<10 Hz) impedance is within the range of 5–7 kOhm and does not depend on frequency; at the "middle" frequencies (0.01–1 kHz) impedance decreases to 0.5–1.5 kOhm, and with further increase of frequency, impedance does not change significantly. During the first 2 h impedance increases over the entire frequency range, then decreasing of impedance occurs.

Quantity of state-space variables of the model (6) for the synthesis of the equivalent circuit of bioelectrical impedance defines the quantity of parallel connections of the elements Ri and Ci. In case of structure analysis of cell suspension, it can be represented as system "fibroblasts-medium". So, the impedance model of fibroblast suspension consists of three elements (Fig. 2). Parameters of the impedance model depend on the time of measurement (Table 1).

Table 1. Parameters of impedance model.

Time	R_0	R_1	C_1,
minute	kOhm	kOhm	mkF
0	0.52	4.48	89.3
30	0.47	5.26	72.2
60	1.04	4.35	40.1
90	1.04	4.68	38.4
120	1.45	5.12	27.4
150	1.02	4.81	40.4
180	0.83	4.45	40.4
210	0.83	4.44	52.1
240	0.83	4.38	52.5

The received dependences of model parameters for different points in time can be used for estimation of cell viability.

4 CONCLUSION

Modeling the bioelectrical impedance of fibroblast suspension by state space method has shown that the impedance model parameters vary over time.

These results allow us to conclude that the proposed method possess a great opportunity for an express estimation of cell viability.

ACKNOWLEDGEMENT

This work was supported by the Ministry of Education and Science of the Russian Federation.

REFERENCES

Grimnes, S. & Martinsen O.G. *Bioimpedance and Bioelectricity Basics*, London: Academic Press, 2000.
Kalakutskiy, L.I. & Akulov, S.A. Modeling of the bioelectrical impedance of blood by synthesis of the equivalent electrical circuits. *IFMBE Proceedings Volume*, 2009, 25/7: 575–577.
Gimsa, JAC-Electrokinetic applications in cell chips: basic understanding and modeling of structural polarization effects. *IFMBE Proceedings* 2012, 37: 1242–124.
Nise, Norman S. *Control Systems Engineering*. Oxford: John Wiley & Sons, Inc., 2011.
Pintelon, R. & Schoukens, J. . *System Identification: A Frequency Domain Approach*. New York: IEEE Press, 2001.
Durbin, J. & Koopman, S. *Series Analysis by State Space Methods*. Oxford: Oxford University Press, 2001.
Eykhoff, P. *Trends and Progress in System Identification*. Oxford: Pergamon Press, 1981.

Information Technology – Wan et al. (Eds)
© *2015 Taylor & Francis Group, London, ISBN 978-1-138-02785-5*

Motility evaluation of amputation rehabilitation patients through simulated abnormal gait analysis

D.J. Moon, J.Y. Kim, H.D. Jung, M.S. Kim & H.H. Choi
Department of Biomedical Engineering, Inje University, Gimhae, Korea

ABSTRACT: We implemented gait with a normal-oriented knee and ankle to obtain properties and electro-myography (EMG) for the simulation of abnormal gait. The simulated gait confirmed that the pattern was the same in normal, reverse pattern, and irregular pattern walking. For each gait type, triaxial accelerometer, pressure sensor, and EMG output data were analyzed. Normal gait and gait of individuals with a knee injury had a stance to swing phase ratio of 64%. The gait of individuals with ankle and knee wound had a ratio of 35–55%. Output of the acceleration sensor was similar for all gaits. For normal gait, the integrated EMG (IEMG) was larger during stance than swing. For the muscle parameters during abnormal gait, a total of three pattern types were confirmed to be the same pattern for normal, reverse pattern, and irregular pattern gait. Each pattern was revealed, regardless of the gait type.

1 INTRODUCTION

The aim of gait is to repeat lower limb motion to move the body forward while maintaining stability. To move the body forward, one lower limb provides support and the other swings forward. This lower limb movement is preceded by their role change, and the repetition cycle in one lower limb is a gait cycle (Perry et al. 2010). The section of the gait cycle supporting the body is the stance phase. The stance phase is divided into two sections: HS (heel strike) where the heel starts touching the ground and TO (toe off) where the big toe starts to detach from the ground. The swing phase is the section from TO to HS.

Individuals show abnormal gait when they suffer from injury or disease. Abnormal gait can be classified by various injured parts. Antalgic gait shortens the stance phase of the affected side's lower limb to reduce the pain. It appears mostly when an individual suffers from external injury to a hip, knee, ankle, foot joint, etc. Arthrogenic gait show excessive plantar flexion and ankle rotation, because of joint stiffening, relaxation, and deformation. Waddling gait from weakening of bilateral hip abductor muscles is characterized by upper body sway (Lee 2012). Leaving these abnormal gaits untreated for a long time may cause complications or additional disease by overtaxing nerves and muscles of other sections. Rehabilitation treatments certainly need to be conducted to prevent additional danger. Various rehabilitation methods have been found to be efficient. Various analysis method are used for efficient gait rehabilitation, and among them electromyography

(EMG) is suitable for evaluating obstacles for normal gait (Hof et al. 2005).

Patient-oriented studies have the danger of additional injury and difficulty to progress. Many studies therefore imitate in experiments with healthy subjects. Kerrigan et al. studied swing phase properties by conducting a virtual experiment of ankylosis gait with healthy subjects. This study showed the possibility of understanding patient-specific mechanism through simulation (Kerrigan et al. 1998). Zajac et al. stated that dynamic simulations help treatment of patients, because they allow obtaining more patient data. They also suggested that experiments with patients can achieve enhanced treatment by obtaining unmeasurable data (Zajac et al. 2003).

In this study, we investigated artificial activities with limitations at the knee and a normal-oriented ankle and obtained gait properties and EMG during abnormal gait. We classified the gait cycle and measured leg acceleration using a triaxial acceleration sensor and a pressure sensor. We confirmed the properties of the quadriceps and biceps femoris thigh muscles using the obtained EMG signals.

2 METHODS

2.1 Experimental setup

While performing a simulated abnormal gait and normal walking, subjects, won the data of EMG and each sensor. Normal gait for the subject was to walk as usual. Abnormal gait simulated patients that could not perform normal gait because of leg injury. For the simulated knee, ankle, and knee and ankle injured patients, each aspect of the normal operation of the damaged joint was assumed impossible. By fixing the ankle and knee using an ankle splint and a knee brace, we constrained movement as in injury.

We measured EMG of biceps femoris and quadriceps, which are used in normal walking, at a 1 kHz sampling rate. Additionally, a triaxial acceleration sensor (AM-3AXIS, NEWTC Inc.) and a pressure sensor (FSR-402, Interlinkelectronics Inc.) were attached and their signals were received by an oscilloscope (WaveRunner 6100A, Lecroy Inc.) for identification and characterization of the gait cycle. The accelerometer was attached to the knee, because it had the larger location radius of the activity. The pressure sensor was attached to the heel and the front of the foot sole to be able to distinguish HS and toe off, which are the critical points to determine the gait cycle.

Twenty male subjects were selected as five people. The average mass and height were 66.8 ± 10.16 kg and 173.2 ± 2.59 cm respectively. The subjects were prohibited from immoderate exercise before the experiment, to prevent interference with the EMG measurements. They practiced gait at their suitable speed. All measurement devices were attached before the experiment.

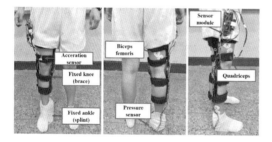

Figure 1.   Mounting position of laboratory equipment.

2.2   Data analysis

The output of the pressure sensor was used to identify the stance and swing phases of the gait cycle. The x-axis of an acceleration sensor was vertical to the ground when it was attached to the subject. It showed a positive value of the acceleration exerted from the ground up. Acceleration in the subject's progression direction had a positive z-axis value.

The obtained EMG signal was analyzed, and IEMG and IAV were calculated. IEMG is the integral of the refined EMG signal over a certain time. It was relevant to display muscle ability. IAV was calculated by dividing the absolute signal value integrated over a certain time period by the number of signals. IEMG and IAV were normalized, and we confirmed the change by gait cycle and analyzed the pattern.

3   RESULTS & DISCUSSION

3.1   Gait cycle analysis

For the normal gait, the ratio between the stance and the swing phase was 64%. This is equal to the ratio that has previously been reported for normal gait (Perry et al. 2010). Also, the gait cycle of simulated knee injury was similar to that of normal gait. The gait ratio of individuals with ankle, and knee and ankle wounds was 35-55% and their stance phase was shorter than that in normal gait. These properties were similar to antalgic gait that is adapted to reduce the pain. The simulated gait was however painless, unlike in patients. However, the stance phase of the leg with the limited joint was shortened because of difficulty in keeping balance.

Although joint movement was restricted during simulated abnormal gait, leg movement according to the gait cycle was similar to normal gait. Output of the acceleration sensor was similar for abnormal and normal gait. When the swing foot touched the ground, the x-axis value increased after an initial decrease at HS. Additionally, the z-axis value seemed to increase in the TO region, because the knee stretched out in front of the body while it was bending before lifting the leg. The z-axis value increased continuously during the swing phase, because the leg moved forward.

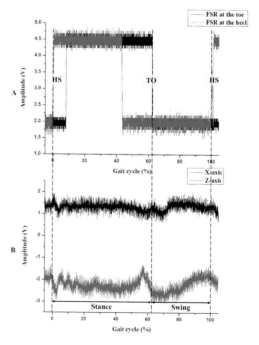

Figure 2.   The output of the acceleration sensor and a pressure sensor at the time of walking.

Table 1. Gait cycle ratio

	Normal gait	Knee injury gait	Ankle injury gait	Knee, ankle injury gait
Stance	60.21%	59.47%	55.90%	55.07%
Swing	39.79%	40.53%	44.10%	44.93%

3.2 EMG analysis

Figures 3 and 4 show the average IEMG and IAV for quadriceps and biceps femoris during normal gait. IEMG was larger during stance than during swing. These patterns were identified in all five subjects. Generally, the IAV pattern was similar to that of IEMG, but at several parts, an irregular pattern was observed. Additionally, the differences between the stance and swing phase were small at the part with a visibly regular pattern. Table 2 lists the muscle parameters for stance and swing of each muscle. IEMG showed a large difference between the values at different parts of the gait cycle. The quadriceps showed a difference of 0.28 and biceps femoris, a difference of 0.23. Meanwhile, for IAV the difference between the values corresponding to the walking cycle was small. Both the quadriceps and the biceps femoris showed a value of 0.04.

For the muscle parameters of abnormal gait, all three pattern types (normal, reverse pattern, and irregular pattern gait) showed the same pattern. Three types of patterns were revealed, regardless of the abnormal gait type. The graph of abnormal gait showed that data of only one subject showed each pattern. Figures 5 and 6 show the same pattern as normal gait. IEMG values of normal walking showed a larger stance value than swing value. However, the difference between the IEMG values by the gait cycle was small. Figures 7 and 8 show the reversed gait pattern with a reversed phase compared to the normal gait pattern. IEMG and IAV values were larger during swing than during stance. Figures 9 and 10 show the irregular gait pattern, where there was no constant pattern according to the gait cycle.

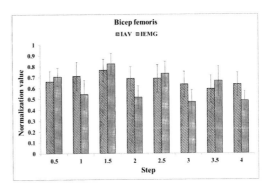

Figure 4. EMG of bicep femoris in normal gait.

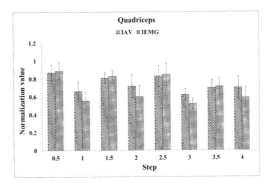

Figure 5. Same as normal gait pattern at abnormal gait EMG of quadriceps.

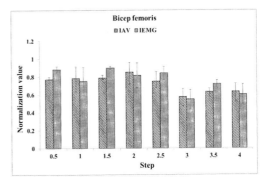

Figure 6. Same as normal gait pattern at abnormal gait EMG of bicep.

4 CONCLUSION

In this study, three types of the simulated abnormal gait and normal gait were compared by analyzing the plantar foot pressure, the knee acceleration, and EMG. We distinguished the stance and swing phase using the plantar foot pressure signal. The knee acceleration and EMG signals of each gait cycle were

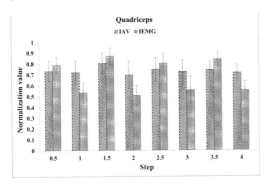

Figure 3. EMG of quadriceps in normal gait.

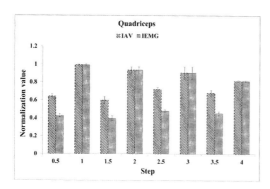

Figure 7.   Reverse as normal gait pattern at abnormal gait EMG of quadriceps.

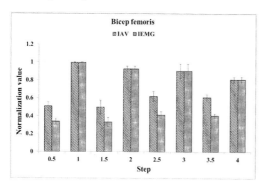

Figure 8.   Reverse as normal gait pattern at abnormal gait EMG of bicep.

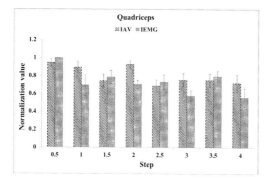

Figure 9.   Irregular pattern at abnormal gait EMG of quadriceps.

analyzed. The abnormal gait had a ratio of stance to swing different to the normal gait. These differences were reflected in significant errors in the muscle activity analysis. Therefore, we propose that the differences in the gait cycle ratio must be taken into consideration in simulated abnormal gait analysis. The muscle activity of simulated abnormal gait showed an

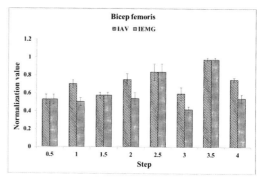

Figure 10.   Irregular pattern at abnormal gait EMG of bicep of quadriceps.

irregular pattern compared to regularly changing the normal gait pattern. The lower the patient's discomfort, the more similar the gait pattern was to normal gait. We considered that this change was useful to the evaluation of rehabilitation exercises. Additionally, quantitative estimation would be possible by quantifying the degree of similarity between normal and abnormal gait patterns.

ACKNOWLEDGMENT

"This research was supported by the MSIP(Ministry of Science, ICT and Future Planning), Korea, under the C-ITRC(Convergence Information Technology Research Center)support program (NIPA-2014-H0401-14-1003) supervised by the NIPA(National IT Industry Promotion Agency)"

REFERENCES

Hof, A.L. Elzinga, H. Grimmius, W. Halbertsma, J.P.K. 2005. Detection of non-standard EMG profiles in walking, *Gait Posture*, 21: 171–177.

Kerrigan, D. C., Roth, R. S. Riley, P. O. 1998. The modelling of adult spastic paretic stiff-legged gait swing period based on actual kinematic data, *Gait & Posture* 7(2): 117–124.

Lee, S.W. 2012. Physical Examination of Arthritis. *The Korean Journal of Medicine* 83(2): 162–173.

Perry, J. Burnfield. M. 2010. *Gait analysis 2nd ed.* SLACK. INC. 3–164

Zajac, F. E. Neptune, R. R. Kautz, S. A. 2003. Biomechanics and muscle coordination of human walking: part II: lessons from dynamical simulations and clinical implications. Gait & Posture 17(1): 1–17.

Information Technology – Wan et al. (Eds)
© 2015 Taylor & Francis Group, London, ISBN 978-1-138-02785-5

Multi-scale binary robust independent feature descriptor

Changqing Yang & Xiaotong Wang
Department of Navigation, Dalian Naval Academy, Dalian, China

ABSTRACT: To solve the problem of the larger memory, time consuming, and lowly matching rate with feature descriptor, a new algorithm was proposed. The algorithm constructs a feature descriptor by fusing the high and low frequency information, the center pixel, and the differential structure local texture feature which extracted in multi-scale. And then, we use hamming distance to measure the final similarity of the descriptor. The experiment conclusively demonstrates that, comparing with SURF, our algorithm takes little memory, and has lower complexity and shorter operation time. At the same time, its matching rate is higher than SURF or identical to SURF.

KEYWORDS: image matching; feature extracting; multi-scale; binary; matching rate

1 INTRODUCTION

The task of finding point correspondences between two images of the same scene or object is part of many computer vision applications. Image registration, camera calibration, object recognition, and image retrieval are just a few.

It has been our goal to develop both a detector and descriptor that, in comparison to the state-of-the-art, are fast to compute while not sacrificing performance. In order to succeed, one has to strike a balance between the above requirements like simplifying the detection scheme while keeping it accurate, and reducing the descriptor's size while keeping it sufficiently distinctive. A wide variety of detectors and descriptors have already been proposed in the literature ([1-5]).

Mikolajczyk, K et al [2] applied the principal component analysis method to remove the redundant, for reducing the descriptor dimension. Calonder, M. et al [3] put forward compact signatures for high-speed interest point description and matching. T. Ojala et al [4] presents a theoretical approach to grayscale and rotation invariant Texture classification based on local binary patterns (LBP) and nonparametric discrimination of sample and prototype distributions. Tuytelaars, et al [5] presents the quantization method combined with PCA. The descriptor in [6], called SIFT for short, computes a histogram of local oriented gradients around the interest point and stores the bins in a 128-dimensional vector (8 orientation bins for each of 4×4 location bins).

Bay, H. et al [4] presents a novel scale- and rotation-invariant detector and descriptor, coined SURF (Speeded-Up Robust Features), by relying on integral images for image convolutions; by building on the strengths of the leading existing detectors and descriptors; and by simplifying these methods to the essential.

In this paper, we construct a new feature descriptor named the multi-scale binary robust independent operator. It extracts the local texture information directly in multi-scale, and it is measured by hamming distance.

2 RELATED THEORY

The algorithm is proposed based on the theory of ID-LBP in the literature [6] and BRIEF in the literature [5]. It constructs the feature descriptor by fusing the high and low frequency information, the center pixel, and the differential structure local texture feature which extracted in multi-scale.

Having the following definition in the area:

$$\tau_1(p;x,y) = \begin{cases} 1 & if \quad p(x) > p(y) \\ 0 & if \quad p(x) \le p(y) \end{cases}$$

$$\tau_2(p;x) = \begin{cases} 1 & if \quad p(x) > \overline{p} \\ 0 & if \quad p(x) \le \overline{p} \end{cases}$$

$$\tau_3(p;y) = \begin{cases} 1 & if \quad p(y) > \overline{p} \\ 0 & if \quad p(y) \le \overline{p} \end{cases} \tag{1}$$

Where $\overline{p}$ is the mean brightness in the area of $S \times S$, $p(x)$ is the smoothing value of $x = (u_x, v_x)$, and $p(y)$ is the smoothing value of $x = (u_x, v_x)$.

After doing the above operations, a binary feature descriptor (named MS-BRIFD) will be constructed. And the feature descriptors of $3 \times n$ dimension are stored in bytes as follows:

$$f_n(p) = \sum_{1 \le i \le n} 2^{i-1}[\tau_1(p;x_i,y_i), \tau_2(p;x_i), \tau_3(p;y_i)] \tag{2}$$

N is selected with 32, 64, and 128 separately. Every feature descriptor has the $l=n/8$ byte, named *MS-BRIFD-L* in this paper.

In the following experiment, we will select the six images of the light group in the image standard database as the test object. The first image is the reference image, and the other image brightness will be changed differently based on the first image. In order to measure the effect of matching algorithm using the Hamming distance, we define the matching rates as follows:

1 We extract D feature points in the reference image, and find their corresponding D matching points in the target image using the homography matrix, then calculate the descriptors of the 2D target points.
2 For each target point in the reference image, we confirm the matched feature point with the minimum Hamming distance.
3 If the target points matched successfully is d, then the matching rate is d/D.

2.1 Region pixels extracting

We construct regional feature descriptor by extracting the coordinates of an N group of pixels (x_i, y_i) using the mathematical distribution method. If the origin of coordinates is in the center of the area, and all the distributions are two-dimensional

distribution, then the sampling process is defined as follows:

1 X and Y are independent and identically distributed as $X, Y \in U(-S/2, S/2)$. The pixel coordinates extracted are close to the image edge mostly.
2 X and Y are independently and identically distributed. $X, Y \in N(0, S^2/25)$. The standard deviation of the normal distribution is $\sigma = S/5$.
3 If $X \in N(0, S^2/25)$, and $Y \sim \in N(x_i, S^2/50)$, the standard deviation of the normal distribution is $\sigma = S/5\sqrt{2}$.
4 X and Y is independent and following the random distribution $(-S/2, S/2)$.

Each of the above four schemes are tested by comparing the matching rate. We use the light group as the test object. The test results show that, the matching rate of scheme (2) and (4) is higher. But the scheme (2) is more stable when in the more difficulty matching condition. So we choose the scheme (2) as the method of extracting pixel.

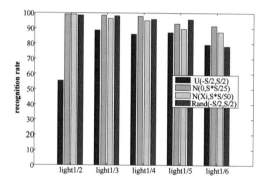

Figure 1. The matching rate of different mathematical distributions.

2.2 Hamming distance distribution

The Hamming distance was used to measure the similarity of the descriptors. In the light group as an example, 5000 randomly pixel matching points are selected in the reference image. Then we confirm the correct matching descriptors in the corresponding image. At the same time, we calculate the every Hamming distance between the correct or error matching points. We can see from Figure 4, the Hamming distance of correct matching point descriptors is stable, but the error matching points are unstable. So that, the Hamming distance can meet the matching requirements.

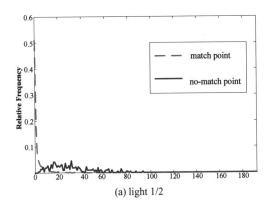

(a) light 1/2

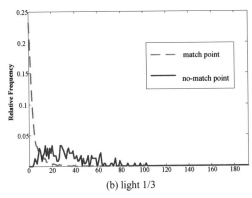

(b) light 1/3

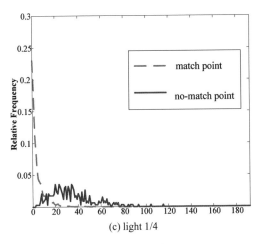

(c) light 1/4

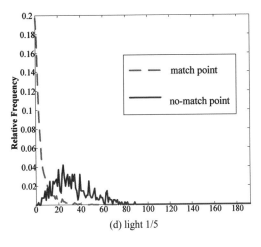

(d) light 1/5

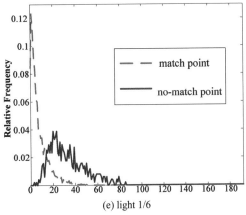

(e) light 1/6

Figure 2.   Hamming Distance Distribution.

(a) light

(b) bikes

(continued)

3   EXPERIMENT

In this section, this algorithm we proposed will be tested. All of the algorithms and experiment in this thesis are based on MATLAB2011 platform. In order to verify the algorithm performance in various aspects, we select four groups of images from the standard image library, as shown in figure 3.

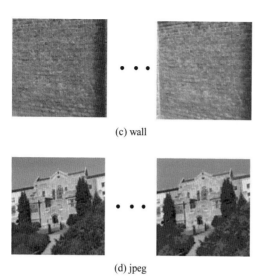

(c) wall

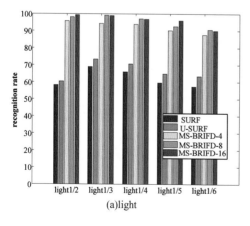

(d) jpeg

Figure 3.　(a is Brightness transform, b is fuzzy transform, c is perspective transform, and d is compression transform).

3.1 *Effect of descriptor size*

In this part, we will analyze the effect of feature descriptor for matching rate. We use the SURF algorithm to extract the feature points at all the tests, to keep consistent with the feature point positions and the number. We construct *MS-BRIFD-4, MS-BRIFD-8, MS-BRIFD-16,* and with n=32, 64, and 128 separately. As shown below:

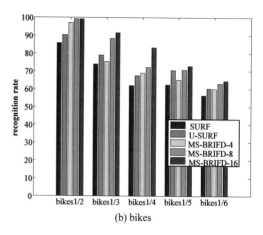

(b) bikes

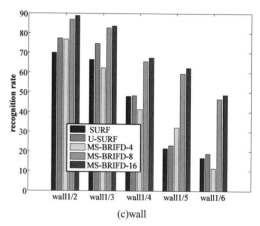

(c)wall

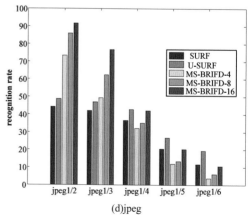

(a)light

Under the condition of the brightness changes, the matching algorithms we proposed were higher than those of SURF and U-SURF algorithm. And under the condition of the fuzzy changes, the matching algorithms approximate the U-SURF algorithm, but even outperforms than the SURF.

(d)jpeg

Figure 4.　Matching rate.

3.2 *Matching rate of rotation*

The MS-BRIFD operator is not considered in the differential direction. In this part, we will discuss the robustness of the MS-BRIFD operator. We use light1 as the test objects. The test condition is 800 pixels which are extracted random with different angles. The test results are shown in the table 1.

Table 1.   Matching rate of rotation.

	1°	10°	30°	90°	120°	180°
SURF	91.17%	76.64%	61.56%	82.39%	63.28%	84.90%
U-SURF	91.74%	79.30%	12.51%	0.17%	0.23%	0.05%
MS-BRIFD	97.08%	44.64%	11.37%	0.14%	0.21%	0.57%

3.3   *Time consuming*

At the same time, we can also use the time consuming to estimate the computational complexity of the algorithm. In this part, we use the consumed time of producing and matching the descriptor as standard to compare with the algorithm of MS-BRIFD, U-SURF, and SURF. The test results are shown in the table 2.

Table 2.   Time consuming.

	SURF	**U-SURF**	**MS-BRIFD**
Descriptor construction time/s	6.81	6.97	0.84
Descriptor matching time/s	3.30	3.28	0.36

4   CONCLUSION

According to the feature descriptor defects, such as computation, memory and others, a new algorithm named the multi-scale binary robust independent operator was proposed. The test results show that, comparing with SURF, our algorithm takes little memory, and has lower complexity and shorter operation time. At the same time, the matching rate of our algorithm is higher than SURF or identical to SURF. But the operator did not consider the direction information, and have the poor Rotating robustness, that will be improved in the following work.

ACKNOWLEDGMENT

This project supported by the National Natural Science Foundation of China **61250006**, and **61273262**.

REFERENCES

[1] Wang Hong, Ji Xiaoqiang, Dai Ming. Improved speed up robust matching algorithm[J]. Infrared and Laser Engineering. 2012, 41(3), pp:811–817.

[2] Mikolajczyk, K., Schmid, C. A Performance Evaluation of Local Descriptors[J]. IEEE Transactions on Pattern Analysis and Machine Intelligence, 2005,27 (10) 1615–1630.

[3] Calonder, M., Lepetit, V., Konolige, K., Bowman, J., Mihelich, P., Fua, P.: Compact Signatures for High-Speed Interest Point Description and Matching[C]// IEEE International Conference on Computer Vision, 2009:357–364.

[4] T. Ojala, M. Pietikäinen, and T. Mäenpää, Multiresolution Gray-Scale and Rotation Invariant Texture Classification with Local Binary Patterns[J], IEEE Transactions on Pattern Analysis and Machine Intelligence, 2002, 24(7):971–987.

[5] Michael Calonder, Vincent Lepetit, Pascal Fua, BRIEF : Binary Robust Indepent Elementary Features[C]// Computer Vision-ECCV 2010 Lecture Notes in Computer Science ,2010:778–792.

[6] Sun Junding, Zhu Shisong, Wu Xiaosheng, Image Retrieval Based on an Improved CS-LBP Descriptor[C]// IEEE 2010 2nd International Conference on Information Management and Engineering(ICIME2010), 2010:115–117.

Information Technology – Wan et al. (Eds)
© 2015 Taylor & Francis Group, London, ISBN 978-1-138-02785-5

New adaptive capacity of audio steganography based on the wavelet transform

Feng Pan, Jun-wei Shen & Yao Zhang
Key Laboratory of Network and Information Security, Engineering University of CAPFXian, China

ABSTRACT: In this article, we took the wavelet packet decomposition method to resolve the audio signals. Then grouped the sub-bands and selected the appropriate groups to be audio carriers according to the marginal bands distortion sensitivity of the packets. We calculated the volume of embedded, selected steganographic position automatically and uniformly according to the sizes of the hidden information and audio carriers. Finally, we improved the encoding algorithm of F5 to embed the information. Experimental results show that the method of high capacity, good imperceptibility, and the computational complexity is low.

1 INTRODUCTION

Today, Steganography research focuses on the image area because the images have more redundancy to hide information. But the images' steganalysis technology is mature, so the image steganography algorithms require high technology and high computational complexity. On the other hand, the proportion of the digital audio files is increasing and there are few statistics models about audio, so audio steganography research is not only necessary but also practical.

MP3stego [1] devised by F. Petitcolas is currently a popular audio steganography tool, but A. Westfeld proposed a steganalysis method[2] based on block length variance is against it. In addition, StegHide, Hide4PGP, Security Suit and other audio steganographys tools use LSB (Least Significant Bit) [3] algorithm in the time domain, but as the bits that are replaced increasing, the imperceptibility is worse and worse.

This paper presents a high-capacity audio steganography algorithm. First, the audio carrier signal makes wavelet packet decomposition, and then adopts a better algorithm to embed secret information in the wavelet coefficients. The experiments show that the method is scientific and effective.

2 AUDIOSTEGANOGRAPHYALGORITHM DESIGN

2.1 Wavelet transform

Wavelet packet decomposition is developed on the basis of Fourier analysis. It is better than the local Fourier transform in that it takes a narrow band of space window at high frequencies, takes wide room windows at low frequencies, which is suitable for handling non-stationary signals. We can understand wavelet packet from the perspective of the average and details. If there is a discrete signal $\{x_1, x_2\}$, respectively, the average (a) and details (d) may be expressed as:

$$a = \frac{x_1 + x_2}{2} \tag{1}$$

$$d = \frac{x_1 - x_2}{2} \tag{2}$$

Discrete-time signal can be expressed as:

$$x_1 = a + d \tag{3}$$

$$x_2 = a - d \tag{4}$$

If x_1 and x_2 are very close and can be representedby $\{a\}$, $\{a\}$ can be regarded as an information signal, d is the lost details.

The nature of the sub-band coefficients after wavelet packet decomposition has the characters of both frequencies and is multi-resolution, which has the advantage for the analysis of each resolution. Dora M. Ballesteros and Juan M. Moreno proposed the chameleon effect [4], which is to embed different details according to different secret information. Cvejic and Seppanen also proposed using LSB embedding method in wavelet domain[5] for audio steganography. But they do not have discriminatory to all the various parts. This affects the character of imperceptibility.

In order to reduce error accumulation, we group audio signal, with 1024 sampling points as a group, and then take 4 layer wavelet packet decompositions, with 16 sub-bands (chart 1). Each sub-band has 64 wavelet coefficients. The energies of 16 sub-bands follow as A, D1, D2, D3 ... D15. A is the minimum frequency resolution information; Di are details, D1 is the lowest resolution detail and has maximum energy of Di. As Di increases, the energy gradually decreases and the frequency sequentially increases. In our experiment, we clear each sub-band in turn, and then restore the original audio signal and calculate the signal to noise ratio (SNR) between them, finally we get Table 1. If the SNR is small, it indicates that the sub-band energy is large, on the contrary, larger SNR, the smaller sub-band energy.

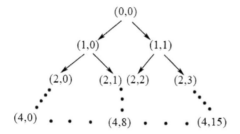

Flow chart 1.　The numbers of the sub-bands.

After repeating embedding, compression, decompression for the audio then extracting the secret information, we find that if the secret information is embedded in the sub-bands of high frequency that could impact the character of imperceptibility. If you embed in the high frequency parts of the Sub-bands, the distortion rate of secret information is high. So we choose the D2-D8as the carriers.

2.2　*New audio steganography algorithm design*

We improve the algorithm of F5[6] which was designed by Andreas Westfeld to suit one-dimensional audio signals. The advantage of this method is that, if we hiden bits secretin formation, we should only change 1 bit audio carrier. This has got a good effect, because our auditory sense is more sensitive.

For example, there are 2 bits secret information $x1$, $x2$. We need 3 bits data carrier like $a1$, $a2$, $a3$, four conditions are as folloqs

If $x_1 = a_1 + a_3$, $x_2 = a_2 + a_3$ make no changes
If $x_1 \neq a_1 + a_3$, $x_2 = a_2 + a_3$ a_1 is changed
If $x_1 = a_1 + a_3$, $x_2 \neq a_2 + a_3$ a_2 is changed
If $x_1 \neq a_1 + a_3$, $x_2 \neq a_2 + a_3$ a_3 is changed

When extracting the information, we need to extract $a1$ $a2$, $a3$ and calculate $x1 = a1 + a3$, $x2 = a2 + a3$.

We give the general form of this algorithm. kbits secret messages will be hidden into nbits ($n = 2^{k-1}$) vector, only 1 bit data changed. We use triples $(1, n, k)$ to describe it. The utilization of the carrier[7] is

$$R = \frac{k}{n}$$

(5)

Embedding efficiency of carrier[7] is

$$E = \frac{2^k \times k}{n}$$

(6)

We use the formula calculate the following Table 2.

In the practical application, we design an algorithm that could auto-choose a more appropriate k via our secret information and carriers' sizes. For example, there is 1000 bits' secret message and the audio signal has 16,000 wavelet coefficients after a total of 4 layers of wavelet packet decomposition. So secret information can be embedded in 7000 (D2-D8) wavelet coefficients. If only embedded in

Table 1.　The energy of sub-bands.

Sub-band	SNR	Energy	Sub-band	SNR	Energy
number	(db)	Code	number	(db)	Code
(4,0)	6.08	A	(4,8)	36.66	D8
(4,1)	18.94	D1	(4,9)	37.87	D13
(4,2)	24.03	D2	(4,10)	37.44	D9
(4,3)	24.64	D3	(4,11)	37.71	D12
(4,4)	30.45	D5	(4,12)	37.51	D10
(4,5)	30.82	D7	(4,13)	38.21	D15
(4,6)	29.51	D4	(4,14)	37.62	D11
(4,7)	30.60	D6	(4,15)	38.05	D14

Table 2.　The rule of the algorithm.

k	n	data utilization embedding	efficiency
		R(%)	E(%)
1	1	100.0	2
2	3	66.7	2.67
3	7	42.86	3.43
4	15	26.67	4.27
5	31	16.13	5.16
6	63	9.52	6.09
7	127	5.51	7.06
8	255	3.14	8.03
9	511	1.76	9.02

the least significant bit of the wavelet coefficients, the utilization R is 14.3% ($1000/7000×100\%$), so according to the Table 2, k cannot exceed 5 for guaranteeing all the secret message embedded.

Furthermore, in order to be against the statistical characteristics, we can produce pseudo random numbers order ri (r_i must be smaller than k) by using logistic chaotic model .The logistic chaotic model[8] is: $r_{n+1} = f(r_n, u) = u \times r_n(1-r_n)$. u is the controlparameter ($0 \le u \le 4$),$r_i \in [0,1]$.If $u > 3.57$, the sequence $r_0, r_1, r_2 \ldots$ are pseudorandom sequence in the interval $0\sim1$. We call the initial value r_0 seed. We can also produce cipher key through the seed to improve the safety.

3 ACHIEVEMENT OF THE ALGORITHM

The algorithm consists of embedding algorithm and extraction algorithm.

The first step of the embedding algorithm is to divide the audio into segments, then use Haar wavelet bases to take 4 layers' wavelet packet decomposition. Get the parameters according to the carrier capacity and the size of message. At last, embed the message by using the improved-rule of F5.

The first step of the extraction algorithm is to divide the audio into segments, then use Haar wavelet bases to take 4 layers' wavelet packet decomposition .Use the same parameters to find the embedding position. At last, extract the message by using the improved-rule of F5.

3.1 The embedding algorithm

The embedding algorithm could divide into 8 steps to achieve;

1 Divide the audio signal into segments (each 1024 sampling points for one part);
2 Every segment uses Haar wavelet bases to take four wavelet packet decomposition, getting 16 sub-bands. Each sub-band has 6 wavelets;
3 Select the seven sub-bands D2–D8 and use the wavelet coefficients of each sub-band to multiply by the scale factor (typically, the maximum of sub-bands), and then convert into binary. coefficients;
4 Select the seeds, then use Logistic chaotic model to generate the secret key to encrypt message, and then convert the cipher-text into binary;
5 Get the parameter k according to the carrier capacity and the size of message;
6 Select the seeds, then use logistic chaotic model to generate the random sequences and deal with

them, getting the positive integer sequences ri (The maximum of ri must be smaller than k). Get length sequences ni according to sequences ri.(You can find congruent relationship in Table 2);
7 Modify one bit of ni in accordance with the encoded-rules to embed;
8 Reconstruct the original signal (Inverse Discrete Wavelet Transform);

You can see it in Flow chart2.

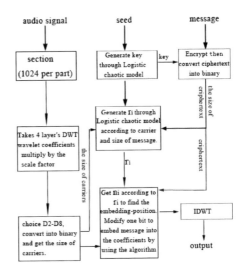

Flow chart2. Embedding algorithm.

3.2 Extraction process

The extraction algorithm could divide into 6 steps to achieve;

1 Divide the audio signal into segments (each 1024 sampling points for one part);
2 Every segment takes four wavelet packet decomposition, getting 16 sub-bands. Each sub-band has 64 wavelet coefficients;
3 Select the seven sub-bands D2-D8, and then converted into binary;
4 Use the same seed to get the same key and the same sequences ri;
5 Get length sequences ni according to sequences ri than search the embedded-position. Extract secret-message using decoded-rules of F5;
6 Decrypt secret-message through key;

You can see it in Flow chart3.

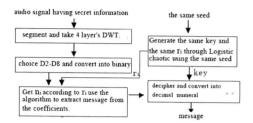

Flow chart3.   Extraction algorithm.

4   EXPERIMENTAL RESULTS AND FUTURE PLAN

We did a lot of experiments by suing different kinds of music such as pops, classical, violin, piano. The experimental results show that the algorithm is universal. We can do experiments and explain them from the aspects of embedding capacity, imperceptibility and undetectability.

4.1   *The embedding capacity*

If the carrier signal uses 44.1 kHz sampling, 16-bit quantization of audio clip, there will be 44,100 samples per second in one channel. After wavelet packet decomposition, there are 44,100 wavelet coefficients and 19,293 (D2–D8) of them could become carrier. If only to modify the least significant bits, the carrier may be embedded 19,293 bit per second. If you modify the minimum 2 to 3 significant bits, there are still good imperceptibility, and the capacity rises to 38586–57879bit per second. Therefore the method has high steganography capacity.

4.2   *Imperceptibility*

Imperceptible damage is caused by distortion of the signal. We usually use it to calculate the distortion level of the SNR, the formula is as follows:

$$SNR(\text{db}) = 10\lg \frac{\sum x^2(n)}{\sum (x(n) - y(n))^2} \qquad (7)$$

In the formula the *x(n)* is original audio signal, *y(n)* is audio signal after embedding secret message .

The experiments compare the SNRs from two dimensions. First, use the same embedding rule in this article, but embedding message in time domain and wavelet domain differently. Severally count the SNRs then compare them. Second, choose the audio carriers at the same domain. One uses the embedding rule in this article to embed the message while the other uses the traditional way. Count the SNRs then compare them.

We show one experiment to explain it. We cut current popular music "wind in the wheat" music clips from 14,262 27,661 ms millisecond, with sampling frequency 44.1 kHZ and sampling bits 16bit. By changing the number of steganographic information, we adaptively select a different k and then get data utilization R(according to Table 2). Also at the same data utilization R, we increase the number of LSBs in per wavelet coefficients. Finally, we get Tables 3–5.

The experiment shows that the SNRs after embedding message in wavelet domain is average 14 percent higher than in time domain. If we use embedding rule in this article, SNRs will increase 7 percent again.

Table 3.   SNRs in wavelet domain by using the embedding rule in this article.

Utilization R	1LSB	2LSBs	3LSBs	4LSBs
R(%)	SNR(db)	(db)	(db)	(db)
100.0	77.32	68.31	60.52	53.92
66.7	79.32	70.52	63.18	56.57
42.86	81.08	73.06	65.75	59.15
26.67	84.51	76.55	69.23	62.66
16.13	88.43	79.87	72.53	66.01
9.52	91.04	83.03	75.64	69.13

Table 4.   SNRs in time domain by using the embedding rule in thisarticle.

Utilization R	1LSB	2LSBs	3LSBs	4LSBs
R(%)	SNR(db)	(db)	(db)	(db)
100.0	68.87	60.07	51.01	40.12
66.7	71.82	63.72	54.21	45.23
42.86	74.66	67.98	59.23	50.21
26.67	77.07	71.22	63.01	54.64
16.13	81.09	74.38	66.98	58.43
9.52	84.07	77.08	70.03	62.55

Table 5.   SNRs in time domain by using the traditional way.

Utilization R	1LSB	2LSBs	3LSBs	4LSBs
R(%)	SNR(db)	(db)	(db)	(db)
100.0	62.33	55.07	43.12	35.92
66.7	65.31	58.34	46.21	38.82
42.86	68.92	61.46	49.92	41.76
26.67	71.27	64.77	52.33	45.15
16.13	73.11	67.25	56.02	48.63
9.52	77.02	70.09	59.69	52.83

4.3 Undetectability

There are not many audio steganography methods, so there are less audio steg analysis methods. J Dittmanna team designed steg analysis algorith [9] against audio steganography of LSBs in time domain. Its judgment flag is r, when r ≥ 1.1,it has audio hidden. We will use these above-mentioned audio data to calculate r. We give a result directly, as in Tables 6–8.

Experimental results show that in the wavelet domain, all the embedded texts are not detected. In time domain, all the embedded texts are detected by using the traditional way. The embedded texts are partially not detected in time domain because we use adaptive embedding rule.

Table 6. The value *r* in wavelet domain by using the embedding rule in this article.

Utilization R	1LSB	2LSBs	3LSBs
R(%)	r	r	r
100.0	1.0674	1.0754	1.0821
66.7	1.0428	1.0501	1.0667
42.86	1.0022	1.0288	1.0421
26.67	0.9587	0.9727	0.9976
16.13	0.9221	0.9321	0.9574
9.52	0.9002	0.9062	0.9121

Table 7. The value *r* in time domain by using the embedding rule in this article.

UtilizationR	1LSB	2LSBs	3LSBs
R(%)	r	r	r
100.0	1.1437	1.1589	1.1766
66.7	1.1228	1.1329	1.1597
42.86	1.1045	1.1128	1.1221
26.67	1.0921	1.1035	1.1197
16.13	1.0844	1.0946	1.1077
9.52	0.9962	1.0182	1.0469

Table 8. The value *r* in time domain by using the traditional way.

Utilization R	1LSB	2LSBs	3LSBs
R (%)	r	r	r
100.0	1.1773	1.1835	1.1989
66.7	1.1672	1.1721	1.1801
42.86	1.1536	1.1602	1.1691
26.67	1.1401	1.1515	1.1597
16.13	1.1229	1.1464	1.1507
9.52	1.1162	1.1291	1.1385

4.4 Results and plan

Experimental results and analysis show that the method has a high capacity, good imperceptibility and low computational complexity and better extraction accuracy.

We will continue to study the above method with lot of experiments to improve embedded-carrier in the future. In addition, we will also design an algorithm for shuffling the embedded-carrier, thereby further improve the imperceptibility.

REFERENCES

[1] Petitcolas F. MP3Stego1.1.18 [CP / OL]. [2010-10-02]
[2] westfel A. Detecting low embedding rates. *FAP, Information Hiding. Noordwijkerhout,* Netherlands: Springer-Verlag, 2003, pp.324–339.
[3] J Fridrich. Feature-based steganalysis for JPEG images and its implications for future design of steganographic schemes. In: *6th Information Hiding Workshop, Toronto, ON, Canada,* 2004.
[4] Dora M. Ballesteros, Juan M. Moreno A "Highly transparent steganography model of speech signals using efficient wavelet masking" *Expert Systems with Applications* 39 (2012) 9141–9149.
[5] N.Cvejic, T.Seppanen, "A wavelet domain LSB insertion algorithm for high capacity audio steganography ", *Digital Signal Processing workshop, 2002 and the 2nd Signal Processing Education workshop, Proceedings of 2002 IEEE* 10th 13–16 October 2002, pp. 53–55.
[6] Westfeld A. F5-a steganographic algorithm high capacity despite better steganalysis. *Lecture Notes in Computer Science,* 2001, 2137 (2): 289 –302.
[7] Fenlin Liu, Jiufen Liu, XiangyangLuo. Digital Image Steganalysis BeiJing:China Machine Press.
[8] Runshenghuang. Chaos and Its application. WuHan:WuHan University Press, 2000.
[9] J Dittmanna. DHesse. Network based intrusion detection to detect steganographic communication channels: on the example of audio data [C]. In: IEEE 6th Workshop on Multimedia Signal Processing, Siena, Italy, 2004:343–346.

Information Technology – Wan et al. (Eds)
© 2015 Taylor & Francis Group, London, ISBN 978-1-138-02785-5

New chaotic image encryption algorithm based on rossler

Shihui Liu & Runhe Qiu

College of Information Sciences and Technology, Engineering Research Center of Digitized Textile & Fashion Technology, Ministry of Education, Donghua University, Shanghai, P. R. China

ABSTRACT: According to the principle and characteristics of chaotic sequence, this paper creates a three-dimensional Rossler chaotic image encryption algorithm, which generates the initial key with the aid of the initial value and the plaintext explicitly, it will obtain iterative sequence when the initial key through the initial chaotic systems, then the original plaintext images iterative sequence encrypted with the method of diffusion and scrambling again. Simulation results and analysis show that the spread of chaos and scrambling encrypted image generated by means of this algorithm, the key space is large, low correlation of adjacent pixels, and strong anti-attack capability. The key space reaches 10^5 in addition to explicit self-generated key.

KEYWORDS: chaotic sequence; initial key; scrambling; diffuse; Rossler

1 INTRODUCTION

With the rapid development of computers and the internet, network transmission becomes easier and faster, more and more information transfer in the network, but with frequent exchange of information, encryption technology has been put on the agenda and becoming important, which to make sure the information of our transmitted and received insecure or not. However, traditional methods of encryption for the large amounts of data can not meet the needs of real-time.In recent years, a novel based on a non-linear chaotic encryption [1–4] has become a hotspot.

As we all know, many factors in a good encryption algorithm should be considered to improve the security of image transmission. In recent years, a number of image encryption algorithm has been proposed: literature [5] proposed an alternating image encryption algorithm, which is generated by means of explicit secondary key, dual chaotic system by scrambling algorithm expanded the key space, and increasing the difficulty of the attacks; literature [6] is encrypted with a disturbance, which improving the security of the algorithm and resisting the plaintext attacks effectively through the secondary scrambling; literature [7] is improving the algorithm, which is designed to replace the pixel values of two rounds of the encryption process. The transformation of alternative encryption formula can be make it complicated. Above all, to design a safe and reliable encryption algorithms, we need to pay attention to as follows:

1. Whether the space of the keys is large enough and we can resist brute-force attacks effectively;
2. Whether scrambling and diffusion of the gray value image generated by the algorithm, and the ciphertext and the adjacent pixel values are distributed uniformly at random, which irrelevant based on the encryption algorithm, thus resist statistical attacks;
3. Whether such encryption algorithms are so sensitive to plaintext and ciphertext that be able to resist the differential attacks effectively.

From the issues above, the literature [8] has been used Rossler chaotic sequence and the shuffle map to the pixel, making encrypt with XOR in pixel values in random order and scrambling pixel location, then to achieve the pixel encryption secondary and the encryption key map once ; While in the literature [9] was proposed a digital image encryption algorithm based on two-dimensional Henon and three-dimensional Rossler chaotic map, this algorithm used a two-dimensional Henon map to encrypted the original image once, then using three-dimensional map of the second Rossler chaos encryptions to ensure adequate security of the encrypted image. This paper proposes a new three-dimensional chaotic image encryption algorithm based on Rossler, the algorithm combine the plaintext and the initial values to obtain the initial key during the encryption process, so the algorithm and the key is not only sensitive to the initial value , but also concerned with plaintext, itcan resist chosen plaintext attacks effectively;

Focused on innovation projects of the Education Commission of Shanghai

Meanwhile, the algorithm is also applied with scrambling and diffusion theory, through the influence of each bit of images spreading to the entire image and airspace for image encryption ,the encrypted image is more complex, so that the algorithm can effectively overcome the defects from anti-known plaintext and chosen-plaintext attacks, so this encryption algorithm is feasible.

2   NEW CHAOTIC IMAGE ENCRYPTION ALGORITHM BASED ON ROSSLER

2.1   *Rossler chaotic map*

Rossler system[10] is a chaotic map, whose dynamics is a three-dimensional differential equations, as shown in Equation (1):

$$
\begin{cases}
\dot{x}_1 = -x_2 - x_3 \\
\dot{x}_2 = x_1 + a x_2 \\
\dot{x}_3 = b + x_3(x_1 - c)
\end{cases}
\tag{1}
$$

Where $\dot{x}_i = d\,x_i\,/\,dt$ represents the system state quantity $x_i(i=1,2,3)$ of the derivative of t times; a, b, c as a system

parameter , when a = 0.2, b = 0.2, c = 5.7, the system is chaotic; namely, given any initial value $(x_{10}\ x_{20}\ x_{30})$, which is generated chaotic iteration series from equation (1).

Since Rossler system is a chaotic system,equation (1) N-1 times of iterations will produce three iteration series, denoted $\{x_{1i}\}, \{x_{2i}\}, \{x_{3i}\}$. And here is used of these three iterative sequences through a certain role in the transformation of the original image, so that the original image becomes into a blurred image, so as to achieve the effect of encryption.Figure 1 shows a flowchart of encryption.

Figure 1.   Encryption flowchart.

2.2   *The initial formation of the key*

In a modern study of chaotic systems, we often just put the initial value as the key, but the key is not effective against differential attacks without any dependence on the plaintext. To this end, this paper will produce the initial key in the form of conversion.

The specific formation process of the key: taking any one of plain image pixels of horizontal, vertical and a sequence of a plurality of arbitrary oblique diagonal, denoted as {R}, {C}, {H}. The sum of these three sequences, respectively, the obtained is denoted by r, c, h. Then given three arbitrary value range between [-1,1] named x_{10}, x_{20}, x_{30}, the number of columns consists of four digits behind the decimal point, sum the last step r, c, h with the initial value, respectively, then modulo of 255 and divided by 255, denoted by $(x_{11}, x_{21}, x_{31},)$ which is the value of the initial state of the system. Since the initial value depends on the plaintext and the given value, so the initial value is highly sensitive.

2.3   *Image scrambling*

An gray image can be represented as a matrix formed M × N, the pixels $\in$ [0 255], confusing the image position is to exchange the image pixel position, so that the original clear image becomes blurred image. So as to achieve the effect of encryption. This image scrambling process as follows.

First, transform the image with a size of M × N pixels into an array of [M × N, 1], from the sequence of iterations $\{x_{1i}\},\{x_{2i}\}$, taking any (M × N) / 2 number (encrypted only specific who know),used these two sets of numbers to transform into form [M × N,1] of the matrix called A; then all the matrix elements are arranged in descending order, obtain a matrix called B ; note the location of the original elements of matrix A corresponds to the position of matrix B, get a position in the array $\{L_i\}$.Here, $\{L_i\}$ represents the index number of B in the sort sequences, then the image gray value matrix [M × N, 1] scrambling according to $\{L_i\}$ sequence number, obtained matrix called T, then transform the matrix T into an array of M × N, called F, F is a matrix obtained by scrambling. If a conversion is not formed deep chaotic image, you can take a different row, column, diagonal values repeat the above process, so that it becomes a completely blurred image in the end.

2.4   *Image diffusion*

The pixels of the image scrambling can produce a blurred image, but it just changed the location of a pixel, it can not change the correlation between two adjacent pixels, so it is easy to be deciphered by the attacker on the basis of the method of comparing pixel values, so just taking the image scrambling encryption cannot be achieved effectively.

Diffusion the image can be used to solve this problem effectively, the encrypted image generated by diffusion can spread each bit effect to the entire image, so that they are almost zero correlation between adjacent picture elements, this can effectively change the statistical properties of the image, improving safety, and against statistical attacks so well. Specific diffusion ideas as follows:

Using the chaotic systems iterative sequence $\{x_{3i}\}$; Transform any sequence number M × N into a new sequence $\{x'_{3i}\}$, and then taking the above scrambling sequence T. First sorting operation(from small to large)is done to $\{x'_{3i}\}$, recording the original sequence element in the new number of columns, and then follow this sequence to T to get I,; Then I converted into a form of M × N, obtain a new matrix called K; Finally, the K treatment and express xor obtain P, P is the ciphertext image we want to get.

2.5 Encryption algorithm process

Step1. Taking any one of the horizontal, vertical and diagonal ramp sequence of plain image pixels, which the elements of each sequence arbitrary, denoted $\{R\}$, $\{C\}$, $\{H\}$. The sum of each sequence were obtained is denoted r, c, h.

Step2. Given the initial parameter x_{10}, x_{20}, x_{30}, between [-1,1], the number of columns consist of four digits behind the decimal point, the last step r, c, h, respectively summed with the initial value, then modulo 255, and divided by 255 to obtain (x_{11}, x_{21},x_{31}) as the initial iteration value (make it as an encryption key).

$$x_{11} = (\mathrm{mod}\ ((r+x_{10}), 255))/255 \qquad (2)$$

$$x_{21} = (\mathrm{mod}\ ((c+x_{20}), 255))/255 \qquad (3)$$

$$x_{31} = (\mathrm{mod}\ ((h+x_{30}), 255))/255 \qquad (4)$$

Step3. Last step we gets the key x_{11}, x_{21},x_{31}, with the formula (1) N-1 times of iterations to obtain $\{x_{11}, x_{12}, x_{13} \dots x_{1n}\}, x_{21}, x_{22}, x_{23} \dots x_{2n}, \{x_{31}, x_{32}, x_{33} \dots x_{3n}\}$.

Step 4. First, transform the image with a size of M × N pixels transform into an array of [M × N, 1], from the sequence of iterations $\{x_{1i}\}$, $\{x_{2i}\}$, taking any (M × N) / 2 number (encrypted only specific who know), used these two sets of numbers to transform into form of the matrix [M × N, 1]; then all the matrix elements are arranged in descending order, obtain a matrix B ; note the location of the original elements of matrix A corresponds to the position of matrix B, get a position in the array $\{L_i\}$.Here, $\{L_i\}$ represents the index number of B in the sort sequences, then the image gray value matrix [M × N, 1] scrambling according to $\{L_i\}$ sequence number, obtained matrix called T.

Step5. Using the chaotic systems, iterative sequence $\{x_{3i}\}$; Transform any sequence number M × N into a new sequence $\{x'_{3i}\}$, and then taking the above scrambling sequence T. First sorting operation (from small to large)is done to $\{x'_{3i}\}$, recording the original sequence element in the new number of columns, and then follow this sequence to T to get I,; Then I converted into a form of M × N, obtain a new matrix called K; Finally, the K treatment and express xor obtain P, P is the ciphertext image we want to get. Algorithm formula as follows:

$$[t(i),L(i)]=\mathrm{sort}(\{x'_{3i}\}) \qquad (5)$$

$$I(i)=T(L(i)) \qquad (6)$$

$$P= f(i) \oplus \mathrm{mod}\ (s * k\ i, 256) \qquad (7)$$

Where $\{x'_{3i}\}$ as chaotic iteration sequence, i (1 2 3 …) is a serial number, T is the result of scrambling sequence, these two sequences are [1, M × N], f is clear pixel matrix, s is an arbitrary constant designed by encrypter, it also can as an intermediate key, P is what we want to get, K is an M × N matrix converted by I, Conversion process is as follows:

> *function b=reshape(a,M,N)*
> *a=1:length(a);*
> *a=a';*
> *b=reshape(a,M,N);*
> *end*

Encryption and decryption process is the reverse process.

3 SIMULATION AND ANALYSIS OF ALGORITHMS

Simulation of this paper was done by using MATLAB to explore the efficiency of this image encryption method. Image of Lena's size is 256 × 256 as the gray image, System parameters a = 0.2, b = 0.2, c = 5.7; initial value input: x_{10} =0.52743, x_{20} = 0.38926, x_{30} =0.65839. The original image and its histogram, cipher image and its histogram shown below.

From the Figure 2a shows a clear image, after the encryption process (Figure 3a) to obtain a chaotic blurred image, from the sensory point of view, it has the effect to play encrypted. Comparing with two image histogram, Figure 2b is non-uniform pixel value, but the ciphertext image (Figure 3b) is the pixel value almost uniformly distributed in the figure, so that it is not easy to find the encryption rule, thereby increasing the difficulty of attack, and the ciphertext has good resistance to statistical analysis; Meanwhile, Figure 4 is obtained by the right image decryption

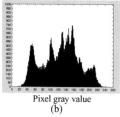

(a)

Pixel gray value
(b)

Figure 2. Plain image and its histogram.

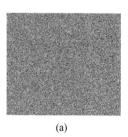

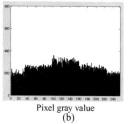

(a)

Pixel gray value
(b)

Figure 3. Ciphertext image and its histogram.

Figure 4. Decrypted image with right keys.

Figure 5: Decrypted image with wrong keys.

method, Figure 5 shows the error plaintext matrix was selected to generate value, it will result in incomplete decrypted, and it will be product blurred image, which shows that this kind of encryption method has a strong dependence and sensitivity on the plaintext. At the same, the difference of the initial value also can not make decryption implementation, which proving the algorithm sensitive to keys again.

As mentioned, a good encryption method needs to consider many factors. In this paper, we also need to analysis the encryption algorithm to see whether it is a viable encryption method, several directions to detect it as follows:

3.1 Test of key sensitive

It does not matter when the initial value will be leaked in this paper, because the text of the iteration of initial (key) is not only related to the initial value, but has a strong relationship with the plaintext. The encrypted plaintext is related to its effect directly, because the

row, column, diagonal and the numbers of the row, column and diagonal's digits we have chosen in plaintext are random, it will cause diffusing and confusing different when choosing a different ranks and even the number of rows or columns of elements, it will result in different deciphering, and increasing the difficulty in deciphering, Figure 6 is the original image, Figure 7 is to choose the number of an error rows, as a result not decrypted properly.

Figure 6. The original image.

Figure 7. Decrypted image with wrong row.

So it is high feasibility of the algorithm.

3.2 Statistical analysis

Getting several pixels from plaintext and ciphertext images horizontal, vertical and diagonal directions randomly, calculated the adjacent pixel correlation with the equation (8-11):

$$r_{xy} = \frac{COV(X,Y)}{\sqrt{D(X)}\ \sqrt{D(Y)}} \tag{8}$$

$$COV(X,Y) = E((X-E(X))\ (Y-E(Y)) \tag{9}$$

$$E(X) = \frac{1}{L}\Sigma_i X_i \tag{10}$$

$$D(X) = \frac{1}{L}\Sigma_i(X_i - E(X)(X_i - E(X)) \tag{11}$$

Variance and covariance of the program are as follows:

Variance procedure:

sum((c(i,j)-mean(c)).^2)/(length(c)-1)

Covariance procedure:

function c=xiefang(a,N)
    for i=1:size(a,N);
      for j=1:size(a,N);
c(i,j)=sum((a(:,i)-mean(a(:,i))).*(a(:,j)
mean(a(:,j))))/(size(a,1)-1);
      end
    end

Where, i, j could be changed when we select different value rows, and columns; correlation between adjacent pixels is shown in TABLE I:

Table 1. Plaintext and ciphertext Lena adjacent pixels of the image correlation.

Adjacent directions	Horizontal	Vertical	Diagonal direction
Plaintext	0.9314	0.9542	0.9105
Ciphertext	0.0106	0.0102	0.0146

According to the Table 1, we can draw the conclusion that the algorithm can be a good method to destroy correlation of pixels adjacent, so that the distribution of the ciphertext pixels randomness.

3.3 Analysis of key space

This paper uses a three-dimensional chaotic system, the initial value of the system as the three original key; Each figure includes an integer and four decimal, the key space is $10^5 \times 10^5 \times 10^5 = 10^{15}$, although the key space of its initial value is not very well, but its plaintext has a significant impact on the key, selecting the value of the different plaintext, it will produce different key, and the sufficient numbers can be produced by plaintext to resist brute-force attack.

3.4 Differential anti-attack capability and noise immunity

According to the principle of image encryption, a good encryption algorithm should be very sensitive to the plaintext. When implementing a differential attack, the attacker usually adjusts plaintext minorly, then attacking by comparing the difference between the minor adjust of the ciphertext and the original ciphertext. For analysis of differential attacks, evaluated by the indexes of formula (12-13) [11] typically.

$$NPCR = \frac{1}{M \times N} \Sigma_{i,j} D(i,j) \times 100\% \quad (12)$$

$$UACI = \frac{1}{M \times N} [\Sigma_{i,j} \frac{abs\ c_{1\ i,j} - c_{2\ i,j}}{255}] \times 100\% \quad (13)$$

Where M, N represent the number of rows and columns of pixels of the image, c_1, c_2 as the pixel between the two images, the difference of their corresponding ciphertext plaintext image is only have one. To test both the value of the Lena image encryption and to give these two values are stabilizing, so the algorithm has a better ability to fight against differential attacks.

However, the transmission of the image will be interferenced by a lot of noise in real life, a good encryption algorithm should have a certain degree of noise immunity, Gaussian white noise added in the ciphertext image in the following, in order to see whether it can recover the original image. The results obtained by simulation as shown, Figure 8 is covering a certain Gaussian white noise, Figure 9 is decrypting the adding noise image, the image decrypted can be seen more clearly, so the anti-noise performance of the algorithm is very well as we can see.

Figure 8. Adding white noise.

Figure 9. Decryption white noise.

4 CONCLUSIONS

Chaotic image encryption is a popular area of security now, the paper proposed a new image encryption method based on a three-dimensional Rossler system, this algorithm is a three-dimensional, which generates values by means of secondary key and making it have a large enough key space; Chaotic sequence generated by the algorithm express a strong dependence on the initial value, so the plaintext is very sensitive; Through performance of simulation and analysis of algorithms shows that the algorithm is simple and has small amount of calculation, but the encrypted ciphertext uniformly distributed random, correlation of adjacent data is low, there is a large key space, easy to crack, better anti-noise performance, and anti-statistics attacks is relatively well, so this encryption algorithm has some applicability.

REFERENCES

[1] Wang Y,Wong K W,Liao X,et al.A new chaos-based fast image encryption algorithm[J].Applied Soft Computing, 2011,11(1):514–522.

[2] Cahit Cokal,Ercan Solak.Cryptanalysis of a chaos-based image encryption algorithm.Physics Letters A,vol.373,pp.1357–1360,2009.

[3] Liquan Chen. contourlet watermarking algorithm based on Arnold scrambling and sing ular value

decoposition. Journal of Southeast University(English Edition([J].2012,28(4):386–391.

[4] WANG Xing-yuan, ZHAO Jian-feng,LIU Hong-jun.A new image encryption algorithm based on chaos[J]. Optics Communications,2012, 285(5):562–566.

[5] Yanfeng Chen, Yifang Li. Alternating mutual scrambling dual chaotic sequence image encryption algorithm [J]., South China University of Technology, 2010, 006(05): 105–121.

[6] Ren Shuai, Gao Chenshi, Dai Qing. High-dimensional chaotic image encryption technique with disturbance [J]. Academic Research, 2011, 0067(03): 67–70.

[7] Xue feng Liu, Hua sheng Zhou. Analysis and improvement of ultra-chaotic image encryption

scheme [J]. Computer Engineering and Applications, 2012, 1002(33): 105–111.

[8] Deng Yue, Guangyi Wang, Yuan Fang. A new image encryption algorithm Rossler chaotic sequence [J]. Hangzhou University Journal of Electronic Science and Technology, 2011, 031(05): 9–12.

[9] Jing Ping, Dinghui Zhang, Yaqi Zhang. Based on Henon mapping and Rossler chaotic map [J]. New Technology, 2012, 093(04): 15–18.

[10] Rossler O E.An equation for continuous chaos.Phys. Letters A,1976,57:397–398.

[11] Guangrong Chen, Yaobin Mao, Chaeles K. Chui.A symmetric image encryption scheme based on 3D chaotic cat maps[J].Chaos,Solitons & Fractals,2004,21(3):749–761.

Information Technology – Wan et al. (Eds)
© 2015 Taylor & Francis Group, London, ISBN 978-1-138-02785-5

Numerical study of heat transfer in bone tumor during radiofrequency ablation therapy

Xiaohui Nie, Qun Nan, Fei Zhai & Xuemei Guo
College of Life Science and Bio-engineering, Beijing University of Technology, Beijing, China

ABSTRACT: Purpose: To simulate temperature distribution of bone tumor during radiofrequency ablation (RFA) by the finite element method, this study would propose a treatment to assist therapy in clinic, and improve the effect of radiofrequency ablation. Method: We employed mode of thermoelectric coupling to achieve ablation size, and got the 50°C temperature curves for ablation treatment producing by applied voltages. Result: When the high temperature ranges from 90°C to 100°C in more than 10 min heating time, the applied voltage is 28 V, created the 47 mm long axis and 34 mm minor axis of ablation. Moreover, by increasing the heating time, the area of ablation was augmenting. Conclusion: It is a benefit for the doctors to make surgical planning and consummate the results from this study, which is shown in the 50°C of temperature distribution of bone tumor and the ablation size.

KEYWORDS: Bone tumor; Radiofrequency ablation; Finite element method

1 INTRODUCTION

Bone tumor was not as common as other types of tumors in the past. However, with better control of the primary tumor and prolonged survival of the patients, the incidence of bone metastases from the primary hepatocellular carcinoma (HCC) has been increasing (Liaw et al., 1989). According to the statistics from American Academy of Neurological and Orthopaedic Surgeons, it always happens to children and youth, which tremendously devastates the entire life quality of patients (Xin Kang et al., 2012). Although, many methods including surgery, thermal therapy, drug treatment, and chemotherapy have been applied to treat various kinds of bone tumors (Rimondi, 2012, Busser, 2011, Andreas, 2011) in the clinic, it is still a challenge to find out the best therapy currently, which can not only alleviate the pain of the patients, but also achieve the ideal treatment effect.

People consider that minimally invasive treatments together with high radiation exposure can accept this challenge and guide the direction of treatment, especially radiofrequency ablation (RFA). It has emerged as a routinely used technique for numerous tumors, such as liver, lung or brain. The mechanism of radiofrequency ablation therapy, based on the thermal coagulation characteristics of biological tissue in the case of high temperature, RF electrode is inserted into the tumor and produce high frequency current. It caused organization charged ions and water molecules friction to shock, and led to rise temperature quickly to coagulate and kill tumor (Zhang Ye, 2012). Currently,

it is one of the most widely used and safest methods in the ablation therapy (Mo Xiaodong 2012).

Previous studies and practices indicated that different parameters of radiofrequency ablation device would led to different ablation results, and it is of great importance to set parameters before therapy. Nevertheless, clinicians always depend on their experiences to estimate the therapeutic effect; we should study how to accurately and completely ablate the tumor by simulating temperature distribution of bone tumor and calculating ablations area during RFA in COMSOL Multiphysics software.

2 METHODS

In this study, the innovation and challenge are that electric field and thermal field will be combined by the Pennes equation (1998). It is followed by:

$$\rho C \frac{\partial T}{\partial \tau} = \nabla \cdot K \nabla T + W_b C_b (T_a - T) + Q_m + Q_r \quad (1)$$

where ρ is the tissue density (kg/m^3), C is tissue specific heat capacity(J/ (kg·°C)), K is the thermal conductivity of tissue (J/ (m·s·°C)), T is the temperature of tissue(°C), τ is the time (s), C_b is the specific heat capacity of blood (J/ (kg°C)), W_b is the blood perfusion rate (kg/ (m^3·s)), T_a is the blood temperature in the heating zone, Q_m is the heat energy of biological tissue (J/(m³·s)), Q_r is the energy of the external heat quantity.

We ignored some parameters in the simulation which is called ideal model, such as blood perfusion rate. So the thermal parameters (Bien, 2010, Ding Jinli, 2013) were assumed to be constant in this simulation as Table 1 showed.

Table 1. Thermal parameters of each material.

Materials	k W/(m·°C)	c J/(kg·°C)	ρ (kg/m³)	F (s/m)
Bone tumor	0.42	4100	1000	0.4
RF electrode Probe	71	21500	132	4×10^6
Insulating shaft	0.026	1045	70	1×10^5

According to clinical application, the water-cooling antenna is widely applied to treat bone tumor in RFA is effective to destroy the tumor cell. Selecting a water-cooling antenna (Dieter Haemmerich, 2003), which has two parts including the 2 cm length antenna tip and the 4 cm length insulating shaft is imposed to the bone tumor (Fig. 1).

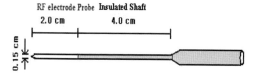

Figure 1. The model of antenna.

3 RESULTS AND DISCUSSION

In the clinical therapy, it is scientific that the highest temperature, generated by RF, is range 90°C to 100°C (Zheng Longpo, 2007, 2008). So we should control the high temperature by means of changing voltage and heating time, which received as more than 10 min in the clinical therapy. On the basis of different voltages and times, the voltage of 28 V, compared with other voltages, was seen as suitable voltage to ablate bone tumor in 960s, which is shown in Fig. 2.

From Fig. 1, we can see that the high temperature in ablation is 95.8°C by applying 28 V voltage on the probe, and when the voltages are 27 V and 29 V, the high temperatures are 89°C and 101°C, which are not involved in the range of right temperature to treat bone tumor in clinic applications. Thus, in allusion to peculiarity of bone tumor, selecting 28 Vvoltage to

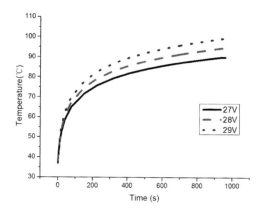

Figure 2. The high temperature of 27V, 28V, 29V.

apply could achieve the aim of best effective ablation by RF in 960 s heating time.

After ablation of simulation, it directly gets the ablation shape at 50°C (F. Rachbauer, 2003), which is regarded as effective temperature, and is considered to completely kill bone tumor and make cell death. Fig. 3 shows that the ablation shape is the ellipse after 960 s, and the long axis and minor axis are 47 mm and 34 mm respectively.

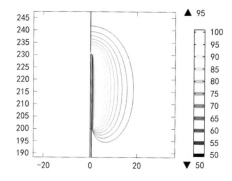

Figure 3. The ablation shape on 50°C (28V).

So if the applied voltage is 28 V and heating time is 960 s, the bone tumor which is the long axis and minor axis are 47 mm and 34 mm could be ablated to kill and destroy the tumor cell.

The influencing factors of RFA are not only the voltage, but also the heating time. With the increase in the heating time, the area of ablation was augmenting. It is shown in Fig. 4.

From Fig. 4, the long and minor axis was amplifying gradually as the heating time was increasing.

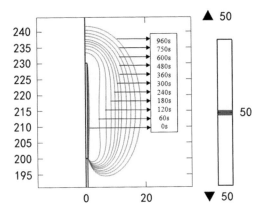

Figure 4. The different ablation with different heating time.

When the heating time was 480 s in applied 28 V, the whole bone tumor, whose long axis and minor axis are 42 mm and 26 mm, was ablation and the tumor cell was death. Attaining the goal of destroying bone tumor which has the 39 mm long axis and 19 mm minor axis, applying voltage of 28 V and setting heating time of 240 s are integrants. If the doctor wants to treat different long and minor axis bone tumor, they can select different heating time in the applied voltage of 28V according to the size of bone tumor.

4 CONCLUSION

To achieve the right high temperature range of 90–100°C in ablation, the specific voltage is 28 V in 960 s. On the basis of simulation ablation, it was shown the accurate value of temperature distribution of bone tumor, which is a benefit for thevdoctors to make surgical planning and consummate the results.

In the future, we will consider mix-material field, because bone tumor is generally in the bone or close to other tissues, like spinal tumor. Moreover, due to hard bone, it is obvious tissue that insertion direction of the antenna is difficult to ablate completely. These problems will be researched in the future.

ACKNOWLEDGMENT

This research is supported by National Science Foundation of China (No. 31070754), Beijing Municipal Commission of Education Project Scientific and Technological Program (KM201410005028), the Importation and Development of High-Caliber Talents Project of Beijing Municipal Institutions for three years (2013-2015)-Nan Qun, and Basic Research Foundation of Beijing University of Technology (X4015999201401).

REFERENCES

Bien, T., Rose, G., & Skalej, M.. FEM modeling of radio frequency ablation in the spinal column. *In Biomedical Engineering and Informatics (BMEI), 2010 3rd International Conference.* 5, 2010, 1867–1871.

Busser, W. M., Hoogeveen, Y. L., Veth, R. P., Schreuder, H. ., Balguid, A., Renema, W. K., & SchultzeKool, L. J. Percutaneous radiofrequency ablation of osteoid osteomas with use of real-time needle guidance for accurate needle placement: a pilot study. *J. Cardiovascular and Interventional Radiology*, 2011, 34(1), 180–183.

Dong, X. M. & Guo H. The development and present situation of adrenal tumors by radiofrequency ablation. J. Journal of Southeast University (Medical Science Edition), 2012, 12, pp. 781–783.

Haemmerich, D., Chachati, L., Wright, A. S., Mahvi, D. M., Lee Jr, F. T., & Webster, J. G.. Hepatic radiofrequency ablation with internally cooled probes: effect of coolant temperature on lesion size. *IEEE Transactions on Biomedical Engineering*, 50(4), 493–500.

Kang, X., Ren, H., Li, J., & Yau, W. P. Statistical atlas based registration and planning for ablating bone tumors in minimally invasive interventions. In Robotics and Biomimetics (ROBIO), IEEE International Conference, 2012, 606–611.

Liaw, C. C., Ng, K. T., Chen, T. J., & Liaw, Y. F. Hepatocellular carcinoma presenting as bone metastasis. *J. Cancer*, 1989, 64(8), 1753–1757.

Li, J. D., Z. H. Juan, N. Qun, L. Y. Jun, & L. Y. Lin. (2013). Surgical planning of microwave thermal ablation for a patient-specific spinal tumor. *J. Journal of Beijing University of Technology*, 39, 1264–1268.

Mahnken, A. H., Bruners, P., Delbrück, H., & Günther, R. W. Radiofrequency ablation of osteoid osteoma: initial experience with a new monopolar ablation device. *J. Cardiovascular and Interventional Radiology*, 34(3), 579–584.

Pennes, H. H. Analysis of tissue and arterial blood temperatures in the resting human forearm. *J. Journal of Applied Physiology*, 1998, 85(1), 5–34.

Po, L. Z. Study on the application of using radiofrequency ablation technology in the treatment of bone tumors. Second Military Medical University, 2007.

Po, L. Z., C. Z. Dong, N. W. Xin, L. Quan, W. Ji, & G. Bo.. Three-dimensional space distribution of thermal conductivity in bone tissue: A thermodynamics finite element analysis. *J. Journal of Clinical Rehabilitative Tissue Engineering Research*, 2008, 12, 7665–7668.

Rachbauer, F., Mangat, J., Bodner, G., Eichberger, P., & Krismer, M. Heat distribution and heat transport in bone during radiofrequency catheter ablation. *Archives of Orthopaedic and Trauma Surgery*, 2003, 123(2–3), 86–90.

Rimondi, E., Mavrogenis, A. F., Rossi, G., Ciminari, R., Malaguti, C., Tranfaglia, C., & Ruggieri, P. Radiofrequency ablation for non-spinal osteoid osteomas in 557 patients. *European Radiology*, 2012, 22(1), 181–188.

Ye, Z., Qian Z., Guo J., Hu G., & Zhao J. Simulation on temperature distribution of effective lesion area for tumor microwave ablation thermotherapy. *J. ACTA Biophysica Sinica*, 2012, 28, 763–770.

Information Technology – Wan et al. (Eds)
© *2015 Taylor & Francis Group, London, ISBN 978-1-138-02785-5*

Parameter estimation for high dimensional ordinary differential equations

Hongqi Xue
Department of Biostatistics and Computational Biology, University of Rochester, Rochester, New York, USA

Patrice Linel
Department of Biostatistics and Computational Biology, University of Rochester, Rochester, New York, USA

Jinhai Chen
Department of Biostatistics and Computational Biology, University of Rochester, Rochester, New York, USA

Ya-xiang Yuan
LSEC, ICMSEC, Academy of Mathematics and System Sciences, Chinese Academy of Sciences, Beijing, China

Hulin Wu
Department of Biostatistics and Computational Biology, University of Rochester, Rochester, New York, USA

ABSTRACT: We consider parameter estimation for high dimensional sparse ordinary differential equation (ODE) systems in this article. We propose to employ the block coordinate descent (BCD) algorithm to solve the underlying optimization problem. Comparisons between the proposed method and existing methods of simulation studies show a clear benefit of the proposed method in regards to the trade-off between computer memory usage and estimation accuracy. We apply the proposed methods to dynamic gene regulatory networks for the yeast cell cycle to further illustrate the usefulness of the proposed approaches.

1 INTRODUCTION

Denote a linear ODE system as follows

$$\begin{cases} X'(t) = AX(t), \forall t \in [t_0, T], \\ X(t_0) = X_0, \end{cases} \qquad (1)$$

where $X(t) = \{X_1(t), \cdots, X_p(t)\}^T$ is a p-dimensional state variable vector, $X'_k(t) = \dfrac{dX_k(t)}{dt}, 0 \leq t_0 < T \leq \infty$, $\mathbf{A} = (\beta_{kj})_{p \times p}$ is an unknown parameter $p \times p$ matrix, and $X(t_0) = X_0$ is the initial value vector. Let A_0 be the true value of A and $X(t, A)$ be the true solutions to Eq. (1) for given A. We usually use notation $X(t)$ to denote $X(t, A_0)$ in this article.

In reality, $X(t)$ cannot be measured exactly and directly; instead, its surrogate Y can be measured. For simplicity, here we assume an additive measurement error model to describe the relationship between $X(t_i)$ and the surrogate Y_i,

$$Y_i = X(t_i) + \epsilon_i, \qquad (2)$$

at random or fixed design time points $t_1, \cdots, t_n$, where the measurement errors $(\varepsilon_1, \cdots, \varepsilon_n)$ are independent with mean zero and a diagonal variance-covariance matrix Σ. Moreover, in the case of random design, assume that the measurement errors are independent of $X(t)$.

Our objective is to estimate the unknown parameters A_0 based on the observations $\{t_i, Y_i\}$. As we know, the sample size is $N = np$ and the number of parameters is p. In practice, p may be large, even larger than n. Then the number of parameters is much larger than the sample size, which will result that we cannot estimate the parameters in Eq. (1). In this article, we consider the high dimensional sparse case, that is, most parameters $\beta_{kj}(k, j = 1, \cdots, p)$ are zeros. Denote $\beta_k = (\beta_{k1}, \ldots, \beta_{kpk})^T$ and $\beta = (\beta_1^T, \ldots, \beta_k^T)^T$ for those nonzero parameters with $M = \max_{k=1}^p pk < \min\{p, n\}$, where M is a bounded constant. Then the sample size is still $N = np$ and the number of parameters is $\sum_{k=1}^{p} pk \leq Mp < np$.

The existing statistical methods for ODE models include the full least squares (FLS) method (Bard, 1974; Li et al., 2005; Xue et al., 2010), the two-stage smoothing-based estimation method (Varah, 1982; Brunel, 2008; Chen and Wu, 2008; Liang and Wu, 2008), the generalized profiling approach (Ramsay et al., 2007) and the Bayesian approaches (Putter et al., 2002; Huang, Liu and Wu, 2006; Donnet and Samson, 2007). Among these methods, we are more interested in the FLS method, where numerical methods such as the Runge-Kutta algorithm have to be firstly used to approximate the solution of the ODEs for a given set of parameter values and initial conditions. Then consequently, the least squares principle (minimizing the residual sum of squares of the differences between the experimental observations and numerical solutions) can be used to obtain the estimates of the unknown parameters.

Eq. (1) is a linear system and exists a close-form solution. In the low dimensional case, it is suitable to apply such close-form solution for the FLS estimation of the coefficient matrix A. However, its computational cost is much higher in the high dimensional case, or even in the moderate dimensional case. In such situation, we have to use its numerical solution. Although the FLS method was the earliest and the most popular method developed for estimating the parameters in ODE models, its theoretical properties have not established until Xue et al. (2010). Xue et al. (2010) proved that the FLS estimator is strongly consistent and asymptotically normal as the maximum step size of the numerical algorithm goes to zero. Moreover, it showed that if such maximum step size is small enough, then the numerical error between the numerical solution and the analytical solution is negligible compared to the measurement error. Global optimization algorithms are often used in the FLS method to obtain an estimation for ODE models, which take great computational cost time and occupy huge computer memory usage.

In this article, we propose to apply the block coordinate direction (BCD) algorithm for solving the above optimization problem, which is also known as nonlinear block Gauss-Seidel method (Ortega and Rheinboldt, 1970, Chap. 7). This algorithm is very simple and easy to implement and has been widely applied for solving large scale optimization problems (Wu and Lange, 2008; Chen and Li, 2013). Its convergence theory has been done by Luo and Tseng (1992), Tseng (2001), Saha and Tewari (2013) among others. In this article, we find that the proposed BCD algorithm for ODEs can save more memory usage than the original FLS method, which is important for high dimensional computational work. Compared with the

two-stage smoothing-based estimation method for ODE models (Chen and Wu, 2008; Liang and Wu, 2008), it can also produce a more accurate estimate. Thus the proposed method has a clear benefit in regards to the trade-off between memory usage and estimation accuracy.

The remainder of this article is organized as follows. In Section II, we introduce the FLS estimation for ODEs and BCD algorithms. In Section III, the theoretical results of the proposed method are established. In Section IV, we present some simulation results to validate the proposed method and compare it with other existing methods. To illustrate the usefulness of the proposed methods, we apply the proposed method to a linear dynamic gene regulatory network based on time course microarray data for yeast cell cycle in Section V. We conclude the article with some discussions in Section VI.

2 ESTIMATION AND ALGORITHM

In this article, we consider a general one-step numerical method. Let $t_0 = s_0 < s_1 < \cdots < s_{m-1} = T$ be grid points on the interval I, $h_j = s_j - s_{j-1}$ be the step size and $h = \max_{1 \leq j \leq m-1} h_j$ be the maximum step size, and

1 Set an initial estimate $\hat{\beta} = (\hat{\beta}_1, \cdots, \hat{\beta}_l)^T$. $j = 0$.

2 If some stopping criterion is satisfied, stop.

3 Obtain the numerical solutions $\tilde{X}(t, \beta), \ldots \tilde{X}(t, \beta)$ by the one-step numerical method introduced in Section 2.1.

4 For $k = 1, 2, \ldots, l$, update β_k^{j+1} by calculating
$$\min_{\beta_k} \Xi(\beta_1^{j+1}, \beta_2^{j+1}, \ldots, \beta_{k-1}^{j+1}, \beta_k, \beta_{k+1}^j, \ldots, \beta_l^j). \quad (5)$$

5 $j = j + 1$, and go to Step 2.

Fig. 1 BCD Algorithm

X_j^h and X_{j+1}^h be the numerical approximations to the true solutions $X(s_j)$ and $X(s_{j+1})$, respectively, which can be typically written as

$$X_{j+1}^h = X_j^h + h\Phi(s_j, X_j^h, X_{j+1}^h, h), \quad (3)$$

where the specific form of Φ depends on the numerical method. The common numerical methods include the Euler backward method, the trapezoidal rule, the r-stage Runge-Kutta algorithm (r is usually between 2 and 5), and so on. Among these algorithms, the 4-stage Runge-Kutta algorithm (Mattheij and Molenaar, 2002, p. 53; Hairer, Nørsett and Wanner, 1993, p.134) has been well developed and widely used in practice. Therefore, we employ the 4-stage Runge-Kutta algorithm as an example in our numerical studies. Following Mattheij and Molenaar (2002, p. 58), the interpolation technique is commonly used if the measurement points (t_i, $i = 1, 2, ..., n$) are not coincident with the grid points (s_j, $j = 1, 2, ..., m - 1$) of the numerical method, and the cubic Hermite interpolation is often adopted. Let $\tilde{X}(t, \beta)$ denote the interpolated numerical solution of $\tilde{X}(t, \beta)$ obtained from the numerical method for given β, then Eq. (2) can be approximately rewritten as $Y(t) \approx \tilde{X}(t, \beta_0) + \in (t)$. The simple numerical solution-based FLS estimator $\hat{\beta}_N$ of β_0 minimizes

$$\Xi(\beta) = \sum_{k=1}^{n} \sum_{i=1}^{n} [Y_k(t_i) - \tilde{X}_k(t_i, \beta)]^2. \quad (4)$$

We can also easily obtain the estimator $\widehat{X}(t) = \widehat{X}(t, \hat{\beta}N)$ for $X(t)$.

The main motivation to use the following BCD algorithm Fig. 1 to solve (4) can be that, at each iteration one block of the coordinates of the current iteration is adjusted in order to minimize (4) along this block coordinate direction. Also, the possible special structure of (4) along this block coordinate direction may be exploited.

Theoretically the number of blocks l can be chosen as the one no more than p^2. However, as will be shown in our numerical simulations, there is a trade-off between the number of blocks and computational costs of BCD method.

3 THEORETICAL RESULTS

In this section, we provide some theoretical results on the BCD algorithm for ODE systems.

Proposition 1. Suppose that (5) is well-defined for β_k with $k = 1, 2, ..., p$ and $\Xi(\beta)$ has a finite number of stationary points. For any stationary point β^*, there exists a neighborhood $A(\beta^*)$ such that for each $\beta \in A(\beta^*)$,

$$\| \beta - \beta^* \| \leq \| \nabla \Xi(\beta) \|. \quad (6)$$

Also, assume that for any $\beta \in A(\beta^*), \Xi(\beta_1, \beta_2, ..., \beta_p)$ is strictly convex with respect to β_k, when the other components of β are held constant. Then $\{\Xi_k(\beta^j)\}$ generated by the BCD method converges at least Q-linearly and $\{\beta^j\}$ converges at least R-linearly to a stationary point. That is,

$$\limsup_{j \to \infty} \| \beta^j - \beta^* \|^{\frac{1}{j}} = v, \quad \text{for some } v \in [0,1). \quad (7)$$

In particular,

$$\| \beta^j - \beta^* \| \leq C v^j, \quad (8)$$

holds for all j, and some constants C, and $v \in [0,1)$.

The proof follows directly from Proposition 3.4 of Luo and Tseng(1993). We can easily get the following result:

Corollary 1. Under the same assumptions as Proportion 1, there exists a neighborhood $A_k(\delta_0)$ of β_0 with radius δ_0 and an integer N_0 for almost every $\{t_i, Y(t_i)\}$, such that if $j > -\dfrac{\ln N}{2 \ln(Cv)}, N \geq N_0$ and $\widehat{\beta^0} \in A_1(\delta_0)$, then $\| \beta^j - \hat{\beta}_N \| \leq C v^j < N^{-1/2}$.

Xue et al. (2012) proved that if the maximum step size h of the q-order numerical algorithm in (3) (see definition in Mattheij and Molenaar, 2002 ; Hairer, Nørsett and Wanner, 1993) goes to zero at a rate faster than $n^{-1/(q^4)}$, the FLS estimator $\hat{\beta}_N$ is strongly consistent and asymptotically normal with the same asymptotic covariance as that of the case where the true ODE solution is exactly known, which means that the numerical error between the numerical solution and the true solution for ODEs is negligible compared to the measurement error. Further, Corollary 1 in this article shows that if the iteration step j is bigger than $O(\ln N)$, then the iteration error can be ignored in comparison with the measurement error. We also note that although a stationary point is not guaranteed to be even a local minimum (it can be a saddle point), in practice the BCD algorithm almost always converges to at least a local minimum.

4 SIMULATION STUDIES

In this section we consider two linear ODE models to test our BCD algorithm, and compare it with the classical FLS method.

Example 4.1: Consider 4-dimensional ODE model:

$$A_0 = \begin{pmatrix} -0.5 & 0.1 & 0 & 0 \\ 0 & 0 & -0.3 & 0 \\ 1 & 0.5 & 0 & 0 \\ 0 & 0.2 & 0 & -1 \end{pmatrix} \text{ and } X_0 = (5, 2, 1, 0)^T.$$

The time interval is $[0, 10]$, and we use 200 linearly equally spaced points of $X(t)$ on this interval as the observed data. That is, $\in_i = 0$ in Eq. (2) for $i = 1, ..., 200$. We use MATLAB ODE solver "ODE45" to represent the solution $X(t, \beta)$ and "lsqnonlin" to solve the underlying nonlinear least square objective function. The stop criterion is one of the following conditions holds at the kth iteration: (a) reach the maximum iteration number $J_{max} = 2000$, (b) relative error $\in^* = \dfrac{\| \beta^j - \beta^* \|_{\infty}}{\beta^*}$ is less than 10^{-6}, or (c) the relative error $\in = \dfrac{\| \beta^{j+1} - \beta^j \|_{\infty}}{\beta^*_{\infty}}$ is less than 10^{-8}. We record the iteration history $\dfrac{\|\beta^j - \beta*\|_\infty}{\beta_\infty^*}$, $\dfrac{\|\beta^{j+1} - \beta^j\|_\infty}{\beta_\infty^j}$ with respect to the iteration step j. In particular, we record the $\lg\dfrac{\|\beta^j - \beta*\|_\infty}{\beta_\infty^*}$ and $\lg\dfrac{\|\beta^{j+1} - \beta^j\|_\infty}{\beta_\infty^j}$. The iterative time is also recorded. The initial iterative value $\beta^{(0)}: \beta_0 + 0.1$ or $\beta_0 + 0.05$. We summary the results in Table 1.

From Table 1, the block partition $(\beta_1\beta_2\beta_3 / \beta_4)$ is optimal among these partings. This can be also verified by our decomposition procedure. In fact, the associated directed graph of β is Figure 2. This graph has two strongly connected components: $v_1v_2v_3$ and v_4. Thus, we may divide β by two blocks: one

consists of all parameters in the first three ODEs, the other contains all parameters in the fourth ODE.

Next, we choose the settings of a two block partition $(\beta_1\beta_2\beta_3 / \beta_4)$ to test the BCD method. The measurement error $\in_{ik} \sim$ Normal $(0, \sigma_i^2)$ in Eq. (2), where σ_i is set as 5%, 10%, 15% or 20% of the standard deviation of $X_i(t)$. The number of observations for each ODE component is 20. The simulation is implemented with 100 replicates.

Table 2 shows the absolute values of biases, standard deviations (STDs), and root mean squared errors

Fig. 2 The directed graph corresponding to β.

(RMSEs) of the parameter estimates averaged over all parameters using the FLS method and BCD method. It shows that the BCD achieves the same level accuracy estimation as the FLS with more computational cost.

Table 1. Computational time and relative error $\dfrac{\|\beta^j - \beta*\|_\infty}{\beta_\infty^*}$

Block	$\beta^{(0)}$	Time (second)		Relative Error	
		BCD	FLS	BCD	FLS
$\beta_1/\beta_2\beta_3\beta_4$	$\beta_0 + 0.1$	82.2665	1.8302	9.3463e-07	1.3902e-09
	$\beta_0 + 0.05$	72.5360	1.5896	9.8332e-07	2.0002e-09
$\beta_1 / \beta_2\beta_3\beta_4$	$\beta_0 + 0.1$	70.0154	1.8302	9.9336e-07	1.3902e-09
	$\beta_0 + 0.05$	63.5562	1.5896	9.4163e-07	2.0002e-09
$\beta_1/\beta_2\beta_3\beta_4$	$\beta_0 + 0.1$	3.1085	1.8302	6.7588e-07	1.3902e-09
	$\beta_0 + 0.05$	2.5261	1.5896	2.5393e-07	2.0002e-09
$\beta_1 / \beta_2\beta_3\beta_4$	$\beta_0 + 0.1$	84.4218	1.8302	9.7976e-07	1.3902e-09
	$\beta_0 + 0.05$	77.7490	1.5896	9.9053e-07	2.0002e-09
$\beta_1/\beta_2\beta_3\beta_4$	$\beta_0 + 0.1$	89.9398	1.8302	9.7878e-07	1.3902e-09
	$\beta_0 + 0.05$	80.3401	1.5896	9.6467e-07	2.0002e-09
$\beta_1/\beta_2\beta_3\beta_4$	$\beta_0 + 0.1$	84.8649	1.8302	9.4067e-07	1.3902e-09
	$\beta_0 + 0.05$	74.4430	1.5896	9.4539e-07	2.0002e-09
$\beta_1/\beta_2\beta_3\beta_4$	$\beta_0 + 0.1$	95.1401	1.8302	9.8859e-07	1.3902e-09
	$\beta_0 + 0.05$	92.4304	1.5896	9.3510e-07	2.0002e-09

Table 2. The mean of computational time, absolute values of biases, standard deviations (STDs), and root mean squared errors (RMSEs) of the parameter estimates averaged over all parameters.

σ	M-time		$\lvert BIAS \rvert \times 10^2$		$STD \times 10^2$		$RMSE \times 10^2$	
	BCD	FLS	BCD	FLS	BCD	FLS	BCD	FLS
5%	2.3215	0.29127	0.19721	0.19725	1.5656	1.5656	1.58	1.58
10%	2.4058	0.29807	0.26153	0.26136	3.1104	3.1101	3.1223	3.1223
15%	2.5731	0.30031	0.27965	0.27946	4.7568	4.7566	4.7676	4.7674
20%	2.9259	0.33131	0.85687	0.85621	6.1568	6.1559	6.2307	6.2297

Example 4.2: Consider a linear ODE system representing a gene model with 41 equations and 205 non-zeros coefficients. This matrix is highly sparse, with all coefficients between [−2, 2].

First, we study the behavior of the BCD method in regard of the FLS performances depending on the initial guess, and to the block partition. These results have been computed on a Linux server of Bi-Xeon X5660 and 48Gb Ram by a Fortran program of the BCD method and least-square method. Table 3 summarizes the result with an initial guess $\beta_0 + 0.1$. Those are preliminary results, and the block partitioning is uniform unless for the last block.

Table 3. Comparison of RSS, time, $\in, \in^*$ E^* and the number of iteration steps for BCD

nb block	RSS	time(seconds)	$\in^*$	iterations
1	8e-7	31	2.e-2	1
2	0.99	1104	0.78	500
8	46.5	1185	2.57	500
20	116	1095	2.57	500
31	156	1257	2.57	500
31 sec	2.6e-5	740	2.3e-3	500
41	420	951	2.37	500

The convergence of the BCD method has been studied (Luo and Tseng, 1992; Tseng, 2001; Saha and Tewari, 2013), and in our particular problem we experienced slow convergence rates which impact the performances. So the valuable aspect is the memory usage constraint, that can be hit by big data problems. In fact, it is well known that the performances of numerical methods are worse when increasing the memory usage, especially in case of swapping. It is then necessary to design algorithms that can deal with the problem with a reasonable memory usage.

The BCD method is particularly well adapted to that situation; for the FLS the code report a memory usage of 63Mb to solve the parameter estimation problem. In the meanwhile, the BCD algorithm uses only 12Mb, which appears to be 5.25 times less than the full approach. Let's remind that here we've taken advantage of the sparsity of the system of ODE, and estimated only the non-zero parameters. For this example of 41 genes, we have 205 parameters, with a maximum of 6 parameters per equation.

Considering that the problem has a memory usage of 10Mb, the estimation of 1681 parameter use 475mb, the estimation of 205 parameters uses 63Mb, and if we remove the 10mb from the 12Mb used by BCD (approx 6 parameters at a time) we find that the usage is around 0.3 Mb per parameter. Extrapolating from this number and taking a limit of 2Gb of memory available on a desktop or laptop, we find that an estimation of 7000 parameters will be necessary to fill up the memory, for such problem the BCD method will take approximately 30Mb (7000 in full matrix which means 84x84, rounding to 100, 100*0.3Mb).

5 REAL DATA RESULTS

We use Genome-wide time course gene expression data for budding yeast *Saccharomyes cerevisiae* throughout the cell cycle were collected by Spellman et al. (1998). The periodic recurring patterns of clusters of genes react on the functioning of mechanisms during the cell divisions. It is of interest to identify such gene regulations. Spellman et al. (1998) carried out a series of experiments to identify all cell cycle related protein-encoding gene transcripts in the genome of Saccharomyes cerevisiae. They used DNA microarrays on samples from yeast cultures that were synchronized by three independent methods. The time course gene expression levels were measured at 18 equally spaced time points during two cell cycles from which 800 genes were identified as cell-cycle related using a numerical score based on a Fourier algorithm (testing periodicity) and a correlation algorithm.

277

First similar to Lu et al. (2011), we used the variable selection on the 41 clusters curve, we obtained a sparsity matrix for the linear ODE system. The parameters used for the estimation of this system are summarized in the terms of residual sum of square (RSS) and computational cost in Table 4 for the FLS method and our BCD approach.

Table 4. Comparison of FLS , BCD, and two-stage smoothing-based method

	FLS	BCD (500it)	Two-Stage
RSS	239	234	355075e7
CPU time (s)	10400	23164	600

Those real data results confirm the results obtained during the simulation studies. In fact, the full least-square is faster and more precise than the BCD method. This real data study is based on the same sparsity matrix used in the simulation study. Also, we tried to estimate the full 41 ODE system, which means optimize 1681 parameters for the system and the 41 initial conditions. After 20000 iterations of the optimization method and 6 days of computation, we obtained a residual sum of square of 9.

We also compare the results obtained by our BCD method with those by the two-stage smoothing-based estimation method (Chen and Wu, 2008; Liang and Wu, 2008) and add the results in the last column of Table 4. We have selected the variable by the two-stage smoothing-based estimated method and used the sparsity matrix to estimate the parameters with BCD method. As expected the convergence is slow, but we manage to get a better estimate than the two-stage smoothing-based estimation method.

For the BCD, we use the strong connect the component to divide the matrix in blocks. This results in 3 blocks, 1 of 39 equations and 2 of 1 equation. For that reason, because the largest block is close to the full matrix, it shows similar performance in terms of residual sum of square. Also, we see that the two-stage smoothing-based estimation method gives a very large estimate of the parameter with a residual sum of square in order of 1e7. Both BCD method and the FLS reduce this RSS to 1000 in the early iterations.

6 DISCUSSION

We have tested our BCD method of simulation and real data and compared the performance with the FLS estimation and the two-stage smoothing-based estimation methods. Though it appears clearly the disadvantage in terms of time for BCD as the iterative process might have a very slow convergence, the method has proven better residual sum of square than the two-stage smoothing-based estimation method and a better memory usage than the FLS method. For very large-scale parameter estimation problem the memory usage is of primary concern, we will be able to address such problems using the BCD method while trading for restitution time.

In simulation studies, we have shown that different block partitions will have different computational cost and produce different accuracy estimation. But it is difficult to get a criterion to find the optimal block, which is still an open problem. It is also interesting to apply the BCD algorithm for variable selection (Tibshirani, 1996; Wu and Lange, 2008) for high dimensional ODEs. In addition, the BCD algorithm adopts in this article can be easily to be extended to nonlinear ODE systems. In Fig. 1, Step 4 of the BCD algorithm tries to update the component estimation in a given order. The random subspace technique may improve the numerical performances.

ACKNOWLEDGMENT

This research was partially supported by the NIH grants HHSN272201000055C, AI087135, and two University of Rochester CTSI pilot awards (UL1RR024160) from the National Center For Research Resources.

REFERENCES

[1] BARD, Y. (1974). *Nonlinear Parameter Estimation.* Academic Press, New York.

[2] BRUNEL, N. (2008). Parameter estimation of ODE's via nonparametric estimators. *Electronic Journal of Statistics*, **2**, 1242–1267.

[3] CHEN, J. and LI, H. (2013). Variable selection for sparse dirichlet-multinomial regression with an application to micro-biome data analysis. *Annals of Applied Statistics*, **7**, 418–442.

[4] CHEN, J. and WU, H. (2008). Efficient local estimation for time-varying coefficients in deterministic dynamic models with applications to HIV-1 dynamics. *Journal of the American Statistical Association*, **103**, 369–384.

[5] DONNET, S. and SAMSON, A. (2007). Estimation of parameters in incomplete data models defined by dynamical systems. *Journal of Statistical Planning and Inference*, **137**, 2815–2831.

[6] HAIRER, E., NØRSETT, S. P. and WANNER, G. (1993). *Solving Ordinary Differential Equations I: Nonstiff Problems*, 2nd Ed. Springer-Verlag, Berlin Heidelberg.

[7] HUANG, Y., LIU, D. and WU, H. (2006). Hierarchical Bayesian methods for estimation of parameters in a longitudinal HIV dynamic system. *Biometrics*, **62**, 413–423.

[8] LI, Z., OSBORNE, M. R. and PRAVAN, T. (2005). Parameter estimation of ordinary differential equations. *IMA Journal of Numerical Analysis*, **25**, 264–285.

[9] LIANG, H. and WU, H. (2008). Parameter estimation for differential equation models using a framework of measurement error in regression models. *Journal of the American Statistical Association*, **103**, 1570–1583.

[10] LU, T., LIANG, H., LI, H. and WU, H. (2011). High-dimensional ODEs coupled with mixed-effects modeling techniques for dynamic gene regulatory network identification. *Journal of the American Statistical Association*, **106**, 1242–1258.

[11] LUO, Z. Q. and TSENG, P. (1992). On the convergence of the coordinate descent method for convex differentiable minimization. *Journal of Optimization Theory and Applications*, **72**, 7–35.

[12] LUO, Z. Q. and TSENG, P. (1993). Error bounds and convergence analysis of feasible descent methods: a general approach. *Annals of Operations Research*, **46**, 157–178.

[13] MATTHEIJ, R. and MOLENAAR, J. (2002). *Ordinary differential equations in theory and practice*. SIAM, Philadelphia.

[14] ORTEGA, J. M. and RHEINBOLDT, W. C. (1970). *Iterative Solution of Nonlinear Equations in Several Variables*, Academic Press, New York.

[15] PUTTER, H., HEISTERKAMP, S. H., LANGE, J. M. and DE WOLF, F. (2002). A Bayesian approach to parameter estimation in HIV dynamical models. *Statistics in Medicine*, **21**, 2199–2214.

[16] RAMSAY, J. O., HOOKER, G., CAMPBELL, D. and CAO, J. (2007). Parameter estimation for differential equations: a generalized smoothing approach. *Journal of the Royal Statistical Society, Series B*, **69**, 741–796.

[17] SAHA, A. and TEWARI, A. (2013). On the nonasymptotic convergence of cyclic coordinate descent methods. *SIAM Journal on Optimization*, **23**, 576–601.

[18] SPELLMAN, P. T., SHERLOCK, G., ZHANG, M. Q., IYER, V. R., ANDERS, K., EISEN, M. B., BROWN, P. O., BOTSTEIN, D. and FUTCHER, B. (1998). Comprehensive identification of cell cycle-regulated genes of the yeast saccharomyces cerevisiae by microarray hybridization. *Molecular Biology of the Cell*, **9**, 3273–3297.

[19] TIBSHIRANI, R. J. (1996). Regression shrinkage and selection via the Lasso. *Journal of the Royal Statistical Society, Series B*, **58**, 267–288.

[20] TSENG, P. (2001). Convergence of a block coordinate descent method for nondifferentiable minimization. *Journal of Optimization Theory and Applications*, **109**, 475–494.

[21] VARAH, J. M. (1982). A spline least squares method for numerical parameter estimation in differential equations. *SIAM Journal on Scientifc Computing*, **3**, 28–46

[22] WU, T. T. and LANGE, K. (2008). Coordinate descent algorithms for Lasso penalized regression. *Annals of Applied Statistics*, **2**, 224–244.

[23] XUE, H., MIAO, H. and WU, H. (2010). Sieve estimation of constant and time-varying coefficients in nonlinear ordinary differential equation models by considering both numerical error and measurement error. *Annals of Statistics*, **38**, 2351–2387.

279

Information Technology – Wan et al. (Eds)
© *2015 Taylor & Francis Group, London, ISBN 978-1-138-02785-5*

Performance analysis of partial-band jamming resistance in cognitive frequency hopping communications

Fei Guo, Jianxin Guo, Wanze Zheng & Yunjun Qi
AFEU, Xi'an, China

ABSTRACT: Aiming at the drawback of the partial-band jamming degrade the FH communications, a cognitive frequency hopping (CFH) system model that combine the cognitive radio with the FH is proposed. The performance of the system under the partial-band jamming is analyzed and the closed-form error probability expression is given. The performance of partial-band jamming resistance in CFH communications and common FH communications are simulated and compared. The theoretic analysis and simulation results show that under the condition of ideal synchronization the CFH can avoid the jamming spectrum, and adjust the hopping frequency dynamically in a clean spectrum hole. Compare with the normal FH communications, the CFH communications have the better performance of jamming resistance.

KEYWORDS: FH communication; Cognitive frequency hopping; Partial-band jamming; Jamming resistance

1 INTRODUCTION

Frequency hopping (FH) communication is defined as the signal carrier is transmitted periodically. For the past few years, FH is widely used in the field of military communications (Yunbin Yan et al., 2011) for its prominent anti-multipath jamming, anti-jamming and anti-fading, and its high ability of improving the communication system anti-interception and anti-jamming. Whereas the electromagnetic environment is growing complex, electronic countermeasures (ECM) increase is diverse. There are various jamming methods to restrict the performance of FH communication. Some of the more major types of jamming are full-band jamming, partial-band jamming, single-tone jamming, multiple-tone jamming and tracking jamming (Zhenyu Na et al., 2009). In comparison with other jamming methods, partial-band jamming power is more concentrated, better fit actual jamming distribution. It is a principal threat to the regular FH communication, and it does more harm to the FH communication. Reference (Yi Zhang et al., 2010) reported that partial-band jamming has a strong impact on the synchronism of the FH communication, and as well increase the error rate. Consequently, the important issues of the FH communication is how to effectively control the partial-band jamming and improve the performance of the FH communication.

In 1999, Dr Joseph Mitol puts forward the conceptual Cognitive Radio (Mitola J 1999).

The CR concept provides a new solution to restrain partial-band jamming. Combining CR technique and FH communication (Kostic Z et al. 2011, Chu et al. 2006), he brings forward the Cognitive Frequency Hopping technique. CFH applies spectrum sensing to FH system. On the face of partial-band jamming, FH system can perceive the spectrum in real time and evade the spectrum in frequency effectively to communicate, thereby promoting the performance of restrain partial-band jamming. At present, studies on CFH focus on system synchronization and equipment compatibility (Geirhofer Stefan et al. 2009, Rongxin Zhi et al.2008, Brown C et al.2005), yet there is less research for the partial-band jamming restraint performance of CFH. Based on the performance of CFH system partial-band jamming restraint, this article applies the theoretical analysis simulation verification, and as well compares the regular FH band jamming restraint to CFH band jamming restraint.

2 SYSTEM ORGANIZATION OF CFH

The essential theories of the CFH are perceiving the electromagnetic environment in real time, to find the spectrum hole (Haykin, 2005) as the transmission band, so that the signal can accomplish FH communication in these bands, thus communication can achieve better performance of jamming resistance. According to the function of CFH, this article designs a specific implementation based on the FH ideas of optimum reception and maximizing the spectrum efficiency, the structure of the main unit and the hopping component recognition unit, the structure is shown in Fig. 1.

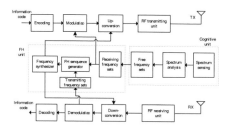

Figure 1.   Sketch map of the structure of CFH.

As shown in the structure diagram, the system mainly includes channel coding, modulation information, the frequency hopping sequence generation, spectrum sensing and a series of modules. In the sending end, the information code encoded was sent into the modulator, cognitive unit gives the available frequency sets through spectrum sensing and spectral analysis, frequency hopping unit generate the available carrier frequency accord to the frequency sets given by cognitive unit and processing the up-conversion for the carrier frequency and modulated signal. In the receiving end, using the same process, demodulation, decoding, the information codes can be restored. This article aims to analyze the performance of partial-band jamming resistance in the cognitive frequency hopping communication system, so all the parts of the system structure principle is no longer in detail.

3   ANALYSIS OF CAPABILITY IN PARTIAL-BAND JAMMING OF CFH SYSTEM

3.1   Mathematic mode of partial-band jamming

Partial-band jamming means that the total jamming power is randomly allocated into the partial frequency hopping bandwidth, compared with broadband noise jamming, the power of the partial-band jamming is centralized. Its mathematic mode can be depicted as follow: partial-band jamming is simulated as gauss noise. Suppose frequency hopping bandwidth is W_s, partial-band jamming bandwidth is W_J, the ratio of the jamming bandwidth and the hopping bandwidth is called jamming factor, marked down as ρ, $\rho = W_J / W_s$, $0 < \rho < 1$. Suppose the power spectrum density of the full frequency band power spectrum density is N_J, the power spectrum density of the partial-band jamming is J, $N_J = J / W_J$. Under the condition of partial-band jamming, jamming bandwidth is W_J, equivalent single power spectral density of the noise is $N_J^{'}$

$$N_J^{'} = \frac{J}{W_J} = \frac{J}{\rho W_s} = \frac{N_J}{\rho} \tag{1}$$

As mentioned before, we know that, under the jamming with the hopping bandwidth, the power, which is $1/\rho$ times of N_J, which is referred to the power spectrum density of the broad band jamming. Thus, partial-band jamming is more efficient in hopping communication.

3.2   Performance analysis of partial-band jamming

Theoretical study shows that (Yading Chen et al., 2007), without jamming, to a normal FH system, modulation mode is BPSK, demodulation mode is coherent demodulation. Bit error rate can be expressed as

$$P_{eAWGN} = \frac{1}{2} \exp\left(-\frac{E_b}{2N_0}\right) \tag{2}$$

where E_b represents signal energy per bit and N_0 represents equivalent single power spectral density of the gauss white noise.

Suppose the system has competed ideal synchronizing, the information bit stream without any channel coding and interleaving. The channel is Gauss white noise. Suppose all channels are available and equal used. In case of the number of FH channel is N, the number of FH channel which is damaged by partial-band jamming is $m = N \cdot \rho$. Suppose the disturbed power J of partial-band jamming is distributed equally in m channels.

Normal FH achieves stochastic FH by FH working list, which is promised in advance, does not change FH working list at will and keeps up communicating in m channels which is jammed. So the error rate of disturbed channel is $P(E_b/(N_0 + N_J / \rho))$, the error rate of undisturbed channels is $P(E_b/N_0)$, the average error rate of normal FH system can be expressed as

$$
\begin{aligned}
P_{pt,com} &= \rho P(E_b/(N_0 + N_J / \rho)) + (1-\rho) \\
&\quad P(E_b/N_0)) \\
&= \frac{1}{2}\rho \exp(-E_b/2(N_0 + N_J / \rho) + \frac{1}{2}(1-\rho) \\
&\quad \exp(-E_b/(2N_0))
\end{aligned}
\tag{3}
$$

In CFH, during that is silent or does not communicate, system detects disturbed frequency by module of detecting frequency real time, avoids choose working channel in disturbed frequency, and communicates in which creates dynamic FH pattern that use undisturbed or low disturbed channels for effectively avoiding biggish error rate.

Suppose CFH chooses N_c channels ($N_{cmin} \leq N_c < N$, N_{cmin} is the least number of channels in which CFH works in gear) from N FH channels as working channels in t, then using chosen channels creates FH pattern and hops in pseudo random. According to the number of disturbed factor, discuss from three conditions.

a. If $0 < \rho < 1 - N_c/N$, $N - m > N_c$, namely the number of undisturbed channel m appeases the number of CFH working channels, CFH must avoid disturbed channels, and has $N - m - N_c$ perfect spare channels.

At this time, the error rate of CFH is equal with the error rate which is in gauss white noise. Its expression is equal with (2).

b. If $\rho = 1 - N_c/N$, $N - m < N_c$, namely the number of undisturbed channel m appeases the number of CFH working channels just right, CFH communicates in avoiding disturbed channels m to a turn. Its expression is also equal with (2).

c. If $1 - N_c/N < \rho < 1$, $N - m < N_c$, namely the number of undisturbed channel m does not appease the number of CFH working channels, CFH chooses N_c channels from $N_c - (N - M)$ bad channels and $N - M$ perfect channels. Due to avoiding $m - \{N_c - (N - M)\} = N - N_c$ bad channels, its error rate makes up two parts, one part is the error rate of $N - M$ perfect channels, another is the error rate of $N_c - (N - M)$ bad channels. So average error rate can be expressed as

$$
\begin{aligned}
P_e &= \left(\frac{N-m}{N_c} \right) \cdot \frac{1}{2} \exp\left(-\frac{E_b}{2N_0}\right) + \left(\frac{N_c - (N-m)}{N_c} \right) \\
&\quad \frac{1}{2} \exp\left(-\frac{E_b}{2(N_0 + N_J/\rho)}\right) \\
&= (1-\rho)N/N_c \cdot \frac{1}{2} \exp\left(-\frac{E_b}{2N_0}\right) + (1 - (1-\rho)N/N_c) \\
&\quad \frac{1}{2} \exp\left(-\frac{E_b}{2(N_0 + N_J/\rho)}\right)
\end{aligned}
\tag{4}
$$

4 SIMULATION OF CAPABILITY IN PARTIAL-BAND JAMMING OF CFH SYSTEM

The performance of partial-band jamming resistance in CFH systems and normal FH systems can be further verified by Matlab simulation. The simulation system

used BFSK modulation, the parameters used in our simulation are listed below: $N = 80$, $W_s = 80$MHz, because the information rate R_s and hopping rate R_c satisfy the relation: $R_s = R_c$, $R_s = 1600$ bps, $R_c = 1600$ hop/s, the minimum number of CFH channels-N_{cmin} is 40.

a. Assuming the interference-to-signal ratio is fixed at 20 dB. The normal FH and CFH are disturbed by partial-band jamming, the jamming factor-ρ is 0.3, 0.5 and 0.6. The bit error rate curve of normal FH and CFH which is adjusted according to the SNR is shown in Fig. 2.

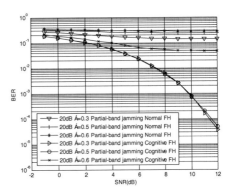

Figure 2. Bit error rate curve of three FHs on different (JSR=20dB).

It can be seen from Fig. 2, when $\rho = 0.3$, $\rho = 0.5$, the number of channels that the jamming did not occupy meets the number of CFH working channels, the bit error rate of CFH system is less than the bit error rate of normal FH system, which shows the performance of partial-band jamming resistance in CFH systems was significantly better than the performance of partial-band jamming resistance in normal FH systems, and this advantage is more obvious as the SNR increases. When $\rho=0.6$, the number of channels that the jamming did not occupy cannot be satisfied with the number of channels CFH, despite the performance of CFH decline, but still better than that of normal FH.

b. Assuming is fixed at 0.3, the interference-to-signal ratio is 0 dB, 10 dB, 20 dB. The bit error rate curve of normal FH and CFH which is adjusted according to the SNR is shown in Fig. 3.

Fig. 3 shows that the bit error rate of CFH and normal FH is substantially independent of the interference-to-signal ratio in the case of the jamming factors in certain, but the bit error rate of CFH can continuously decrease as the SNR increases, while the bit error rate of normal FH is essentially unchanged.

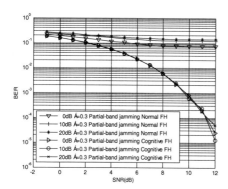

Figure 3. Bit error rate curve of three FHs on different ρ(JSR=20dB).

The CFH system has outstanding anti-interference performance.

5 CONCLUSION

In this article, the performance of partial-band jamming resistance in CFH systems has carried on theoretical analysis and performance simulation under the different jamming factors and different JSR. Theoretical analysis and simulation results show that, compared with normal FH, CFH which can effectively avoid partial-band jamming when the number of channels that the jamming did not occupy meets the number of CFH working channels has better performance in partial-band jamming resistance. So CFH system can effectively solve the problem of partial-band jamming and has great application value.

REFERENCES

Yunbin Yan, Houde Quan & Peizhang Cui. Simulation and Analysis of Jamming Pattern in GMSK Frequency-Hopping Radio. *Computer Measurement & Control*, 2011, 19(12), 3082–3084.

Zhenyu Na, Zhihe Gao & Qing Guo. Anal yses of Typical Jamming Performance on Frequency Hopping Communication System. *Science Technology and Engineering*, 9(8), 2009, 2072–2076.

Yi Zhang, Jing Zeng & Zhe Xue. Analysis of the Anti-jamming Capabilities in the Adaptive Frequency-hopping. *Journal of Projectiles, Rockets, Missiles and Guidance*, 2010, 30(1), 179–181.

Mitola J. Cognitive radio-making software ra dios more personal. 1999, *IEEE Personal Communications*, 6(4), 13–18.

Kostic Z, Maric I & Wang X. Fundamentals of dynamic frequency hopping in cellular systems. *IEEE J. Sel. Areas Commun*, 2001, 19(11), 2254–2266.

Chu, W.Hu & G.Vlantis. Dynamic frequency hopping community. *online, IEEE802.22 Working Group, Technical proposal submitted to IEEE 802.22*, 2006, pp. 1023–1032.

Geirhofer Stefan, Sun John Z, Tong Lang & Sadler Brian M. Cognitive frequency hopping based on jamming prediction: theory and experimental results. *Mobile Computing and Communications Review*, 2009, 13(2), 49–61.

Rongxin Zhi, Luyong Zhang & Zheng Zhou. Cognitive Frequency Hopping. *Cognitive Ratio Oriented Wireless Networks and Communications, 2008. CrownCorn:IEEE Press*, 2008, pp. 1–4.

Brown C & Vigneron P. J. Robust. Frequency Hopping for High Data-rate Tactical Communications. *in Proc. ST-054 Symposium on Military Communications, Rome*, 2005, 18–19.

Haykin S. Cognitive radio :brain-empowered wireless communications. *IEEE Journal on Selected Areas in Communications*, 2005, 23(2), 201–220.

Yading Chen, Xiaohui Liu, Yufan Cheng & Shaoqian Li. The architecture and anti-jamming test of ultra-fast frequency hopplatform. *Journal of Electronics & Information Technology*, 2007, 29(9), 2191–2194.

Information Technology – Wan et al. (Eds)
© 2015 Taylor & Francis Group, London, ISBN 978-1-138-02785-5

Predictive analysis of outpatient visits to a grade 3, class A hospital using ARIMA model

Yunming Li, Nianyi Sun, Fan Wu, Chi Zheng & Yong Li
Chengdu Military General Hospital, Chengdu, Sichuan Province, China

ABSTRACT: Forecasting hospital outpatient visits is useful for outpatient management strategy. An ARIMA model of outpatient visits was established using a time series forecasting model in Predictive Analytics Software (PASW). The total number of outpatient visits between January 2010 and March 2014 was 3,036,000, with a mean annual growth rate of 24.07%. Compared with the actual data for January 2010 to March 2014, the relative error of the data forecast was 4.11%, suggesting a good fit and accurate forecasting. Based on data for January 2010 to March 2014, the number of outpatient visits predicted for years 2014 and 2015 was 1,132,000 and 1,295,000, respectively. The ARIMA model in PASW software is easily manipulated, enabling good fitting and accurate prediction of outpatient visits to a large general hospital. In addition, the results were easy to understand. ARIMA could be popularized as a tool for the short-term forecasting of variables with seasonal fluctuations.

1 INTRODUCTION

Statistical forecasting has been an indispensible tool for the analysis of hospital-related information that can provide objective evidence for hospital management strategies(Xiong, 2002, Li, 2012). The ability to forecast trends in hospital outpatient visits, by accurately evaluating the dynamic changes in outpatient visits and fitting these data to a rational statistical model, is extremely useful for the appropriate allocation of human, financial and material resources and for future planning (Yuan et al., 2005). Over the past three decades, the autoregressive integrated moving average (ARIMA) model (also known as the Box-Jenkins model) has been a popular linear model for time series forecasting, and it is now widely used in the field of econometrics(Shen, 2014, Wang et al., 2014). In the field of bio-medicine, ARIMA has been used mainly in the forecasting of infectious diseases(Guan, 2013, Zou et al., 2013), although it has also found preliminarily application in the prediction of outpatient volume(Zhou et al., 2011a, Liang et al., 2006, Xiang and Chen, 2009, Zhou et al., 2008, Zhou et al., 2011b). In order to obtain objective information that would facilitate the planned expansion of the outpatient building of a grade 3, class A (first-class) general hospital in China, the present study was designed to collect the data for outpatient visits during the last three years, and establish an ARIMA model of the outpatient visits by time series forecasting, using Predictive Analytics Software (PASW). The effect of the model was evaluated by forecasting future outpatient visits during the subsequent two years.

2 MATERIALS AND METHODS

2.1 Data source

Data were retrieved from the outpatient medical records of the No. 1 Military Medical Project, between January 1, 2010 and March 31, 2014. Information concerning the monthly local outpatient visits over a total of 51 months was summarized.

2.2 Forecast model

Data for the monthly outpatient visits were imported into PASW software (IBM, USA)(Li and Luo, 2010), and the yearly and monthly data defined using the Define Dates dialog box. The Time Series Modeler was used, with the outpatient volume data selected as the dependent variable, and the yearly and monthly data as the explanatory variable. The model was chosen to be an Expert Model (full model including ARIMA), and the predicted value as well as the upper and lower limits were saved. The forecasting time in the Options dialog box was selected to be December 2015. The outpatient visits for the first quarter of 2014 were forecast based on data from 2010 to 2013. The accuracy of the model was analyzed using the relative error and absolute error, with an absolute error < 5% taken to indicate good accuracy(Zhou et al., 2011a, Liang et al., 2006, Xiang and Chen, 2009, Zhou et al., 2008). Stationary-R^2 and R^2 were used as indicators of goodness of fit, with values > 0.75 considered to indicate good fitting by the model. The outpatient visits for years 2014 and 2015 were predicted on the basis of the data for the period January 2010 to December 2013.

3 RESULTS

3.1 General information concerning the outpatient visits

As shown in Figure 1, the cumulative total number of outpatient visits during the statistical interval was 3,277,000, with 3,036,000 occurring from 2010 to 2013; the mean annual growth rate was 24.07%. The patients seen during these visits included 1,463,000 males (44.65%) and 1,814,000 females (55.35%), with a male-female ratio of 0.81. The mean age of all the patients was 40.36 ± 19.32 years, with 8.42% of patients aged ≤ 10 years, 6.08% aged 11–20 years, 15.86% aged 21–30 years, 18.60% aged 31–40 years, 21.31% aged 41–50 years, 14.02% aged 51–60 years, and 15.70% aged ≥ 61 years. There were 1,884,000 medical patients (57.49%) and 1,393,000 surgical patients (42.51%), with a medical-surgical ratio of 1.35:1. In addition, insured patients accounted for 18.11%, and patients consulted by high-ranking professionals or experts accounted for 30.10%.

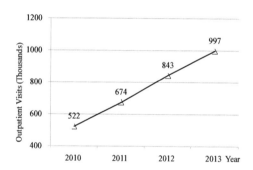

Figure 1.   Outpatient visits for the period 2010 to 2013.

3.2 The forecasting model and evaluation of outpatient visits

Based on the data for the period January 2010 to December 2013, the ARIMA model estimated the following parameters: Alpha (Level), 0.055431; Gamma (Trend), 0.000001; and Delta (Season), 0.001000. The stationary-R^2 and R^2 values were 0.783 and 0.951 (i.e., > 0.75), respectively, suggesting that the model fitted the data well. As shown in Table 1, the number of outpatient visits forecast for the first quarter of 2014 was 254,000, compared with an actual value of 240,000. In other words, the actual number of outpatient visits was within the prediction interval; the absolute error was 9,882, while the relative error was 4.11% (i.e., < 5%), indicating a good forecasting ability (Table 1).

Table 1.   The ability of the arima model to forecast the number of outpatient visits during the first quarter of year 2014.

Time	Actual no. of visits	Forecast no. of visits	Absolute error	Relative error (%)
January	73821	80880	954	1.29
February	71396	77013	2408	3.37
March	95072	96590	6520	6.86
First quarter	240289	254483	9882	4.11

3.3 Short-term forecasting of outpatient visits

Based on the data for the period January 2010 to March 2013, the ARIMA model estimated the following parameters: Alpha (Level), 0.098007; Gamma (Trend), 0.000002; and Delta (Season), 0.000178. The respective stationary-R^2 and R^2 values were 0.792 and 0.950 (i.e., > 0.75). The ARIMA model was used to forecast the number of outpatient visits for the period April 2014 to December 2014, and for the year 2015. The number of outpatient visits for 2014 was forecast to be 1,132,000, with lower and upper limits of 1,057,000 and 1,207,000, respectively (Fig. 2). The value forecast for 2015 was 1,295,000, with lower and upper limits of 1,191,000 and 1,399,000, respectively (Figure 2).

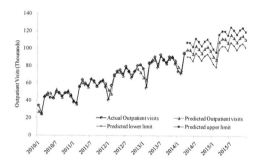

Figure 2.   The actual number of outpatient visits and the number forecast by the ARIMA model for the period 2010 to 2015.

4 DISCUSSION

An important aspect in the management of outpatient departments in a large general hospital is the accurate forecasting of the number of outpatient visits, since this helps to guide the use of outpatient consulting rooms, the allocation of medical staff and the formulation of management objectives for various clinical departments. In a recent preliminary study, the ARIMA model was applied to the forecasting of outpatient visits with reportedly good results(Zhou et al., 2011a, Liang et al., 2006, Xiang and Chen, 2009, Zhou et al., 2008, Zhou et al., 2011b). The main findings of the present study, which utilized time series forecasting with an ARIMA model in PASW software to make two-year predictions of outpatient visits, was that the model was easy to develop and use, fitted the dataset well, and had good forecasting ability. Furthermore, the results were easy to understand, making ARIMA a useful tool for short-term forecasting of variables showing seasonal fluctuation (such as outpatient visits, inpatient visits and surgical visits to a hospital).

The statistical interval in the present study (from year 2010 to 2013) corresponded with a major reform of the medical system in our country. The implementation of the new rural co-operative medical system and the medical insurance system for urban residents resulted in a rapid increase in the number of emergency and outpatient visits over the country as a whole. For example, compared with the year 2010, the total number of emergency and outpatient visits in 2011 rose by 7.67% nationwide, with increases of 11.69% in Beijing, 5.27% in western areas of China, and 7.07% in Sichuan Province. The national increase in emergency and outpatient visits to general hospitals was 11.08%. The hospital in the present study is a grade A, tertiary general hospital with a mean annual increase in outpatient visits of 24.07%, higher than that reported nationally, and higher than the values for western areas of China, Sichuan Province, and general hospitals nationwide. On the one hand, this indicates that the hospital is in a period of rapid development, with good outpatient management and exploitation of the expanding market, by virtue of the new medical reform policy. On the other hand, these data also suggest that the emergency and outpatient visits to first-class hospitals and second-class hospitals in western areas (Sichuan Province) have risen more slowly. The rapid increase in the number of patients attending high-level hospitals has yet to settle, requiring the attention of the relevant health administration departments. Unlike previous investigations into outpatient visits, which rarely considered demographic data(Yuan et al., 2005, Zhou et al., 2011a, Liang et al., 2006, Xiang and Chen, 2009, Zhou et al., 2008, Zhou et al., 2011b), the present study collected the medical records of almost 3,000,000 outpatients to provide preliminarily information concerning the male-female ratio and age of the patients, the characteristics of the departments attended, and the nature of the medical fee payments. These data are potentially important to managers of hospital outpatient departments, as they may help to guide the decision making process.

The ARIMA model used to forecast outpatient visits in the present study utilized PASW software, to which the SPSS software package (originally by SPSS Inc., Chicago, USA) was renamed following its acquisition by IBM in 2009(Li and Luo, 2010). Previous forecasting with ARIMA models had been accomplished using SPSS, SAS and Eviews software(Zhou et al., 2011a, Liang et al., 2006, Xiang and Chen, 2009, Zhou et al., 2008, Zhou et al., 2011b), but this required the data analyst to have a profound knowledge of statistical theory: the ARIMA model was ultimately determined by drawing the self-correlation and partial correlation functional diagrams, a complex procedure requiring interpretation of the results by a skilled professional. In contrast, the time series forecasting model of PASW software allows utilization of the 'Expert Models' setting to evaluate the fitness of the evaluation model through simple parameter settings. The results are easy to understand, confirming the feasibility of this ARIMA model for the forecasting of outpatient visits.

ACKNOWLEDGMENTS

This work was supported by the Chengdu Military General Hospital fund for cultivation of research-oriented talents (2011YG-C12, 2013YG-B021) and the Health Bureau of Sichuan Province (120566).

REFERENCES

National Health and Family Planning Commission, China Health Statistics Yearbook 2011 [Online]. Available: http://wsb.moh.gov.cn/htmlfiles/zwgkzt/ptjnj/year2011/index2011.html [Accessed 6.30 2014].

National Health and Family Planning Commission, China Health Statistics Yearbook 2012 [Online]. Available: http://wsb.moh.gov.cn/htmlfiles/zwgkzt/ptjnj/year2012/index2012.html [Accessed 6.30 2014].

Guan, L. 2013. Application of ARIMA Model in Predicting Incidence Trend of Nosocomial Infection. *Practical Preventive Medicine* 20(10): 1247–1249.

Li, Z. 2012. *Xijing Information Division Manual*. Xi'an: Fourth Military Medical University Press.

Li, Z. & Luo, P. 2010. *PASW / SPSS Statistics Statistical analysis tutorial*. Beijing: Electronic Industry Press.

Liang, G., Liu, Y. & Deng, S. 2006. Applications of ARIMA model on predictive workload of outpatient department. *Chinese Journal of Hospital Statistics* 13(1): 24–26.

Shen, Q. 2014. ARIMA model-based analysis of China's total retail sales of social consumer goods. *China chain* 4(2): 61–62.

Wang, Z., Yang, Y., Zhang, F., et al. 2014. The application of ARIMA model in the per capita income of rural households. *Joural of inner mongolia agricultural university* 35(1): 165–169.

Xiang, Q. & Chen, P. 2009. Construction and application of an ARIMA model for predicting the number of outpatient visits in general hospitals. *Journal of Southern Medical University* 29(5): 1076–1078.

Xiong, J. 2002. *Hospital management and medical statistics.* Beijing: People's Medical Publishing House.

Yuan, J., Gao, X., Shi, Y., et al. 2005. Survey and analysis of distribution of outpatients registration and visiting time. *J Fourth Mil Med Univ* 26(1): 83–85.

Zhou, Q., Chen, Q., Ye, X., et al. 2011a. A Forecast of Hospital Outpatient Number Based on ARIMA Model. *Chinese Medical Record* 12(1): 51–52.

Zhou, X., Sun, L. & Zhou, Z. 2011b. Use Eviews software for mental health Outpatient visits conducted ARIMA model prediction. *Chinese Journal of Health Statistics* 28(6): 687–688.

Zhou, Z., Lu, H. & Zou, Y. 2008. Time series analysis by ARIMA interfering model to forecast amount of outpatient. *Time series analysis by ARIMA interfering model to forecast amount of outpatient* 15(2): 110–112.

Zou, Y., Li, H., Chen, Y., et al. 2013. Dynamic analysis of patients with viral encephalitis by using ARIMA model combined with circle distribution. *Dynamic analysis of patients with viral encephalitis by using ARIMA model combined with circle distribution* 14(10): 784–787.

Information Technology – Wan et al. (Eds)
© *2015 Taylor & Francis Group, London, ISBN 978-1-138-02785-5*

Purification and research of fish oil

Hua Liu & Li Zhao

College of Life Sciences, Jiangxi Science and Technology Normal University, Nanchang, China

ABSTRACT: Our paper presents the optimum conditions of fish oil purification, and compared the anti-oxidation effect of some kinds of antioxidants on fish oil. And all the raw material was offcut of grass carp. The fish oil of grass carp was obtained by enzymolysis, and we took the treatment of degumming, deacidification, decoloration, and deodorization to obtain refined fish oil. In order to compare the anti-oxidation effect of antioxidants on fish oil, we used the POV and TBARS as an index. The results show that the optimum conditions were: for enzymolysis the addition of trypsin was 1.5%, the optimum pH and temperature were 8.0 and 50°C; the 80% phosphoric acid was used to degumming, and the volume of addition was 1%;for deacidification we used 3.65% sodium hydroxide, and the volume of addition was 1.5%; and the activated carbon and activated clay were used to decoloration, the ratio of activated carbon and activated clay was 1:20 and the volume of addition was 2.1%. And the anti-oxidation effect of PG was better other antioxidants we used.

1 INTRODUCTION

Grass carp is one of the most important aquatic products in China, which contributes 9.5% of Chinese catches (Kasankala 2007). The fish resources of our country are very abundant, and China is one of the superpower freshwater fish production countries in the world. Recently, with the increase of Fish product diversification, and with the expansion of the breeding technology and application, the output of grass carp increased continuously, and also the annual production capacity is relatively high. However, a large amount of waste was generated by fish processing industry from grass crap, which includes viscera, scales and bones. The by-product during freshwater fish processing was about 40%-50%. It is reported that about 131 million tons of fish was produced in 2000 in the world, and nearly 74% was consumed by human (Arvanitoyannis 2008). Therefore, there are a lot of grass carp byproduct to be processed, and these wastes are dumped may cause pollutions and emit an offensive odor. According to many studies, the byproducts of fish are rich in fish oil and many nutrient contents, such as protein and unsaturated fatty acid (Rubio-Rodríguez 2008, Bougatef 2008) . But the rescues are not well used. It's not only a serious waste of resources and also caused a great pollution of the environment. DHA and EPA are two kinds of unsaturated fatty acid; they have many physiological actions, such as cancer prevention, anti-inflammatory, inhibit platelet aggregation, and improve brain function (Menendez 2006, Chen 2000, Kato 2007, Wall 2010, Lau 1993, Camuesco 2006, Palakurthi 2000, Siddiqui 2005).

In addition, supplementation of fish oil during pregnancy may decrease the incidence rate of food and IgE-associated eczema allergy, and supplementation of fish oil during pregnancy also can improve neuro-developmental outcomes (Furuhjelm 2009, Bouwens 2009). And as well known that fish oil has a beneficial effect on human immune cells. The study reported that a high intake DHA and EPA can alter the gene expression which related to an inflammatory reaction, and induce the expression of anti-inflammatory gene (Tofail 2011). Furthermore, supplementation of fish oil has beneficial effects in hyperlipidemia, and also in some other heart diseases, for example, regular daily ingestion of fish oil can prevent dysfunction, which caused by pressure overload conditions (Eslick 2009, Duda 2009) and reduces potentially risk of fatal ventricular arrhythmias (Leaf 2005).

With improved food safety awareness, as a kind of health and safety functional foods, fish oil become more and more popular. But the DHA and EPA are easily oxidized, oxidation is an essential reaction in foods that can cause the spoilage of fish oil, and also can affect the nutritive value. Therefore, to prevent the oxidation of fish oil and guarantee the nutritive value is the most important thing during processing. There are many methods to prevent the fish oil from oxidation, and the main methods are: stored in vacuum or vacuum with nitrogen charging, cold storage, keep in a dark place, microencapsulation (Keogh 2006, Heinzelmann 1999, Baik 2004, Heinzelmann 2000), packing with special material (Kagami 2006) and add antioxidants (Nieto 1993, Maqsood 2009, Seppanen 2010, Huber 2009, Kindleysides 2012, André 2010).

But considering from practicability and price, the addition of antioxidants is more convenient, effective, economical and practical compare with other methods.

Therefore, the main objective of this study was to obtain crude oil by enzymolysis from grass carp, then after the process of degumming, deacidification, decoloration, and deodorization to obtain refined fish oil. And the oxidative stability of VC, VE, BHT and PG on fish oil was detected, in order to obtain the optimal antioxidant to prevent the fish oil from oxidation.

2 MATERIALS AND METHODS

2.1 *Material and instruments*

By-products of grass carp were obtained from Jiangxi Dong Hai Food Co., Ltd. (Jiangxi, China). Sodium hydroxide, hydrochloric acid and potassium hydroxide were obtained from Beijing Mr. Lai treasure technology Co., LTD (Beijing, China). Trypsase and pepsase were purchased from Novozmes enzyme preparation Co. Other chemicals and reagents used in this study were of analytical grade. HH-4 digital display constant temperature water bath pot was purchased from Jintan Branch Analysis Instrument Co. Ltd and rotary evaporators were the product of Shanghai Ailang Instruments Co., Ltd.

2.2 *Production of crude fish oil*

Fresh by-products of grass carp were washed and minced by meat grinder, and make sure that by-products were completely minced. Then 2 kg ground meat was put into 1000mL distilled water, the ratio of ground meat and distilled water was 1:1. The pH was adjusted to 8-9, and trypsin was added at 45 °C, the volume of the addition was 1.5%. After 4 h of enzymolysis, the mixture was put into seperate funnel to obtain the supernatant liquid, and centrifuged at 3500r/h for 5 min, and then the supernatant liquid was washed three times. The crude oil was stored at 4°C.

2.3 *The treatment of degumming*

The phosphoric acid was added into 1% crude oil, the final concentration were 0%, 10%, 20%, 40%, 80%, and reacted at 80°C for 5 min, and stirring during the reaction. After reaction samples were centrifuged at 3500r/h for 5 min, the supernatant liquid was washed three times.

2.4 *The treatment of deacidification*

As described above, after the treatment of degumming, samples were heated at 65°C by thermostat water bath, and 1mol/L NaOH was added, the final concentrations were 0.0%, 0.2%, 0.5%, 1.0%, 1.5%, 2.0%.

Then samples were centrifuged at 4500r/h for 15 min when the temperature rose to 75°C, keep the supernatant liquid, and washed three times, then centrifuged again at 4500r/h for 10 min, the supernatant liquid was what we wanted.

2.5 *The treatment of decoloration*

After the treatment of deacidification, the activated carbon and activated clay were added to samples at 75°C, and the addition of activated carbon and activated clay was: 10%, 0%; 0%, 10%; 5%, 0%; 0%, 5%; 2%, 0.1%; and 4%, 0.2%. During the reaction stirring was needed. Then samples were filtrated after the reaction last for 30 min.

2.6 *The treatment of deodorization*

After all the treatment described above, glacial acetic acid was added into samples, which was easily volatilized. Then samples were deodorized by rotary evaporators at 80-90°C. The time of deodorization were 30, 60, 90 min respectively.

2.7 *The test of oxidization*

VE (0.2%), VC (0.2%), BHT (0.02%), PG (0.01%) were added into fish oil respectively, and stored at 80 °C. The TBARS and POV were detected on 1, 2, 3, 4 and 5 day.

3 RESULTS AND DISCUSSION

3.1 *The yield of fish oil by degumming*

As it was shown in Fig 1, the concentration of phosphoric acid has great effect on fish oil yield. With the concentration increased, thought the yield of fish oil decreased, but the peroxide value was decreased (data not shown), and the yield still kept at 99.1%, it means that using high concentrations of phosphoric acid for degumming has good effect of fish oil.

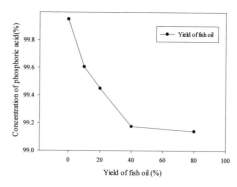

Figure 1. Yield of fish oil after degumming by phosphoric acid.

3.2 The yield of fish oil by deacidification

Deacidification is the most important process during the production of refined fish oil. For the process of deacidification we used NaOH in different concentration. And from Fig 2, we can see that with the concentration of NaOH increased, the yield of fish oil decreased, but the peroxide value was decreased. The reason of this phenomenon is the neutral oil of fish oil was saponificated by high concentration of NaOH, it caused the decrease yield. And the result indicated the optimum condition for deacidification is adding 1.5% NaOH.

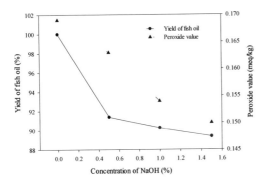

Figure 2. Yield and peroxide value of fish oil after deacidification.

3.3 The yield of fish oil by c

We used activated clay and activated carbon for the decoloration of fish oil. The main purpose of decoloration is to eliminate natural pigment and the color that produces during chemical changes. And the optimum condition of decoloration was shown in table 1.

Table 1 The effect of activated clay and activated carbon on yield and peroxide value of fish oil

Activated clay (%)	Activated carbon (%)	Peroxide value (meq/kg)	Yield (%)
10	0	0.014	84
0	10	0.024	94.3
5	0	0.019	86.583
0	5	0.02	95.254
2	0.1	0.021	85.146

As Table 1 shown, the optimum condition of decoloration is added 2% activated clay and 0.1% activated carbon, with this condition, though the yield was decreased at this condition, but the peroxide value was also decreased, and the fish oil was colorless.

3.4 The results of deodorization

Due to the smell of fish oil makes people feel comfortable, so it's necessary to eliminate the smell of stench. We used glacial acetic acid for deodorization, and found the optimum time of reaction. The results were shown in table 2. As the result shown, when the time of reaction last 30 min, the stench was still evident, and the viscosity was obviously. And after reacting 60 min, the odor of stench became slightly. But when the time last to 90 min, the odor became normal and the viscosity almost had no change. The result indicated that the optimum reaction time is 90 min.

Table 2.   The effect of react time on deodorization.

Time (min)	Viscosity	Odor
30	obviously	stench
60	slightly	slight stench
90	no-change	normal

3.5 The effect of antioxidant on fish oil

For the effect of antioxidant of fish oil, the VE (0.2%), VC (0.2%), BHT (0.02%), PG (0.01%) were used, samples were stored at 60°C for a week, then the value of TBARS and POV were detected to evaluate the antioxidant effect of antioxidants which were described above.

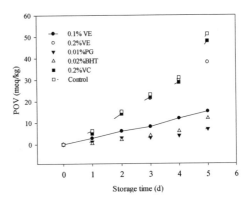

Figure 3.   Result of POV with the protection of antioxidants.

The effect of antioxidants on the value of POV was shown in Fig. 3. As the result shown, with the storage time increasing, the value of POV has a sharp increase. But with the protection of antioxidants, the values of POV were lower than control. And from Fig. 3 we can see that 0.1% VE, 0.02% BHT, and 0.01% PG were obviously inhibited the oxidation. When the samples were stored for 5 days, the POV value of control was increased to 51.221, but the values of VE, BHT, and PG were 15.232, 12.003 and 6.982 respectively. And within these three antioxidants the effect of PG was better than others.

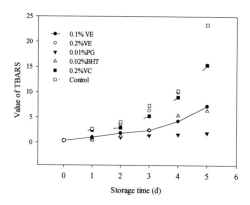

Figure 4. Result of TBARS with the protection of antioxidants

As Fig. 4 shown, with the increase of storage time, the value of TBARS was also increased. But as the results shown, the effect of antioxidants was obviously compared to control, especially PG. When the storage time last to 5 days, the value of TBARS was 6.982, it's obviously lower than the value of control, the value of control was 23.341. And compared to other antioxidants, PG shown the best effect for inhibition of oil oxidation. From the result we can figure that the effect of VE and VC were not remarkable. The oxidation resistance ability of different antioxidant for fish oil were 0.01% PG>0.02% BHT>0.1% VE>0.2% VE>0.2% VC>control. The result indicated that PG can be used as antioxidant in storage of fish oil.

4 SUMMARY

The main purpose of the study was to obtain refined fish oil from by-products of grass carp. And the optimum conditions were studied. And 1.5% trypsin was added to hydrolysis, and for the process of degumming, 1% phosphoric acid was added, and the optimum final concentration was 80%. And as the result shown, for the process of deacidification the optimum condition was adding 1.5% NaOH. And for the process of decoloratio, the ratio of 1:20 (activated clay and activated carbon) shown a great effect. The antioxidant effects of several antioxidants were studied and PG shown the best effect at low concentration.

Due to the fish oil has lot of benefits for human health, and with the rapidly increasing demand for fish oil (Péron 2010, there will be a great approach to make waste profitable, if the manufacturers could make good use of fish by-products. Besides, it also can protect the environment from pollution and expand the application prospect of fish in health food industry.

REFERENCES

André C., I. Castanheira, J. Cruz, P. Paseiro, & A. Sanches-Silva (2010). Analytical strategies to evaluate antioxidants in food: a review. *Trends Food Sci Tech. 21*, 229–246.

Arvanitoyannis I.S., & A. Kassaveti (2008). Fish industry waste: treatments, environmental impacts, current and potential uses. *Int J Food Sci Tech. 43*, 726–745.

Baik M.Y., E. Suhendro, W. Nawar, D. Mcclements, E. Decker, & P. Chinachoti (2004). Effects of antioxidants and humidity on the oxidative stability of microencapsulated fish oil. *J AM Oil Chem Soc. 81*, 355–360.

Bougatef A., N. Nedjar-Arroume, R. Ravallec-Plé, Y. Leroy, D. Guillochon, A. Barkia, & M. Nasri (2008). Angiotensin I-converting enzyme (ACE) inhibitory activities of sardinelle (Sardinella aurita) by-products protein hydrolysates obtained by treatment with microbial and visceral fish serine proteases. *Food Chem. 111*, 350–356.

Bouwens M., O. Rest, N. Dellschaft, M. Bromharr, L. Groot, J. Geleijnse, M. Muller, & L. Afman (2009). Fish-oil supplementation induces antiinflammatory gene expression profiles in human blood mononuclear cells. *AM J Clin Nutri. 90*, 415–424.

Camuesco D., M. Comalada, A. Concha, A. Nieto, S.Sierra, J. Xaus, A. Zarzuelo, & J. Galvez (2006). Intestinal anti-inflammatory activity of combined quercitrin and dietary olive oil supplemented with fish oil, rich in EPA and DHA (n-3) polyunsaturated fatty acids, in rats with DSS-induced colitis. *Clinl Nutr. 25*, 466–476.

Chen Z.Y., & N. Istfan (2000). Docosahexaenoic acid is a potent inducer of apoptosis in HT-29 colon cancer cells. *Prostag Leukotr Ess. 63*, 301–308.

Duda M. K., K. O'shea, A. Tintinu,W. Xu, & R. Khairallah (2009). Fish oil, but not flaxseed oil, decreases inflammation and prevents pressure overload-induced cardiac dysfunction. *Cardiovasc Res. 81*, 319–327.

Eslick G.D., P. Howe, C. Smith, R. Priest, & A. Bensoussan (2009). Benefits of fish oil supplementation in hyperlipidemia: a systematic review and meta-analysis. *Int J Cardiol. 136*, 4–16.

Furuhjelm C., K. Warstedt, J. Larsson, M. Fredriksson, M. Bottcher, K. Falth-Magnusson, & K. Duchen (2009). Fish oil supplementation in pregnancy and lactation may decrease the risk of infant allergy. *Acta Paediatr. 98*, 1461–1467.

Heinzelmann K., & K. Franke (1999). Using freezing and drying techniques of emulsions for the microencapsulation of fish oil to improve oxidation stability. *Colloid Surface B. 12*, 223–229.

Heinzelmann K., K. Franke, B. Jensen, & A. Haahr (2000). Protection of fish oil from oxidation by microencapsulation using freeze-drying techniques. *Eur J Lipid Sci Tech. 102*, 114–121.

Huber G. M., H. Vasantha Rupasinghe, & F. Shahidi (2009). Inhibition of oxidation of omega-3 polyunsaturated fatty acids and fish oil by quercetin glycosides. *Food Chem. 117*, 290–295.

Kagami Y., S. Sugimura, N. Fujishima, T. Kometani, & Y. Matsumura (2006). Oxidative stability, structure, and physical characteristics of microcapsules formed by spray drying of fish oil with protein and dextrin wall materials. *J Food Sci. 68*, 2248–2255.

Kasankala L.M., X. Yao, W. Yao, S. Hong, & Q. He (2007). Optimization of gelatine extraction from grass carp (Catenopharyngodon idella) fish skin by response surface methodology. *Bioresource Technol. 98*, 3338–3343.

Kato T., N. Kolenic, & R. Pardini (2007). Docosahexaenoic acid (DHA), a primary tumor suppressive omega-3 fatty acid, inhibits growth of colorectal cancer independent of p53 mutational status. *Nutr Cancer. 58*, 178–187.

Keogh M. K., B. O'Kennedy, J. Kelly, M. Auty, P. Kelly, A. Fureby, & A. Haahr (2006). Stability to Oxidation of Spray-Dried Fish Oil Powder Microencapsulated Using Milk Ingredients. *J Food Sci. 66*, 217–224.

Kindleysides S., S. Quek, & M. Miller (2012). Inhibition of fish oil oxidation and the radical scavenging activity of New Zealand seaweed extracts. *Food Chem. 133*, 1624–1631.

Lau C.S., K. Morley, & J. Belch (1993). Effects Of Fish Oil Supplementation On Non-Steroidal Anti—Inflammatory Drug Requirement In Patients With Mild Rheumatoid Arthritis—A Double-Blind Placebo Controlled Study. *Rheumatology. 32*, 982–989.

Leaf A., C. Albert, M. Josephson, & D. Steinhaus (2005). Prevention of fatal arrhythmias in high-risk subjects by fish oil n-3 fatty acid intake. *Circulation. 112*, 2762–2768.

Maqsood S., & S. Benjakul (2009). Comparative studies of four different phenolic compounds on in vitro antioxidative activity and the preventive effect on lipid oxidation of fish oil emulsion and fish mince. *Food Chem. 119*, 123–132.

Menendez J.A., A. Vazquez-Martin, S. Ropero, R. Lupu, R. Colomer, & J. Trueta (2006). Exogenous supplementation with ω-3 polyunsaturated fatty acid docosahexaenoic acid (DHA; 22: 6n-3) synergistically enhances taxane cytotoxicity and downregulates Her-2/neu (c-erbB-2) oncogene expression in human breast cancer cell. *Eur J Cancer Prev. 14*, 263–270.

Nieto S., A. Garrido, J. Sanhueza, L. Loyola, G. Morales, F. Leighton, & A. Valenzuela (1993). Flavonoids as stabilizers of fish oil: an alternative to synthetic antioxidants. *J AM Oil Chem Soc. 70*, 773–778.

Palakurthi S.S., R. Flückiger, H. Aktas, A. Changolkar, A. Shahsafaei, S. Harneit, E. Kilic, & J. Halperin (2000). Inhibition of translation initiation mediates the anticancer effect of the n-3 polyunsaturated fatty acid eicosapentaenoic acid. *Cancer Res. 60*, 2919–2925.

Péron G., J. Mittaine, B. Gallic (2010). Where do fishmeal and fish oil products come from. An analysis of the conversion ratios in the global fishmeal industry. *Marine Policy. 34*, 815–820.

Rubio-Rodríguez N., S. de Diego, S. Beltrán, I. Jaime, M.T.Sanz, & J.Rovira (2008). Supercritical fluid extraction of the omega-3 rich oil contained in hake (Merluccius capensis–Merluccius paradoxus) by-products: Study of the influence of process parameters on the extraction yield and oil quality. *J Supercrit Fluid. 47*, 215–226.

Seppanen C.M., Q. Song, & A. Csallany (2010). The antioxidant functions of tocopherol and tocotrienol homologues in oils, fats, and food systems. J AM Oil Chem Soc. 87, 469–481.

Siddiqui R.A., M. Zerouga, M. Wu, A. Caetillo, K. Harvey, G. Zaloga, & W. Stillwell (2005). Anticancer properties of propofol-docosahexaenoate and propofol-eicosapentaenoate on breast cancer cells. *Breast Cancer Res. 7*, 645–654.

Tofail F., I. Kabir, J. Hamadani, F. Chowdhury, S. Yesmin, F. Mehreen, & S. Huda (2011). Supplementation of fish-oil and soy-oil during pregnancy and psychomotor development of infants. *J Health Popul Nutri. 24*, 48–56.

Wall R., R. Ross, G. Fitzgerald, & C. Stanton (2010). Fatty acids from fish: the anti-inflammatory potential of long-chain omega-3 fatty acids. *Nutr Rev. 68*, 280–289.

Information Technology – Wan et al. (Eds)
© *2015 Taylor & Francis Group, London, ISBN 978-1-138-02785-5*

Recent advances in camera tracking and virtual reality

Wenju Zhou & Huosheng Hu
School of Computer Science and Electronic Engineering, University of Essex, Colchester, UK

Peter James Smith
Vitec Videocom Ltd, Bury St Edmunds, Suffolk, UK

ABSTRACT: This paper presents a comprehensive literature review and market evaluation of camera tracking and its related applications in the broadcast industry. The current commercial camera tracking hardware and software are outlined. The key market players for developing these systems are also introduced. Moreover, the applications of the camera tracking are discussed, and popular tracking software is compared. By surveying the current industry on broadcast camera tracking and VR, it is clear that the prospect of the camera tracking and VR business are broad and many advanced automatic camera tracking systems will be developed for VR within the broadcast markets soon.

KEYWORDS: Camera Tracking, Virtual Reality, Optical Tracking sensors, Laser Scanners

1 INTRODUCTION

Camera tracking refers to a camera that follows a person or an object that is physically moving. Previously, the camera tracking can be realised by using a dolly or pedestal with the mounted camera, which then be manually pushed on the road or on rails [1]. Recently, more sophisticated methods have become possible for camera tracking. For example the use of vision or laser based systems alongside robotic devices. These advancements have generated interest in this area for researchers, who constantly evolve new techniques and algorithms to enhance tracking technologies[2]. Some of the novel sensors used in the modern tracking applications include: lasers, ultrasonic and optical methods. As the research interest grows in this area, so does the development potential for new camera tracking sensors and products.

Camera tracking provides accurate, repeatable smooth movements. This is important when a camera shot needs to be one consistent, continuous move. Camera tracking applications can be seen in early Hollywood [3], through to modern television broadcasts. It is particularly apparent in major sporting events, and news broadcasts, where the tracking data is used to produce VR or AR output renderings. Increasing demand in these areas has led to huge demands and opportunities in the camera tracking domain.

The rest of this paper is organised as follows. Section I presents current research trend in the camera tracking and virtual reality. Section III collects typical

hardware system sensor components. In Section IV, two tracking structures are described, namely the camera mounted and the environmental mounted. Section V outlines several popular algorithms for camera tracking used in VR and AR. This includes tracking methods, classifications, and currently available software. There are several camera tracking software in current market can be found in Section VI. Finally, a brief conclusion is given in Section VII.

2 CURRENT RESEARCH TRENDS

Television broadcasters and film makers require virtual or augmented elements to their programmes. This requires highly accurate and repeatable positional measurements. This is the main driving factor for researching virtual reality and camera tracking. We searched using key words "Virtual Reality", "Camera Tracking" and "Virtual Reality + Camera Tracking" from "Google Scholars website". The results from these searches represent the current research interest in these subject areas. These results suggest technology trends and business opportunities. Figure 1 and Figure 2 show the results for the searches of "Virtual Reality" and "Camera Tracking", respectively. Figure 3 shows the result of searching using word "Virtual Reality + Camera tracking".

These results highlight that although the trend for separate use of tracking and virtual reality are decreasing, the joint search results show that combined, the research interest is growing. In conclusion

Figure 1. "Virtual Reality" at google scholar.

Figure 2. "Camera Tracking" at google scholar.

Figure 3. "Virtual Reality + Camera Tracking" at google scholar.

we can suggest this may be due to a shift from the theoretical to practical domain of the technologies.

3 CAMERA TRACKING HARDWARE

Camera tracking systems provide real-time positional data that must be both accurate and repeatable. They must also provide this data synchronously with minimal jitter[4]. This includes when stationary and moving. This positional data can be collected from many types of sensors which can be classified as optical, laser scan, or ultra-sonic based etc. Due to the constant decline in cost of cameras it is now becoming feasible to consider image processing based techniques for the purposes of camera tracking. The tracking system methods can be classified as pedestal mounted or environmentally mounted. The following sections provide details on both of these techniques.

3.1 Optical tracking sensors

This is a technique that uses optical sensors, e.g. CCD/CMOS cameras. Optical tracking sensors can be classified as RGB (Red-Green-Blue), RGB-D (Depth) and thermal camera types. Although broadcast cameras could be used to acquire image information for camera tracking, their control, complexity and high cost prevent this from being practical for this application. Therefore, independent cameras are preferred. The independent tracking cameras should be flexible, easily controllable, programmable, and of low cost.

RGB cameras are commonly used in photography and filming. These types of camera are often used in conjunction with advanced algorithms for environmental mapping purposes also. Figure 4 below shows the tracking of a women's face using a RGB camera and advanced recognition and tracking algorithm. Lumenera [5], PX4 [6] and Surveyor[7] are the companies that can provide this type of product.

Figure 4. RGB camera with advanced algorithm tracking a women's face.

RGB-D cameras combine RGB camera and depth sensors that were originally designed for capturing human motion. As they provide depth information combined with the RGB image, they attract researchers of computer vision as well as robotic vision apply them to human machine interaction and SLAM. Some of them even mount RGB-D sensor on camera tracking system to navigate the robot autonomously[8]. The depth image is shown in Figure 5, the depth data information is denoted by different colours.

Figure 5. Depth data information image obtained by a RGB-D camera.

An infrared thermography camera or more commonly known as an infrared camera is a device that forms an image using infrared radiation, similar to a common camera that forms an image using visible light[9]. Instead of the visible light camera, which works in wavelengths ranging between 450nm and

750 nm, infrared cameras operate in wavelengths as long as 14,000 nm (14μm). Thermal cameras are also used in camera tracking systems to take thermal images or video. Specification of typical thermal cameras can be found from various companies such as Optris[10], Thermoteknix[11], Micro-Epsilon[12] and FILR[13].

3.2 *Laser scanners*

Laser-Scan was founded in 1969 by three academics from the Physics Department at the University of Cambridge[14]. The laser scanner captures surfaces in the form of a three dimensional scatter plot. They provide quick, precise, and cost-effective three-dimensional scanning of scenes and objects. The fields of application include; map generation, surveying of rooms and spaces, building surveillance, tunnel or mine shaft inspection, etc. Therefore the laser scanners are suitable for use as a broadcast camera tracking system. Figure 6 shows an image built by a laser scan, scatter plot. The companies can provide specification products include SICK[15], Phoenix[16], Fraunhofer[17] and FARO[18].

Figure 6.   Laser scan image.

3.3 *Ultrasonic tracking sensor*

Ultrasonic sensors work on a principle similar to that of radar or sonar which evaluates attributes of a target by interpreting the echoes from a transmitted radio or sound wave, respectively. Ultrasonic sensors generate high frequency sound waves and evaluate the echo which is received back by the sensor. Sensors calculate the time interval between sending the signal and receiving the echo to determine the distance to an object. Therefore, ultrasonic sensors can be used to detect movement of targets and to measure the distance to targets in many automatic navigation systems. Because ultrasonic sensors use sound rather than light for detection, they work in applications where photoelectric sensors may not. Specifications of typical products can be found from SICK[15], Bakatronocs [19], Senix [20] and TOMTOP[21].

4   TRACKING STRUCTURE

There are two main types of tracking structures. The first is target mounted and the second environmentally mounted. There are many advantages and disadvantages to these systems. In this section a list of companies that provided these technologies is presented.

4.1 *Sensors mounted on the target system*

Most of the sensors mounted on the target system use vision-guided forms. These applications include the camera and laser scanner sensors. As cameras become cheaper and more affordable, camera navigation systems have become commercially viable in many more target tracking applications. One such example of a famous target mounted tracking system maybe the NASA Mars exploration rover, and the ESA ExoMars rover as can be seen in website link[21]. You can see an artist's impression of the Mars rover, an example of an unmanned land-based vehicle. The stereo cameras mounted on top of the Rover used for tracking.

Another example application is a laser scanner mounted on the commercial forklift truck [22]. The navigational forklift truck is a self-tracking system. The pallets are loaded on and off of the fork lift robot using Aurigas laser guided system. Comparing "human forklift trucks" to the Auriga laser guided unit, the Auriga is by far safer, and cheaper to operate. Auriga shuttles are available in many configurations and can be adapted to a wide variety of tasks. Laser guided vehicles like this can operate as individuals or as part of a collaborative team controlled by centralised system that can maximise efficiencies by pre planning the work of the collective group.

4.2 *Sensors mounted in the environment*

One of the most well-known environmentally mounted sensor systems is GPS. GPS is a space based system were satellites of known orbits with known time stamps transmit information to receivers on earth that can calculate accurately the position (with 10 approximately meters accuracy) of any location on the planet. This requires unobstructed line of sight to the satellites. For this calculation a minimum of four satellites is required by the receiver (because the earth is spherical) the higher the number of satellites the higher the accuracy of the calculated position[23]. This system is used around the world for military, civil and commercial applications. It is maintained by the United States government and is freely accessible to anyone with a GPS receiver.

Another example of environmentally mounted cameras is a team of collaborative robots for playing a football competition. The images are captured by the environmental cameras and the recorded images are processed in real-time by a backend PC. This PC provides instructions to the robots to move towards the identified target, a ball. Many tactics are applied to what robot should move and how far to optimise the chance of the team scoring a goal collectively[24].

The mark-based motion tracking system, VICON, is also the environmentally mounted cameras system, which has been deployed to track a team of quad-rotor helicopters in order to provide their individual positions in the space for motion control [27]. The cameras mounted on the walls have illumination rings and each helicopter has embedded retro-reflective marks for motion tracking. These form part of the track and control system that maintains the quad-rotor array in real-time.

5 ALGORITHMS FOR CAMERA TRACKING

In camera tracking, not only the sensors hardware are required, but also the software for tracking is indispensable. This section will address the algorithm of camera tracking.

5.1 *Tracking methods and classification*

According to the process of camera tracking, the algorithm of camera tracking should include two stages, first stage is target match and second is path planning. Figure 7 shows the camera tracking process. The aim of "target match" is finding the tracking target quickly, which is operated on one image. The process means "who is the target". Path planning is to find the optimal motion path to control camera from the positions of targets in several images (frames). This process implies "where is the target". These two processes are required to adopt different algorithms. Some commonly used methods are described as follow.

Figure 7.　Flow chart for camera tracking.

5.2 *Target match*

One main category of gradient based methods is to use shape/contour to represent objects. Gavrila [25] presented a contour based hierarchical

chamfer matching detector for pedestrian detection. Lin et al. [26] extended this work by decomposing the global shape models into parts and constructing a hierarchical tree for the part templates.

Another main category is to use the statistical summarization of the gradients. For example, Lowe [27] introduced the well-known SIFT descriptor for object recognition as shown in Figure 8. Bay et al. proposed SURF[28], which is a much faster scale and rotation invariant interest point descriptor. Dalal and Triggs [29] used the Histogram of Oriented Gradient (HOG) descriptor in training SVM classifier for pedestrian detection. Satpathy et al. [30] proposed a quadratic classification approach on the subspace of Extended Histogram of Gradients (ExHoG) for human detection, which alleviates the problem of discrimination between a dark object against a bright background and vice versa inherent in Histogram of Gradients(HG). Maji et al. [31] also demonstrated promising results using the multi-resolution HOG descriptor and the faster kernel SVM classification. Fan et al. [32] described a part based deformable model based on the multi-resolution HOG descriptor for pedestrian tracking.

Figure 8.　SIFT object recognition.

5.3 *Colour feature*

To increase the discriminative, colour feature descriptors have been proposed [33], which are robust against certain photometric changes. Weijer et al. [34] applied an error propagation analysis to the hue transformation. The analysis shows that the certainty of the hue is inversely proportional to the saturation. Color invariants had been also used as an input to the SIFT descriptor [35], which leads to a CSIFT descriptor that is scale-invariant with respect to light intensity. More detailed performance evaluation of color descriptors can be found in [36].

5.4 *Texture feature*

Texture is a measure of the intensity variation of a surface which quantifies properties such as smoothness and regularity [37]. In recent years, increasing interest is paid on investigating image's local patterns for better detection and recognition [38]. Especially, local patterns that are binarized with an adaptive threshold provide state-of-the-art results on various topics, such as targets recognition and segmentation algorithms as shown in Figure 9 [39].

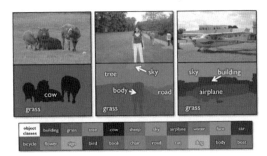

Figure 9.   Targets recognition and segmentation based on texture features.

Figure 10   Humans detection and tracking.

Ojala et al. developed a very efficient texture descriptor, called Local Binary Patterns (LBP) [40]. The LBP texture analysis operator is defined as a grayscale invariant texture measure, derived from a general definition of texture in a local neighbour-hood. The most important property of the LBP operator is its tolerance against illumination changes. Many variants of LBP have been recently proposed, including Local Ternary Patterns (LTP) [41] and multi-scale block LBP (MB-LBP) [42].

5.5   *Spatio-temporal feature*

Local space-time features have recently become a popular representation for action recognition and visual detection. Several methods for feature localization and description have been proposed in the literature and promising results were demonstrated for action classification and various detection tasks [43]. Ke et al. [44] studied the use of volumetric features for event detection in video sequences. They generalized the notion of 2D box features to 3D spatio-temporal volumetric features, which is an extension of the Haar-like features[45]. To characterize local motion and appearance, Ivan et al. computed coarse histograms of the oriented gradients (HOG) and optic flow (HOF) accumulated in space-time neighbourhoods of detected interest points [46].

5.6   *Multiple features fusion*

The feature fusion scheme typically achieves boosted system performance or robustness, which attracts much attention of researchers from image and video retrieval, visual tracking and detection [47]. Figure 10 shows the humans detection and tracking.

Tuzel et al. [48] utilized the covariance matrix as the descriptor for human representation. Their method can encode the gradients' strength, orientation and position information in symmetric positive definite covariance descriptors which lie on a Riemannian manifold. The main disadvantage of the covariance matrix lies in that the operations through

Riemannian geometry are usually time-consuming. Hong et al. [49] proposed a novel descriptor called Sigma Set. Compared with the covariance matrix, Sigma Set is not only more efficient in distance evaluation and average calculation, but also easier to be enriched with first order statistics. Recently, biological features also received a lot of attention from robust tracking research [50]. The biological features tried to mimic human beings' biological vision mechanism in order to achieve robust recognition.

6   CONCLUSION

In this paper, comprehensive literature review and market evaluation of camera tracking and its related applications are presented. In order to obtain the impressive visual effects to audience, VR technology is usually used in the broadcast fields. Therefore, the broadcast cameras are controlled to track targets accurately and record real-time positional data, from sensors and relation software for obtaining the targets precise position. Based on this survey, it is clear that broadcast camera tracking and VR systems have many potential applications within the broadcast markets.

ACKNOWLEDGMENT

This work was supported by KTP (knowledge transfer partnerships) Project, UK (grant numbers KTP167).

REFERENCES

[1] B. Tracking, "http://www.raindance.org/10-best-tracking-shots/," *[Online; accessed 21-May 2014]*.

[2] R. Jagathishwaran, K. Ravichandran, and P. Jayaraman, "A Survey on Face Detection and Tracking," 2014.

[3] C. Tracking, "http://www.thefoundry.co.uk/products/plugins/cameratracker/," *[Online; accessed 24-May-2014]*.

[4] S. S. Dessai, A. Hornung, and L. Kobbelt, "Automatic data normalization and parameterization for optical motion tracking," *Journal of Virtual Reality and Broadcasting*, vol. 3, pp. 981–992, 2006.

[5] Lumenera, "http://www.lumenera.com/," *[Online; accessed 21-May 2014]*, 2014.

[6] PIX4, "http://pixhawk.org/," *[Online; accessed 21-May 2014]*, 2014.

[7] Surveyor corporation, "http://www.surveyor.com/stereo/," *[Online; accessed 21-May 2014]*, 2014

[8] A. S. Huang, A. Bachrach, P. Henry, M. Krainin, D. Maturana, D. Fox, and N. Roy, "Visual odometry and mapping for autonomous flight using an RGB-D camera," in *International Symposium on Robotics Research (ISRR)* 2011, pp. 1–16.

[9] R. Gade and T. B. Moeslund, "Thermal cameras and applications: a survey," *Machine Vision and Applications* vol. 25, pp. 245-262, 2014.

[10] Optris PI Infrared Cameras, "http://www.optris.co.uk/infrared-cameras," *[Online; accessed 21-May 2014]*, 2014.

[11] Thermoteknix, "http://www.thermoteknix.com/," *[Online; accessed 21-May 2014]*, 2014.

[12] Micro-Epsilon, "http://www.micro-epsilon.co.uk/download/products/cat--thermoIMAGER-TIM--en.pdf," *[Online; accessed 27-May-2014]*, 2014.

[13] FLIR, "http://www.flir.com/cs/emea/en/view/?id=42103," *[Online; accessed 21-May 2014]*, 2014.

[14] Laser-scan, "http://www.laser-scan.com/demo/laser-scan-history/," *[Online; accessed 21-May 2014]*, 2014.

[15] SICK, "https://www.mysick.com/eCat.aspx?go=FinderSearch&Cat=Row&At=Fa&Cult=English&FamilyID=344&Category=Produktfinder&Selections=34390,34243," *[Online; accessed 21-May 2014]*, 2014.

[16] Phoenix, "http://www.phoenixse.com/media/pdf/PhoenixCatalogue.pdf," *[Online; accessed 27-May-2014]*.

[17] 3D-Scanner, "http://www.3d-scanner.net/," *[Online;accessed 17-May-2014]*.

[18] FARO, "http://www.faroasia.com/resource-centre/registeredUserFrm.php," *[Online; accessed 27-May-2014]*.

[19] Bakatronics, "http://www.bakatronics.com/," *[Online;accessed 27-May-2014]*.

[20] "SENIX, "http://www.senix.com/pdf_files/senix-data-sheet-tspc-21s.pdf," *[Online; accessed 24-May-2014]*, 2014. 2014.

[21] "TOMTOP, "http://www.tomtop.com/," *[Online; accessed 17-May-2014]*, 2014.

[22] "OCME, "http://www.ocme.com/website/," *[Online;accessed 17-May-2014]*, 2014.

[23] W. i. GPS, "http://www.loc.gov/rr/scitech/mysteries/global.html," *[Online;accessed 24-May-2014]* 2014.

[24] F. robot, "http://en.wikipedia.org/wiki/Robot_Football," *[Online; accessed 27-May-2014]* 2014.

[25] D. M. Gavrila, "Pedestrian detection from a moving vehicle," in, *Computer Vision—ECCV 2000, ed: Springer*, 2000, pp. 37–49.

[26] Z. Lin, L. S. Davis, D. Doermann, and D. DeMenthon, "Hierarchical part-template matching for human detection and segmentation," in, *Computer Vision, 2007. ICCV 2007. IEEE 11th International Conference on* 2007, pp. 1–8.

[27] D. G. Lowe, "Distinctive image features from scale-invariant keypoints," *International journal of computer vision*, vol. 60, pp. 91-110, 2004.

[28] H. Bay, A. Ess, T. Tuytelaars, and L. Van Gool, "Speeded-up robust features (SURF)," *Computer vision and image understanding*, vol. 110, pp. 346–359, 2008.

[29] N. Dalal and B. Triggs, "Histograms of oriented gradients for human detection," in, *Computer Vision and Pattern Recognition, 2005. CVPR2005. IEEE Computer Society Conference on*, 2005, pp. 886–893.

[30] A. Satpathy, X. Jiang, and H.-L. Eng, "Human Detection by Quadratic Classification on Subspace of Extended Histogram of Gradients," 2014.

[31] S. Maji, A. C. Berg, and J. Malik, "Efficient classification for additive kernel SVMs," *Pattern Analysis and Machine Intelligence, IEEE Transactions on*, vol. 35, pp. 66–77, 2013.

[32] X. Fan, S. Mittal, T. Prasad, S. Saurabh, and H. Shin, "Pedestrian Detection and Tracking Using Deformable Part Models and Kalman Filtering," *Journal of Communication and Computer*, vol. 10, pp. 960–966, 2013.

[33] L. Bretzner, I. Laptev, and T. Lindeberg, "Hand gesture recognition using multi-scale colour features, hierarchical models and particle filtering," in, *Automatic Face and Gesture Recognition, 2002. Proceedings. Fifth IEEE International Conference on* 2002, pp. 423–428.

[34] J. Van De Weijer, T. Gevers, and A. D. Bagdanov, "Boosting color saliency in image feature detection," *Pattern Analysis and Machine Intelligence, IEEE Transactions on*, vol. 28, pp. 150-156, 2006.

[35] A. E. Abdel-Hakim and A. A. Farag, "CSIFT: A SIFT descriptor with color invariant characteristics," in, *Computer Vision and Pattern Recognition, 2006 IEEE Computer Society Conference on*, 2006, pp. 1978–1983.

[36] G. J. Burghouts and J.-M. Geusebroek, "Performance evaluation of local colour invariants," *Computer vision and image understanding* vol. 113, pp. 48–62, 2009.

[37] R. Fergus, P. Perona, and A. Zisserman, "Object class recognition by unsupervised scale-invariant learning," *Computer Vision and Pattern Recognition, 2003. Proceedings. 2003 IEEE Computer Society Conference on, 2003*, in, 2003, pp. II-264-II-271 vol. 2.

[38] P. Jackman and D.-W. Sun, "Recent advances in image processing using image texture features for food quality assessment," *Trends in Food Science & Technology*, vol. 29, pp. 35–43, 2013.

[39] J. Shotton, J. Winn, C. Rother, and A. Criminisi, "Textonboost for image understanding: Multi-class object recognition and segmentation by jointly modeling texture, layout, and context," *International journal of computer vision*, vol. 81, pp. 2–23, 2009.

[40] T. Ojala, M. Pietikainen, and T. Maenpaa, "Multiresolution gray-scale and rotation invariant texture classification with local binary patterns," *Pattern Analysis and Machine Intelligence, IEEE Transactions on*, vol. 24, pp. 971–987, 2002.

[41] X. Tan and B. Triggs, "Enhanced local texture feature sets for face recognition under difficult lighting conditions," *Image Processing, IEEE Transactions on*, vol. 19, pp. 1635–1650, 2010.

[42] S. Liao, X. Zhu, Z. Lei, L. Zhang, and S. Z. Li, "Learning multi-scale block local binary patterns

for face recognition," in, *Advances in Biometrics*, ed: Springer, 2007, pp. 828–837.

[43] H. Wang, M. M. Ullah, A. Klaser, I. Laptev, and C. Schmid, "Evaluation of local spatio-temporal features for action recognition," in, *BMVC 2009-British Machine Vision Conference,* 2009.

[44] Y. Ke, R. Sukthankar, and M. Hebert, "Efficient visual event detection using volumetric features," in, *Computer Vision, 2005. ICCV 2005. Tenth IEEE International Conference on* 2005, pp. 166–173.

[45] P. Viola and M. Jones, "Rapid object detection using a boosted cascade of simple features," in, *Computer Vision and Pattern Recognition, 2001. CVPR 2001. Proceedings of the 2001 IEEE Computer Society Conference on,* 2001, pp. I-511-I-518 vol. 1.

[46] I. Laptev, M. Marszalek, C. Schmid, and B. Rozenfeld, "Learning realistic human actions from movies," in, *Computer Vision and Pattern Recognition, 2008. CVPR 2008. IEEE Conference on,* 2008, pp. 1–8.

[47] H. Yang, L. Shao, F. Zheng, L. Wang, and Z. Song, "Recent advances and trends in visual tracking: A review," *Neurocomputing,* vol. 74, pp. 3823–3831, 2011.

[48] O. Tuzel, F. Porikli, and P. Meer, "Human detection via classification on riemannian manifolds," in, *Computer Vision and Pattern Recognition, 2007. CVPR'07. IEEE Conference on,* 2007, pp. 1–8.

[49] X. Hong, H. Chang, S. Shan, X. Chen, and W. Gao, "Sigma set: A small second order statistical region descriptor," in, *Computer Vision and Pattern Recognition, 2009. CVPR 2009. IEEE Conference on,* 2009, pp. 1802–1809.

[50] J. Fan, Y. Wu, and S. Dai, "Discriminative spatial attention for robust tracking," in, *Computer Vision–ECCV 2010,* ed: Springer, 2010, pp. 480–493.

Information Technology – Wan et al. (Eds)
© *2015 Taylor & Francis Group, London, ISBN 978-1-138-02785-5*

A method for the towed line array sonar's port-side/starboard-side discrimination

Zhaopeng Xu, Weiliang Zhu, Bo Li & Bin Li
Navy Submarine Academy, Qingdao, China

ABSTRACT: Focused on the problem of port-side/starboard-side discrimination when the towed line array sonar is straight, the target function whose parameters are target' bearing and range is established according to the theory of time domain beam-forming in the near-field zone. Target' bearing and mirror fountainhead' bearing could be discriminated based on the changing characters of target function' grey-scale picture when the array shape distorts. Thus a new method which could solve the problem of port-side/starboard-side discrimination is acquired. The validity of this method is validated by simulations.

1 INTRODUCTION

The array of towed line array sonar is usually a single line array, which is working in a passive way. It could acquire bearing of target, but not distance. Besides, the array is a line array which is made up of many non-directional hydrophones, so if there is a target at the port-side/starboard-side, there must be a mirror fountainhead at the starboard-side/port-side with the same relative bearing due to the directivity of cylindrical symmetry. The target's bearing and the mirror fountainhead' bearing could not be discriminated, that is the problem of port-side/starboard-side discrimination [1].

Generally, the low frequency target and towed line array sonar are in the near-field zone due to the big acoustic aperture of towed line array sonar, so the sound wave is spherical wave. Above condition is the precondition of the research.

Focused on the problem of port-side/starboard-side discrimination, a method is put forward. First, the frequency of target could be acquired after spectrum analysis based on the signals received from the each unit of towed line array sonar. Second, the delay time of signal received by each array unit could be acquired compared with the benchmark array unit. Third, the target function whose parameters are target' bearing and range is established according to the theory of time domain beam-forming in the near-field zone. Then two-dimensional searching is carried out based on the two parameters, target' bearing and mirror fountainhead' bearing could be discriminated based on the changing characters of target function' grey-scale picture when the array's shape distorts. This is the whole process of the method which could solve the problem of port-side/starboard-side discrimination.

2 FREQUENCY OF TARGET

Supposing the number of array units is M. In time domain, the signal received by each array unit is

$$X(t) = [x_1(t), x_2(t), \cdots, x_M(t)] \tag{1}$$

The frequency of target could be acquired after Fourier transform. In order to improve the quality of signal, signal $x_m(t)$ ($m = 1, 2, \cdots, M$) should be handled beforehand, $x_m(t)$ is divided to continuous N parts, each part and proximate part overlap each other in half. After Fourier transform, in frequency domain the signal is

$$Y(f) = [y1(f), y_2(f), \cdots, y_M(f)] \tag{2}$$

In order to improve the quality of signal, the Eq. (2) should be changed to Eq. (3)

$$Y(f) = \left| \sum_{m=1}^{M} y_m(f) \right| \tag{3}$$

Then the power density spectrum could be calculated, and the frequency f_m could be acquired.

3 PORT-SIDE/STARBOARD-SIDE DISCRIMINATION

Supposing signal $s(t)$ transfers from target to array unit 1, which is a benchmark array unit at the velocity v, compared to signal received by array unit 1, the signal received by array unit m is

$$x_m(t) = s(t - t_m) + q_m(t) \tag{4}$$

$q_m(t)$ is additive Gaussian noise, t_m is time delay which is affected by position of array unit. Fig. 1 shows the situation. Array unit 1 is the center of coordination, θ is the bearing of target. For convenience, the towed line array sonar is put on the negative half part of coordination in the y direction. The position of array unit m is (x_m, y_m), so t_m is

$$t_m = \left(R - \sqrt{(R\sin(\theta) - x_m)^2 + (R\cos(\theta) - y_m)^2}\right)\Big/v \quad (5)$$

R is the distance of target.

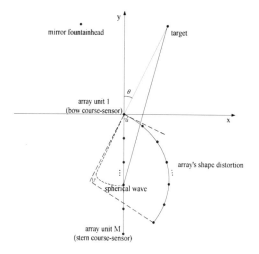

Figure 1.   Situation and array shape distortion picture.

Time domain beam-forming is the most effective method to calculate the bearing of acoustic target. After weighting as well as time delay and summation, $B(t)$ is

$$B(t) = \sum_{m=1}^{M} a_m x_m(t - t_m) \quad (6)$$

a_m is weighting coefficient. The target function $M(\theta, R)$ whose parameters are target' bearing and range is established .

$$M(\theta, R) = \left| \sum_{m=1}^{M} \sum_{n=1}^{N} y_{mn}(f_m) e^{i(2\pi f_m n\delta t + \varphi_m)} \right| \quad (7)$$

$$\varphi_m = 2\pi f_m \sqrt{(R\sin(\theta) - x_m)^2 + (R\cos(\theta) - y_m)^2}\Big/v \quad (8)$$

$y_{mn}(f_m)$ is the result of the part n signal received by array unit m after Fourier transform. δt is the sampling interval. ϕ_m is phase delay of signal received by array

unit m compared with array unit 1. Two-dimensional searching is carried out related to distance and bearing based on Eq. (7), then the maximum value of $M(\theta, R)$ is found. The parameters related to this value are the target's distance and bearing.

When the towed line array sonar is straight, the sonar likes a symmetrical axis, $M(\theta, R)$ could be the maximum value when the position is target or mirror fountainhead according to the structural features of function in the Eq. (8) as well as symmetrical features of function $\sin(\theta)$ and $\cos(\theta)$, Fig. 1 shows the situation. But the symmetrical features will be broken, when the towed line array's shape distorts [2]. Then the maximum value of $M(\theta, R)$ is exclusive, that is the position of target, not the position of mirror fountainhead, meanwhile the problem of port-side/starboardside discrimination is solved.

4   SIMULATION AND ANALYSIS

Supposing the number of array units is 101. The distance between each proximate unit is 2.5 m. Signal is sinusoidal with single frequency, the frequency is 100 Hz. SNR is -3 dB, the distance of target is 8km, bearing of target is 45°. Sampling frequency is 512 Hz and N is 2 during power density spectrum analysis. Fig. 2 shows the result. Obviously the frequency is 100 Hz.

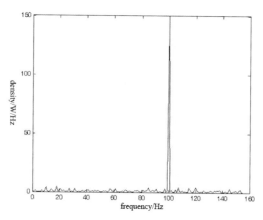

Figure 2.   Received signal's power density spectrum picture.

Then two-dimensional searching is carried out based on the f_m, θ in searching area is from 0° to 359°, R in searching area is from 5 km to 10 km. The result picture is showed in grey-scale mode in Fig. 3.

When towed line array sonar distorts, the position of each array unit could be known according to [2] and [4].

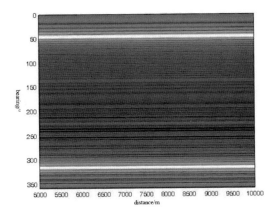

Figure 3. Grey-scale picture of target function when the towed line array sonar is straight.

In this simulation, supposing the position of each array unit could be acquired according to the bow course-sensor and stern course-sensor in Fig. 1. For convenience, supposing the shape of array is typical arc when the array shape distorts. In this simulation the result of bow course-sensor is 333° and the result of bow course-sensor is 23°, Fig. 4 shows the result.

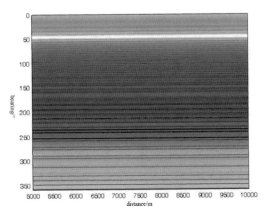

Figure 4. Grey-scale picture of target function when the array's shape distorts.

The analysis about result of simulation are as follows:

1 According to Figs. 3 and 4, $M(\theta, R)$ is sensitive to the change of parameter θ, and if θ does not change, $M(R, \theta)$ is affected by the change of R slightly.

2 According to Fig. 3, if the towed line array sonar is straight, the value of $M(\theta, R)$ is greater than other positions when θ is 45° or 315°. The two values are nearly the same, so it could not discriminate which side the true target is.

3 According to Fig. 4, if the array's shape distorts, the value of $M(R, \theta)$ is constant when θ is 45°. But the value of $M(R, \theta)$ becomes smaller obviously. So the bearing of target is 45°, not 315°.

4 Monte Carlo simulation is done 20 times when the array' shape distorts. Target's bearing and distance could be acquired from simulation. The bearing of target is 45°, this value is precise. The distance of target is 8300 m, this value is not precise, but acceptable.

5 CONCLUSION

Focused on the problem of port-side/starboard-side discrimination, the target function whose parameters are target' bearing and range is established, and then two-dimension searching is carried out. Target' bearing and mirror fountainhead' bearing could be discriminated based on the changing characters of target function' grey-scale picture when the array shape distorts. Thus the problem of port-side/starboard-side discrimination is solved. The validity of this method is validated by simulations.

REFERENCES

[1] He Xin-yi, Zhang Chun-hua, Li Qi-hu. Rough introduction of the towed linear array sonar and port/starboard discrimination methods. *Ship Science and Technology*, 2006, 28(5):9–14.
[2] He Xin-yi, Jiang Xing-zhou, LI Qi-hu. The array shape distortion of the towed line array and port/starboard discrimination. *Acta Acoustic*, 2004, 29(5):409–413.

Information Technology – Wan et al. (Eds)
© *2015 Taylor & Francis Group, London, ISBN 978-1-138-02785-5*

Comprehensive early warning system of coal and gas outburst

Qing He
China Coal Technology Engineering Group Chongqing Research Institute, Chongqing
Chongqing, China

ABSTRACT: The comprehensive early warning system of coal and gas outburst is a complicated system project, which has blended together the coal mine safety theory, the information technology, and the early warning technology. The paper discusses the principle, logic structure, and implementation process of gas outburst-warning devises, the software and hardware structure of the warning system, therefore, offers a theoretical thought and methods for the forecasting of coal and gas outburst. Practice proves that the comprehensive early warning system has a great significance in predicting advance calamity, forecasting, and reducing accidents.)

KEYWORDS: Gas Hazards; Comprehensive Early Warning System; Trend Warning; Status Warning

1 INTRODUCTION

Coal and gas outburst is an extremely complicated gas dynamic phenomenon, apparently showing that enormous amount of coal and rock mass is broken and thrown with gas emission in a very short time.[1] China is one of the countries that suffers from the most serious coal and gas outburst disasters. China's situation of outburst prevention is very serious because of coal mining depth increased every year with an average annual increase of 10–30 m.[2] The risk and complexity of coal and gas outburst is increasingly serious because of the ground stress, gas pressure, and content increased constantly. Meanwhile, the difficulty of outburst prevention and controlling is increasing too. The key for controlling coal and gas outburst is prediction and warning for hidden outburst hazard and accident, and then take effective control measures to eliminate the danger before the accident occurred. Early warning mechanism is the integration of early warning technology and crisis management system that can evaluate the possibility and harmfulness of hazard according to the monitoring information and collected data. Therefore, it can raise the alarm at an early age to prompt decision-makers to take pre-control countermeasures and prevent the accidents to a great extent.

Comprehensive Early Warning System (CEWS) for coal and gas outburst is a new technology and method to guarantee coal mine safety. It is also a complicated system, which is related to mine security theory, information technology, and early warning technical system. CEWS can send warning alarm in a primitive period to ensure managers to adopt pre-control measures and guarantee the security of system. To research and develop the early warning theory and

technology of coal and gas outburst and to build up comprehensive early warning system have become an effective approach and trend of gas preventing.

2 COMPREHENSIVE EARLY WARNING PRINCIPLE OF COAL AND GAS OUTBURST

There are many complex reasons and factors in coal and gas outburst. With the continuing movement of the work face, the related factors, such as the gas geology and coal seam occurrence, are in the dynamic changes of time and space. The outburst usually happens suddenly and turns to be accidents, which leads to have a major destruction. All kinds of relevant materials and information about the coal mine site and coal and gas outburst, which are scattered in the space have various information forms, are mastered in different functional departments. The realization of coal and gas outburst warning is based on the premise that we get real-time information, central management, and intelligence analysis.

The basic principle of coal and gas outburst early warning is: with the help of information and computer and network technology and based on the comprehensive and timely collection of the static and dynamic safety information and center control and information share, we make full use of the existing technical equipment and the management methods to make an intelligent, real-time and dynamic analysis about the working face dangerous condition and development trend with the development of the spatial mining and early warning analysis model, and to remind managers to take measures to prevent outburst, strengthen management and remove hidden dangers according to the automatically given warning information by the results of analysis.

3 COMPREHENSIVE EARLY WARNING METHODS OF OUTBURST

Comprehensive Early Warning of outburst mainly adopts three methods: Index Early Warning, Statistics Early Warning, and Model Early Warning.

3.1 Index Early Warning

CEWS sets up an early warning index system, monitors and analyzes a comprehensive index of early warning, and compares it with the pre-warning index. Finally, it sets out signals at different levels. Coal and gas outburst risk prediction and effect inspection more use predictor of critical common law, comprehensive index K D determination, single index method, R index method, and method of drilling index, etc. factors that influence the coal and gas outburst are buried depth, geological structure, gas content and operation technics, etc. Gas content and tectonic soft coals are outburst prediction indexes of the early warning system of coal and gas outburst.

3.2 Index Early Warning

According to the statistical results of relevant safety information, CEWS combines the statistical patterns of outburst. When the statistical results and changes satisfy the supposed statistical patterns and mode, CEWS will set out signals at different levels.

3.3 Index Early Warning

On the basis of the relational model of outburst danger and safety information, which is established at the preprocessing stage, CEWS builds up early warning models of outburst, such as Artificial Neural Network Mode 1 (Figure 1). With the help of the monitoring safety information, it takes model calculation and outputs the warning results at different levels.

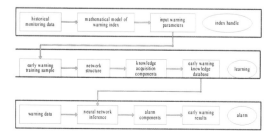

Figure 1. Early warning model of coal and gas outburst using the neural network.

4 IMPLEMENTATION PROCESS OF EARLY WARNING SYSTEM

According to the colliery safety monitoring system, maintenance and management of LAN and functional departments, and in order to realize the prediction of coal and gas outburst, we need to monitor the main dangerous resources of the coal and gas outburst to get the safety information, such as coal seam hosting, geological structure distribution, roadway layout, space position of face, information from gas-geology around the face and the information of stress concentration, gas emission, daily forecast, the performance of outburst prevention measures, effect test of outburst prevention measures, forecast equipment and violation, etc. The safety information is divided into static and dynamic safety information; the former is obtained by the digitalization process of mines, while the later is obtained by the monitoring system and LAN[3]. All of the information is mastered by the different departments, and we need to build a centralized, comprehensive database to share and manage and renew information, identify and analyze the safety information and calculate the related warning index, make grade division through various warning rules in the database. Information identification, analysis, and judgment should be completed by warning servers and professional analysis software, and the warning results should be issued by network, sound and light alarm, message and so on. The implementation process is shown in Figure 2.

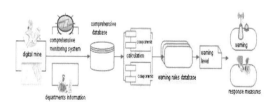

Figure 2. Implementation process of early warning for coal and gas outburst.

5 HARDWARE STRUCTURE DESIGN OF EARLY WARNING SYSTEM

The hardware structure of warning system is composed of monitoring system servers, office local area network, warning server, computer terminals, and notification appliance. According to the warning database management process, we integrate the related data down the mine form the different functional departments with the help of professional software, and construct the comprehensive warning platform of coal and gas outburst on the basis of realizing the centralized management and sharing of the safety information. Hardware is the carrier and environment support of warning software, whose logical structure is structured by data input, data transfer, data analysis, data output, etc. The main hardware equipments mainly include warning server, office local area network, office local area network, mobile notification appliance and so on [4].

Monitoring system server mainly provides real time information, i.e., security environment monitoring parameters. The office local area network mainly manages the transport of important dynamic security information that cannot be monitored, and the issue of warning results. A warning server mainly provides large database storage, safety information analysis, and warning instructions. Computer terminal takes on the warning subsystem operation, dynamic security information inputting, warning results show and history information querying of warning. The mobile notification appliance is used for sending out warning results and information to the mine technicians of safety management. Combining the above hardware equipment and environment, the hardware platform of the warning system is constructed, as shown in Figure 3.

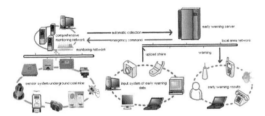

Figure 3.   Framework of hardware platform.

6   SOFTWARE STRUCTURE DESIGN OF EARLY WARNING SYSTEM

The warning software system of the coal and gas outburst is the kernel of the comprehensive early warning [5], which is composed of the gas geological information system, analysis of the gas emission system, dynamic management and analysis of outburst prevention, ventilation information management and dynamic network solution, management of geological measurement information, and Comprehensive analysis of early warning system, as depicted in Figure 4.

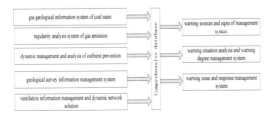

Figure 4.   Composition of comprehensive early warning software system.

6.1   Geological survey information management system

To measure basic parameters (gas content, gas pressure contour, etc.) of gas in coal seam, make a research on gas occurrence regularity in coal seam and geological distribution, set up strip dividing of advanced risk of face, making out dynamic gas geological map of mining face designing, tunneling, mining period. Gas pressure contour automatic update effect drawing is seen in Figure 5.

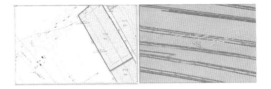

Figure 5.   Gas pressure contour automatic update effect drawing.

6.2   Analysis of gas emission

On the basis of the monitoring system of KJ90 seamless connection, CEWS can query and statistic gas monitoring data and curve, analyze gas emission during coal cutting, shooting and drilling, etc. Through the automatic conversion with the forecasting index of V30, KV, wave attenuation and coefficient, etc. CEMS can realize several indicators and continuous dynamic analysis of working face outburst prediction, as shown in Figure 6.

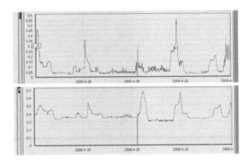

Figure 6.   Dynamic analysis system diagram of gas emission law.

6.3   Dynamic management and analysis of outburst prevention

Outburst prevention information (working face gas geological information, mining information, borehole data, Index distribution information sharing, etc), intelligent design of outburst prediction and drilling measures can be managed. Correlate analysis of forecasting index and its influence can also be performed. The system can analyze the measures of outburst prevention, early warning of outburst danger, and over mining, as shown in Figure 7.

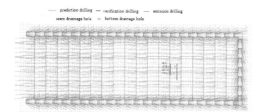

Figure 7. Dynamic management and analysis system of outburst prevention.

6.4 Ventilation information management and dynamic network solution

Master mine workings' ventilation parameters through determination and analysis. Establish stable evaluation model calculate of ventilation network and calculate the ventilation network dynamically. The system is shown in figure 8.

Figure 8. Ventilation information and dynamic network solution system.

6.5 Management of geological measurement information

Digital mine based on 3D GIS, can compile and output mine chart, manage geological drilling data, and wire information. It can manage mining working face information, inquiry and statistic geological measurement information, as shown in Figure 9.

6.6 Comprehensive analysis of early warning

The comprehensive early warning system integrates with all kinds of information to analyze outburst risk of working face[7], and provide the client with basic instruments of early warning information. It mainly includes early warning server and early warning management platform. The early warning server collects basic information input by each system in time.

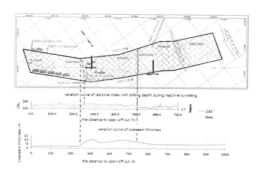

Figure 9. Gas geological measurement information system.

According to pre-determined rules, it analyzes outburst risk from actual risk state (state warning) and developing trend (trend warning). The early earning management platform is engaged in making out working face warning rules, setting expression form of early warning information, inquiring information and browsing relative mining sketch.

- Warning sources and signs management system

According to the organizational structure and work contents of coal mine functional departments, the early warning system establishes warning sources and signs management system as shown in Figure 10. Warning sources management system includes acquisition, transfer, and a collection of security information. The system uses a semi-automatic way to obtain and process information. Monitoring data collected automatically through the coal mine safety monitoring system, the production and management data are manually entered by corresponding functional departments. The warning sign analysis system, as an important link in the early warning process and by deeply excavating the basic data and safety information on the comprehensive early warning database, screens and recognizes effective information and calculates the corresponding early warning indicators.

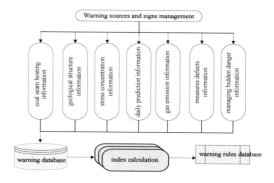

Figure 10. Management system of warning sources and signs.

310

- Warning situation analysis and warning degree management system

The structure of warning situation analysis and warning degree management is shown in Figure 11. The comprehensive early warning system consists of two parts: the warning management platform and early warning services. The warning management platform is mainly used to choose and custom warning rules and display forms of warning degree, query, and analyze warning results and information, checks the operating status of the early warning system, and gives warnings and suggestions of operating faults. Comprehensive early warning services include two parts: data preparation and implementation of early warning. Data preparation is to confirm the basic information and connection of monitoring equipment prior to the outburst warning of the working face, and then analyzes, in the mining process, its dangerous condition and trend based on early warning indicators, rules and models.

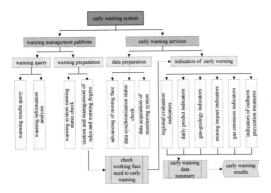

Figure 11. Architecture of warning situation analyze and warning level manage.

The warning degree directly sends dynamic safety signals to the user's working face. The display forms of warning degree employ two formats: the degree output based on a comprehensive model and the single factor time-order degree output[8]. The degree output based on a comprehensive model, in accordance with the dangerous scales of coal and gas outburst, demonstrates the status and trends of warning degree in words and in color respectively as shown in Table 1. For each result of warning degree, descriptive terms are given to explain the warning condition and analyze the basis in accordance with sources, signs, and conditions of early warning[9].

- Warning issue and response management system

The warning issue system is mainly used for sending warning signals to relevant departments and personnel when the early warning degree amounts to the set value,

Table 1. Emergency level of early warning.

Type of warning results	Grade of warning results			
Status warning	*Normal*	*Attention*	*Dangerous*	*Catastrophe*
	Green		Orange	Red
Trending warning				

and customizing different warning issue measures according to requirements of warning degree. When the warning results of working face are at the "normal" level, the warning issue system will show the warning degree in green indicating no warning and will not send out the warning information. When the warning results of working face are at the "abnormal" level, the system will send different signals according to different user types. For general users, the system will show the warning degree in brunet fonts and colors, and issue warning through the web page. For main users such as the technical personnel of outburst prevention, warning results will be issued by sound, photoelectric signals and short messages through the warning management platform. For key users such as leaders of coal mines and persons in charge of outburst prevention, the warning will be issued in real time by short message services. The warning message sending system is shown in Figure 12.

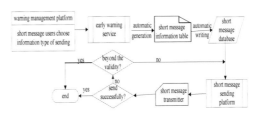

Figure 12. Logical structure of warning message sending system.

The warning response management system consists of two parts: daily monitoring and risk management. The daily monitoring system exercises special monitoring and control over the signs of accidents, which is realized through the warning sources and sign management system and the warning situation analysis system. The risk management system is employed to take measures when outburst risks exist or disasters occur on the working face. First, by using the risk management system, it sends warning information to different departments and leaders in accordance with the warning degree information. Then, on the basis of the overall measures taken and the effects of outburst-removing measures, in particular, it reanalyzes the safety status and issues new warning results in order to determine whether mining can be allowed. The risk management system structure is shown in Figure 13.

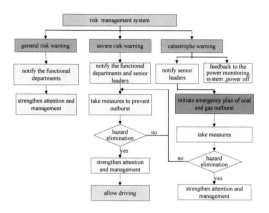

Figure 13. Structure of emergency management for early warning.

7 CONCLUSION

The warning system of coal and gas outburst is a new method and a new means to guarantee safety in coal mines, and also a complicated and systematic project, which combines the safety theory of coal mines, the information technology, GIS technology and the early warning theory. On the basis of a systematic study of the specific characteristics and laws of mining disasters, it employs safety monitoring systems of coal mines and local network conditions and adopts a series of professional analysis software as its carrier. As a technological system of early warning that accords with mining production, it is of great realistic significance for forecasting and reducing accidents of coal and gas outbursts.

ACKNOWLEDGMENT

This work was financially supported by the national science and technology supporting plan of the twelfth five-year of China (2012BAK04B01).

REFERENCES

[1] Chongqing Research Institute of China Coal Technology & Engineering Group Corporation. Research report on early warning technique of ventilation gas hazards [R].2010.In Chinese.
[2] Zhao Xusheng, Zou Yunlong. Accident cause analysis and countermeasure of coal and gas outburst nearly two years of our country[J]. Mining Safety and Environmental Protection. 2010.37(1):84–87.
[3] Liu Cheng, Zhao Xusheng, The comprehensive solutions of gas disaster warning technology and computer system construction. Mining Safety and Environmental Protection. 2009.36(8): 60–64.
[4] Liu Xianglan, Zhao Xusheng etc.Research on early warning system for coal and gas outburst[A]. ICMHPC 2010, 560–568.
[5] Administration of National Security http: //www .chinasafety.gov.cn.
[6] Zhao Xusheng, Kang Xiaoning,etc.Coal and gas outburst comprehensive warning system and warning methods[P]. CN101550841,2009-10-07.
[7] Hou Shaojie, Forecast model of coal and gas outburst indexes and testing software design[J]. Coal Mine Machinery 2010.31(3):71–73.
[8] Guo Deyong, Early warning of coal and gas outburst by GIS and neural network[J]. Journal of University of Science and Technology Beijing, 2009.31(1):15–18.
[9] Zhao Xusheng. Study on comprehensive early warning methods for coal and gas outburst[D]. Shandong University of Science and Technology, 2012.

Information Technology – Wan et al. (Eds)
© 2015 Taylor & Francis Group, London, ISBN 978-1-138-02785-5

Signal processing technology in fault diagnosis

Rui Gao & Xiao Yu
Qingdao University of Science and Technology Automation and Electronic Engineering Qingdao,China

ABSTRACT: An internal exception occurs when machinery is usually accompanying vibration increases and changes in work performance. According to statistics, 60% are reflected by the above mechanical failure through vibration. Without downtime and disintegration you can pass judgement on the nature of the degree of deterioration and failure by means of mechanical vibration signal measurement and analysis. Fault diagnosis and signal processing are meant to prevent accidents and ensure the safety of people and equipment and raise the level of equipment management and promote reform as well as to improve the efficiency of the equipment maintenance system.

Diagnosis method of signal analysis and processing collect status information through the machinery and equipment placed on the sensors. Analysis and processing extract on the situation and trouble-free operation of the device. Its key technology is the method of signal analysis and processing. There are main domain, frequency domain, frequency domain, time-frequency analysis.

KEYWORDS: *Troubleshooting; signal processing; bearing failure; envelope spectrum analysis*

1 INTRODUCTION

With the rapid development of science and technology in modern industry, the production equipment is increasingly large-scale, integrated, high speed, automation and intelligent, and the equipment is becoming more and more important in the production. The management of equipment also put forward higher requirements. Failure will cause enormous economic loss and may lead to serious social harm. Managing advanced production equipment plays an important role in improving their reliability and effectiveness and reducing equipment repair costs and the production cost of enterprises and improves the competitiveness of enterprises.

Fault diagnosis technology has made significant progress from theory to practice since seventy times develops, which provides an important basis for the management and repair equipment, significant economic and social benefits. Because of a variety of equipment fault diagnosis object diversity, complexity there are a lot of theories and methods to further improve. In this paper, we mainly research on signal processing technology in fault diagnosis.

Signal analysis and processing method of diagnosis acquire mechanical equipment state information through the sensors on the mechanical equipment. And then we can obtain the results by analyzing the process and the running situation of the equipment and judge whether there are faults, fault occurrence and development status. The key technology is the analysis method of signal processing. There are time domain, frequency domain, cepstrum domain, time–frequency analysis.

Vibration characteristic parameters are the peak to peak, mean square amplitude, mean, variance and standard deviation, three moment, four moment, waveform factor. These feature parameters can be used for on-line monitoring due to measurement comparison directly and can be used as feature various other diagnosis methods of extraction parameters, auxiliary diagnosis.

Frequency domain analysis is mainly through a transformation, the vibration signal from time domain to frequency domain, a feature extraction method, the main treatment method with estimation method and modern spectrum estimation method of classical spectrum.

Classical spectral method includes periodogram method, correlation method and some other improved algorithm. It intends to the modern law of maximum entropy spectral estimation and the ARMA method and the minimum variance method. Classical spectral method can be calculated quickly by FFT. It has a clear physical meaning, but the spectral resolution is very low and the amount of data is needed. In addition, the variance of performance is not so good. The modern spectrum analysis method has higher resolution and the requirement of data quantity is rather less, but prone to distortion and has low signal-to-noise ratio.

2 MATHEMATICAL MODELING BEARING FAULT DIAGNOSIS

2.1 Analysis of rolling bearing defects in mathematics

Electric power enterprises will directly result in the abnormal motor mechanical equipment shutdown and it influences the equipment and personal safety and reduces the equipment operation efficiency. AC motor fault analysis and diagnosis, making corresponding treatment, have a very important role in judging and handling equipment fault and hiding trouble to the safe and stable operation of equipment.

Fig. 1 D is for rolling bearings for rolling bearing diameter, D is the diameter, α is the contact angle, Ra is the radius of gyration A point, Rb is the radius of gyration at B point.

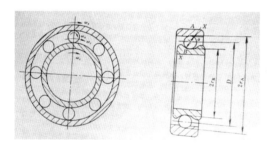

Figure 1. Motor bearing rolling element profile no assumption by the deformation of the available.

$$R_a = \frac{D}{2}\left(1 + \frac{d}{D}\cos\alpha\right) \tag{1}$$

$$R_b = \frac{D}{2}\left(1 + \frac{d}{D}\cos\alpha\right) \tag{2}$$

This makes the ω_r outer raceway (ring) the rotational angular velocity, angular velocity ω_a as the inner raceway, ω_c for the angular velocity of the mass center of the ω_s rolling, rolling body for rotation, f_r, f_a, f_c rotation frequency of each. The inner raceway and rolling body without the assumption of sliding contact points in the A available.

$$R_a\omega_c + \frac{1}{2}d\omega_s = R_a\omega_a \tag{3}$$

By the outer raceway and rolling body without the assumption in sliding contact points available.

$$R_b\omega_c - \frac{1}{2}d\omega_s = R_b\omega_r \tag{4}$$

It can be determined by solving the resulting equations simultaneously.

$$\omega_s = \frac{2}{d}\frac{R_a R_b}{R_b + R_a}(\omega_a - \omega_r) \tag{5}$$

$$\omega_c = \frac{(R_a\omega_a + R_b\omega_r)}{R_b + R_a} \tag{6}$$

The angular velocity is expressed as a circular frequency form, and (1) (2) two type substitution, then two formula can be written as

$$f_s = \frac{D}{2d}(f_a - f_r)\left(1 - \frac{d^2}{D^2}\cos^2\alpha\right) \tag{7}$$

$$f_c = \frac{1}{2}\left[f_a\left(1 + \frac{d}{D}\cos\alpha\right) + f_r\left(1 - \frac{d}{D}\cos\alpha\right)\right] \tag{8}$$

2.2 The rotation frequency characteristic of rolling body defect

The rotation characteristic frequency of rolling body has some defect. When the rolling body has only one defect and is shown in Fig. 1 in the X–X interface. Because of the defect, the each rolling body rotate in a week, and the impact on the hob once. When the defect increases, the impact becomes more obvious. Defect impact frequency and rolling body should be relative, so it is called a characteristic frequency of rolling defects. It is denoted by f_b. Its value is:

$$f_b = f_s = \frac{D}{2d}(f_a - f_r)\left(1 - \frac{d^2}{D^2}\cos^2\alpha\right) \tag{9}$$

When the bearing outer ring is stationary, $f_a = 0$

$$-\frac{D}{2d}f_r\left(1 - \frac{d^2}{D^2}\cos^2\alpha\right) = f_b \tag{10}$$

Here appears minus because of the speed trend contrary to the assumed direction. We add the absolute value:

$$\left| \frac{D}{2d}(f_a - f_r)(1 - \frac{d^2}{D^2}\cos^2\alpha) \right| = f_b \qquad (11)$$

When $f_a = 0$

$$\frac{D}{2d} f_r(1 - \frac{d^2}{D^2}\cos^2\alpha) = f_b \qquad (12)$$

When the inner ring outer ring movement is still, $f_r = 0$

$$\frac{D}{2d} f_a(1 - \frac{d^2}{D^2}\cos^2\alpha) = f_b \qquad (13)$$

3 ENVELOPE SPECTRUM SIGNAL PROCESSING TECHNOLOGY IN ROLLING BEARING FAULT DIAGNOSIS

3.1 Envelope spectrum analysis

The line spectrum can be used to form the amplitude modulation signal demodulation. In this case, the high frequency carrier signal frequency modulation signal separates. The line can be created by different methods. This paper introduces it through the modulation and demodulation, frequency filtering, envelope spectrum rectifying.

Rectified signal is through the half wave rectifier which is a full wave rectification or edge grinding. This means that the separation is through demodulation and rectifying the carrier frequency band. Low pass filtering will inhibit the carrier frequency and the other is through rectification that is caused by high frequency signal. Rectification of bandpass filter ensures only the carrier frequency demodulation. The low frequency signal is rectified and the other cannot pass through the low pass filter low frequency signal, which can be effectively suppressed.

Fig. 2 shows rectifying process, including the line spectrum:

Half wave rectifier can demonstrate the envelope spectrum image well. Using sine function is an example:

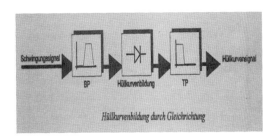

Figure 2. Line spectrum obtained by the rectification process include.

$$s(t) = s_o \sin(2\pi f_d t) \qquad (14)$$

Fig. 3 shows that the sine signal and square wave signal are multiplied by the half wave rectifier.

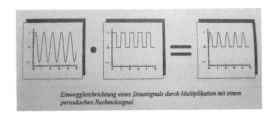

Figure 3. Sine and square wave signals are multiplied by the half-wave rectifier.

In the half wave rectification process, all the signals are filtered out. The following is the mathematical derivation of the sinusoidal half wave rectifier.

The following frequency change type is:

$$s_{gl}(f) = \left((so[\delta(f+f_d)+\delta(f-f_d)])*\left[si\left(\frac{\pi f}{2f_d}\right) \circ III(\frac{f}{f_d})\right]\circ e^{-j\pi\frac{f}{2f_d}}\right) \qquad (15)$$

Among them

$$III(\frac{f}{f_d}) = \delta(f-f_d)*III(\frac{f}{f_d}) = III(\frac{f-f_d}{f_d}) \qquad (16)$$

$$sgl(f) = \left(so\left[si\left(\frac{\pi(f+fd)}{2fd}\right)+si\left(\frac{\pi(f-fd)}{2fd}\right)\right]\circ III(\frac{f}{f_d})\right)\circ e^{j\pi\frac{f}{2f_d}} \qquad (17)$$

3.2 Envelope spectrum analysis of rolling body at fault

In the raceway, each of the inner and outer rings is likely to cause the fault of rolling. So the pulse repetition frequency in most cases is two times bigger than the rolling bearing rolling frequency. On this assumption, the rolling body is always around an axis of rotation and the shaft on the rail failure appears. This is only applied to single axis rotation, such as cylinder, piston, etc. Proved by experiments, the ball damage is irregular, so a damage diagnosis is very important.

The following is the bearing fault mathematical analysis and envelope spectrum:

$$x(t) = \left(a_w(t) \bullet \left(n \bullet 2 \bullet f_w \bullet III(2f_w t) \right) \right) * g(t) \tag{18}$$

4 HYDRAULIC SERVO SYSTEM SIGNAL PROCESSING

Hydraulic servo system is based on an automatic control system of hydraulic transmission principle. The hydraulic servo systems have fast response, load stiffness, large control power and other unique advantages in industry. National defense industry has widely used. With the rapid development of automation technology, the system starts toward the high power, high efficiency and high integration development. Control of hydraulic servo system has become mature development of science and technology and the control has experienced the classic PID control, adaptive control. At present, the fuzzy control and neural networks control of nonlinear control technique have been well applied in servo control system.

Control of hydraulic servo system is influenced by many factors, such as the external load, environmental factors, and the inherently nonlinear systems. Using the classical control theory, the control strategy can achieve the desired requirements. However, in recent years with the machining accuracy and the increased response speed requirements control of hydraulic servo system requirements have been raised to a new level. In this paper, a mathematical model is developed for the hydraulic servo system. The control theory of linear quadratic optimal control in two applications in the hydraulic servo system and weighted matrix affects the system dynamic performance optimization.

4.1 The mathematical model of hydraulic servo system

Hydraulic servo system is composed of hydraulic pump, hydraulic control valves, actuators, control object.

Ignoring the hydraulic system leakage, excluding the elastic case, the following is the model of the hydraulic system of mathematics.

$$Q_m = D_m \omega + \frac{V_1}{\beta} \cdot \frac{dp}{dt} = D_m \frac{d\theta_m}{dt} + \frac{V_1}{\beta} \cdot \frac{dp}{dt} \tag{19}$$

$$pD_m = J \frac{d^2\theta_m}{dt^2} + B \frac{d\theta_m}{dt} + C\theta_m + T_L \tag{20}$$

Hydraulic control valve displacement is proportional to the input signal:

$$X_s = k_2 u \tag{21}$$

$$Q_b = K_q \alpha \tag{22}$$

Hydraulic cylinder displacement and the control valve are proportional to the displacement:

$$X = k_3 X_s \tag{23}$$

The displacement of the hydraulic cylinder with variable pump swashplate angle is proportional to:

$$X = k_1 \alpha \tag{24}$$

The inherent frequency of the system and the damping ratio are expressed as:

$$\omega_n = \sqrt{\frac{\beta D_m^2}{V J_1}} \tag{25}$$

$$G(s) = \frac{\theta_m}{u} = \frac{k_q k_2 k_3 / D_m k_1}{s \left(\frac{s^2}{\omega_n^2} + 2 \frac{\xi_n}{\omega_n} s + 1 \right)} \tag{26}$$

4.2 Two linear quadratic optimal control theory

Make $\lambda(t) = K(t)X(t)$

The $\lambda(t)$ problem is transformed into solving the matrix $K(t)$ differential equation, $\lambda(t) = K(t)X(t)$ is substituted into the above formula. Function matrix $K(t)$ foot (Riccati equation) is:

$$K(t) = -K(t)A(t) - A^T(t)K(t) +$$
$$K(t)B(t)R^{-1}(t)B^T(t)K(t) - Q(t) \tag{27}$$

$G(t) = -R^{-1}(t)B^T(t)K(t)$ is the optimal feedback gain matrix.

4.3 Design and simulation of two quadratic optimal controllers for linear

The formula can get the system state space expression:

$$\begin{bmatrix} \dot{x_1} \\ \dot{x_2} \\ \dot{x_3} \end{bmatrix} = \begin{bmatrix} 0 & 1 & 0 \\ 0 & 0 & 1 \\ 0 & -\omega_n^2 & -2\xi\omega_n \end{bmatrix} \cdot \begin{bmatrix} x_1 \\ x_2 \\ x_3 \end{bmatrix} + \begin{bmatrix} 0 \\ 0 \\ K_4\omega_n^2 \end{bmatrix} U$$

$$Y = \begin{bmatrix} 1 & 0 & 0 \end{bmatrix} \begin{bmatrix} x_1 \\ x_2 \\ x_3 \end{bmatrix}$$

The natural frequency of the system:

$$\omega_n = 250, \xi_n = 0.4,$$

$$k_q k_2 k_3 / (D_m k_1) = 100, R = 1, q_1 = 5, q_2 = 0, q_3 = 0.$$

The general weighted matrix Q, R is set to:

$$Q = \begin{bmatrix} q_1 & 0 & 0 \\ 0 & q_2 & 0 \\ 0 & 0 & q_3 \end{bmatrix}, R = 1$$

4.4 Simulation analysis

Matlab programming can get the optimal feedback gain matrix $G(t) = \lfloor 2.2361 \quad 0.0097 \quad 3.2183 \times e^{-5} \rfloor$. The step response of the system as is shown in Fig. 4.

In order to study the weighted matrix Q parameters influence on the optimal controller, we change the Q parameter. The $q_1 = 5, q_2 = 1, q_3 = 1$ optimal feedback gain matrix is:

$$G(t) = \begin{bmatrix} 2.2361 & 0.9913 & 5.3213 \times e^{-4} \end{bmatrix}$$

The step response of the system is shown in Fig. 6. The state of the system x_1, x_2, x_3 has different weighting matrix parameters under the condition of step response curve as shown in Figs. 5 and 7.

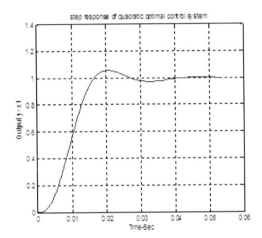

Figure 4. Step response of quadratic optimal control system.

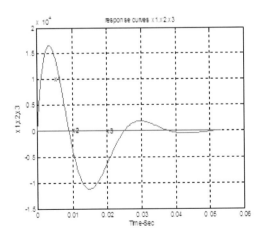

Figure 5. Response curves of $x1, x2, x3$.

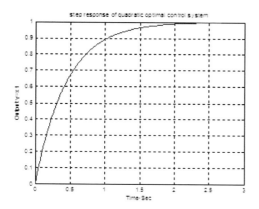

Figure 6. Step response of quadratic optimal control system.

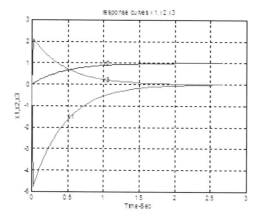

Figure 7. Response curves of x1,x2,x3.

ACKNOWLEDGMENT

First and foremost, I would like to show my deepest gratitude to my supervisor Dr. Zhongdong Liu a respectable responsible and resourceful scholar who has provided me with valuable guidance in every stage of the writing of this thesis. Without his enlightening instruction and impressive kindness and patience, I could not have completed my thesis.

I would also like to thank all my teachers who have helped me to develop the fundamental and essential academic competence.

Last but not least, I would like to thank all my friends, for their encouragement and support.

REFERENCES

[1] Tetsuhiko Yamamoto, Shin-ichi Hanada, Kunihiko Nakazono of inverted pendulum evolved by GA with rough evaluation. 1995,61(591), pp. 4276–4281.
[2] John H. Davies. MSP430 MicrocontrollerBasics. USA:Newnes, 2008 Spong M W,Block D. The penduot: a mechatronic system for control research and education. USA:New Orleans, 1996, pp. 256–278.
[3] D.boulahbal, G. Farid, F. Ismail. Amplitude and phase wavelet maps for the detection of cracks in geared systems. Mechanical Systems and Signal Proces sing,1999, 13(3), pp. 423–430.
[4] Chris Nagy.Embedded System Design using MSP430 series. USA:Newnes, 2003.
[5] Murphy Eva, Colm Slattery. Ask The Application Engineer--33 All About Direct Digital Synthesis. Analog Dialog, 2004, 38, pp. 1–5.
[6] Spong M W.The swing uo control problem for the Acrobot. Control Systems Magazine. 1995, 15(1), pp. 49–55.

Information Technology – Wan et al. (Eds)
© *2015 Taylor & Francis Group, London, ISBN 978-1-138-02785-5*

Screening and identification of osteosarcoma cell aptamers

Hua Wang
Dep. of Orthopaedics, Affiliated Hospital, Beihua University, Jilin, Jilin, China

Ji Liang
Reproductive Medicine Center, The First Hospital of Jilin University, Changchun, Jilin, China

Na Sun, Qilai Wang, Yuan Gao & Wei Xia
School of Laboratory Medicine, Beihua University, Jilin, Jilin

ABSTRACT: In this study, we reported that oligonucleotide ligands single strand DNA (ssDNA or aptamer) binding to osteosarcoma cells with high affinity and specificity had been screened and identified through the systematic evolution of ligands by exponential enrichment (SELEX) technology. During the study, the osteosarcoma (U2-OS) cells were used as positive target cells and fibrosarcoma cells as negative-screening cells. The enrichment of selected ssDNAs was monitored by fluorescence spectroscopy scanning technology. Screening was ended when fluorescence spectra did not show obvious shift between the eleventh and thirteenth round of selection. The ssDNA pools from the thirteenth round of selection were cloned. The ability of recognizing and binding to the cells from clinical cases with osteosarcoma had been identified. Results show that a novel method used for diagnosis of osteosarcoma has been explored successfully. Fluorescence scanning technology has high sensitivity, selectivity and the advantages of wide linear range.

1 INTRODUCTION

Small oligonucleotide ligands of DNA or RNA have been extensively researched and used in the medical and other bioscience fields [1–5]. Because small oligonucleotide ligands of DNA or RNA can fold into 3D structure and recognize and interact with target, shch as amino acids [6,7], peptides [8], bacteria [9], proteins [10–12], and a whole cells [13–18], they show abilities of selective binding and affinity same as antibodies. Molecules of oligonucleotide DNA or RNA can be chemically synthesized and modified to incorporate some functional groups, such as biotin, amino and carboxyl which do not affect the recognition and binding of the aptamers to the target. In addition, oligonucleotide ligands of DNA or RNA have low toxicity, immunogenicity and blood residence time compared with antibodies [14]. Therefore, small oligonucleotide ligands of DNA or RNA have been used for probers or aptamers of tumour marker, such as platele derived growth factor (PDGF), vascular endothelial growth factor (VEGF), human epidermal growth factor receptor 3(HER3). Some aptamers were screened from whole cancer cells, such as acute lymphoblastic leukemia cells, liver cancer cells and small cell lung cancer [14]. Aptamers were usually generated by SELEX. Capillary electrophoresis-SELEX [19–20], complex target-SELEX

[21,22], Subtractive-SELEX [23] and cell-SELEX have been developed according to the purpose of research and application. Cell-SELEX technology includes the incubation of the ssDNA or RNA library pool with target cells. After incubation, the sequences binding to the target cells were recovered from cells by heating the complex of cells-ssDNA. Sequences which had been recovered were amplified by PCR to generate a new ssDNA or RNA pools for next round of selection. This process was repeated until the sequences that specifically recognize the target cells were enriched.

Osteosarcoma is the most common bone cancer in children [24,25]. Age at diagnosis is around 15. Boys and girls are just as likely to get this tumor until the late teen years, when it occurs more often in boys. Osteosarcoma is also common in people above 60 years, and the cause is not known. In some cases, osteosarcoma runs in families. At least one gene has been linked to an increased risk. This gene is also associated with familial retinoblastoma. This is a cancer of the eye that occurs in children. At present, diagnosis of osteogenic sarcoma is dependent on the analysis by synthesis of imaging, serology and pathology. This way of analysis by synthesis needs longer time for diagnosis. Therefore, a novel diagnosis method needs to be developed. Due to abnormal expression of cancer-related genes and the changing proteins structures [26], as same as other cancer cells, osteosarcoma cells were endowed with some changes in morphology and biological behavior. More complicated epitopes on the surfaces of cells would be shown than normal cells. So it is possible of using cell-SELEX technology to screen sequences of small oligonucleotide ligands of DNA that recognize and bind to osteosacoma cells with high specificity and affinity.

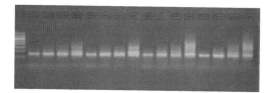

Figure 1. Agarose gel electrophoresis of PCR products with various cycles and template volume of selected DNA library pool.Lane M :100bp ladder; 6, 8,10 and 12 are cycles of PCR; 5, 10, 15 and 20 mean percent of the DNA template.

2 RESULTS

2.1 *Optimization of PCR condition*

After incubation of cells with ssDNA library pools, complex of ssDNA-cells was produced, and needs to be separated by boiling and centrifugation. Supernatant containing ssDNA was amplified by PCR to produce dsDNA, which were used for generation of a new ssDNA library pool. To ensure PCR products dsDNA without unspecific amplification, the number of cycles and volume of amplified DNA selected pool in the PCR reaction system need to be optimized. The image of agarose gel electrophoresis shows the optimization results as shown in Fig. 1. It is confirmed that 6 cycles and 20% of template show the band clearly and without unspecific amplification, so we finished the preparation of dsDNA by PCR under same condition. Results are shown in Fig. 2.

2.2 *Fluorescence spectrometry assay*

In order to assess the process of enrichment, the selected ssDNA pools from rounds of third, sixth, ninth, eleventh, twelveth and thirteenth were respectively labeled with FITC and determined by fluorescence spectrometry. Emission spectra are shown in Fig. 3. Sequences of oligonucleotide DNA from the thirteenth round selection were cloned into DH5a with pGM-T vectors. Plasmids containing sequences selected were further checked by PCR. Positive

plasmids with 76nt DNA sequences were sequencing and preserved.

2.3 *Affinity assay*

Binding three sequences to three samples from patients with osteosarcoma without pathological types to confirm the specificity and affinity of oligonucleotides ssDNA selected with the samples from clinical cases, resulte as shown in Fig4. Imaging of agarose gel electrophoresis shown sequences from No.2, 6 and 11 plasmids are efficient aptamers for recognizing and binding to sample1; Sequences of No.2 and 11 can also bind to sample2; Sequences of No.6 and 11 can also tie together with sample3.

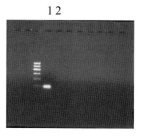

Figure 2. A agarose gell electrophoresis of the PCR product band clearly and without unspecific amplification. Lane 1: 100 bp ladder; lane 2: PCR product.

3 DISCUSSION

SELEX technology is an ideal strategy to discover new markers of tumour. Aptamers of DNA or RNA generated by this technology show their characteristics of higher recognition capability, selective binding and affinity likened to antibodies. Other special features of aptamers are that they can be synthesized, modified and preserved for a longer time. Labeled or unlabeled aptamers could be detected with methods of fluorescence, colorimetry or agrose gel electrophoresis. Because aptamers are stable at room temperature, they can be immobilized in microchip or microfluid using early stage diagnosis of tumors. Technologies of histological section-SELEX, composite target-SELEX, chip-SELEX, live cell-SELEX and other SELEX being developed largely propel the application of nucleotides aptmers. On the basis of the theories described, it is possible to generate any aptamers aiming at various targets. Therefore, we tried to screen specific sequences of oligonucleotides DNA as diagnosis probers thereby recognizing and binding to osteosarcoma by SELEX. During selection, we paid attention to optimization of PCR condition for generation of ssDNA pools needed by subsequent round of selection. 6–8 cycles and 10–20% volume DNA template was desired for PCR. Fibrosarcoma cell line had been used for negative selection to fill out the sequences that bind to the cells of both osteosarcoma (U-2 OS) and fibrosarcoma. The negative selection accelerated the process of enrichment of oligonucleotides ssDNA library pools.

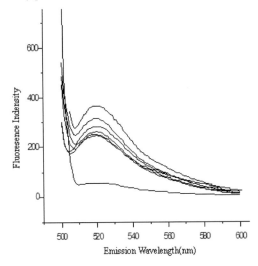

Figure 3. Fluorescence spectrometry assay to assess the binding of selected ssDNA pools to osteosarcoma cells. From top to bottom is round 3; round 6; round 9; round 11; round 12, round 13 and control.

The process of enrichment of oligonucleotides ssDNA library pools was usually monitored by flow cytometry. In this study, we tried to use fluorescence spectrometry for binding assays. Fluoresence spectrophotometer excitation wavelength scanning range is generally 190 nm–650 nm, emission wavelength range is 200 nm–800 nm. In the nonsynchronous scanning fluorescence measurement, fluoresence signal intensity is a function of the emission wavelength. Results showed that it was an effective method to assess the process of enrichment via fluorescence spectrometry assays. After screening of oligonucleotides ssDNA library pools binding to U2-OS cells, the selected ssDNA library pools were incubated with cells from patients with osteosarcoma without pathological details. Sequences binding to cells were recovered and amplified by PCR. Selective binding and affinity are assessed according to the agarose gel electrophoresis image. No doubt that we have generated the oligonucleotides probers for recognizing and binding to sample cells from clinical cases successfully. But these results also suggest that surface of cancer cells with intricate epitope would show difference of affinity with same sequence of aptamers. We need to classify these sequences selected to the different families matching the pathological types.

4 MATERIAL AND METHODS

4.1 Cell lines

Osteosarcoma U-2 OS cells for positive selection was maintained in McCoy'5A medium containing 10% fetal bovine serum. Fibrosarcome cells for negative selection were grown in DMEM medium containing 10% fetal bovine serum. Cells and culture medium were all purchased from Boster bio-engineering limited company.

4.2 Primers and library

Primers and library sets 5′-atccagagtgacgcagca-40nt-t ggacacggtggcttagt-3′; 5′-FITC-atccagagtgacgcagca-3′ and 5′-bitin-actaagccaccgtgtcca-3′ had been synthesized and purified (Shanghai Songon Biotech Co.) according to Ref. [14]. 40nt means random region of four deoxyribonucleotides.

4.3 Positive and negative selection

Positive selection: Osteosarcoma U-2 OS cells were cultured in 100 mm × 20 mm (first round selection) plates 24 h before using at a density of 5×10^6 cells (first round selection) in McCoy'5A medium containing 10% fetal bovine serum. After 24-h culture, cells were washed three times with washing buffer [1 l of DPBS containing 4.5 g of glucose and 5 ml of 1 M $MgCl_2$]. After washing, cells were incubated with cold ssDNA library pool in 1 ml of binding buffer [1 l of DPBS containing 4.5 g of glucose, 1 g BSA, 100 mg tRNA (Invitrogen) and 5 ml of 1 M $MgCl_2$] at 37 °C, 50 r/min shaking for 1 h. (10 nmol of ssDNA library for the first round of selection, 1000 nmol of that for the subsequent round of selection). Then, cells were washed three times with washing buffer, and one time with DNAse-free water. Detached cells from culture dish by cell scraper and transfer cells into a 1.5 ml tube. Boil tube at 95 °C for 10 min and collect supernatant by centrifugation at 13,000 × g for 5 min at 4 °C. Oligonucleotides ssDNA sequences in the supernatant were amplified by PCR.

Negative selection: Fibrosarcoma cells were cultured in 60 mm × 15 mm plates 24 h before using at a density of 3×10^6 cells in DMEM medium containing 10% fetal bovine serum. After 24-h culture, cells were incubated with cold oligonucleotides ssDNA library pool from positive selection (1000 nMol final concentration) in 600 µl of binding buffer, shaking for 40 min at 37 °C, 50 r/min. After incubation, collect the binding buffer from culture dish. This liquid that contains sequences of ssDNA only recognizing and binding to U-2 OS cells was amplified by PCR to generate new ssDNA pool for next round of positive selection. Negative selections were finished by the fifth and eighth round of selections.

4.4 Optimization of PCR condition

Optimization of PCR condition is required for each round of selection, such as 4 tubes are prepared with and 100 μl mixture in each tube. Each PCR mixture contained 50 μl 2 × Taq MasterMix (Taq polymerase, dNTP and buffer) (Tiangen Biotech Co. LTD.), 0.5 μM concentration of each primer. 5 μl , 10 μl, 15 μl and 20 μl of ssDNA templates separately in 4 tubes. H$_2$O was used for supplementary to total volume of 100 μl. Mix thoroughly in each tube and divide 100 μl of mixture into four PCR tubes evenly. PCR was performed for 6, 8, 10 and 12 cycles with denaturing at 95 °C for 30 s, annealing at 56 °C for 30 s, elongation at 72 °C for 30 s, followed by one cycle at 72 °C for 3 min. After amplification, load of 15 μl of PCR products in each lane of 3% agarose gel, electrophoresis at 100 V. Optimization of PCR condition was confirmed from agarose electrophoresis gel image, as shown in Figs. 1 and 2.

4.5 Sense ssDNA library pools generation

PCR products dsDNA from each round of selection needed to generate a new sense ssDNA library pool which is used for next round of selection. The methods are as follows: unlabeled sense primers and biotin-labeled antisense primers were used for PCR to generate biotin labeled dsDNA. Added 200 μl of streptavidin sepharose beads suspension (HP 17-5113-01 GE Healtheare) into 3 ml chromatographic column, washed the beads with 3–4 ml of 1× DPBS, loaded the PCR products to beads, washed beads with 3 ml of 1 × DPBS, added 550 μl of 200 mM NaOH, collected the eluate containing unlableted sense strand DNA(new ssDNA pool). This new ssDNA pool was purified, quantified and dried.

4.6 Monitor the progress of screening

Osteosarcoma U-2 OS cells were cultured with McCoy'5A medium containing 10% fetal bovine serumin in 60 mm × 15 mm plates 24 h before using at a density of 3 × 10^6 cells. Cells were washed three times with washing buffer, incubated with cold oligonucleotides ssDNA library pool in 600 μl binding buffer (1000 nMol final) at 37 °C, 50 r/min shaking for 40 min. After incubation, cells were washed three times with washing buffer, soaked with 300 μl binding buffer, and then scraped and transferred into a 1.5 ml tube. Boil tube at 95 °C for 10 min, collect supernatant by centrifugation at 13,000×g for 5min at 4 °C (3K30 SIGMA).

260 μl supernatant was collected. 114 μl supernatant was used for generation of new ssDNA pool for subsequent selection. 150 μl supernatant was used for amplification by PCR with FITC-labeled sense primers and biotin-labeled antisense primers. FITC-labeled sense stand DNA pool was collected via streptavidin sepharose chromatographic column as described, and monitored by fluorescence spectrophotometer (Agilent).

4.7 Cloning and sequencing

On the basis of the fluorescence spectrometry results, ssDNA pools from the thirteenth round of selection were cloned using pGM-T kit (VT302-02, TIAGen, Co.) according to the procedure as described by the manufactures. Plasmids from white bacterial colony were extracted and amplified by PCR with unlabeled primers, and performed agarose gel electrophoresis. Plasmids of positive clones were selected for sequencing (Shanghai Shengon Co.) .

4.8 Affinity assay

Primary cells from three patients with osteosarcoma without pathological types were cultured in McCoy'5A medium containing 10% fetal bovine serum in 60 mm × 15 mm plates. Sequences from three plasmids were amplified by PCR with double-labeled primers, antisense strands were removed via streptavidin sepharose beads, and sence strand DNA were colected for following use: the third generation cells (3 × 10^6) of osteosarcoma were incubated with sence strand DNA in 600 μl binding buffer (1000 nmol final) at 37 °C, 50 r.p.m./ min shaking for 40 min. After incubation, sence strand DNA was separated from cells by the method described previously, and amplified by PCR. PCR mixture contained ddH$_2$O 7.25 μl, 2 × Taq MasterMix 12.5 μl, each primer 1 μl, supernatant (ssDNA templte) 3.25 μl in a total volume of 25 μl. PCR was performed for 8 cycles with 30 s of denaturing at 95 °C, annealing for 30 s at 56 °C, extension at 72 °C for 30, followed by one cycle at 72 °C for 3 min. After amplification, loaded 15 μl of PCR product in each lane of 3% agarose gel and performed electrophoresis at 100 V.

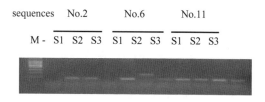

Figure 4. M : 100 bp ladder; – : negative control; S1~3: samples from three clinical cases; Sequences of No. 2, No. 6 and No. 11 were chosen from plasmids cloned.

REFERENCES

[1] Wallis, M. G. *In vitro* selection of a viomycin-binding RNA pseudoknot. *Chem. Biol.* 1997, 4:357–366.

[2] Williams, K.P.et al. Bioactive and nuclease-resistant 1-DNA ligand of vasopressin. *Proc.Nail Acad. Sci.U. S. A.*, 1997, 94: 11285–11290.

[3] Tuerk, C., MacDougal, S. & Gold, L. RNA pseudoknots that inhibit human immunodeficiency virus type 1 reverse transcriptase. Proc.Nail Acad. Sci.U. S. A., 1992, 89, 6988–6992.

[4] Tuerk, C., Gold, L. Systematic evolution of ligands by exponential enrichment:RNA ligands to bacteriophage T4DNA polymerase. *Science,* 1990, 249:505–510.

[5] Bruno, J. G. & Kiel, J.L. *In vitro* selection of DNA aptamers to anthrax spores with electrochemiluminescence detection. *Biosens. Bioelectron.* 1999, 14, 457–464.

[6] Geiger, A., Burgstaller, P., von的人Eltz, H., Roeder, A.& Famulok, M. RNA aptamers that bind 1-arginine with sub-micromolar dissociation constannts and high enantioselectivity. *Nucleic Acids Re*s. 1996, 24, 1029–1036.

[7] Harada, K. & Frankel, A.D. Identification of two noval arginine binding DNAs. *EMBO J.* 1995, 14, 5798–5811.

[8] Nieuwlandt, D., Wecker, M. & Gold, L. In vitro selection of RNA ligands to substance P. *Biochemistry* 34, 5651–5659.

[9] Hamula, C.L.A., Zhang, H., Guan, L.L., Li, X.-F. & Le, C. Selection of aptamers against live bacterial cells. *Anal.Chem.* 80, 7812–7819.

[10] Che, H., McBroom, D.G., Zhang,Y.-Q., Gold, L., & North, T.W. Inhibitory RNA ligand to reverse transcriptase from feline immunodeficiency virus. *Biochemistry,* 1996, 35, 6923–6930.

[11] Blank, M., Weinschenk, T., Priemer, M. & Schluesener, H. Systematic evolution of a DNA aptamer binding to rat brain tumor microvessels, selective targeting of endothelial regulatory protein pigpen. *J. Biol. Chem.* 2001, 276, 16464.

[12] Cerchia, L. et al. Neutralizing aptamers from wholecell SELEX inhibit the RET receptor tyrosine kinase. PLoS Biol. 2005, 3, 0697–0704.

[13] Shangguan, D. et al. Aptamers evolved from live cells as effective molecular probes for cancer study. Proc. Natl. Acad. Sci. U. S. A. 103, 11838–11843 (2006).

[14] Sefah, K., Shangguan, D. et al. Development of DNA aptamers using Cell-SELEX. *Nature Protocols.* 2010, 5(6):1169

[15] Hicke, B.J. et al. Tenascin-C aptamers are generated using tumor cells and purified protein. *J. Biol. Chem.* 2001, 276, 48644–48654

[16] Shangguan, D. et al. Identification of liver cancerspecific aptamers using+ whole live cells. *Anal. Chem.* 2008, 80, 721–728.

[17] Chen, H.W.et al. Molecular recognition of small-cell lung cancer cells using aptamers. *ChemMedChem* 2008, 3:991–1001.

[18] Sefah, K.et al.Molecular recognition of acute myeloid leukemia using aptamers. Leukemia 23,235–244(2009).

[19] Mosing, R.K., Mendonsa, S.D., Bowser, M.T. Capillary electrophoresis-SELEX selection of aptamers with affinity for HIV-1 reverse transcriptase. *Anal. Chem.* 77(19):6107– 6112.

[20] Axelroda, Nicholas,P., Giustinia and Steven, W. Suljaka. Capillary electrophoretic development of aptamers for a glycosylated VEGF peptide fragment. *Analyst* 135(11):2945–2951.

[21] Morris, K.N., Jensen, K.B., Julin, C.M., Weil, M. & Gold, L. High affinity ligands from In vitro selection: complex targets. Proc. Natl. Acad. Sci.U. S. A. 95, 2902–2907(1998).

[22] Shamah SM, Healy JM, Cload ST. Complex target SELEX. *Acc.Chem.Res.*2008, 41:130–138.

[23] Wang, C.et al. Single-stranded DNA aptamers that bind differentiated but not parental cells: substractive systematic evolution of ligands by exponential enrichment. *J. Biotechnol.* 2003, 102: 15–22.

[24] Lerner, A., Antman, K,H. Primary and metastatic malignant bone lesions. In: Goldman L, Schafer AI, eds. Goldman's Cecil Medicine. 24th ed. Philadelphia, PA: Elsevier Saunders. Chapter 208(2011).

[25] Gebhardt, M,A, Springfield, D,, Neff, J,R. Sarcomas of bone. In: Abeloff, M.D., Armitage, JO, Niederhuber JE.et al.eds. Abeloff's Clinical Oncology. 4th ed. Philadelphia, PA: Elsevier Churchill-Livingstone. Chapter 96(2008).

[26] Tang, A. et al. Selection ofaptamers for molecular recognition and characterization of cancer cells. *Anal. Chem.* 2007, 79, 4900–4907.

*This project is supported by Education Commission of Jilin Province, China [2011], No. 132.

Information Technology – Wan et al. (Eds)
© 2015 Taylor & Francis Group, London, ISBN 978-1-138-02785-5

Self quotient image based on facial symmetry

Liqiao Hu
College of Information Sciences and Technology, Engineering Research Center of Digitized Textile & Fashion Technology, Ministry of Education, Donghua University, Shanghai China

Runhe Qiu
College of Information Sciences and Technology, Engineering Research Center of Digitized Textile & Fashion Technology, Ministry of Education, Donghua University, Shanghai China

ABSTRACT: In order to improve the recognition rate of facial image under the complex illumination conditions and improve the speed of running, a novel face recognition method based on Self Quotient Image and human face priori symmetry theory is proposed. In the paper, firstly, we analyzed the Gamma transformation algorithm and applied it to improve the contrast of the image, Furthermore; we elaborated the facial symmetry theory base on parity decomposition theory and extracted the half even symmetrical facial image for face recognition system. Finally, we use a quotient image algorithm to half even face image for training and testing. The experimental results of Yale B show that the proposed method can eliminate the effects of illumination, strengthen the robustness of uneven illumination and achieved a better recognition effect.

KEYWORDS: face recognition; self quotient image; Gamma algorithm; face image parity decomposition

1 INTRODUCTION

Face recognition technology is an important research topic in the field of computer vision and pattern recognition and have a broad application prospect in the field of identity authentication, intelligent surveillance, human-computer interaction [1]. In recent years, face recognition technology has got great development, face recognition system under standard conditions have reached a satisfactory level, however, when applied to the actual environments, due to the influence of illumination, pose the complexity, diversity and other factor expression, the effect is often unsatisfactory [2]. This article mainly talks about improving the recognition rate of face recognition system in the side of the complex illumination.

In order to solve these problems and improve the recognition rate under the complex illumination, a commonly used method is illumination normalization before recognition. There were lots of typical illumination normalization methods such as histogram equalization, homomorphism filtering, Gamma transformation method and so on [3]. These methods only compressed the dynamic range of gray from the view of image enhancement to improve the image contrast. Neither eliminates the influence

of illumination nor recovers facial features of the exposure and shadow areas. Many milepost type algorithms have been proposed after introducing the Lambertian model and the Retinex algorithm to resolve the illumination problem, which analyzed the relationship between light and face image. For example, W.L. etc., firstly using Discrete Cosine Transform to transform the face image to DCT area and eliminating low-frequency components associated with the light by setting Low-frequency coefficients to zero. Then get the illumination normalized face image by inverse cosine transform [4]. Apart from above algorithm, there were lots of effective approaches do well in eliminating the influence of illumination on the human face images, such as X.M. etc., proposed multi scale Retinex with color restoration(MSRCR) algorithm base on Retinex theory [5]. H.T. et al. proposed Self Quotient Image base on Quotient Image algorithm [6]. X.G. et proposed Morphological quotient image algorithm base on SQI and Wavelet Transform [7]. S.C. proposed Multilevel Quotient Image algorithm [8]. However, experimental results from H.T show that the face feature of some partial shade and exposed areas still difficult to restore. Unlike the above methods, Y.N etc., introducing face image parity decomposition approach and extract even symmetrical facial image base on function parity decomposition [9]. Y.F. proposed Reconfigure method base on face bilateral mirror symmetry [10]. This method solved the issue generated by uneven light of half face at a certain extent. However, such approach using the mirror symmetry of the face has great limitations.

Aimed for the above problem, this article will adopt priori facial symmetry as a constraint and extract half even symmetrical facial image feature which less susceptible to interference. Then I adopt half even symmetrical facial image as sample to accept self quotient image algorithm, finally, reconstructing the half face image for training and testing.

2 QUOTIENT IMAGE

2.1 Retinex model

The Retinex algorithm decomposes the image into the reflectance image and the illuminant image as equation 1 [11].

$$I(x, y) = R(x, y)L(x, y) \qquad (1)$$

where I(x, y) is original image, L(x, y)is illuminant image, R(x, y) is the reflectance image, which reflected the reliable information of the object such as texture. I(x, y) * F is used to get the illuminant image L(x, y) and calculated with a Gaussian function for a smoothed version. And then according to the Retinex algorithm, we can get equation 2 as below.

$$R(x,y) = \frac{I(x,y)}{L(x,y)} = \frac{I(x,y)}{I(x,y) * F(x,y)} \qquad (2)$$

2.2 Self quotient image

H.T. etc proposed SQI algorithm after analyzing the relationship of Quotient Image and Retinex model, The SQI was defined by a face image I(x, y) and its smoothed version L(x, y) as equation 3.

$$Q(x,y) = \frac{I(x,y)}{I(x,y) * F(x,y)} \qquad (3)$$

In the equation 3, where * is the convolution operation and F(x, y) is the smoothing kernel which is a weighted Gaussian filter. SQI algorithm can be calculated from one image without pre-training which was very important for real world applications.

Figure 1 denotes the effect image of SQI algorithm.

Original-image

SQI-image

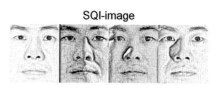

Figure 1. The effect image of SQI.

2.3 Gamma self quotient image

Figure 1 denotes that there are some high-frequency noises in the original shaded area, the reflected image extracted from the original image is not very satisfactory. So do some illumination compensation preprocessing to the original image is very necessary. Gamma transformation is a nonlinear enhancement method which can extend darker pixels, compress brighter pixels and improve image contrast by compressing the dynamic range. We can adopt it to eliminate the influence of illumination in a certain degree.

Equation of Gamma transformation as below:

$$F' = F^{\gamma} \qquad (4)$$

Original-image

Gamma-image

Gamma-SQI-image

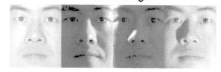

Figure 2. The effect of gamma conversion.

In the equation 4, where ã is Gamma coefficient, in this paper, we adopt Gamma transform as pre-processing method and set 0.2 as the Gamma coefficient for the original image before running SQI algorithm. Figure 2 shows the original images and the effect images of Gamma transformation to compare the effects of Gamma-SQI.

3 PARITY DECOMPOSITION OF FACE IMAGE

3.1 Parity decomposition of function

Any function f(x) can be decomposed into an even function and an odd function as equation 5

$$f(x) = f_e(x) + f_o(x) \tag{5}$$

In the equation 5, where $f_e(x)$ is even function, $f_o(x)$ is odd function. And $f_e(x)$, $f_e(x)$ can be expressed as equation 6[9]:

$$f_e(x) = \frac{f(t) + f(-t)}{2}, \qquad f_o(x) = \frac{f(t) - f(-t)}{2} \tag{6}$$

3.2 Parity decomposition of face image

X.W etc., bring function decomposition ideologies into human face area [12], He concluded that an image can be decomposed into an even image and an odd image as equation 7:

$$\begin{cases} A = A_e + A_o \\ A_e = \dfrac{A + A_m}{2} \\ A_o = \dfrac{A - A_m}{2} \end{cases} \tag{7}$$

In the equation 7, where A, A_m, A_e, A_o respective expressed as an original face image, mirror symmetry image, even symmetrical facial image, odd symmetrical facial image, we can see them in figure 5.

In the figure 3, even symmetrical facial image is an average face image generated by original image and the mirror symmetry image; it contains the main structure of the human face. At the same time, even symmetrical facial image own better stability and less susceptible to interference, however, odd symmetry face image more susceptible to interference such as uneven illumination. We choose an even symmetrical facial image A_m as sample of face recognition image. More importantly, we choose half even symmetrical facial image to test due to it is completely symmetrical and save running time.

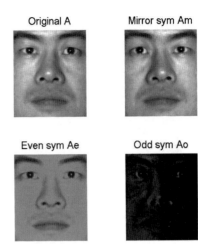

Figure 3. Decomposition image of human face.

Figure 4, 5 respectively denotes the effect of half even symmetrical facial image from the same person, but different illumination and the same illumination but different person.

3.3 Flow-chart of the proposed method

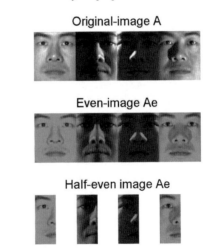

Figure 4. Half even image of different illumination.

According to the above analysis, this paper proposes a novel algorithm, which combines human face priori symmetry theory and SQI classical algorithm. Specific steps as below:

Step1: Adopting Gamma transformation algorithm as Pre-processing to improve the contrast of the original image

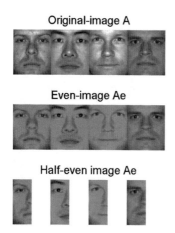

Original-image A

Even-image Ae

Half-even image Ae

Figure 5. Half even image of different person.

Step2: Exact half even symmetrical facial image to eliminate the asymmetric information caused by outside interference such as uneven illumination.

Step3: Exact texture feature by SQI algorithm.

Step4: Reconstruct half even symmetrical facial SQI image as sample to train and test.

Figure 6 denotes the flow chart of the proposed system as below.

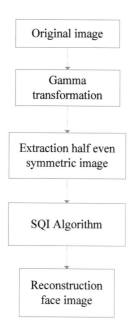

Figure 6. Flow chart of proposed system.

4 EXPERIMENT RESULTS AND ANALYSIS

In order to verify the recognition performance of the proposed algorithm under the complex illumination conditions, we adopt most popular Yale face database B for experiments. Yale Face Database B includes 10 subjects, each with 64 frontal images under different light conditions. All the images are manually cropped to size 168x192 for removing the backgrounds effect.

These images are divided into five subsets according to the angle between the light source direction and the camera axis: subsets 1(0°~12°), subsets 2 (13°~25°), subsets 3(26°~50°), subsets 4(51°~77°), subsets 5) >77°) [13],The details are shown in table 1:

Some examples of face images in each subset are shown in figure 7, and figure 8 denotes the effect image of the proposed method. At the same time, we

Table 1. Five image subsets according to light source direction.

	Subsets 1	Subsets 2	Subsets 3	Subsets 4	Subsets 5
Light Direction (°)	0~12	13~25	26~50	51~77	>77
Number (piece)	70	120	120	140	190

compared three algorithms (SQI, Reconfigure-SQI, Our method) effect images to highlight our methods in Figure 9.

In the Figure 9, Recog-SQI method is combine reconfigure-method and SQI algorithm.

To further validate the recognition performance of the proposed method. We test Yale face database B, with PCA algorithm to extract features and classify these features with k-Nearest Neighbor algorithm. We use three methods such as the original image, Average face and SQI to compare with the proposed method.

The experiments choose subset 1 as training sets and the others as testing sets respectively. Table 2 shows the evaluation results of all methods.

In the Table 2, we were told that the recognition rate is affected by illumination seriously. Words "None" in Table 2 on behalf of the original image without any pre-processing, in this column, the recognition rate is the lowest. And the proposed method outperforms other tested methods. So we concluded that the proposed method in this paper can strengthen the robustness of uneven illumination, and the performance of the system is increased.

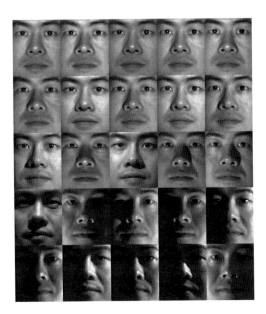

Figure 7. Examples of face images Yale B database.

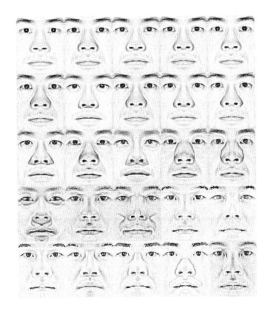

Figure 8. The effect image of proposed method.

5 CONCLUSION

In this paper, we proposed a novel face recognition method which is robust to illumination variation.

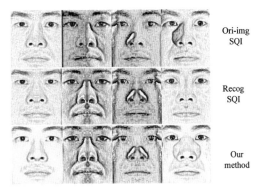

Ori-img
SQI

Recog
SQI

Our
method

Figure 9. The effect images of four methods.

Table 2. The evaluation results of all methods.

	subset 2	subset 3	subset 4	subset 5	average
None (%)	96.67	83.33	37.14	15.26	58.1
Avg-even (%)	99.17	98.33	84.29	66.32	87.03
SQI (%)	98.33	99.17	95.71	92.11	96.33
Our Meth (%)	98.33	98.33	94.14	99.61	97.6

Our method is based on SQI and human face priori symmetry. In the pre-processing stage, Gamma transformation is applied to improve the contrast of the original image. Then adopt half even symmetrical facial image as sample to accept SQI algorithm. Finally, we reconstructed the half even face image for training and testing. According to the results of a series of experiments, we conclude that our method can eliminate the effects of illumination, strengthen the robustness of uneven illumination and achieved a better recognition effect.

REFERENCES

[1] X.M. WANG, X.Y. FANG, J.G. LIU. Complex adaptive face image illumination enhancement. Computer Engineering and Applications, 2011, 47(2):15–18.
[2] S.F. LIANG, Y.H. LIU, LS, Li. Face recognition algorithm to extract the wavelet transform and LBP log domain wavelet transform and LBP log domain feature. Journal of Signal Processing, 2013, 29(9).
[3] H. HAN, S.G. SHAN, W. GAO. A use of facial symmetry illumination normalization method. Journal of Computer Research and Development, 2013, 50(4):767–775.

[4] W.L. CHEN, S.Q.WU. Illumination compensation and normalization for robust face recognition using discrete cosine transform in logarithm domain. IEEE Trans on Systems ,Man and Cybernetics : Part B, 2006, 36(2):458–466.

[5] X.M. WANG. Multi scale Retinex image enhancement algorithm improved. Journal of Computer Application, 2010, 30(8):2091–2093.

[6] H.T. WANG, Z.Q. ZHI, Y.S. WANG. Face recognition under varying conditions using self quotient image. Proc of the 6th IEEE International Conf on Automatic Face and Gesture Recognition. Piscataway, NJ: 2004. 819–824.

[7] X.G. HE, J. TE, L. F. Wu, Y. Y. ZHANG. Morphological quotient image based illumination normalization algorithm. Journal of software, 2007, 18(9):2318–2325

[8] S.C. SHI, CH. YANG, T. WANG, X.Q. LV. Face image enhancement based on self quotient image. Computer Engineering and Applications, 2013, 49(13):142–144.

[9] Y.N. WANG, JB SU. Symmetry description and application of symmetry based on LBP features in face recognition.

[10] [Y.F.ZHANG, Eliminating variation of face symmetry .proc of the 4th international conf on audio and video based biometry person authentication.2003 523–530.

[11] Z.H. LIU. Illumination on face feature extraction algorithms under varying. NANGJING, 2011. 21–22

[12] X.W. HUANG. Application of block faces parity symmetry in face recognition. GUANGDONG. 2013.

[13] X.M. WANG. Variable light detection and recognition of human face.. SHANGHAI 2010.

Information Technology – Wan et al. (Eds)
© 2015 Taylor & Francis Group, London, ISBN 978-1-138-02785-5

Simplified classification in multimodal affect detection using vocal and facial information

Yunyun Wei & Xiangran Sun
Communication University of China, Beijing, China

ABSTRACT: Affect detection is an interesting and challenging problem, where human mental states are recognized by the computer through speech, facial expressions, physiology and other modalities. It requires not only accurate machine learning and classification techniques, but also lots of training data and emotional models. In this paper a novel approach is presented to recognize the emotional expressions of human using vocal and facial features from voices, facial visual and thermal images. We also propose a simplified classification which is composed of Hidden Markov Model (HMM) and Support Vector Machines (SVM). It saves computation time considering real-time processing and achieves high performance. Experimental results show that our proposed method for affect detection has a good accuracy of emotion recognition.

1 INTRODUCTION

1.1 *Affect detection*

Affect detection or emotion recognition plays an important role in the interaction of multimedia and social media, especially in rapidly developing digital media. It is necessary and helpful to import emotional taste into the course of communications between human and digital media, including video game, videoconference, instant messenger (such as Twitter, Wechat and QQ) and intelligent personal assistant (such as Siri and Google Now). Digital media should be able to interact in a friendly manner with a human. If digital media understand the emotional state of a human and responds the need of a person, our society will be more enjoyable in the future. The goal of our research is to develop software and hardware architecture which can perceive human feelings or mental states.

The communication between humans and digital media need to become more natural and more human-like. This is particularly meaningful in the design of video game and intelligent personal assistant, where emotional intelligence is of great significance. The human sensory system uses multimodal analysis of multiple communication channels to interpret face-to-face interaction and to recognize another party's emotion. The psychological study (Mehrabian 1968) indicated that humans mainly rely on vocal intonation and facial expression to judge someone's emotional state. Hence, automatic affect detection systems should at least make use of the information from voice and face to achieve a reliable and robust result. Thus, considering multiple modalities may give a more complete description of the expressed emotion

and generally tends to lead to more accurate results than unimodal techniques (Metallinou & Narayanan 2010).

Recently a number of techniques have been developed for affect detection and studies tend to use features from audio-visual, speech-text, dialog-posture (Pantic & Rothkrantz 2003). Most of these studies have applied single classifiers, such as support vector machines, k-nearest neighbors, Bayesian network etc. with single and multiple modalities. However, finding a single classifier that works well for all modalities and individuals is difficult (Sebe et al. 2005). Combining classifiers is thought to provide more accurate and efficient classification results. Instead of just one classifier, a subset of classifiers (aka base classifiers) can be considered along with the best subset of features for the best combination (Kuncheva 2004). Moreover, a certain base classifier may do well on a certain modality, but it is challenging to generalize one classifier for multiple channels or modalities. Because a single classifier cannot perform well when the natures of features are different, we use two combinations of classifiers with a subset of features may provide a better performance (Utthara et al. 2010). Combining classifier can be suitable in emotion studies for affect detection or classification where features contribute from multiple modalities and individuals in varying environmental setups. It is a serious problem that several classifiers would take more time and system resource.

1.2 *Overview of simplified method*

In this paper, we present a new simplified method to recognize the emotional expressions using vocal and facial information. We use Microsoft's Kinect integrated multi-array microphone, RGB camera and depth sensor as voice sensor, image sensor and infrared sensor, so we can obtain voice data, visual images and thermal images. Depending on the thermal images observed by infrared ray, the recognition of facial expressions is not influenced by lighting conditions. For voices, we use such prosodic parameters as pitch signals, energy, and their derivatives as extracted features. For facial expressions, we use feature parameters of thermal images in addition to visible images as extracted features. After the processing of feature selection, these features are trained by our simplified classification for recognition. Emotions are described with three continuous-valued emotion primitives, namely valence, activation, and dominance in a 3D emotion space. In our simplified classification we firstly separate the features from vocal and facial information into the positive and negative nature of emotions in the term of valence using HMM. Secondly, we recognize five types of emotions, including surprise, happy, neutral, sadness and anger by use of SVM. It saves computation time considering real-time processing and boosts performance.

2 PROCEDURE OF AFFECT DETECTION

Figure 1 shows the examples of male face visual image and thermal image. On the whole, the temperature measurement by thermal images is not affected by the color of skin, background, darkness and lighting condition, resulting in that face and its characteristics are easily extracted from the captured image containing face and its scene.

 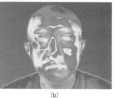

(a)   (b)

Figure 1. Example of face image: (a) visual image, (b) infrared image.

Figure 2 illustrates the procedure for recognizing the emotional state contained in voices, visual and thermal images. The procedure is comprised by feature extraction, feature selection and classification. In order to recognize emotional states in both vocal and facial expressions, we need to utilize the information to extract emotional feature parameters. Firstly, we analyze vocal expressions which contain emotional information, including six kinds of feature

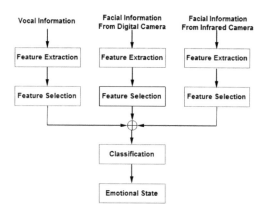

Figure 2. Overview of the procedure for affect detection using voices, visual and thermal images.

parameters. We also extract key feature parameters from facial expressions of both visible and thermal images as well as emotional feature extraction from vocal expressions.

3 EMOTIONAL FEATURE EXTRACTION AND SELECTION

3.1 Facial feature extraction and selection

In the recent years, many studies have been performed using an ordinary digital camera to resolve the issues of recognizing emotional states of human through facial expressions. But those studies still don't seem to get the perfect achievement since among various facial expressions there is a slight difference in terms of characteristic features for the gray level distribution of visual images using a digital camera. Thus, applying thermal distribution images to facial expression recognition using infrared cameras combined with ordinary images can achieve better performance.

In our experiment we take two timing positions to acquire face images from ordinary camera and infrared camera where the first and the second ones are the maximum voice parts respectively. When a face image is given in the computer, it is necessary to extract face-parts areas correctly, which will be important for better recognition performance. Some face-part areas in an ordinary visual image are extracted by exploiting both the information on corresponding thermal image and the simulated annealing method for template matching of the block-region including the eyes. Then the face-parts areas which consist of three areas in the visual image and six areas in the infrared image respectively are extracted. In the next step, we generate different images between the averaged neutral face image and the test face image in the extracted face-parts areas to perform a discrete cosine transformation (DCT). For infrared image, some face-parts areas are also extracted through segmentation of the image, and then DCT is performed.

Because the feature selection was performed separately for all individual modalities, we used SVR-SFFS technique for facial feature selection respectively from digital camera and infrared camera (Kanluan et al. 2008). SFFS is the Sequential Forward Floating Search technique, and SVR is a regression method based Support Vector Machines (SVM). This method is a good way of choosing the best subset of features and used for discarding redundant, noisy features. Firstly, the DCT coefficients between 10% and 90% quintiles were normalized to a scale of [0, 1]. Then, they were sorted in descending order according to their correlation to the manual labels of the emotion primitives.

Since the DCT coefficients had the largest absolute value of correlation coefficient, we took the coefficient as SVR-SFFS. The correlation coefficient is between the manual reference and the estimates for each emotion primitive as a function of the feature set. The number of features for each emotion primitive was the number that yields the lowest mean linear error, which were retained for the estimation. Given our analysis, we selected the corresponding coefficients for the feelings such as surprise, happy, neutral, sadness and anger respectively. In our experiments we extracted several coefficients of the feature, respectively which are from the eye region image in non-overlapping 50×50 blocks of pixels. We found that these relevant features are DCT coefficients extracted from the facial image regions from both low-frequency and high-frequency bands of various blocks. Our experiments showed inducing more features may represent more irrelevant information and hardly improve the estimation results.

3.2 Vocal feature extraction and selection

The prosody plays an important role as an indicator of the characteristics of acoustic emotion (Waibel 1986), so many studies on automatic affect detection (Kanluan et al. 2008) used prosodic features to implement emotion recognition. In our experiment we extracted prosodic features from the fundamental frequency (pitch), the energy contours of the speech signals, and the derivative elements of these contours. The pitch signals in the vocal regions are smoothed by a spline interpolation. Before using HMM to train the features, we use discrete duration information to reduce the effect of a speaking rate. The following statistical characteristics can be calculated from the above signals: mean value, standard deviation, median, maximum, minimum, as well as the difference between the maximum and minimum, and differences between the quartiles. In order to acquire the temporal characteristics of acoustic emotion, we utilize speech rate, pause to speech ratio, as well as mean and standard deviation of the speech duration, and mean and standard deviation of the pause duration. In accordance with classical studies (Waibel 1986) we also used spectral characteristics in 13 sub-bands derived from the MFCCs. Totally, 137 acoustic features were extracted. They were normalized to the range [0, 1].

To improve the system performance and decrease the number of acoustic features, the Sequential Forward Floating Search technique is applied during the procedure of feature selection. It is enough and efficient to use 20 features for each of the emotion primitives. Even though more features are involved, it is hard to improve the experimental results.

4 SIMPLIFIED EMOTION CLASSIFICATION

The preceding sections show the extraction and selection of vocal and facial information. Here, we describe the classification of these features in our architecture of affect detection.

Hidden Markov models have been widely used for many classification and modeling problems. One of the main advantages of HMMs is their ability to model non-stationary signals or events. Dynamic programming methods allow one to align the signals so as to account for the instability. However, the main disadvantage of this approach is that it is very time-consuming since all of the stored sequences are used to find the best match. It uses the transition probabilities between the hidden states and learns the conditional probabilities of the observations given the state of the model. An HMM is given by the following set of parameters:

$$\lambda = (A, B, \pi)$$
$$a_{ij} = P(q_{t+1} = S_j | q_t = S_i), 1 \leq i, j \leq N$$
$$B = \{b_j(O_t)\} = P(O_t | q_t = S_j), 1 \leq j \leq N \quad (1)$$
$$\pi_j = P(q_1 = S_j)$$

where A is the state transition probability matrix, B is the observation probability distribution, and π is the initial state distribution. The number of states of the HMM is given by N. It should be noted that the observations (O_t) can be either discrete or continuous, and can be vectors. In the discrete case, B becomes a matrix of probability entries, and in the continuous case, B will be given by the parameters of the probability distribution function of the observations.

These selected features can be directly processed by dynamic modeling as HMM. Yet, streams usually need to be synchronized for this purpose (Wimmer & Matthias 2008). The facial feature vector t_v describes the currently visible face and it is assembled from the structural and the temporal features mentioned. The time series T_v is constructed from a sequence of t_v sampled at frame rate. It is established in a certain amount of time, which is determined by speech processing. These vocal feature vectors t_a and the time series T_a are constructed from sampling t_a over a certain amount of time. We therefore prefer the application of functional f to the combined descriptors $T_c = [T_v, T_a]$ in order to obtain a feature vector $x \in R^d$, see Equation 2. As opposed to the time-series data, this feature vector is of constant dimension d, which allows for an analysis with standardized techniques.

$$f : T_c \rightarrow x \qquad (2)$$

SVM is popular based on their compatibility and performance in many applications (Nguyen et al. 2005). It is simple to implement, span a variety of machine learning theories and techniques to make SVM suitable in combined classifiers for addressing the diversity of features and subject variability. Then we classified the feature vector x using SVM with polynomial Kernel and a multi-class discrimination strategy. We proposed an SVM algorithm to build the classification.

Because the main advantage of SVM is that it has limited training data, it has very good classification performance. The classification of SVM is done by using the following formula.

$$\langle w \cdot x \rangle + b_0 \geq 1, \forall y = 1$$
$$\langle w \cdot x \rangle + b_0 \leq -1, \forall y = -1 \qquad (3)$$

where (x, y) is the pair of training set and $y \in \{+1, -1\}$. $\langle w \cdot x \rangle$ represents the inner product of w and x whereas b_0 refers to the condition. SVM employs the linear kernel function which is given by the formula below:

$$Kernel(x, y) = (x \cdot y) \qquad (4)$$

The selection of the most relevant features is as important as the choice of an optimal classifier. It saves computation time considering real-time processing and boosts performance as some classifiers are susceptible to high dimensionality. In order to avoid NP-hard exhaustive search recommended in (Schuller et al. 2005), we employ classification error as an optimization criterion by using Sequential Forward Floating Search (SFFS) with SVM as a wrapper. As a vocal and facial super vector is constructed, we can select features in one pass to point out the importance of vocal and facial features, each. The optimal number of features is determined according to the highest accuracy throughout the selection.

5 EXPERIMENTS

In this section, we present the experimental results for detecting five types of emotions, including surprise, happy, neutral, sadness and anger. First, we capture the voice, visual and thermal images simulating five emotional states acted by two male subjects. We assemble 20 samples per each emotional expression per a subject as training data and 20 as test data. Table I shows the average recognition rate is 73% in five emotional states (except for no answers).

Our proposed method can be applied to any Chinese word without the preparation for the special word. In Table 1, the main reason for misrecognition

Table 1. Recognition accuracy for emotional state with vocal information.

		Input Emotion With Vocal Information				
		Surprise	Happy	Neutral	Sad	Anger
Output	Surprise	70				20
	Happy		95			
	Neutral			65		
Emotion	Sadness				65	
	Anger	10			20	70
	No Answer	20	5	35	15	10

334

Table 2. Recognition accuracy for emotional state without vocal information.

		Input Emotion With Vocal Information				
		Surprise	*Happy*	*Neutral*	*Sad*	*Anger*
	Surprise	50				30
Output	*Happy*		75			
	Neutral			75		
	Sadness				50	
Emotion	*Anger*	10			30	65
	No Answer	40	25	25	20	5

Table 3. Recognition accuracy for emotional state without facial expression.

		Input Emotion With Vocal Information				
		Surprise	*Happy*	*Neutral*	*Sad*	*Anger*
	Surprise	50	20			
Output	*Happy*		30			20
	Neutral			60	40	
	Sadness				50	
Emotion	*Anger*	40				40
	No Answer	10	50	40	10	40

of the facial expression was the misrecognition of vocal feature. Using the training data for each state at whole speech timing, which is just speaking one word, we made each averaged neutral state of face image. In order to apply the proposed method to any speaker, we need to prepare six averaged neutral face images for each state: one image before speaking and five images in speaking the word. If the difference of the averaged neutral face image in speaking the word, we use the averaged neutral face image for the vocal feature resulting in six averaged neutral face images for each state.

We applied the proposed method for one state who spoke any Chinese word, after preparing the learning data for some special words. The mean accuracy of facial expression recognition in expressing one of the emotions of surprise, happiness, neutral, sadness and anger was 63% without vocal information showed in Table II. Table III shows the mean accuracy of the emotion recognition without facial expression was 46%, even worse than without vocal information. In the experiment, the mean accuracy of the emotion recognition with facial expression and vocal information is 73%.

Because there was increased recognition caused by using the vocal information as learning data to help recognizing a facial expression, the affective detection accuracy could be made better by improving the facial expression recognition and the speech recognition accuracy.

6 CONCLUSION

In this work, a novel approach to recognize emotions by use of a simplified emotion classification constructed with HMM and SVM is presented. The proposed work is able to process vocal and facial information from voice sensor, image sensor and infrared sensor. Our simplified classification saves computation time considering real-time processing and boosts performance. Among the five expressions, happy expression has been recognized with an accuracy of 95%. Expressions neutral and sad cannot be distinguished very well. Hence the future work aims to apply more extracted features from facial expressions in this work and also considering more number of emotions.

REFERENCES

Mehrabian, A. 1968. Communication without words. *Psychol. Today, vol.2: 53–56.*

Metallinou, A. Lee, S. & Narayanan, S. 2010. Decision level combination of multiple modalities for recognition and analysis of emotional expression. *in Proc. of ICASSP, Dallas, Texas: 2462–2465.*

Pantic, M. & Rothkrantz, L. J. M. 2003. Toward an affect-sensitive multimodal human-computer interaction. *Proceedings of the IEEE, Special Issue on human-computer multimodal interface, 91(9): 1370–1390.*

Sebe, N. Cohen, I. Gevers, T. & Huang, T. S. 2005. Multimodal Approaches for Emotion Recognition: A Survey. *Proc. SPIE, 5670: 56–67.*

Kuncheva, L. I. 2004. Combining pattern classifiers: Methods and algorithms. *Wiley-Interscience.*

Utthara, M. Suranjana, S. Sukhendu, D. & Pinaki, C. 2010. A Survey of Decision Fusion and Feature Fusion Strategies for Pattern Classification. *IETE Technical Review, 27: 293–307.*

Kanluan, Ittipan, Michael Grimm & Kristian Kroschel, 2008. Audio-visual emotion recognition using an emotion space concept. *16th European Signal Processing Conference, Lausanne, Switzerland.*

Waibel, A. 1986. Prosody and Speech Recognition. *Doctoral Thesis, Carnegie Mellon Univ.*

Wimmer & Matthias, 2008. Low-Level Fusion of Audio, Video Feature for Multi-Modal Emotion Recognition. *VISAPP (2): 145–151.*

Nguyen, T. Li, M. Bass, I. & Sethi, I. K. 2005. Investigation of Combining SVM and Decision Tree for Emotion Classification. *Seventh IEEE International Symposium on Multimedia, Irvine, California, USA: 540–544.*

Schuller, B. Mueller, R. Lang, M. & Rigoll, G. 2005. Speaker independent emotion recognition by early fusion of acoustic and linguistic features within ensembles. *In Proc. Interspeech, Lisboa, Portugal. ISCA.*

Information Technology – Wan et al. (Eds)
© *2015 Taylor & Francis Group, London, ISBN 978-1-138-02785-5*

Simulation and analysis of the in-pipe robot based on the software of Pro/E and ADAMS

Hongmei Xu & Xinbo Ma
College of Engineering, Huazhong Agricultural University, Wuhan, China

Qinghong Liu
Mechanical Engineering School, Xi'an Jiaotong University, Xi'an, China

ABSTRACT: Taking the in-pipe robot with inclined wheel as the research object, the software of ADAMS was adopted to simulate its performance. The basic principles and performance requirements of the in-pipe robot were briefly discussed. The Pro/E software was used to establish the geometric model of the in-pipe robot, which was then imported to the ADAMS software for the simulation. On the basis of the virtual prototype model of the in-pipe robot, the effects of stiffness coefficient and preload of the spring on the crawling speed and bearing capacity of the in-pipe robot were discussed in depth. Additionally, the crawling speed and bearing capacity of the in-pipe robot with inclined wheel were compared with that of the conventional in-pipe robot. The simulation results show that the telescopic link is in favor of increasing the bearing capacity of the in-pipe robot. The performance of the in-pipe robot can be greatly improved by adjusting the preload of the spring and optimizing the structure of the robot.

1   INTRODUCTION

As a smart device, the pipeline robot can creep along the inner or outer surface of pipes, carrying one or various sensors and operating equipment, such as posture and position sensors, CCD camera, ultrasonic sensor [1], pipe-cleaning device, pipe-welding device and simple manipulator. Thus the pipeline robot can inspect, maintain and clean pipes without pipeline damaging. The in-pipe robot with inclined wheel is easily controlled, which can advance and retreat freely, stable and reliable.

In addition to this, the connecting rod of the robot is characterized for its high scalability and flexibility, which can enhance the bearing capacity and self-adaptability of the in-pipe robot. Relevant parameters and structure design of the in-pipe robot have great effects on its performance such as bearing capacity and speed. In order to improve the performance of the physical prototype and reduce the test times, the relationship between performance and parameters or structure is simulated and analyzed by the software of ADAMS.

The virtual prototype of the in-pip robot is established to carry out motion analysis on pipeline robot. Generally there are two research methods: (1) a geometric model is built by using the drawing software such as Pro/E and UG, and then imported to the simulation software such as ADAMS and ANSYS;

(2) The establishment of the geometric model and simulation analysis of the in-pipe robot are both completed based on the simulation software. On account of relatively complex structure of pipeline robot, the first method was adopted. Furthermore, Pro/E behaves itself with stronger processing functionality for the three-dimensional model, advanced design philosophy and simple practical operation [2]. As the special mechanic dynamic analysis software, ADAMS is easy to carry out statics, kinematics and dynamics analysis of the virtual machine. For these reasons, a geometric model of the pipeline robot with inclined wheel was firstly established by the software of Pro/E, and then imported to ADAMS to construct the model of the robot [3]. The effects of the relevant parameters and structure design of the in-pipe robot on its performance were investigated on the base of the virtual prototype. The results can provide preferences for the prototype development of the pipeline robot.

2   GETTING STARTED

In general, the in-pipe robot is mainly composed of mechanical system, operating system and functional module. Fig.1 shows an image of the simplified model of the in-pipe robot with inclined wheel, which consists of pipeline vehicle, link mechanism and guide mechanism [4].

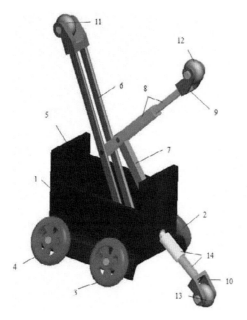

Figure 1. Simplified model of the in-pipe robot with inclined wheel.

1, 2, 3, 4 - driving wheel; 5 - base; 6, 7 - rocker; 8 - telescopic connecting rod; 9, 10 - wheel support; 11, 12, 13 - small wheel; 14 - telescopic boom.

As the main part of in-pipe robot, the pipeline vehicle is made up of base and driving wheels. Driving wheels are fitted on either side of the base, and the side plate of base is tilted, which is different from traditional wheel in-pipe robot. Consequently, it is suitable for in-pipe robot to crawl in circular pipe, and the static friction between the wheels and tube wall is raised, and the stability of the pipeline robot is also improved.

Link mechanism is mainly composed of rocker 6, telescopic rod, rocker 7, wheel support, small wheel 11 and 12, and spring which is used for connecting rocker 7 and base. Connecting rod uses telescopic structure. Namely, two parts are connected together through another spring, and relative movement between the two parts may happen. Because of spring force, a small wheel of the connecting rod is always glued to the tube wall. The movement of four-bar mechanism is determined by the diameters of pipe. In other words, pipe diameters decide the location of each part. When the pipe comes in a smaller size, the barycenter of link is lowered, conversely, elevated. Therefore, the in-pipe robot has the property of self-adaptability.

Guide mechanism mainly comprises a telescopic arm, wheel support and small wheel. Telescopic arm is connected to the base through a globe hinge. Wheel support is fixed on the telescopic arm, and rotates about the axis of telescopic boom, which forms universal wheel with the small wheel. Relative movement between two sections of telescopic boom will happen on corners, which is analogous to the working principle of the air cylinder. Altering length and adjusting the direction of guide mechanism conduce to robot crawling in elbow pipes.

Taking the simplified model of the in-pipe robot with inclined wheel as the research object, the impacts of related structure and parameters on the bearing capacity and crawling speed of robot are mainly underlined. Additionally, performance comparison between the novel in-pipe robot and traditional wheel in-pipe robot is made. The influence of telescopic structure and dip angle of driving wheel on its performance is also discussed.

3 SIMULATION ANALYSIS

Importing the above-mentioned model of in-pipe robot with inclined wheel into ADAMS and adding the related constraints, the virtual prototype model of pipeline robot is established as shown in Fig.2.

Figure 2. Virtual prototype model of in-pipe robot.

According to the design requirements of in-pipe robot, the in-pipe robot must have a greater bearing capacity, namely having enough traction force and high and stable crawling speed. In the model of in-pipe robot, the speed of the pipeline robot is mainly determined by the speed of driving wheels. Speed change of driving wheels has a certain effect on stability. Magnitude of traction force is mainly decided by the static friction force between the wheels and the wall, thus the influence factors of maximum static friction force are same to that factor of maximum traction force. Therefore, velocity, bearing capacity and superior performance of in-pipe robot with inclined wheel is discussed in the article.

3.1 Velocity analysis

There are many factors that affect the speed of in-pipe robot. Most of them are preload and stiffness coefficient of spring. Accordingly, the influence of preload and stiffness coefficient of spring between connecting rocker 7 and based on speed is discussed. When the stiffness coefficient and other parameters remain unchanged, velocity curves of mass center with different preload are measured, shown in Fig.3. When the preload of spring is constant, velocity curves of mass center with stiffness coefficient are measured and shown in Fig. 4.

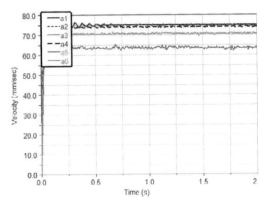

Figure 3. Velocity curves of center of mass under different preload.

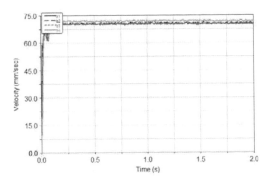

Figure 4. Velocity curves of center of mass under different stiffness coefficient.

Obviously, preload of spring have a certain effect on the velocity. In Fig.3, curve a1, a2, a3 and a4 coincide with each other substantially, indicating that the speed is almost invariable when the preload is in a certain range. Curve a5 and a6 are significantly lower than the above-mentioned four curves, revealing that the speed of in-pipe robot decreases inversely when the preload increases. Because the movements of

in-pipe robot mainly depend on the static friction force between the wall and wheels, the driving wheels roll forward in the pipe and drive the in-pipe robot under driving force. However, when the preload is too large, it is not just rolling friction that exists between the wheels and the wall, but sliding friction which results in the decreased crawling speed of in-pipe robot. In addition, excessive preload usually leads to obvious fluctuation of the speed curve and affects the service life of spring significantly.

It can be seen from Fig.4 that the stiffness coefficient of spring has a little effect on the velocity of pipeline robot when the preload remains constant. When the stiffness coefficient is smaller, speed fluctuation emerges. Speed fluctuation arises from many factors, including the gravity of the robot, friction between the wall and wheels, impulse force and inertia moment of driving wheels. In addition to this, the model of in-pipe robot is not symmetrical, leading to uneven bearing force. In a word, preload of spring have more influence than stiffness coefficient on speed. The spring should have a certain rigidity to ensure the preload. It is necessary to take into account the preload and the stiffness coefficient of spring synthetically when choosing the spring.

3.2 Bearing capacity analysis

Bearing capacity, namely drag force of pipeline robot, mainly depends on the maximum static friction force. The influencing factors of the maximum static friction force include positive pressure and static friction coefficient between the wheel and wall. So it can change the maximum drag force by altering the static friction coefficient or positive pressure. Obviously, bearing capacity increases with the friction coefficient. Positive pressure between the wall and wheels is mainly determined by gravity and the force of a spring between the rocker 7 and the base. Tensile force, resulting from the spring acting on rocker 7, has a certain effect on bearing capacity. Therefore, it is beneficial for optimizing the in-pipe robot with an inclined wheel to investigate the influence of preload on bearing capacity.

In order to measure the maximum drag power of pipeline robot, it is essential to append an auxiliary spring connecting the base with fixed point. Fig.5 shows the virtual prototype of in-pipe robot to measure the maximum drag force. The maximum pressure of auxiliary spring represents the bearing capacity. So when pipeline robot's speed and the friction coefficient remains unchanged, the influence of preload on the bearing capacity of pipeline robot can be discussed by changing the preload and measuring the force of auxiliary spring. Fig.6 shows curves that force acts on the auxiliary spring under various preloads.

Figure 5. Virtual prototype of in-pipe robot to measure maximum drag force.

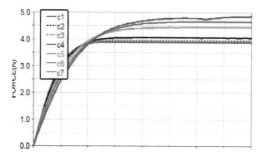

Figure 6. Force acting on the auxiliary spring curve under different preload.

Fig. 6 shows the force acting on the auxiliary spring under different preload. The curve c1, c2, c3 and c4 almost coincide well with each other, suggesting that the preload has a smaller effect on the bearing capacity of in-pipe robot when the preload is small. Since the preload is smaller, wheels do not have full contact with the wall. Thus the pressure and maximum static friction force between the wheel and wall is mainly determined by gravity. The maximum drag force does not change significantly with the increase of preload. When the preload is relatively great, speed of the in-pipe robot will decline, leading to unstable stress of auxiliary spring and weakened increasing trend of its maximum pressure. When the preload is appropriate, the wheel contacts with the wall completely, and the maximum pressure of auxiliary spring increases with preload. Namely, bearing capacity of auxiliary spring increases with preload. In practice, the preload of spring can be adjusted to guarantee enough bearing capacity of in-pipe robot. When the in-pipe robot is moving in vertical pipeline, it is particularly important to act suitable preload.

3.3 *Performance comparisons between in-pipe robot with inclined wheel and traditional wheel in-pipe robot*

Since the in-pipe robot with inclined wheel adopts leaning wheel, it has more superior performance

than a traditional wheel in-pipe robot with the axes of driving wheel perpendicular to the vertical plane. According to the reference [5], the larger the actual wheel tread is, the greater the equivalent friction coefficient between wheel and tube wall is. When the driving wheel is inclined, the actual wheel treads off in-pipe robot will become greater. Therefore, the equivalent friction coefficient between driving wheel and the wall increases, giving rise to the bearing capacity of the pipeline robot.

The virtual prototype of the in-pipe robot with traditional wheel is set up. Performance comparisons between the in-pipe robot with inclined wheel and that traditional wheel are made as shown in Fig.7, Fig. 8, Fig.9 and Fig.10.

Maximum pressure of an auxiliary spring of in-pipe robot with inclined wheel is greater than that with traditional wheel exerting the same static friction coefficient and constraints. Namely, in-pipe robot with inclined wheel has the greater bearing capacity than that with traditional wheel. The reason lies in the fact that the actual wheel track of in-pipe robot

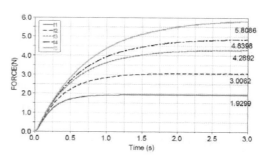

Figure 7. The force acting on the auxiliary spring of traditional wheel in-pipe robot.

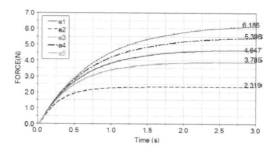

Figure 8. The force acting on the auxiliary spring of in-pipe robot with inclined wheel.

with inclined wheel is larger than that of in-pipe robot with traditional wheel. Greater bearing capacity can be obtained by adjusting the angle of inclined wheel.

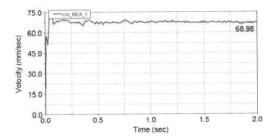

Figure 9. Velocity curve of the center of traditional wheel in-pipe robot.

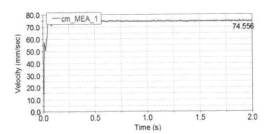

Figure 10. Velocity curve of the center of inclined wheel in-pipe robot.

In the view of crawling speed, the crawling speed of in-pipe robot with inclined wheel is higher and more stable than that of a pipeline robot with traditional wheel when the other parameters remain unchanged and only the inclination angle is changed. In both cases, nominal radius of pipeline robot has nothing in common, but the actual radius of driving wheel is different. In other words, the actual radius of in-pipe robot with inclined wheel is larger than that of pipeline robot with traditional wheel. When the driving wheel is in pure roll, the linear velocity of its center can be calculated as follows: $v=r*\omega$. Linear velocity increases with the actual radius at the constant rotate speed. In short, the in-pipe robot with inclined wheel is superior to the in-pipe robot with traditional wheel.

In addition, the connecting rod of in-pipe robot uses a telescopic structure, which is designed to ensure the small wheels of the rocker and connecting rod contact with the wall and give rise to the bearing capacity of pipeline robot. On the base of moving characteristics of four-bar linkage, if the connecting rod of planar four-bar mechanism does not adopt telescopic structure, it cannot guarantee that two small wheels of the rocker and the connecting rod simultaneously contact with the wall. Contact between two small wheels and wall relates to the inner diameter of pipeline.

In order to validate the relationship between the structure design of telescopic connecting rod and excellent performance of in-pipe robot, maximum drag force of the above two different types of

connecting rod was measured, as shown in Fig.11. Curve m and curve n represents force acting on auxiliary spring without and with telescopic structure respectively.

Maximum value of curve n is greater than that of curve m, suggesting that telescopic structure of the connecting rod contributes to increasing the bearing capacity. For the telescope structure, two small wheels can completely contact with the wall, which increases the positive pressure between the pipeline robot and

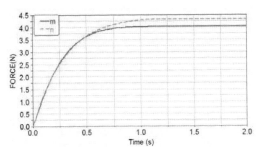

Figure 11. The force acting on an auxiliary spring curve under two different types of connecting rods.

wall. As a result, the maximum static friction force between pipeline robot and wall is raised. Thus the in-pipe robot has greater maximum drag force. In brief, greater bearing capacity and better adaptability to pipe diameter of pipeline robot can be attributed to the telescope structure of connecting rod.

4 CONCLUSION

The simulation results show that the preload of spring has an influence on bearing capacity and speed stability. When preload is in a specific range, crawling velocity basically remain the same; when the preload is relatively large, crawling speed of in-pipe robot is decreased slightly. When the prestressing force is very small, maximum tractive force changes little with its varying; when a prestressing force makes small wheels, full contact with the shell of pipe, maximum tractive force increase with preload, namely, there is a positive correlation between them. They also show that the stiffness coefficient of spring has little impact on the magnitude and stability of crawling velocity. Moreover, a comparative analysis revealed that the in-pipe robot with inclined wheel has greater carrying capacity and more stable and higher speed than in-pipe robot with traditional wheel, when the robot haves suitable parameters. And the telescope structure of the connecting rod improves the bearing capacity of the pipeline robot.

341

ACKNOWLEDGMENT

This research was supported by the Specialized Research Fund for the Doctoral Program of Higher Education of China "Research on the prediction and optimization mechanism of the acoustic performance for vehicle-use sound packages based on the hybrid method" under grant Nos. 20110146120008, National Innovation Experiment Program for University Students"Prediction and experimental study on the acoustic performance of vehicle-use sound packages based on FE-SEA hybrid model " under grant Nos. 1210504013, and Student's Science & Technology Innovation Fund of Huazhong Agricultural University under grant Nos. 11028.

REFERENCES

Yi, D., Wang, R.H, Jiang C.Y. 2009.Development of pipe-line robot and application in packaging field, Packaging and Food Machinery, 27 (3):128–130

Tan, X.S., Zhu, J.B., Yue, G.Y. 2004.Extrapolation—the Chinese version of practical training of mechanical design. The first edition, Beijing: Posts and Telecommunications Press.

Zhang, M.Y., Han, X.Q.2011. Design and Simulation of Pipeline Robot Based on Pro/E and ADAMS, Machinery Design & Manufacture. 7:22–24.

Yu, D.Z.2011. Analysis of working principle of micro pipe robot and parameter optimization, Machine Tool & Hydraulics, 39(7):77–81

Chang, Y.L., Fan, L.H., Gao, Sheng.2007.Parameters optimization of wheel pipeline robot based on genetic algorithm, Machine Design Research, 23(6):51–64.

Information Technology – Wan et al. (Eds)
© 2015 Taylor & Francis Group, London, ISBN 978-1-138-02785-5

SLM-companding algorithm for reducing the peak-to-average power ratio of OFDM systems

Ying Lin, Haiyan Chen, Xijun Zhang, Jianbin Xue

School of Computer and Communication, Lanzhou University of Technology, Lanzhou, Gansu,China

ABSTRACT: Since one of major problems of OFDM-based systems is high peak-to-average power ratio (PAPR) of its transmitted signal, many PAPR reduction techniques and combined schemes with individual techniques have recently been developed. The cause of high PAPR in IEEE 802.16d OFDM system is investigated. Based on the existed SLM and companding algorithms to reduce the PAPR, the best combined rules in algorithms are studied. The performances of combined algorithms (SLM-Companding) are analysed. Simulation results show that SLM-Companding is the better combined algorithm. It can reduce the PAPR efficiently, and has lower computational complexity and restrain noise as well.

1 INTORDUCTION

Orthogonal Frequency Division Multiplexing (OFDM) technique has been adopted in European digital audio/video broadcasting (DAV/DVB), wireless local/metropolitan area network standards (WLAN/WMAN) and Long Term Evolution (LTE), due to high spectral efficiency and excellent anti-multipath fading capability.

However, a large peak-to-average power ratio (PAPR) is a major disadvantage of OFDM, and thus various PAPR reduction techniques have been proposed for OFDM systems, such as clipping [Han et al.2005], block coding [Jiang et al.2008], partial transmit sequence (PTS) [Park et al.2007], and selected mapping (SLM) [Ku et al.2010]. Among these methods, SLM has been well investigated because of its good PAPR reduction performance.

In literatures, many methods have been proposed to solve the high PAPR problem [Ku et al.2010,Chen et al.2010,T.jiang et al.2008,Byung et al.2012]. At present, there are three classical methods to reduce high PAPR: predistortion technology is the first method[Han et al.2005, Jiang et al.2008, Park et al.2007]. This typical method is clipping and companding. The characteristic of this algorithms is cutting down the signal amplitude directly, which is irreversible and simple, but leads to great error easily. Coding technology is the second method[Chen

et al.2010,T.jiang et al.2008,Byung et al.2012]. Such as block coding, Golay complementary sequence and Reed-Muller codes. Its basic principle is using various types of encoding to avoid the appearance of OFDM symbols with high peak power. But these algorithms have high computational complexity. Distortionless method is the third one, which reduces the probability of high PAPR appears through linear transformation . Selected Mapping(SLM), Partial Transmit Sequence(PTS) and partition dummy sequence insertion (PDSI) are typical algorithm.

This article is based on the IEEE 802.16 d agreement of OFDM transmission system. The reasons of high PAPR and the existing main PAPR reduced algorithm are studied in this paper. Yet, using only one algorithm can't achieve the demand effectively. This paper proposes and evaluates a new joint algorithm of companding and SLM. The joint algorithm uses a special airy function and is able to achieve ideal reducing effect. This joint algorithm is a good compromise algorithm.

2 SYSTEM DESCRIPTION

In OFDM system, a fixed number of successive input data samples are taken from a quadrature-amplitude modulation (QAM) or quadrature- phase-shift keying (QPSK) constellation. Let

This work is supported by the National Natural Science Foundation of China (61362034), Natural Science Foundation of Gansu Province, China (1310RJYA020, 1212RJYA033)

$X = [X(0), X(1), X(2), \cdots, X(N-1)]^T$ represent the data sequence to be transmitted in an OFDM symbol with N subcarriers. The time domain complex baseband OFDM signal can be represented as:

$$s(t) = \frac{1}{\sqrt{N}} \sum_{k=0}^{N-1} X_k e^{j2\pi f_k t}, \quad 0 \le t \le T_S \tag{1}$$

where T_S is the duration of the OFDM symbol[Byung et al.2012]. In the discrete-time domain, the nth element of the over-sampled OFDM symbol vector $x = [x_0, x_1,, \cdots, x_{JN-1}]^T$ can be expressed as:

$$s_n = s(n\Delta t) = \frac{1}{\sqrt{N}} \sum_{k=0}^{N-1} X_k e^{j2\pi f_k (n/JN)}, 0 \le n \le JN-1 \tag{2}$$

where J is the over-sampling factor, which can be achieved by performing a JN-point Inverse Fast Fourier Transform (IFFT) operation to X with $(JN-1)$ zero-padding in its middle. s_n is the nth signal component in OFDM output symbol, X_k is the kth data modulated symbol in OFDM frequency domain, and N is the number of subcarriers . According to the central limit theorem, when N becomes large, both the real and imaginary parts of $s(t)$ become Gaussian distributed, each with zero mean and a variance of $\frac{E[|s(t)|^2]}{2}$, and the amplitude of the OFDM symbol follows a Rayleigh distribution. Consequently it is possible that the maximum amplitude of OFDM signal may well exceed its average amplitude. In general, practical hardware (e.g. A/D and D/A converters, power amplifiers) has finite dynamic range, therefore the peak amplitude of OFDM signal must be limited.

The PAPR of the transmitted OFDM signal can be defined as:

$$PAPR_{[s_n]} = 10\log \frac{\max_n \{|s_n|^2\}}{E\{|s_n|^2\}} \quad (dB) \tag{3}$$

where E[•] is the expected value operator.

Moreover, since the PAPR is a random variable, a more helpful description for it is the complementary cumulative distribution function (CCDF), which can be formulated as the probability that the PAPR of s exceeds an assigned threshold γ_0, i.e,

$$CCDF_s(\gamma_0) = \Pr\{PAPR_s > \gamma_0\} \approx 1 - (1-(e^{-\gamma_0}))^{JN} \tag{4}$$

3 METHOD DESCRIPTIONS

3.1 Companding technique

The basic idea of companding transformation is high power signal compressed, and the small power signal amplified, which can make the signal's average power remains the same, also can reduce the PAPR [David et al.2011].

Now, a better companding algorithm using a smooth function is proposed, namely the airy special function. The companding function can be expressed as:

$$f(x) = \beta \cdot sign(x) \cdot [airy(0) - airy(\alpha \cdot |x|)] \tag{5}$$

where $airy(\alpha)$ is the airy function of the first kind. α is the parameter that controls the degree of companding and ultimate PAPR. β is the factor adjusting the average output power of the compander to the same level as the average input power:

$$\beta = \sqrt{\frac{E[|x|^2]}{E[|airy(0) - airy(\alpha \cdot |x|)|^2]}} \tag{6}$$

where $E[•]$ denotes the expectation.
The decompanding function is the inverse of $f(x)$:

$$f^{-1}(x) = \frac{1}{\alpha} \cdot sign(x) \cdot airy^{-1}\left[airy(0) - \frac{|x|}{\beta}\right] \tag{7}$$

This algorithm is flexible in adjusting its specifications simply by changing the value of α in the companding function.

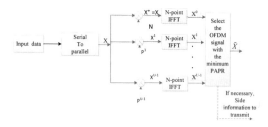

Figure 1. Block diagram of selective mapping (SLM) technique for PAPR reduction.

3.2 SLM technique with embedded SI

Figure 1 shows the block diagram of selective mapping (SLM) technique for PAPR reduction. Here, the input data block $\mathbf{X} = [X[0], X[1], \cdots, X[N-1]]$ is multiplied with U different phase sequences $P^u = [P_0^u, P_1^u, \cdots, P_{N-1}^u]^T$, where $P_v^u = e^{j\varphi_v^u}$ and $\varphi_v^u \in [0, 2\pi)$ for $v = 0, 1, \cdots, N-1$ and $u = 0, 1, \cdots, U$, which produce a modified data block $\mathbf{X}^u = [X^u[0], X^u[1], \cdots, X^u[N-1]]^T$. IFFT of U independent sequences $\{X^u[v]\}$ are taken to produce the sequences $\mathbf{x}^u = [x^u[0], x^u[1], \cdots, x^u[N-1]]^T$, among which the one $\tilde{\mathbf{x}} = x^{\tilde{u}}$ with the lowest PAPR is selected for transmission[Yong et al.2010], as shown as:

$$\tilde{u} = \arg\min_{u=1,2,\cdots,U} (\max_{n=0,1,\cdots,N-1} |x^u[n]|)$$

In order for the receiver to be able to recover the original data block, the information (index u) about the selected phase sequence P^u should be transmitted as a side information(SI). The implementation of SLM technique requires U IFFT operations. Furthermore, it requires $\lceil \log_2 U \rceil$ bits of side information for each data block[He et al.2012].

However, SLM scheme needs to transport the SI which taking up valuable frequency band. In order to overcome this defect, we choose to embed the SI in the signal without sending SI separately.

Set the subcarrier number OFDM for N, In order to calculate simply, we choose four special values as φ. The value of the corresponding phase sequence is

$$\varphi_i = [\pi/2, \pi, 3\pi/2, 2\pi]; P_i = [j, -1, -j, 1]$$

due to the phase period T = 4, number of cycles is $T_X = N/4$, all cycle T_X can be divided into odd number of cycles T_O and Even number of cycle T_E :

$$\begin{cases} am(T_O) + am(T_E) = T_X \\ am(T_O) = am(T_E) = T_X/2 \end{cases}$$

am (•) is the number of cycle in there.

Set the data on the kth subcarrier is X, according to the value of the phase sequence P_k in X, decided whether to rotate other subcarrier data phase, the following rules is shown in table I.

In this table ,we know that the φ is the the original phase of ith subcarrier. φ_i is the phase of the subcarrier after processing .

According to the phase sequence value j and -j, deciding even or odd cycle phase number of cycles corresponding data have the rotation of $\pi/4$. Then we send SI embedded in the data directly through the

Table 1. The rules of phase sequence rotation

φ_i	$P_k = 1$	$P_k = -1$	$P_k = j$	$P_k = -j$
$i \in T_E$	φ	$\varphi + \pi/4$	φ	φ
$i \in T_O$	φ	$\varphi + \pi/4$	$\varphi + \pi/4$	$\varphi + \pi/4$

above phase rotation. Using the same method to cycle extension can determine the phase sequence.

At the receiving end, checking and dealing with the adjacent two cycles or an odd one even cycles period of data can detect the sideband information and recovery the send information.

3.3 Proposed joint algorithm of companding and SLM

In this section, we introduce a new SLM-companding combined algorithm. The SLM-companding can reduce PAPR performance compared to as SLM only. Moreover, once we use the SLM before clipping, the clipping noise would be less than the one using the clipping technique alone. This is one of the main benefits of the combined scheme of clipping and SLM techniques. On the other hand, computational complexity can be further reduced if the two methods are combined. As shown in Figure 2, it should be noted that the processed data is defined in the time domain after SLM, but clipping algorithm is defined in the frequency domain. Therefore, when combined in the algorithm, we must join the FFT operation to meet the transformation from time domain to frequency domain[David et al.2011,Zhan et al.2011,Md. et al.2013].

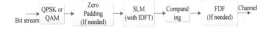

Figure 2. The block diagram of joint algorithm of SLM-companding.

The essence of SLM is adding phase rotation factor to the data sub-carrier vectors, then choose the minimum PAPR of OFDM symbols and phase combination as a transport sequence. To reach the purpose of reducing PAPR, we must find the optimal phase rotation factor at the expense of complex calculation. However, the essence of companding is clipping the signal exceeds a certain limit and amplifying small power signal. If we make companding firstly and then do SLM algorithm, the effect of companding would be destroyed during the signal reconstruction of data subcarriers. Otherwise, if we do the companding after SLM algorithm, which not only retains reducing effect of SLM algorithm, but also embodies the advantage of companding algorithm.

The BER performance of the companding scheme with the SLM technique is theoretically analyzed over fading channels. Bit error rate of a SLM-based OFDM system can be represented as follows:

$$P_{b,SLM} \oplus P_b \cdot (1 - P_{e,SI}) + 0.5 P_{e,SI} \tag{8}$$

where P_b is the average bit error probability with M-ary signaling, $P_{e,SI} = P_b \cdot [\log_2 V]$ is the probability of the erroneous detection of side information (SI) and V is the number of phase sets for the SLM. The performance of the combined scheme of companding and SLM techniques can be obtained by simply putting the $\overline{P_b}$ of companding technique. However, once we use the SLM before companding, the companding noise would be less than the one using the companding technique alone. This is one f the main benefits of the combined scheme of companding techniques. Therefore, (8) is only valid for the combined technique when the channel noise is much larger than the companding noise, and (8) is rewritten as follows:

$$\overline{P}_{b,SLM} \approx \overline{P}_b \cdot (1 - \overline{P}_{e,SI}) + 0.5 \overline{P}_{e,SI} \tag{9}$$

4 SIMULATION RESULTS AND ANALYSIS

In this section, the performance of the proposed scheme was evaluated in terms of the reduction in PAPR and the characteristic of CCDF through numerical simulations. The OFDM system used in the simulation based on IEEE802.16d protocol , which consists of 64 QAM-modulated data points. The size of the FFT/IFFT is 256, meaning a 4× oversampling.

We can see from Figure 3 that signal PAPR reduced effect mainly depends on the companding factor α in the companding algorithms. CCDF characteristics of Signal PAPR after clipping algorithm is shown in figure 3. With the increasement of α, the reduction effect of signal PAPR become better.

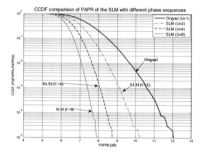

Figure 4. CCDF comparison of PAPR of the SLM with different phase cycle.

In Figure 4, we can see the extent of SLM for the reduction of signal PAPR depends on the number of the subsequence SSN (Sub Sequence Num) in PTS algorithm. With the addition of the block number N, PAPR reducing effect is remarkable, but the computational complexity is increase.

We can also see from Figure 4, when the CCDF occurrence probability is 10^{-2}, U = 4 SLM method reduce the PAPR of OFDM system value is about 7.8 dB, while U = 16 reduce the PAPR of OFDM system is nearly 7 db. With the increasement of the U value, PAPR performance in the OFDM system is gradually improved, and the probability of large peak signal appears more and more low. But this method needs to compute $U - 1$ sets of additional IFFT, receiver needs to know the selected vector random phase sequence. On the other hand, a lot of sideband information must

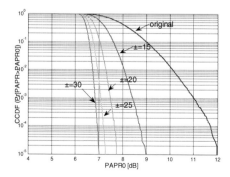

Figure 3. Complementary cumulative distribution function of original and companded.

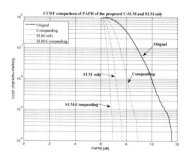

Figure 5. CCDF comparison of PAPR of the proposed ESLM and C-SLM.

be sent in the process of data transmission, which takes up a large part of the bandwidth.

In the case of the setting same parameters, we choose factor α in the companding algorithms equals to 20, while phase cycle in SLM algorithms is 8. we can see from Figure 5 that SLM-Companding algorithm to reduce the PAPR value is more effective than using SLM only. One more observation from the simulation is: the new algorithm is flexible in adjusting its specifications simply by changing the value of α in the companding function.

5 CONCLUSIONS

In this paper, the performance of the companding scheme with the SLM technique is theoretically analyzed and compared with simulation results over fading channels. The performance of combined scheme is analyzed with various clipping ratios, phase sets for SLM, and modulation schemes over flat and frequency selective fading channels. Based on the results of analysis, therefore, one can design the effective clipping scheme with the SLM technique for the PAPR reduction of OFDM-based systems.

REFERENCES

Byung Moo Lee and Youngok Kim, 2012.Performance Analysis of the Clipping Scheme with SLM Technique for PAPR Reduction of OFDM Signals in Fading Channels. Wireless Personal Communications 63: 331–344.

Chen, J.C, 2010. Partial transmit sequences for PAPR reduction of OFDM signals with stochastic optimization techniques. IEEE Transactions on Consumer Electronics 56(3): 1229–1234.

David Solomon Raju Y, Sridhar V, 2011. Efficient OFDM PAPR reduction method using linear phase variation. International Journal of Elecrtonics Communication and Computer Engineering 2:1–5.

Han, S. H, Lee, J. H, 2005. An overview of peak-to-average power ratio reduction techniques for multicarrier transmission. IEEE Wireless Communications 12(2):56–65.

He Xuansen and Zeng Qingfang, 2012. Embedding and detection method of SI based on SLM. Journal of Electronic Measurement and Instrument 26: 150–154.

Jiang, T., & Wu, Y, 2008. An overview: Peak-to-average power ratio reduction techniques for OFDM Signals. IEEE Transactions on Broadcasting 54(2): 257–268.

Ku, S. J., Wang, C. L., & Chen, C. H, 2010. A reduced-complexity PTS-based PAPR reduction scheme for OFDM systems. IEEE Transactions on Wireless Communications 9(8):2455–2460.

Md. Mahmudul Hasan, 2013. VLM Precoded SLM Technique for PAPR Reduction in OFDM Systems. Wireless Personal Communications 73:791–801.

Park, D. H., & Song, H. K, 2007. A new PAPR reduction technique of OFDM system with nonlinear high power amplifier. IEEE Transactions on Consumer Electronics 53(2): 327–332.

T. Jiang and Y. Wu, 2008. An overview: peak-to-average power ratio reduction techniques for OFDM signals. IEEE Trans. Broadcast 54(2): 257–268 .

Yong Soo Cho and Jaekwon Kim, 2010. MIMO-OFDM Wireless Communications with MATLAB , John Wiley & Sons (Asia) Pte Ltd.

ZHAN Yanyan, 2011. Study on Reducing the PAPR of OFDM Systems. Journal of Shenyang Ligong University 30(5):6–10.

Information Technology – Wan et al. (Eds)
© *2015 Taylor & Francis Group, London, ISBN 978-1-138-02785-5*

Spectrum sensing method via signal denoising based on sparse representation

Yulong Gao, Youxiang Zhu & Yongkui Ma
Communication Research Center, Harbin Institute of Technology, Harbin, China

ABSTRACT: The detection performance of traditional spectrum sensing with energy method in cognitive radio is restricted to the signal noise rate (SNR) of the received signal. The detection probability with constant false alarm rate decreases sharply because of the SNR wall. The sparse representation of signal can reveal the inherent property of the ideal signal. Using this feature, we can remove most of the noise added into the ideal signal. The sparse representation for denoising can be directly used if the sparse basis is known. If not, the K-SVD dictionary learning algorithm is adopted to build the sparse redundant dictionary. After denoising based on sparse representation, the noise energy falls down as well as the SNR increases greatly. As a consequence, compared with traditional energy detection, the sensing performance of new method improves apparently, especially, when the detection effect turns bad as the SNR falls down gradually.

KEYWORDS: spectrum sensing; signal denoising; sparse representation; K-SVD; dictionary learning

1 INTRODUCTION

It has been generally recognized that utilization of radio spectrum by licensed wireless systems, for example, TV broadcasting, aeronautical telemetry, is quite insufficient [2]. In particular, at any given time and spatial region, there are frequency bands without signal occupancy. That is, licensed signal transmissions are sparse in time domain for a given frequency band. There has been recent interest in improving spectrum utilization by permitting secondary usage using cognitive radios [1,2]. Cognitive radios use spectrum sensing to detect if frequency bands are vacant or not, and determine cognitive signal transmissions on such frequency bands to satisfy regulatory constraints of avoiding interference to authorized users.

The traditional method of spectrum sensing is energy detection; it is a sub-optimal and noncoherent detection method. The most advantage of energy detection is the short sensing time and its simplicity in implementation. But there are also several drawbacks, such as low signal noise rate (SNR) situations, and the threshold setting. The detection performance decreases sharply when the SNR falls down. With varying conditions on the channel between an authorized user and a cognitive radio user due to Gaussian noise with zero-mean, spectrum sensing can become extremely hard. To improve the performance of these detection techniques in such conditions, signal denoising based on sparse representation is proposed in this paper.

Using sparse representations for denoising of signals has drawn a lot of research attention in the past decade or so. The majority of signals in nature are with some certain characteristics, and can be sparse decomposed over a specific orthogonal basis or over-complete dictionary. On the contrary, Gaussian white noise with zero-mean is random at any time; it cannot be sparse decomposed in any orthogonal basis or dictionary. As a consequence, it is hard to separate the noise from the signal in time domain, when Gaussian noise is added into a signal, but once it is sparse decomposed over a specific orthogonal basis or over-complete dictionary, we can distinguish the signal from the noise easily. The SNR improves obviously after the signal denoising and the performance of the spectrum sensing promotes with it.

This paper is organized as follows: In Section 1, spectrum energy sensing and signal sparse representation is introduced. In Section 2, a specific spectrum sensing method: energy detection is presented.

Supported by the National Natural Science Foundation of China (NSFC) (61301101)

In Section 3, we review the signal denoising based on sparse representation and present the dictionary learning algorithm: K-SVD, when the sparse basis is unknown in advance. The simulation experiment is executed to prove the effect of the method in Section 4. Concluding remarks are given in Section 5.

2 SPECTRUM ENERGY SENSING

Energy detection is a noncoherent signal detection technique used in spectrum sensing extensively. To measure the signal power in a fixed frequency band in time domain, a band pass filter is used to the received signal and the power of the signal samples is calculated to the measurement. Figure 1 shows the detection process.

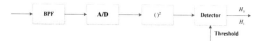

Figure 1.   The process of energy detection.

The detection is the test of the following hypotheses:

$$y(n) = \begin{cases} w(n), & H_0 \\ s(n)+w(n), & H_1 \end{cases} \quad (1)$$

where $s(n)$ is the sample of the target signal, $y(n)$ denotes the received signal detected by CR user. The noise sample $w(n)$ is assumed to be Gaussian white noise with zero mean and variance σ_w^2. The decision statistic for detector is as follows:

$$Z = \sum_{n=1}^{2TW} y(n)^2 \quad (2)$$

where $2TW$ is the length of the receive signal, T is the observation time, and W is the frequency band. For the convenience of discussion, the received signal is normalized by the noise variance, $w'(n) = w(n)/\sigma_w$, $s'(n) = s(n)/\sigma_w$. The decision statistic for energy detector is as follows:

$$Z = \sum_{n=1}^{N} y'(n)^2 \quad (3)$$

Hence,

$$Z = \begin{cases} \sum_{n=1}^{2TW} w'(n)^2, & H_0 \\ \sum_{n=1}^{2TW} (s'(n)+w'(n))^2, & H_1 \end{cases} \quad (4)$$

The normalized noise sample $\{w'(n), n = 1, 2...2TW\}$ is a standard independent and identically distributed Gaussian random variable. Therefore, in the absence of an authorized user, the statistic is the central chi-square distribution with degree of freedom (DOF) $2TW$. When the authorized user exists, the statistic follows noncentral chi-square distribution with the same degree of freedom. Noncentral parameter is as follows:

$$\delta = \sum_{n=1}^{2TW} s'(n)^2 = \sum_{n=1}^{2TW} \left(\frac{s(n)}{\sigma_w}\right)^2 = \frac{\sum_{n=1}^{2TW} s(n)^2}{\sigma_w^2} = \frac{2TWP_s}{P_n} = 2TW\gamma \quad (5)$$

where γ is the signal noise rate (SNR).

The performance of detection is measured by two parameters P_d and P_f. Each pair is associated with a particular threshold γ that tests the decision statistic

$$Z > \lambda \ decide \ signal \ present$$
$$Z < \lambda \ decide \ signal \ absent \quad (6)$$

Then P_d and P_f can be evaluated as follows:

$$P_d = P(Z > \lambda|H_1) = Q_u(\sqrt{\delta}, \sqrt{\lambda}) \quad (7)$$

$$P_f = P(Z > \lambda|H_0) = \frac{\Gamma(u, \frac{\lambda}{2})}{\Gamma(u)} \quad (8)$$

where $\Gamma(.)$ and $\Gamma(.,.)$ is the Gamma function and incomplete Gamma function. $Q_u(.,.)$ is the general Marcum Q function. u is the product of T and W. λ is the preset threshold.

In case that the signal power is unknown, the false alarm probability P_f is set to a specific constant first, the threshold λ can be determined with equation (8). Then, for the fixed number of samples $2TW$, the detection probability P_d can be evaluated by substituting the threshold in (7).

3 SIGNAL DENOISING BASED ON SPARSE REPRESENTATION

The majority of signals in nature has its own inherent property in different aspects, and can be sparse decomposed in a specific orthogonal basis or redundant dictionary. On the contrary, Gaussian white noise with zero-mean is random in time domain; as a consequence, it cannot be sparse decomposed in any orthogonal basis or dictionary. When Gaussian noise is added into a signal, it is hard to identify them in time domain, but once it is sparse decomposed in a specific orthogonal basis or over-complete dictionary, we can distinguish the signal from the noise easily. Large proportion of the noise energy has been removed away after we get the sparse representation. Hence, the detection probability with constant false alarm rate improves obviously after using the denoising signal.

3.1 Sparse representation

We address the classical signal denoising problem: an ideal signal x is corrupted by an additive Gaussian white noise n with zero-mean and standard deviation σ_n, and the observed signal y is generated with equation (9)

$$y = x + n \tag{9}$$

Assume that $x \in R^N$ has sparse representation over a specific orthogonal basis (known in advance) or over-complete dictionary, we get the denoising model as follows:

$$\hat{\alpha} = \arg\min_{\alpha} \|\alpha\|_0 \quad s.t. \quad \|y - \Psi\alpha\|_2^2 \leq \delta \tag{10}$$

And we get the restoration of signal $\hat{y} = \Psi\hat{\alpha}$.

We desire to design an algorithm that can remove the noise from y, getting the original signal x as close as possible.

3.2 Choosing the sparse basis or atoms for the sparse representation

The sparse representation ability of the signal is closely related to the selection of the sparse basis or dictionary. If we have known the sparse domain of the signals received in advance, the effect of the sparse representation would be the best. And we can get the sparse representation of the signal over the known orthogonal basis. But in the overwhelming majority of cases, the sparse domain is unknown in advance. The best practice of sparse representation is to decompose the signal over a redundant dictionary.

3.2.1 Know the sparse domain in advance
Imagine that the ideal signal is sparse over the orthogonal basis $\Psi = [\varphi_1 \ \varphi_2 \ ... \ \varphi_N]$, $\langle \varphi_i, \varphi_j \rangle = 0$, $i \neq j$, $\varphi_i, i \in [1,2...N]$ is a $N \times 1$ vector which is mutually orthogonal.

The ideal signal is decomposed as follows:

$$x = \Psi\alpha \tag{11}$$

α is the sparse representation of the ideal signal whose sparsity is K, defined by $\|\alpha\|_0 = K$.

Similarly, the noise cannot be decomposed sparsely in any orthogonal basis, but we still decompose the noise over these basis.

$$n = \Psi\upsilon \tag{12}$$

While υ is the decomposition of the noise n over the orthogonal basis which is not sparse.

Hence, we get the received signal

$$y = x + n = \Psi\alpha + n = \Psi(\alpha + \upsilon) = \Psi\beta \tag{13}$$

As a consequence, we choose the largest number of K from β, namely β', then recover the signal by the orthogonal basis Ψ

$$y' = \Psi\beta' \tag{14}$$

y' is the denoising signal used for spectrum sensing.

3.2.2 Sparse representation over a learned dictionary
The sparse basis derived from orthogonal decomposition. However, as the signal decomposition is based on the basis function, it is hard to adaptively adjust for different signal during the reconstruction process, which leads to unsatisfactory sparsification results. Therefore, it is necessary to search for better sparse basis adaptively according to the property of target signal.

Along with the development of signal processing technology, non-orthogonal decomposition arouses more and more attention. Mallat and Zhang propose the idea of sparse decomposition with redundant dictionary based on wavelet analysis [8]. Aharon and Elad combine singular value decomposition (SVD) and K-means clustering algorithms together and then put forward a over-complete dictionary training method K-SVD based on sparse decomposition [9]. For a group of given training signals, the over-complete dictionary of sparse representation can be learned self-adaptively under the sparse constraint condition using K-SVD method. The dictionary resulted from this method has good sparsification result for the signals in the trained group. And it is also effective for the signals which keep similar property with the trained ones.

The main idea of K-SVD dictionary learning algorithm is the alternatively application of the pursuit of L1-norm sparse constraint and SVD algorithm. The dictionary is updated simultaneously with sparse decomposition coefficients. Consequently, it is possible to adaptively adjust for a set of signal.

Arranging L training signals column vectors w_i to the matrix form $W = \left[w_i\right]_{i=1}^{L}$, defining the initial trained dictionary and the sparse coefficients as

$D = [d_m]_{m=1}^{M}$ and $X = [x_l]_{l=1}^{L}$, respectively, we can describe the K-SVD algorithm model as follows:

$$\min_{D,X} \|W - DX\|_2 \quad s.t. \quad \forall i \; \|x_i\|_0 \leq K \tag{15}$$

where K represents the given sparse level.

The steps to solve equation (15) consist of dictionary initialization, sparse coding, dictionary iteration, and iteration termination condition.

After we get the learned dictionary D, the received signal can be decomposed over it by the sparse representation with Orthogonal Matching Pursuit (OMP).

$$y = x + n \approx D(\alpha + \upsilon) = D\beta \approx D\beta' = y', \|\beta'\|_0 = T_0 \tag{16}$$

The iterations of the OMP algorithm or the sparse level of β' is usually higher than K according to the redundancy of dictionary to get enough energy of the ideal signal as the dictionary is over-completed.

3.3 Signal denoising

Imagine that the sparse representation of the signal contains most percentage of the original signal, namely

$$y' = x + n' \tag{17}$$

Then, the noise energy removed can be estimated as follows:

$$E_{denoising} \approx E_n - E_{n'} = E_y - E_{y'} \tag{18}$$

Therefore, the energy of the noise after sparse representation is $E_{renew\_n} = E_n - E_{denoising}$, thus, the noise variance turn out to be $\sigma_{renew\_n}^2 = E_{renew\_n}/N$.

$\sigma_{renew\_n}^2$ ought to be smaller than σ_n^2, as we remove most part of the noise using the sparse representation, and then spectrum sensing solution can be gained by the process in Section 2.

After the sparse representation for signal denoising, the large percentage energy of the noise has been eliminated from the received signal. Using the denoising signal in spectrum sensing, the detection probability improved largely compared with traditional direct energy detection.

4 EXPERIMENTS

In this section, signal denoising based on sparse representation is used in spectrum energy sensing.

As the ideal signal is sparse in some specific domain, it is sensible to assume that the signal is sparse over DCT basis. To confirm the sparsity of the signal, we generate random location and amplitude α in DCT basis first, and transform it into time domain as follows:

$$x = \Psi\alpha, \|\alpha\|_0 = K \tag{19}$$

Where x and α are all $N \times 1$ vectors, the number of nonzero value in α is K, the value and the location is randomly generated, Ψ is a $N \times N$ matrix whose columns are DCT atoms.

Hence, the signal x is sparse over DCT basis. Adding the noise according to the SNR in time domain, we get the actual received signal.

In our simulations, the length of the signal N is 64 and the sparsity $K=2$, the adding noise produced every time is determined by the energy of the ideal signal and the SNR we set.

We distinguish the simulation in terms of if the sparse domain is known in advance or not.

4.1 The sparse domain known in advance

To show the denoising performance of the sparse representation, the signal with 3 dB Gaussian noise is shown in Figure 1. Using signal denoising based on sparse representation (DCT basis), we get the signal which has removed most of the noise, as the Figure 2 shows.

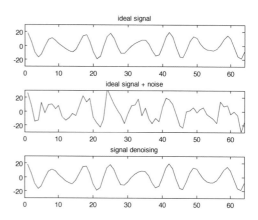

Figure 2. The time domain of (a) ideal signal (b) ideal signal + noise (c) signal denoising using orthogonal basis.

As we can see from the Figure 2, the denoising signal is almost close to the ideal signal.

Then we use the denoising signal in spectrum sensing in different SNR circumstance, setting the false alarm probability $P_f = 0.05$, and get the threshold λ in (8). Finally, detection probability improvement is shown in Table 1, respectively.

A promotion in detection probability is obvious when the SNR is between -7 dB and -2 dB, as we can see from Table 1. There is no promotion in detection probability when the SNR is higher than -1 dB, as the detection probability has already reached 1.

Table 1. Improvement in detection probability.

Improvement in detection probability						
SNR	-7	-6	-5	-4	-3	-2	-1
Pd	0.26	0.37	0.45	0.55	0.78	0.90	1
Pd′	1	1	1	1	1	1	1

Table 2. Improvement in detection probability.

Improvement in detection probability						
SNR	-7	-6	-5	-4	-3	-2	-1
Pd	0.26	0.37	0.45	0.55	0.78	0.90	1
Pd′	0.88	0.98	0.99	1	1	1	1

4.2 Unknown sparse domain

To get the learned over-complete dictionary, we choose the DCT dictionary (64 × 256) to be the initial trained dictionary. Then arrange L=100 training signals just as the ideal signal we get in the last section.

After we get the learned dictionary D through K-SVD algorithm, then decompose the signal with 3 dB Gaussian noise. As the dictionary is redundant (64 × 256), we choose the iterations of the OMP algorithm $T_0 = 3K = 6$ here.

$$y \approx D\beta', \quad \|\beta'\|_0 = T_0 \qquad (20)$$

Transforming the T_0 big coefficient β' back into the time domain, $y' = D\beta'$, we get the denoising signal as the Figure 3 shows.

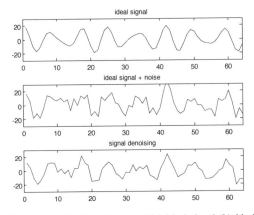

Figure 3. The time domain of (a) ideal signal (b) ideal signal + noise (c) signal denoising using learned dictionary.

As we can see from the Figure 3, the noise energy added to the signal reduce greatly after denoising by sparse representation.

Using the denoising signal in spectrum sensing in different SNR circumstance, the detection probability improvement is shown in Table 2 respectively.

From the Figure 2 and Figure 3, it can be seen that the denoising based on sparse representation is effective. Most of the noise is removed away after the reconstruction through sparse representation. Meanwhile, the improvement in detection probability with constant false alarm rate is apparent as we can see from Table 1 and Table 2, when the SNR is between -7 dB to -2 dB.

Certainly, with the change in the observation time (number of samples), the detection probability will change. What is more, the sparsity of the ideal signal has some influence on the effect of the denoising.

5 CONCLUSION

In this paper, a spectrum sensing method via denoising based on sparse representation is proposed. The effect of denoising through sparse representation is significant, as a consequence, the performance for spectrum sensing improves obviously. This meaningful improvement is promising in some environment where the SNR is low enough to detect using the traditional spectrum sensing method.

ACKNOWLEDGMENT

This work was supported by the National Natural Science Foundation of China (NSFC) under Grant 61301101.

REFERENCES

[1] D. Cabric, I. D. O'Donnell, M. S.-W. Chen and R. W. Brodersen, "Spectrum sharing radios," IEEE Circuits and Systems Magazine, pp. 30–45, 2006.
[2] Federal Communications Commission—First Report, and Order and Further Notice of Proposed Rulemaking, "Unlicensed operation in the TV broadcast bands," FCC 06–156, Oct. 2006.
[3] Y. L. Polo, Y. Wang, A. Pandharipande, and G. Leus, "Compressive Wide-Band Spectrum Sensing," IEEE International Conference on Acoustics, Speech, and Signal Processing, pp.2337–2340, 2009.
[4] Zhai Xuping, Pan Jianguo, "Energy-Detection Based Spectrum Sensing for Cognitive Radio," IET Conference on Wireless, Mobile and Sensor Networks, pp. 944–947, 2007.
[5] Li C, Ma L, Wang Q, Zhou Y, Wang N, "Construction of sparse basis by dictionary training for compressive

sensing hyperspectral imaging," International Conference on Acoustics, Speech, and Signal Processing, pp.1442–1445, 2013.

[6] A. Koutrouvelis. A. Harma. "Compressive Sensing in Footstep Sounds, Hand Tremors and Speech Using K-SVD Dictionaries." The 18th International Conference on Digital Signal Processing, 2013, July 1–3.

[7] Urkowitz. "Energy detection of unknown deterministic signals," Proceedings of the IEEE, VOL. 55, NO. 4,pp, 523–526, APRIL 1967.

[8] Mallat, S. G., Zhang, Z. "Matching Pursuits with Time-frequency Dictionaries," IEEE Transactions on Signal Processing, vol.41, no.12, pp.3397–3415, 1993.

[9] Aharon, M., Elad, M., Bruckstein, A. "K-SVD: An Algorithm for Designing Overcomplete Dictionaries for Sparse Representation," IEEE Transactions on Signal Processing, vol.54, no.11, pp.4311–4322, 2006.

[10] Fadel. F. Digham, M. Alouini, K. Simon, "On the energy detection of unknown signals over fading channels," IEEE Transactions on Communication, vol.55, no. 1, pp.21–24, 2007.

Information Technology – Wan et al. (Eds)
© *2015 Taylor & Francis Group, London, ISBN 978-1-138-02785-5*

Study of cardiac hemodynamics based on collecting ECG signal

Li Ke & Yue Zhao
Institute of Biomedical and Electromagnetic Engineering,
Shenyang University of Technology, Shenyang, Liaoning, China

Qiang Du
School of Medical Devices,
Shenyang Pharmaceutical University, Shenyang, Liaoning, China

ABSTRACT: The cardiac hemodynamic has a close relationship with human health, so the study of cardiac hemodynamic becomes crucial. The main purpose of this paper is to extract cardiac hemodynamic parameters based on human's ECG signal to study cardiac hemodynamic. Collecting ECG signal from 10 people and designing the method of baseline correction and noise removal processing. Then marking ECG signal feature points for further calculation of cardiac hemodynamic parameters: Heart rate (HR), stroke volume (SV), cardiac output (CO), cardiac index (CI). Results show that main cardiac hemodynamic parameters can be obtained by ECG signal, and also can be used as a function parameter evaluation of human body, and have certain reference significance for health check.

1 INTRODUCTION

Cardiovascular disease has become one of the highest morbidity and mortality disease. The scientific and rational treatments of cardiovascular disease are from the correct diagnosis, so the monitoring of cardiac function and hemodynamic is more significant. Cardiac hemodynamic is closely related to human body health, if blood flow of heart underpowered may cause shock, cardiac failure, myocardial infarction and other major diseases, resulting in breathing difficulties in patients with cardiac failure, and symptoms and physical signs are lack of specificity (Cheng, 2001). Early hemodynamic monitoring can be measured cardiovascular pathological and physiological changes.

In order to obtain cardiac hemodynamic parameters, pulmonary artery floating catheter method is widely used clinically. Pulmonary artery floating catheter method is invasive and risky and not suitable for repeated use, the demand of operation technology level is higher, the applicable scope is limited (does not apply to critically ill patients, mild patients and healthy people) as well as the monitoring cost is high, some studies have shown that invasive check can result damage (Rhedes et al., 2002, Hnsky & Vincent, 2005) and the application is limited (Hong, 2005).

Impedance cardiogram method is noninvasive to evaluate cardiovascular function and one of cardiovascular hemodynamic applications in human impedance. Currently widely used is the Kubicek theory – chest is regarded as cylinder model (Sise et al., 1981), Kubicek improved formula (Bernstein & Lemens, 2005) to calculate the SV, but the technology is not clinically widely used.

Ultrasonic cardiogram is divided into M type, Doppler, two-dimensional and three-dimensional echocardiography (Wang, 2009), in the first 3 kinds of echocardiographic measurements of left ventricular ejection fraction, apex biplanar Simpson method of measurement results are more reliable, its accuracy is only less than three-dimensional echocardiography (Wang, 2009). But due to technical and cost reasons, such as three-dimensional echocardiography is still cannot replace two-dimensional echocardiography in measuring left ventricular ejection fraction. The twote-dimensional echocardiography has certain limitations yet (Wang, 2009): (1) the image is affected by acoustic window condition and two-dimensional image quality; (2) accuracy decreases by ventricular aneurysm formation and segmental ventricular wall motion abnormalities of left ventricular shape deformation; (3) measurement for the same patient, will be introduced by the operator of different errors. The above three methods in calculating cardiac hemodynamic parameters are insufficient, body surface electrocardiogram (ECG) is simple measurement and widely used clinically, therefore, on the basis of the ECG signal to calculate the cardiac hemodynamic parameters is faster and more convenient. This article introduces a set of system: collecting ECG signal, processing and analysis. Calculate cardiac hemodynamic parameters by processing the body surface ECG signals, and then analyze the parameters. Collecting method is noninvasive, simple and inexpensive and can be used in the calculation of cardiac hemodynamic parameters, finally achieve the goal of monitoring hemodynamics.

2 COLLECTING ECG SIGNAL

2.1 *ECG signal collecting system*

The ECG signal collecting system is developed by our laboratory independently. Collect ECG signal of experiment subjects. Collecting lead ECG signal: left upper limb of the human body with positive electrode, right upper limb of the human body with negative electrode, record the potential difference between left arm and right arm, right lower limb to ground. Electrode placement is as shown in Fig. 1(a), ECG display part adopts VB interface in PC window, as shown in Fig. 1(b), is a more stable ECG signal.

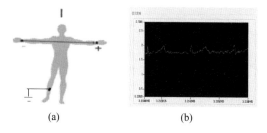

(a)   (b)

Figure 1.   (a) Electrode placement; (b) PC window display.

2.2 *Ecg signal processing*

As shown in Fig. 2(a) for the original ECG signal, the data is with the baseline drift and noise interference. The collected ECG signal requires pretreatment of baseline correction and filtering to calculate the related parameters of cardiac hemodynamics.

2.2.1 *Remove the baseline drift*
Baseline drift seriously affect the ECG signal analysis and processing. Baseline drift is an important influence factor in ECG signal amplitude measurement and morphological analysis. Talking, laughing, and mobile situation of experiment subject will produce a great deal of baseline drift in the process of collecting ECG signal, so the inhibition of baseline drift is the precondition of correct information of ECG. This article selects db3 wavelet to the original ECG signal in 9 layer decomposition (Ren & Yang, 2010), with the increase of scale, time resolution decreases, high frequency signal reduces gradually, the rest of

the signal approximates baseline drift component. Before the signal reconstruction process, the wavelet coefficients to be properly selected, abandon the low-frequency wavelet coefficients, all the original value of the high frequency coefficients are reserved, only the low frequency coefficient is zero, so that can correct baseline drift in the ECG signal, treatment effect as shown in Fig. 2(b), through compared with the original ECG signal, after baseline drift correction, signal value calibrates from 2.4 mV to 0 mV.

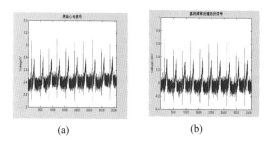

(a)   (b)

Figure 2.   (a) Original ECG signal; (b) Remove the baseline drift.

2.2.2 *Remove the noise and feature point mark*
In the process of collecting, ECG signals will contain noise because of subject, movements. Electromyographic signal noise may come from human skin, so wipe some physiological saline or some coupling agent on electrode placements to reduce noise. Larger noise will seriously affect the calculation of cardiac hemodynamic parameters, so after baseline correction, ECG signal is needed to remove the noise. This article uses the Butterworth low-pass filter to remove the noise of ECG signals after baseline correction. Result of removing noise is shown in Fig. 3(a), through compared with the ECG signals after baseline correction, high frequency noise is filtered obviously, and it is advantageous for marking the feature points of ECG signal.

After low pass filtering processing of ECG signals, mark P wave, QRS complex and T wave of the feature points in the tag, computing the amplitude of the three waves and the R–R period, which are used in calculating cardiac hemodynamic parameters. The results of marking is shown in Fig. 3(b), each feature point is clear and accurate, on the basis of accurate marked feature points can make calculation of R–R period and amplitude more accurate.

2.2.3 *The calculation of hemodynamic parameters*
Heart rate: Heart rate is averaged about 75 times/min (60–100 times/min) in normal adults. In this paper, the calculation of heart rate is based on the electrocardiogram, with each average R–R period to calculate

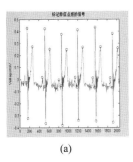 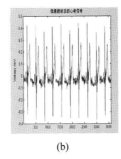

| (a) | (b) |

Figure 3. (a) Low pass filtered ECG signal; (b) Feature points mark.

the number of the R wave per minute, to determine the heart rate, unit: times per minute (bpm).

Left ventricular ejection fractions: volume quantity percentage of stroke volume in ventricular end-diastolic. Patrick (2005) invented the formula according to the body surface ECG computing *LVEF* in 2005, as in

$$LVEF = \beta 1 \times avR + \beta 2 \times age \tag{1}$$

In Eq. (1), *LVEF* represents left ventricular ejection fraction, $\beta 1$ and $\beta 2$ are constant parameters, the value of $\beta 1$ is 0.264, the value of $\beta 2$ is 0.645, *avR* represents the amplitude value (mV) sum of R wave and S wave in ECG signal, which are determined by the amplitude results of marking feature points of ECG signal, *age* is the actual age of participants. Eq. (1) has been proved that can be used to estimate the ejection fraction. Ejection fraction can also be calculated by stroke volume, as in

$$EF = SV(ml) / VEDV(ml) \times 100\% \tag{2}$$

EF represents ejection fraction, *SV* represents stroke volume of the heart, *VEDV* represents the volume of ventricular end-diastolic. Normal volume of left ventricular end-diastolic is about 145 ml. When calculating the parameters, the volume of left ventricular end-diastolic is estimated for 145 ml. Calculate stroke volume according to Eq. (2).

Stroke volume: the volume of blood that injected by one side of the ventricular in a heart beat, referred to as "stroke volume". In this paper, stroke volume is calculated by ejection fraction with equation (2), ejection fraction is calculated by Eq. (1).

Cardiac output: the volume of blood that injected by one side of the ventricular per minute, referred to as "cardiac output", as in

$$CO = HR \times SV \tag{3}$$

CO represents the cardiac output, the unit is L/min, *HR* represents the heart rate, *SV* represents stroke volume in Eq. (3).

Cardiac index: in terms of surface area of every square meter of cardiac output, as in

$$CI = CO / SA \tag{4}$$

CI represents cardiac index, the unit is L/(min·m²), *CO* represents the cardiac output, *SA* represents the human body surface area, Eq. (5) is suitable for calculating surface area of the Chinese people (Hou, 1985):

$$SA = 0.0061 \times H + 0.0128 \times W - 0.1529 \tag{5}$$

SA represents the human body surface area; *H* represents a person's height, the unit is cm; *W* represents a person's weight, the unit is kg. The body surface area of normal medium height adult is 1.6–1.7 m².

Then calculating the parameters with the processed ECG signals according to the above equations.

3 THE RESULT OF THE EXPERIMENT AND ANALYSIS

3.1 *The experiment subjects*

Experiment subjects were healthy, height range from 153 cm to 190 cm, weight range from 45 kg to 83 kg, age range from 23 to 30. Four girls: height range from 153 cm to 160 cm, weight range from 45 kg to 67 kg, age range from 23 to 24; six boys, height range from 165 cm to 190 cm, weight range from 56 kg to 83 kg, age range from 24 to 30.

3.2 *The experimental process*

In the process of experimental subjects were resting flat, wiped some physiological saline and some coupling agent on electrode placements, then red electrodes placed on the right upper limb, yellow electrodes placed on the left upper limb, black electrodes placed on the right lower limb. Then observed the ECG signal of PC display window and began collecting when the ECG signal was smooth. The experiment subjects remained resting flat, no talking, no moving in collecting process. Collected two sets of data, 3 min of every subject each time, the sampling frequency was 360 Hz, the data were stored in TXT format.

3.3 The experimental results

Processed ECG signals with baseline drift correction and low pass filtering, and then marked ECG signal feature points, calculated the amplitude and R–R period between the waveform, finally got cardiac hemodynamic parameters HR, SV, CO, CI in girl and boy groups respectively, the results are shown in Tables 1 and 2:

Table 1. Cardiac hemodynamic parameter calculation results of girl group.

Serial number	Height (cm)	Weight (kg)	HR (bpm)	SV (mL)	CO (L/min)	CI L/ (min·m²)
Data 1	153	67	62.5	93.7	6.1	3.4
Data 2	154	56	80	80.5	6.4	3.9
Data 3	160	49.5	66.4	90.8	6.1	3.8
Data 4	160	45	93.4	69.5	6.5	4.3

Table 2. Cardiac hemodynamic parameter calculation results of boy group.

Serial number	Height (cm)	Weight (kg)	HR (bpm)	SV (mL)	CO (L/min)	CI L/ (min·m²)
Data 5	170	65	77.7	76.5	5.9	3.2
Data 6	169	57.5	77.4	60.5	4.7	2.7
Data 7	168	67.5	71.5	82.9	5.9	3.2
Data8	190	75	65.7	58.1	3.8	2.1
Data 9	165	56	64.3	70.1	4.5	2.7
Data10	178	83	60.3	95.7	5.8	3.4

3.4 The experimental results analysis

Normal range of HR is from 60 to 100 times/min; normal range of SV is from 60 to 80 ml; normal range of CO is from 4.5 to 6.0 L / min in quiet state of healthy adult, woman's CO is lower 10% than man's with the same weight; normal range of CI is from 3.0–3.5 L / (min • m²) in quiet state (Zhu, 2009).

The results of a girl group can be seen from Table 1. Data 1 is overweight; SV is much higher; Data 2 is a bit over weight, the various parameters are slightly higher. Data 3 is body symmetry, SV is higher than the normal range, the other parameters are within the normal range; Data 4 is relatively thin. HR, CI and CO are a bit higher. The results of boy group can be seen from Table 2: Data 5 is body symmetry and in good health, parameters are within the normal range; Data 6 is relatively thin, CI is lower; Data 7 is body symmetry, SV is a bit higher, other parameters are normal; in addition to HR is low, the other parameters have larger errors in Data8; Data 9's CI is lower, the other parameters are normal; Data 10 is over weight,

SV is a bit higher, the other parameters are normal. In general most cardiac function parameters are within the normal range.

4 CONCLUSION

Based on the laboratory of ECG signal collecting system to complete the collection of ECG signal, designed to remove baseline drift and noise in the process of data processing and marked feature points of ECG signal, completed calculation and analysis of the cardiac hemodynamic parameters: HR, SV, CO, CI, the results of calculation are almost in the normal range. From signal collection and signal processing to the analysis of the results, it proves the feasibility of the whole system. The system can be taken as a reference to the evaluation of human body function parameter, which has certain reference significance for health check.

REFERENCES

Bernstein, D.P. & H.J. Lemens. Stroke volume equation for impedance cardiography. *Medical & Biological Engineering & Computing*, 2005, 43(4), 443–50.

Cheng, Y.L. The diagnosis and treatment of the elderly heart failure. *Journal of the Chinese Elderly Medical Journal* 2001, 20(2), 146-150.

Hnsky, M.R. & J.L. Vincent (2005). Let us use the pulmonary artery catheter correcdy and only when we need it. *Critical Care Medicine*, 2005, 33 (5), 1119–1122.

Hong, H., X.J. Jin, & C.Z. Pan. A noninvasive cardiac hemodynamic monitor with echocardiography to detect cardiac function index correlation analysis. *Chinese Journal of Medical Apparatus and Instruments*, 2009, 33(5), 328-331.

Hou, Z.L. (1985). Chinese Medical Encyclopedia: Physiology. Shanghai, Shanghai Science and Technology Publishing, 167.

Patrick, R.K.. Left ventricular ejection fraction can be derived from simple ECG measurements. *J Am Coll Cardiol.*, 2005, 45(3), SupplA118A.

Ren, J. & L.X. Yang. Based on the wavelet transform coefficients of ecg signal baseline drift noise removal method. *Journal of Medical and Health Care Equipment*, 2010, 31(11), 24–26.

Rhedes, A., I.U. Cmack, & P.J. Newman. A randomized controlled trial of the pulmonary artery catheter in critically ill patients. *Int Cam Med.*, 2002, 28(3), 256–264.

Sise, M.J., P. Hollingsworth, & J.E. Brimm. Complications of the flow directed pulmonary artery catheter. A prospective analysis in 219 patients. *Critical Care Medicine*, 1981, 9, 315–318.

Wang, X.F. Echocardiography. Beijing, People's Medical Publishing, 2009, 998.

Zhu, D.N. Echocardiography. People's Medical Publishing House, 978, 2009.

Information Technology – Wan et al. (Eds)
© 2015 Taylor & Francis Group, London, ISBN 978-1-138-02785-5

The noise reduction in indirect-conduct electrocardiograph detection implements by electric potential integrated circuit sensor

Hao Sun
City College, Kunming University of Science & Technology, Kunming, China

Wenjing Hu
Applied Technology College, Dalian Ocean University, Dalian, China

Wei Chen
Biomedical Engineering Department, Zhejiang University, Hangzhou, China

ABSTRACT: We present a gel-free, indirect-contact ECG detector with a couple of Electric Potential Integrated Circuit (EPIC) sensors. EPIC is a capacitive sensor which detects the coupling capacitance between skin and sensor surface and so does not rely on ohmic contact to the body for measuring bio-electrical signals. For a capacitive sensor, the power frequency noise will heavily interface the weak electrophysiological signals. Here, we design a Driven-Right-Leg (DRL) system coupled capacitively to the body unlike the conventional ECG configuration which directly coupled the DRL signal to the patient's skin. Result show the method can effectively reduce the 50Hz power line noise to -40dB when the DRL feedback gain had chosen to 100, and the conductive fabric area will heavily influence the quality of ECG signals, in our experiment, the optimum conductive fabric area is 10cm×10cm.

1 INTRODUCTION

Recently, the home test biomedical diagnostic instruments for healthcare and early diagnosis are more and more popular. It is crucial that these devices must be safety and convenience when use at home without professional surveillance. For this purpose, many approaches had been proposed by researchers, including the so called Indirect-Contact ECG (IDC-ECG). (Harland 2003, Chi 2010). Compared to the conventional ECG extraction system, the most remarkable feature of IDC-ECG lies about it is a gel-free, indirect-contact or dry-contact system which excludes the risk of electric shock for heart and convenient to wearing (Lim 2006, Sullivan 2007). It is promising that IDC-ECG has a wonderful prospect in home diagnostic and healthcare (Jung 2012, Baek 2012). Nevertheless, the drawback of IDC-ECG is the intrinsic noise induced by power line which coupled the 50Hz power frequency into ECG, this limit the application of IDC-ECG.

This paper shows a capacitive coupling Driven-Right-Leg (DRL) configuration for power line noise reduction. Result show the method can effectively reduce the 50Hz power line noise.

2 INDIRECT-CONDUCT ECG DETECTION

IDC-ECG is a state-of-the-art ECG extraction technology which has the ability to measure ECG without direct skin contact. The pursuit of monitoring ECG in any case such as during exercise, sleeping or working, driven the researchers dedicated into the IDC-ECG technology. YG Lim and his coworkers design an IDC-ECG device for monitoring ECG during sleep that is adequate for long-term use is provided. The large conductive textile sheet was used in their design (Lim 2007). Tsu-Wang Shen has given an ultra-wearable smart sensor system combines electrocardiogram (ear-lead ECG) which measures the ECG signals from an ear to an arm. They use metal shell of earphone as the lead of 3-lead ECG detection circuits (Figure 1). According to their research, the ear-lead is a linear combination with regular ECG lead I and III (Shen 2008). Chong-Rong Wu develop an indirect-contact ECG measurement system with their well-designed movable electrode. His experiment shows the signal quality of movable electrode is higher than the one of the fixed electrodes when moving forward and backward as well as the turning left and right (Wu 2013).

Figure 1. Ear-lead dry electrode placement of ear-lead ECG.

Figure 2 shows the architecture of IDC-ECG system which include a couple of EPIC sensor, capacitive coupling Driven-Right-Leg (DRL) electrode, the instrumental amplifier and filters.

Here, we use the PS25201A as the capacitive coupling sensor which brought from Plessey Semiconductors Ltd. Figure 3 shows the photograph of PS25201A. The PS25201A is an ultra high impedance solid state ECG sensor with high input resistance, typically 20GΩ and low input capacitance, typically 15pF.

The EPIC sensor is highly sensitive to the potential variation of our skin, which capacitively coupled with sensor; the potential variation had been amplified by adaptive feedback amplifier module inside the sensor before it was sent to the differential amplification circuit. The adaptive feedback techniques can both lower the effective input capacitance of the sensing element (Cin) and boost the input resistance (Rin). The technique is suitable for monitoring ECG signals with a favorable frequency response characteristic.

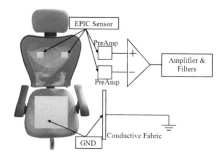

Figure 2.   The architecture of IDC-ECG system.

The IDC-ECG detection circuit implements by 'capacity ground' which without the need to contact the skin of tester directly. In our experiment, we cover the test-chair with a sheet of conductive fabric, show in (Figure 2), which connects to the detection circuit's ground for ensuring the high performance of 'capacity ground'.

Figure 3.   The photograph of PS25201A.

3   THE NOISES COME FROM POWER LINE

There are a large number of researchers had studied the noise reduction method of the ECG detection system (Leonhardt 2008, Chamadiya 2009, Chi 2010). We simplify the schematic of the IDC-ECG detector's common-mode noise come from the power line, see in (Figure 4). Here, the key point we concerned focus on the noises capacitively coupled from the power line. Here, V_N is the power voltage, V_B represents the potential of the human body, Z_N means the impedance between power line and our body, Z_G is the impedance between our body and ground, Z_{E1}、Z_{E2} represent the impedance of two EPIC sensor, Z_{A1}、Z_{A2} represent the input impedance of the amplifier for EPIC signal enhancement, V_{A1}、V_{A2} represent the output voltage of the two amplifiers.

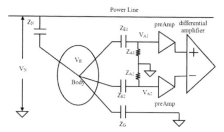

Figure 4.   The schematic of the IDC-ECG detector's common-mode noise come from the power line.

It is very common that using the differential amplification circuit to extract ECG signals. Theoretically, the differential amplification circuit can reduce the common-mode noise effectively, but in practice, the effect of the noise reduction often not as well as we expect. It is caused by the large difference of impedance between the two channels of input signals which mainly come from the impedance difference of tow EPIC sensor which caused by the interference coupled in body.

$$V_D = V_{A1} - V_{A2} = \left(\frac{Z_{A1}}{Z_{E1} + Z_{A1}} - \frac{Z_{A2}}{Z_{E2} + Z_{A2}} \right) \times V_B \quad (1)$$

$$V_D = V_B \times \frac{\Delta Z_E}{Z_A} \quad (2)$$

where, $Z_A = (Z_{A1} + Z_{A2})/2$, $\Delta Z_E = Z_{E1} - Z_{E2}$. Equation.1 describes the two EPIC sensor's differential noise V_D (which caused by the common-mode noise V_B) by the asymmetry of two EPIC impedance. When $Z_E \ll Z_A$, Equation. 1 can be substitute by Equation.2. It can be concluded from Equation 2 that V_D is in direct proportion to the noise of V_B and $\triangle Z_E$ and in inversely proportional to Z_A.

4 DRIVEN RIGHT LEG (DRL) CIRCUITS

Equation 2 illustrates the key point of reducing the differential noise of ECG signal is to restrain the common-mode noise (Lim 2010). One of the most effective methods of common-mode noise reduction is the DRL circuit (Driven-Right-Leg Circuit). In the DRL configuration, the common side of EPIC sensors connects the right leg and the system ground, see (Figure 5). From the viewpoint of the common-mode noise side, the impendence between human body and systems will be sharply reduced, thanking the inverse amplification circuit which drives the common-mode noise.

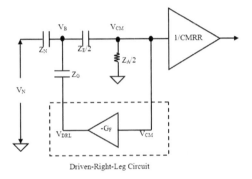

Driven-Right-Leg Circuit

Figure 5. The common-mode noise model, including Driven-Right-Leg circuits.

$$V_{CM} = \frac{Z_A Z_G V_N}{(Z_A + Z_E)Z_G + ((1+G_F)Z_A + Z_E + 2Z_G)Z_N} \quad (3)$$

Equation 3 describes the effect of DRL on the reduction of common-mode noise, where V_{CM} represents the common - mode component. It is obvious that increasing the G_F (Gain of the DRL) can reduce the common-mode noise on the system. In addition, increasing the Z_G (impendence between our body and system ground) also can reduce the V_{CM} (Lee 2009).

In practice, we use the INA321 as differential amplifier which drive the DRL with its input and connect to a big sheet of conductive fabric between our body and system ground through the inverse amplifier.

5 RESULT AND DISCUSSION

In our experiment, the ECG signal come from INA321 output is filtered by a band filter with frequency range from 0.5Hz to 200Hz and sampled with the sample rate of 1 KHz. The subject wore a cotton sweat suit with thickness less than 0.5mm. The influence of G_F and the conductive fabric area to ECG signal quality had been studied.

The influence of G_F to ECG waveform shows in (Figure 6). Here, we connect ground with a piece of wire, to exclude the interference of the ground contact area.

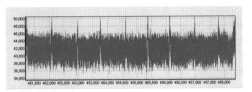

(a) Passive ground

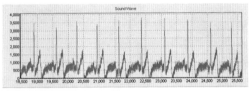

(b) G_F is 10

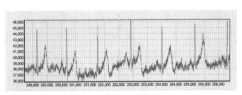

(c) G_F is 100

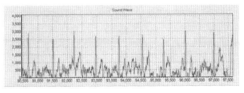

(d) G_F is 500

Figure 6. The influence of G_F to ECG waveform.

It can be concluded from (Figure 5) that capacitive coupling DRL circuit can effectively reduce the power line noise when the DRL feedback gain less to 100, the SNR is -40dB, it matches the compute result of (Eq.3), but the SNR will reduce when the G_F greater than 100, this is caused by the noise of power line's non-common-mode component. Given this, the G_F of DRL circuit commonly set to 100.

The influence of conductive fabric area to ECG waveform shows in (Figure 7). Here, we set G_F to 100. It is obviously that enlarging ground contact area of the body can effect reduce the noise, but oversize the area contact to the body will induce to ECG signal distortion.

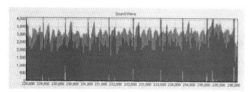

(a) Passive ground

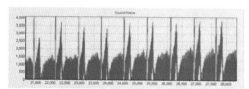

(b) Connect ground with a piece of wire

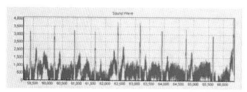

(c) Connect ground with a piece of 1cm×1cm conductive fabric

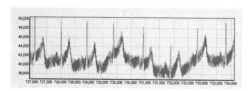

(d) Connect ground with a piece of 10cm×10cm conductive fabric

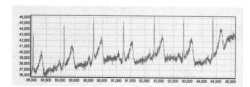

(e) Connect ground with a piece of 30cm×30cm conductive fabric

Figure 7. The influence of conductive fabric area to ECG waveform.

Therefore, the DRL circuit with large contact area and high gain is helpful in increasing the SNR. But the oversize ground contact area can induce the ECG signal distortion and when the gain of the DRL higher than 100, the power line's non-common-mode component will influence the ECG signals.

ACKNOWLEDGMENT

This study is supported by the National Science Foundation of Yunnan Province, China (No. 2009ZC051M) and General Research Program of Zhejiang Medicine and Health (No. 2013KYA084).

REFERENCES

Baek, H. J., Chung, G. S., Kim, K. K., & Park, K. S. (2012). A smart health monitoring chair for nonintrusive measurement of biological signals. Information Technology in Biomedicine, IEEE Transactions on, 16(1), 150–158.

Chamadiya, B., Heuer, S., Hofmann, U. G., & Wagner, M. (2009, January). Towards a capacitively coupled electrocardiography system for car seat integration. In 4th European Conference of the International Federation for Medical and Biological Engineering (pp. 1217–1221). Springer Berlin Heidelberg.

Chi, Y. M., & Cauwenberghs, G. (2010, June). Wireless non-contact EEG/ECG electrodes for body sensor networks. In Body Sensor Networks (BSN), 2010 International Conference on (pp. 297–301). IEEE.

Harland, C. J., Clark, T. D., & Prance, R. J. (2003). High resolution ambulatory electrocardiographic monitoring using wrist-mounted electric potential sensors. Measurement Science and Technology, 14(7), 923.

Jung, S. J., Shin, H. S., & Chung, W. Y. (2012). Highly sensitive driver health condition monitoring system using nonintrusive active electrodes. Sensors and Actuators B: Chemical, 171, 691–698.

Lee, K. M., Lee, S. M., Sim, K. S., Kim, K. K., & Park, K. S. (2009, September). Noise reduction for non-contact electrocardiogram measurement in daily life. In Computers in Cardiology, 2009 (pp. 493–496). IEEE.

Leonhardt, S., & Aleksandrowicz, A. (2008, June). Non-contact ECG monitoring for automotive application. In Medical Devices and Biosensors, 2008. ISSS-MDBS 2008. 5th International Summer School and Symposium on (pp. 183–185). IEEE.

Lim, Y. G., Kim, K. K., & Park, S. (2006). ECG measurement on a chair without conductive contact. Biomedical Engineering, IEEE Transactions on, 53(5), 956–959.

Lim, Y. G., Kim, K. K., & Park, K. S. (2007). ECG recording on a bed during sleep without direct skin-contact. Biomedical Engineering, IEEE Transactions on, 54(4), 718–725.

Lim, Y. G., Chung, G. S., & Park, K. S. (2010, August). Capacitive driven-right-leg grounding in indirect-contact ECG measurement. In Engineering in Medicine and

Biology Society (EMBC), 2010 Annual International Conference of the IEEE (pp. 1250–1253). IEEE.

Shen, T. W., Hsiao, T., Liu, Y. T., & He, T. Y. (2008, November). An ear-lead ECG based smart sensor system with voice biofeedback for daily activity monitoring. In TENCON 2008–2008 IEEE Region 10 Conference (pp. 1–6). IEEE.

Sullivan, T. J., Deiss, S. R., & Cauwenberghs, G. (2007, November). A low-noise, non-contact EEG/ECG sensor. In Biomedical Circuits and Systems Conference, 2007. BIOCAS 2007. IEEE (pp. 154–157). IEEE.

Wu, C. R. (2013). Design of an Indirect-Contact ECG Measurement System Using the Movable Electrode.

Information Technology – Wan et al. (Eds)
© 2015 Taylor & Francis Group, London, ISBN 978-1-138-02785-5

The land use change in Haikou city based on RS and GIS

Feng Xia Wang
College of Tourism in Hainan University, Hainan

Wei Hong Liu
Information Center of Land and Resources, Department of Zhejiang Province, Zhejiang

ABSTRACT: The paper analyzes the time and spatial change of land use from 2000 to 2010 in Haikou city by using the RS and GIS technologies and their professional softwares and draws conclusions as follows: (1) The spatial distribution. The residential areas, and construction land and forest land scope present a tendency of spreading outward. The distribution of unused land gradually reduces. No change in the cultivated land and water. (2) The time aspect: Residential areas and construction land, cultivated land and forest land in the area of three classes show a trend of rising, while water and unused land area gradually reduce. (3) Land use transfer aspect: The area of unused converted to other land use types is the largest , followed by the cultivated land and forest land, and the area water change is the least.

KEYWORDS: land use change; Haikou City; RS; GIS; land transfer matrix

1 INTRODUCTION

With the rapid development of the global economy, resources, environment, population and social problems have become increasingly prominent, causing the global environment changing constantly. As the main cause of the earth's land surface change, land use and land cover change (LUCC) is an important aspect of global environmental change. The rapid development of economy intensifies people to utilize the land resources, causing the land cover changed to some extent. It leads to the global and even regional environment change. Therefore, in the 1990s, the International Geosphere Biosphere Program (IGBP) and the International Human Dimensions Program (IHDP) formally put forward the LUCC research program, designed to monitor, explanation, simulation and prediction of the world and different regional scale of land use and land cover changes. Land use/cover research as a hot topic of current research on global environmental change, especially the research on how to allocate land resources scientifically and reasonably based on the limited land resources, and at the same time coordinate the global natural environment and social economic development to realize the goal of sustainable development and has been a top priority in the study of land. Land use refers to the use of land, it is the human that according to the natural characteristics of the land and certain economic and social purpose, take a series of biological and technical measures for long-term or periodic land management and governance reform activities. Land cover refers to the state of the land surface coverage, not only including the surface vegetation, but also various artificial mulch and transform objects, and is the total sum of the earth's surface vegetation cover and artificial mulch. Land use and land cover have a close relationship. Land use is the external driving force of land cover change. Land cover, in turn, affects the way of land use, they both form a unified whole in the surface.

Before the analysis of land use cover changes, the standard classification should be determined firstly. According to the actual situation of the land use in Haikou, a new RS classification system of land use has been built, including five classifications (Table 1).

Table 1. The RS classification system of land use cover.

First Classification	Code
Farmland	1
Forest	2
Water	3
Urban and village construction	4
Unused land	5

Financial supports from Hainan Province Nature Science Fund (413122) and National Soft Science Project (2013GXS4D140)

2 STUDY AREA

Haikou city is located in north latitude 19° 32′ -20° 05′, longitude 110° 10′ -110° 41′, located in the northern part of Hainan Island, Flat terrain, slightly heart-shaped landscape basic is divided into the northern coastal plain, central order along the Yangtze river region, eastern, southern tableland, lava in western region; East from DaZhiPo old town village, west to west township south village, 60.6 km distance, South from Dapo Town five car Village, north to the sea, 62.5 km distance, the city's land area of 2304.84 km². Haikou city is located in the low latitude tropical edge, belongs to the tropical maritime monsoon climate, annual sunshine time is long, the radiation energy is large, the average annual sunshine hours 2000 hours, solar radiation can be up to 11 to 120 000 Card, annual average temperature 24.2°C, annual average rainfall of 1684 mm, the average annual evaporation is 1834 mm, the average relative humidity is around 85%, southeast and northeast wind is given priority to all the year round, the annual average wind speed 3.4 m/s; north Linhai, sea area of 830 km², a coastline of 131 km , sea water average temperature 25°C.

3 DATA COLLECTION AND DATA PREPARATION

3.1 Remote sensing data

The Remote Sensing data of the study are mainly Landsat TM images (Figure 1). The Landsat TM data are composed with scenes. Every scene equals to the area of 185 km × 170 km on the ground. The scanning period of Landsat is 16 days. The Landsat TM data, which are the most widely used earth observation data, are used mostly in the fields of resource exploration, environmental monitoring and land cover investigation. In this study, 6 of 7 bands, in Visible Band and Near Infrared Band, are used: TM1-5 and TM7, with the space resolution of 30 m.

3.2 Ancillary data

The ancillary data are also important in the remote sensing interpretation. The classification result combined with the geographic information data is better than simply using the spectral features. At present, the method of the information extraction of land cover is mainly the combination of analysis the remote sensing data, Geographic Information System (GIS) and on-the-spot investigation. It is very useful to remote the precision of remote sensing interpretation of data merging of multiplex data in the GIS environment.

The main ancillary data include: (1) special data: DEM data with 10m grid; (2) Normalized Difference Vegetation Index(NDVI): NDVI=(NIR-R)/(NIR+R), here NIR is Near-Infrared Waveband(TM4), R is Red Waveband(TM3); (3) Vegetation Coverage fcover: fcover = (NDVI-NDVIsoil)/(NDVIveg-NDVIsoil), here NDVIsoil is the NDVI value of bare soil or non-vegetation soil, NDVIveg is the NDVI value of fully-vegetation soil; (4) Map of soil type and map of geomorphologic type.

4 HAIKOU CITY LAND USE CHANGE FROM 2000 TO 2010

4.1 Data processing

4.1.1 The data extraction

The data between 2000 and 2010 of the Haikou city Landsat5 TM multi spectral remote sensing image and administrative map were used for this study. As the image data is Subband data, it needs above all use ERDAS9.2 do the combination of band, and then use the ERDAS9.2 software do geometric correction and strip the image processing. The pretreatment of data, through the software Mosaic Image function, do stitching Image and get two mosaic map. Then using ArcGis software Extract by Mask function, based on the scope of the administrative map of Haikou city, extracting Haikou city remote sensing image, the end result is shown in Figure 1.

Figure 1. The extraction of 2000 and 2010 results figure.

4.1.2 The data preprocessing

Since the process of obtaining remote sensing images is influenced by atmospheric scattering reflection or the weather, the image inevitably with noise or bad visual effect is not good, such as low contrast, image blur, and sometimes the problems of necessary information are not prominent. In order to improve the image quality, increase the image visual effect, and highlight the required information, you need use image enhancement techniques and remote sensing image processing software to enhance the image quality. This passage by using ERDAS9.2 software, through histogram equalization, haze reduction, jump solution with radiation enhancement and Sharpening, focusing on the spatial enhancement function, did two Landsat5 TM image of Haikou city of 2000 and 2010 enhancement processing, the results are shown in Figure 2.

Figure 2. The reusult of 2000 and 2010 image processing.

4.1.3 The supervised classification

Different kinds of feature of spectral information and spatial information characteristics have differences. According to the differences, all the pixels in the image can be divided into several categories basing on its property that is the image classification. Through classification, the actual object can be recognized from the image, and then extract the feature information. The supervised classification is one of the commonly used methods for remote sensing image classification. Supervised classification is based on remote sensing image of the object class attribute, and it has a priori knowledge on the basis of selecting all to distinguish from the image of the various features of the sample, building templates, and then reusing software for automatic identification. According to the different characteristics of Haikou city land resources and landscape types, and combined with field investigation, Haikou city landscape pattern is divided into waters, residential areas and the construction land, forest land, cultivated land and unused land 5 classes in this article, and then by using ERDAS9.2 software to select sample, establish classification template, and evaluate whether the classification template already meet the needs of research. This article is passing the probability matrix method for testing, and the classification error matrix is greater than 85%, which meets the needs of the analysis, then it is supervised classification. After finishing the classification and then evaluating its accuracy, the result also meet the needs of the classification. The ultimate land landscape classification result is shown in Figure 3.

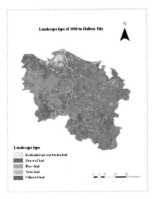 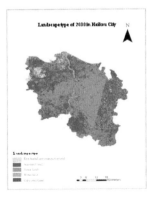

Figure 3. The Haikou classification reusult of 2000 and 2010.

4.2 The haikou land use present situation analysis in 2000–2010

4.2.1 Haikou land use analysis in 2000

According to the five types of land use types, in 2000, Haikou city land use types are mainly the use of forest land and farmland, the use of residential areas and the construction land and water is less, whereas unused is relatively more. The distribution of land use in Haikou city in 2000, the cultivated land are mainly distributed in the southeast of Haikou city, and occupied the main land use types. Forest land is mainly distributed in the northwest of Haikou city, thus, the occupying proportion is bigger too. Residential areas and the construction land are mainly distributed in the north of Haikou city, its distribution is less. The water is mainly distributed in NaDuJiang basin, the southeast is also of sporadic spread some small lakes.

4.2.2 Haikou land use analysis in 2010

According to a 2010 Haikou city land use classification diagram, forest land and farmland are the two ways of land use, which still occupies the main part of the Haikou city land utilization, almost distributing in the whole city area, which illustrates vegetation coverage rate is higher in Haikou city, in addition, the use of water has little change, change is bigger of residential areas, construction land and unused land of land-use types, residential areas and construction land area increase larger, and unused land area is less and amount more. For the distribution of land use in 2010, in addition to the north and northwest of Haikou city, forest land and farmland almost covered the whole city area; Residential areas and the construction land are mainly distributed in the north and northwest; Nandujiang is the mainly waters through Haikou City; unused land is less and distribution more scattered.

4.3 Haikou land use comparative analysis in 2000–2010

4.3.1 Residential areas and construction land comparative analysis

Through the use of ArcGis software, the spatial distribution of different land types can be demonstrated by contrast, the change of Haikou city's residential areas and construction land is greater in 2000 and 2010, Not only on the number of rapid increase, on the spatial distribution, also on the constantly expanding in 2000, it accounts for only a small area of the north, up to 2010, it becomes a major way of land use of north and northwest. The decade of change is obvious.

4.3.2 The unused land comparative analysis

By comparing the two periods of unused land present status (Figure 3) in 2010, unused land dropped sharply, compared with 2000, especially in northern and central and eastern regions, reduced particularly acute.

4.3.3 The woodland comparative analysis

In 2000, Haikou city forest land is mainly distributed in the northwest region, other parts of the distribution are less. In 2010, they are mainly distributed in the central and southeastern regions. The change tendency is, woodland area of northwest in reducing trend, while, central and southeast area of forestland area constantly on the rise.

4.3.4 The water land comparative analysis

Waters land use change is slow. NanDuJiang is basically as the main river and runs through the north and south direction of Haikou city. But some of the land area of the waters is constantly changing, and it can be seen from the figure NanDuJiang waters use in the east from 2010 to 2000, has increased a lot. Rarely, changes occur in the western regional water land.

4.3.5 The farmland comparative analysis

From the two periods of cultivated land present status Figure 3, it can be seen that cultivated land areas and the spatial distribution of the two were very weak and are the main land use types and distribution range is wide. It can be seen that cultivated land in the southeast area of is over more than in the northwest.

5 THE QUANTITATIVE ANALYSIS OF HAIKOU LAND USE CHANGE FROM 2000 TO 2010

5.1 Land transfer matrix

Analysing the land use transfer matrix of Haikou city in 2000 and 2010 by ArcGIS10 software, get the data of land use transfer matrix of Haikou city in 2000 and 2010 (Table 2).

Table 2. Land use transfer matrix of haikou city in 2000 and 2010 unit: Ha.

2010 / 2000	Settlement and construction land	Unused land	Arable land	waters	forest land	in total
Settlement and construction land	-	16.44	123.38	7.44	66.85	214.11
Unused land	263.44	-	385.6	2.15	211.61	862.80
Arable land	237.7	59.96	-	5.85	221.26	524.77
waters	103.08	10.37	47.25	-	28.66	189.36
forest land	221.63	32.67	388.54	0.82	-	643.66

5.2 Land transfer matrix analysis

From Table 1, it can be seen in 2000, the land use type of Haikou city, the transformation of unused land area is the most, followed by the cultivated land and forest land, and transformation of water area at least. Among them, the unused land is mainly converted into residential areas, cultivated land and forest land and construction land, be transferred to the water area is less; Cultivated land mainly transferred to cultivated land, forest land and settlement and construction land; Woodland mainly transferred to cultivated land, forest land and settlement and construction land, which is also explained in 2000 the cultivated land, forest land and residential areas and the construction land area changes in Haikou city large, and growth tendency. From the perspective of the land use type of 2010, settlement and construction land mainly come from the unused cultivated land and forest land from the 2000 years, cultivated land mainly comes from the forest land and unused land and farmland from the 2000 years, forest land mainly comes from the cultivated land, forest land and unused land.

land, the area waters transformed least, the unused transformation of land use types, mainly residential areas and construction land, forest land and farmland, in 2010, land use types in Haikou city, increase area is more residential areas and construction land, forest land and cultivated land. This suggests that from 2000 to 2010, are mainly the increase of residential areas and construction land area and unused land area decreases.

Haikou city has unique natural resources and land resources, which bring economic benefits at the same time, also with a certain social and environmental benefits. Although the Haikou city land use types in terms of spatial distribution and structural composition, relative to the more balanced, but the use of land resources structure is not reasonable enough, intensive land use has yet to be perfected. This requires the development of economy and tourism industry at the same time, planning and reasonable use of land resources, to the sustainable development of both environment and the natural environment and land resources.

6 CONCLUSION AND PROSPECT

Through the above analysis, from 2000 to 2010, it can be seen,with the rapid development of the economy and improvement of people's material life level, especially in recent years and the vigorous development of tourism industry in Haikou city of Hainan province and the construction of the Hainan International Tourism Island, significant changes have taken place in Haikou city land use types, mainly reflects in:

1 On the spatial distribution, settlement and construction land scope expanding constantly, mainly distributed in the north in Haikou city from 2000 to 2010 gradually expand outward, extending to the northwest and central and eastern regions; Unused land is mainly distributed in the northern region in 2000, by 2010, the distribution gradually reduces; Forest land began to spread from northwest to southwest and the south; while the spatial distribution of arable land and water changes less.

2 In time, not only showing changes of the spatial distribution, but also reflected in the number. In all the land use type, residential areas and three types of construction land, cultivated land and forest land in the area of the class shows a rising trend, and water and unused land use area presents gradually decreased tendency.

3 In the transfer of land use from 2000 to 2010, the land use type in the Haikou city, the unused converted to other land use types of the largest area, followed by the cultivated land and forest

ACKNOWLEDGMENT

Thank Z.H, Y.XY and J. CX for the figures making and other works. Thank all the authors of the reference articles.

REFERENCES

[1] Turner IIBL. Land use and Land cover change (LUCC): Science/Research plan. IGBP Reports, N035, 1995-6-9.

[2] Chen Youqi, Yang Peng. The land use / land cover change research progress. Journal of economic geography, 2001, 21 (1): 0095–0100.

[3] Shi Mingle. International LUCC research present situation,problems and future prospect. Journal of Shangrao Normal University, 2004, 24 (3): 0089–0094

[4] Shili Li, Tang Guotao, Dong Xiansheng, Wang Sijie. The domestic and foreign land use / land cover change research. Guangxi Agricultural Sciences, 2008, 23 (3): 0042–0044.

[5] Ni Shaoxiang. In recent years our country land use / Cover Change Research Progress — land cover change and its environmental effects. planet publishing house, 2002:7–8.

[6] Xiao Duning Li Xiuzhen. The progress and outlook of contemporary landscape ecology. Journal of geographical science, 1997, 52 (4): 356–364.

[7] Guo Jinping. discipline integration of landscape ecology and Chinese landscape ecology outlook. Geographical Science, 2003,23 (6): 277–281.

[8] Liu Delin, Li Bicheng. The Loess Plateau on the yellow small watershed land-use landscape pattern analysis.

Journal of Surveying and Mapping, 2014,39 (01): 0078–0082.

[9] Zheng Xinqi, Fu Mei Chen. Landscape pattern spatial analysis technology and its application. Beijing: Science Press, 2010:92–146.

[10] P.G.Silva, J.R.Santos, Y.E.Shimabukuro, P.E.U.Souza. Change Vector Analysis technique to monitor selective logging activities in Amazon. IEEE, 2003, pp. 2580–2582.

[11] Myneni R B, Hoffman S, Knyazikhin Y, et al. Global product of vegetation leaf area and fraction absorbed PAR from year one of MODIS data. Remote Sensing of Environment, 2002, pp. 214–231.

[12] ZHANG Bao-lei, SONG Meng-qiang, ZHOU Wan-cun. Exploration on Method of Auto-classification Based on GIS for the Main Ground Objects of the Three Gorges Reservoir Area. Remote Sensing Technology and Application, 2006, pp. 71–76. (in Chinese)

[13] CAO Xue-zhang, ZUO Wei, SHEN Wen-ming. "Remote Sensing Analysis Changes in Land Cover in the Three Gorges Reservoir Region", Rural Eco-Environment, 17(4):6–11 (2001). (in Chinese).

[14] ZHANG Bao-lei, ZHOU Wan-cun, MA Ze-zhong. "Expolorations on Method of Auto-Classification for the Main Types of Ground Objects in the Three-Gorge Region", Resources and Environment in the Yangtze Basin, 445–449 (2005). (in Chinese).

[15] ZHANG Yuan-xing, ect. "The Application Study on the RS Technology in the Land Detailed Investigation", Anhui Agricultural Science Bulletin, (3), 52–53 (2007). (in Chinese).

[16] ZHANG Bao-lei, SONG Meng-qiang, ZHOU Wan-cun. "Exploloration on Method of Auto-classification Based on GIS for the Main Ground Objects of the Three Gorges Reservoir Area", Remote Sensing Technology and Application, 21(1):71–76 (2006). (in Chinese).

[17] WANG Feng-xia, ZHOU Wan-cun. The Method Study on the Land Use/Land Cover Changes in Three Gorges Reservoir Region Based on RS and GIS, 2008 Proceeding of Information Technology and Environmental System Sciences, ITESS'2008, pp. 593–596.

[18] WANG Feng-xia, ZHOU Wan-cun. The Land Use/Land Cover Changes in Three Gorges Reservoir Region Based on RS and GIS, International Symposium on Computer Science and Computational Technology, 2008, pp 525–530.

Information Technology – Wan et al. (Eds)
© *2015 Taylor & Francis Group, London, ISBN 978-1-138-02785-5*

TD-LTE-based E-government IOT access network design

Jiandong Yang & Nong Si
Electronic Information and Control Engineering College

Pu Wang
Beijing University of Technology, Beijing, China

Daoxin Chen & Xiaochun Xi
CAPINFO, Beijing, China

ABSTRACT: In this Paper, a TD-LTE-based E-government IOT (Internet of Things) access network design solution has been put forward based on the actual needs of the Beijing municipal E-government wireless IOT by comparing the characteristics of the fixed broadband wireless technology, McWiLL broadband wireless technology and TD-LTE technology; this design solution has started from the structure characteristics of the TD-LTE network, detailed the design targeting the usage and distribution of the frequency resources, simultaneously designed the typical scenario coverage at such four different regions as densely-populated urban, general urban, suburban and rural areas, and then estimated the scale of base stations accordingly to have designed a set of access network integrating the same and different frequency networking solutions for the Beijing municipal E-government wireless IOT.

1 INTRODUCTION

The Internet of Things (IOT) is the core area, triggering a new wave in the information industry after computer, Internet and mobile communications. In order to accelerate the IOT construction, promote the development of the Beijing IOT and enhance the level of urban management and emergency capacity, the Beijing municipal government has been attaching great importance to the IOT industry, with a decision made to push forward the demonstration construction of the Perception of Beijing, actively develop the IOT industry as the city symbolic project, take the lead to carry out the IOT infrastructure construction in China, construct a private wireless IOT for Beijing E-government to provide a unified, safe, wide and standard sensing information transmission channel for city management and emergency applications, avoid repeated construction of sensing network in different departments, reduce the cost of the IOT construction and applications, make full use of the limited spectrum resources to ensure the security of the sensing information and thus lay a solid foundation for rapid popularization of the IOT applications and fast development of the IOT industry in Beijing.

The wireless E-government IOT private network is an important part of the E-government IOT infrastructure constructed by Beijing municipal government based on Beijing cable E-government private network, with wireless base stations, access optical fibers and other facilities supplemented to provide a safe and reliable transmission channel for sensing information transmission, gathering, processing and distribution of various IOT applications in Beijing. This network has a unified standard for sensing information access, able to complete identity authentication of the sensing information and achieve secure transmission, and able to satisfy the IOT application requirements of all governments at all levels, as well as the relevant committees, offices and bureaus in Beijing for urban safety operation management and emergency management.

2 COMPARISON AND SELECTION OF WIRELESS E-GOVERNMENT IOT TECHNOLOGY

In the three-layer architecture of the E-government IOT, the network layer bears the function of reliable transmission, namely, it has access to the Internet through a variety of communication networks to reliably interact and share all kinds of perceptual information anytime and anywhere, and manage the application and perceptive equipment. The network layer mainly includes the access network, transmission network, core network, service network, network management system, and business support system (Shao & Sun 2013). The IOT network layer must have greater throughput and higher security.

Table 1. Comparison of technical indexes.

	Fixed broadband wireless technology	McWiLL broadband wireless technology	TD-LTE technology
Duplex mode	TDD FDD HFDD	TDD	TDD
Net peak rate	15Mbit/s (5MHz)	15Mbit/s (5MHz)	25Mbit/s (5MHz)
Channel bandwidth	1.25–20MHz	1–20MHz	1.4MHz,5MHz 10MHz,20MHz
Applicable frequency band	Less than 6GHz	1.8GHz 400MHz 3300MHz	Around 2GHz
Multi-access mode	TDMA OFDMA	CS-OFDMA	OFDMA SC-FDMA
Modulation system	QPSK 8PSK 16QAM 64QAM Support AMC	QPSK 8PSK 16QAM, 64QAM Support AMC	QPSK 8PSK 16QAM 64QAM
Multi-antenna support	AAS MIMO SDMA	Smart antenna MIMO	Smart antenna
Typical coverage	1–5km	1–3km	0.3–3km

In recent years, the global communication technology features rapid progress with each passing day, the cellular mobile communication has developed into the third generation mobile communication technology, TDSCDMA, WCDMA and CDMA2000 have been widely applied, and the fourth generation mobile communication technology is also in its fast development. According to the characteristics of the wireless E-government IOT, such as massive data transmission and reliable security, we have compared the performance indexes and relevant contents of the fixed broadband wireless access technology (FBWA), McWiLL wireless broadband access technology and TD-LTE technology, making every effort to find out a technology more suitable for transmission of the wireless E-government IOT data.

Based on comprehensive consideration of the networking capacity, professional proficiency, commercial applications, industrial chain maturity, policy support and cost, etc. of such three kinds of technology, the TD-LTE technology and fixed broadband wireless access technology both can meet the needs of the current network business, the TD-LTE technology is more advantageous because it is a widely-supported international standard and

fostered by the national industrial policy, so it boasts a wide prospect of applications and plays a stronger leading and exemplary role. Therefore, the TD-LTE technology is selected as the technology to carrying the wireless E-government IOT private network.

3 TD-LTE NETWORK STRUCTURE

The wireless E-government IOT is regarded as the access network of E-government private network cable infrastructure and the wireless nodes can achieve connectivity by means of a cable private network. The IOT wireless base stations will have handy access to the municipal-level private network convergence nodes (PE nodes) via optical fiber links (Hu, Zhu, Meng, & Yupeng. 2010). According to the analysis concluded in the previous section, preference will be given to the TD-LTE technology as a technical route for construction of base stations, with the structure of the wireless access network as shown in Figure 1:

TD-LTE adopts a flat IP-based network structure, with IP transmission applied to the interface between network nodes. In the figure above, each network element has its main features including:

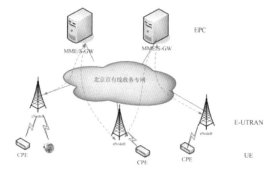

Figure 1. TD-LTE wireless network structure.

1 E-UTRAN (Evolved Universal Terrestrial Radio Access Network) is composed of eNodeB, whose main functions include establishment, management and release of the control and user planes during the process of users' communications, as well as management of partial wireless resources.

2 Core network (Evolved Packet Core): responsible for UE control and bearing establishment. MME (Mobility Management Entity): responsible for handling of the control signaling and paging between UE and CN, as well as control information distribution and load control to ensure the NAS signaling safety and mobility management; P-GW (PDN Gateway): UE IP distribution, QoS warranty, billing and IP

packet filtering; S-GW (Serving Gateway): used as mobility control anchors and downlink data cache in time of UE switching between cells(Qiu, Wu, & Zhou 2012).

4 TD-LTE ACCESS NETWORK DESIGN

4.1 *Frequency resource use and allocation*

The TD-LTE system is such a system based on orthogonal frequency division multiplexing (OFDM) and orthogonal frequency division multiple access (OFDMA) multi-carrier modulation technology, able to distinguish between users through time or frequency sub-band and able to use the same and different frequency networking solutions (Yang, Bi,

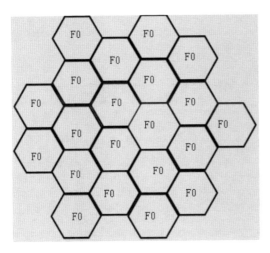

Figure 2. Same frequency networking.

Dong, & Zhang 2012).

1 Same frequency networking solution
In the same frequency networking solution, all cells throughout the network use the same frequency and such relevant technology as smart antennas can be used to reduce interference, but the signal at the edge of cells will also be affected.

The same frequency networking has such advantages as follows:
1 Highest spectrum utilization, able to save frequency resources
The TD-LTE system uses a time-division duplex working mode, in need of no paired uplink and downlink frequency bands or protective isolation frequency between the uplink and downlink carrier waves. The same frequency networking solution can maximize the advantage of the TD-LTE system

with high spectrum efficiency and save operators' frequency resource costs.
2 Frequency simplification planning
In times of network design, construction and expansion, the frequency planning is very simple.
3 Higher cell peak rate
Relative to the different frequency networking, the same frequency networking can get a higher cell peak rate because each sector can dispatch all frequency resources within the system bandwidth, and in theory, when the frequency reuse factor is 1, the cell peak rate is three times that when the frequency reuse factor is 3.

2 Different frequency networking solution
If there are three frequency points available, the S1x3x1 cellular network mode can be adopted for networking to reduce the same frequency interference, improve the network capacity and meet the IOT requirements for reliable transmission of massive data, as shown in Figure 3. The different frequency networking solution has its advantage as follows: it can cause relatively small interference.

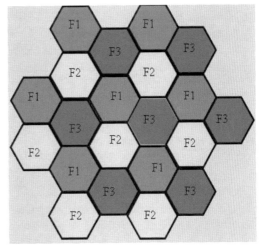

Figure 3. Different frequency networking.

The TD-LTE system does not have the same frequency interference in the local cell and the interference mainly comes from the neighboring cell using the same frequency. The different frequency networking can have different frequencies kept between its service cell and its most adjacent cell, using the space transmission distance to isolate the cells using the same frequency to reduce the same frequency interference as much as possible.

Compared with the same frequency network, the cell carrier interference ratio (C/I) capacity has been greatly improved by means of different frequency networking. This means that:

1 In the same coverage, users' communication quality will be more stable, and to obtain the same frequency resource digits, users will enjoy a higher transmission rate. At the same time, the users at the edge of the coverage can harvest a higher peak rate.

2 Relatively larger coverage
Based on the enhancement of the C/I capacity and improvement on the rate of edge users, the base station can achieve wider coverage to save network investment accordingly.

3 Frequency point planning
In order to better meet the operational needs of the E-government wireless network and make the financial investment give play to greater effectiveness, Beijing Municipality plans to apply for 20 MHZ bandwidth continuously available in the frequency band of 1427-1525 MHZ for construction of E-government IOT data private network.

The TD-LTE wireless system features more flexible individual frequency bandwidth, with 5M, 10M and 20M available. In order to improve the spectrum utilization efficiency and the overall network capacity, it is recommended to use the mode of 20 MHZ same frequency networking:

This networking mode has such advantages as follows (Wang, Zhang, & Zhao 2012):

1 Its performance indicators are as shown below: The uplink frequency spectrum efficiency is 0.7bps/Hz/sector in time of the same frequency networking, so the same frequency networking features higher spectrum efficiency, with the 20 MHZ bandwidth fully utilized throughout the network.

2 The LTE technology is one of today's most advanced wireless communication technologies,

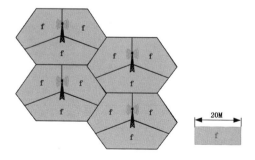

Figure 4. 20 MHZ same frequency networking.

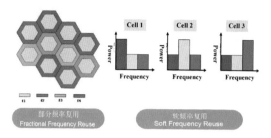

Figure 5. Schematic of fractional frequency reuse and soft frequency reuse.

featuring higher spectrum efficiency, which is higher than the system throughput rate achieved by means of 3G and other wireless technologies. Moreover, in the real production environment, its performance will be reduced somewhat due to the influence of various interference factors.

3 The same frequency networking technology has been used to improve the system throughput, which can satisfy the needs for medium and long-term development of the IOT and other wireless E-government services. In the same frequency networking mode, there is frequency interference between adjacent cells, but TD-LTE has solved such problem by using the ICIC (Inter Cell Interference Coordination) technology.

4 The basic idea of the inter-cell interference coordination (ICIC) is to limit the use of resources in the inter-cell coordinated manner, including limiting which time and frequency resources are available or restricting their transmit power at certain time-frequency resources. ICIC can be classified into three categories from the resource coordinated manner: FFR (Full Frequency Reuse), SFR (Soft Frequency Reuse) and FFR (Fractional Frequency Reuse), among which SFR achieves a good balance between spectrum utilization and scheduling complexity. SFR is mainly used in the industry now.

4 Time Slot Allocation Scheme
TDD separates the receiving and transmitting channels with time. In the TDD mobile communication system, reception and transmission adopt different time slots at the same frequency as the carrier bearer. The channel resources in single direction are not continuous in time, and the time resources are allocated in the two directions. TD-LTE supports different uplink and downlink allocations which can be adjusted according to the type of service to meet the nonsymmetrical service needs (Liu, Hu, & Zhu 2010).

The TD-LTE standard supports the configuration of 5ms and 10ms. The 10ms radio frame consists of two half-frames of 5ms, each of which consists of five sub frames of 1ms. The regular subframe consists of two time slots of 0.5ms. The special subframe consists of DwPTS, GP and UpPTS, and support 5ms and 10ms DL & UL switching point cycles.

TD-LTE uplink and downlink can be configured in many ways:

Wherein the S sub-frame (the special subframe) consists of DwPTS (downlink pilot time slot), GP (up and down conversion interval) and UpPTS (uplink pilot time slot). For the time slot allocation scheme of the special subframe, see Table 3.

Considering that the data is mostly uplink in the practical application of the networking to increase the uplink data service capacity and the maximum coverage distance of the system, the network time slot allocation adopts the 5ms frame structure configured 0 (1 downlink: 3 uplinks), and the special sub frame adopts the configuration of 5 (3:9:2).

4.2 *Typical Scenario Covering Design*

The choice of the TD-LTE IOT covering solutions depends primarily on the distribution of users and services, service volume, service rate, environmental conditions and facilities. According to the current needs, the TD-LTE outdoor coverage is only considered (Li 2009).

1 Dense urban areas

Table 2. Uplink and downlink ratio of TD-LTE in different frame periods.

Uplink-down link configuration	Downlink to-Uplink Switch-point	Subframe number									
		0	1	2	3	4	5	6	7	8	9
0	5ms	D	S	U	U	D	D	S	U	U	D
1	5ms	D	S	U	U	D	D	S	U	U	D
2	5ms	D	S	U	D	D	D	S	U	D	D
3	10 ms	D	S	U	U	D	D	D	D	D	D
5	10 ms	D	S	U	D	D	D	D	D	D	D
6	5ms	D	S	U	U	U	D	S	U	U	D

Table 3. Uplink and downlink time slot allocation scheme of the special sub-frame.

Special subframe configuration	Routine CP			Extended CP		
	Downlink pilot time slot	Downlink-to-Uplink Switch-point periodicity	Uplink pilot time slot	Downlink pilot time slot	Downlink-to-Uplink Switch-point periodicity	Uplink pilot time slot
0	3	10	1	3	8	1
1	9	4	1	8	3	1
2	10	3	1	9	2	1
3	11	2	1	10	1	1
4	12	1	1	3	7	2
5	3	9	2	8	2	2
6	9	3	2	9	1	2
7	10	2	2	-	-	-
8	11	1	2	-	-	-

Dense urban areas are the coverage areas that the TD-LTE IOT needs to focus on at the initial stage of building the network. Dense urban areas are typically characterized by high-rise buildings, the majority of which have a height of 30 meters (10 floors) or more. There are lots of high-rise buildings with more than twenty stories, among which the low-rise and midrise buildings are located; some buildings are huge, and some have one or more layers of underground commercial facilities or parking; the roads are wide, and the streets of some areas are narrow. The planning and setting of the base stations in the dense urban areas should be combined with the topography, environment, building characteristics of the coverage areas for reasonable layout. Generally, the wireless environment of the areas is very complex with large expansion potential for the future. The distributed BBU + RRU system combined with dual polarized antenna 4 + 4 is proposed to achieve the outdoor macro cell coverage. Using BBU + RRU mode with RRU installed near the antenna enables continuous coverage, reduces the room requirements, increases the installation layout flexibility and accelerates the speed of the project. The 8 double antenna beam forming technology is a major technology for TD-LTE network construction. The use of eight antennas can effectively improve the throughput performance of the uplink-downlink for the TD-LTE system and the cell coverage. S1/1/1 can be configured at the initial stage of the network construction, and can be flexibly expanded as needed in the future. In some hot service spots difficult to be covered by macro cell or blind-spot areas, the subsequent use of integrated base station can be made based on the network construction optimization conditions to achieve fine coverage. The integrated base station supports outdoor installation without room, adopting dual antenna and supporting the output of 20W per channel.

2 General urban areas

The building density in general urban areas is small, and there are less large buildings. Clear streets or green areas tend to separate the buildings in general urban areas. The population density is low, and the phone traffic is relatively small. For the general urban scene, use the BBU + RRU distributed system combined with 4 +4 dual-polarized antenna to achieve the outdoor macro cell coverage with the S1/1/1 configuration of the initial stage of the network construction. The frequency applications and blind area coverage are the same with the dense urban areas.

3 Suburbs

In the rural environment, buildings are sparse, and most of them are low-rise buildings. Thus,

relatively speaking, the scene is more open, the wireless environment is not very complex, and the user density is not very high. The type of the base station in the suburbs should be selected based on the situation of the local real traffic. Omnidirectional or sector macro base station may be selected to meet the coverage and capacity requirements. BBU + RRU base station combined with 4 +4 dual-polarized antenna is recommended to meet the requirements with 8 beams forming providing greater coverage.

4 Rural areas

In the rural environment, the buildings are more sparse compared to the suburbs, and mostly are low rise buildings. The scene is open with the simple radio environment and low user density. The type of the base station in the suburbs is recommended to be selected based on the situation of the local real traffic. Omnidirectional or sector macro base station may be selected to meet the coverage and capacity requirements. BBU + RRU base station combined with 4 +4 dual-polarized antenna is

Table 4. Estimates of the number of base stations.

Cell radius	1.1km	1.56km	3.72km	Number of base stations
Number of base stations for Phase	221	0	0	221
Number of base stations for Phase	221	366	0	587
Number of base stations for Phase	221	366	152	739

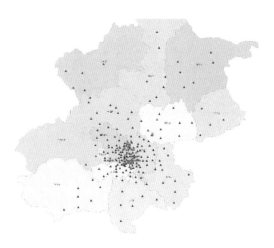

Figure 6. Site location map of the base stations.

recommended to provide greater coverage (Wang & Wang 2012).

4.3 Estimating the size of the base station

According to TD-LTE technical specifications and the actual coverage results in a production environment, each base station can achieve the outdoor wireless coverage within a radius of 5 miles. Based on the link budget simulation and comprehensive consideration of the actual environmental conditions in Beijing and the required transmission rate requirements, the coverage radius within the Fourth Ring Road and the central region of each county and its suburbs is designed for 1.1km to ensure the signal quality of the cell boundary. The Phase II single base station coverage radius is calculated at 1.56km. Taking into account the characteristics of wireless networks and the general needs of the IOT service related to urban management, the Phasebase station density is much lower than the Phase and Phase II projects, and the single base station covers a radius of 3.72km.

During the site planning and design, first of all, use as far as possible the available resources such as the wired-government network rooms, 800 MB radio government network base station rooms and county government offices for site planning and construction in line with the principle of the cellular network; secondly, for the operators which have the site, with the conditions permitting, consider sharing their infrastructure resources such as rooms, towers and external power; finally, considering the new construction, the owners rooms should be rented. During the site planning and construction, try to avoid duplication construction and waste of resources, make full use of all available resources that can be used to maximize the input-output ratio and to achieve the construction of an intensive society. Part of the base station locations for the preliminary design is shown in Figure 6:

5 CONCLUSION

Through careful study of the actual needs of the wireless government IOT in Beijing, we contrasted fixed broadband wireless technology, McWiLL broadband wireless technology and TD-LTE technology according to the needs; ultimately we chose TD-LTE technology as the networking technology for the wireless government IOT; herein we designed an access network design program with the same frequency networking combined with the different frequency networking for the wireless government IOTin Beijing based on the characteristics of TD-LTE technology to provide various departments in Beijing with an IOT data transmission network with large data transmission and high security.

The access network of the wireless government IOT constructed based on this design program solved the problem in the middle and long-distance transmission of data obtained from the perception layer. The data can be connected to the government IOT in Beijing via the e-government cable network in Beijing so as to provide a full coverage, easy and convenient access network for the city management and emergency command for the government departments at all levels in Beijing and to meet the application requirements of the IOT in city safe operation and emergency management from all levels of governments in Beijing and the relevant bureaus.

REFERENCES

Hu, H., Q. Zhu, F. Meng, & L. Yupeng. (2010). Td-lte networking strategy research. *Mobile Communication. 5*, 49–53.

Li, X. (2009). Td-lte wireless network coverage characteristics analysis. *Telecommunications Sciences. 1*, 22–26.

Liu, B., H. Hu, & Q. Zhu (2010). Id-lte wireless network planning research. *Telecommunications Engineering Technology and Standardization. 1*, 16–20.

Qiu, J., Q. Wu, & Y. Zhou (2012). Td-lte core network epc introduction strategy and networking solution research. *Systems and Programs 5*, 83–88.

Shao, C. & G. Sun (2013). Friction factors in open channels. *IOT Principles and Industrial Applications. 15*, 318.

Wang, X., Y. Zhang, & X. Zhao (2012). Td-lte key technologies and network planning strategy. *Telecom Engineering Techniques and Standardization 7*, 1–5.

Wang, Y. & Z. Wang (2012). A unified bar-bend theory of river meanders. *Wireless Sensor Networks. 1*, 162–165.

Yang, J., D. Bi, J. Dong, & W. Zhang (2012). Td-lte optimization networking strategy analysis. *Communication & Information Technology. 3*, 55–57.

Information Technology – Wan et al. (Eds)
© *2015 Taylor & Francis Group, London, ISBN 978-1-138-02785-5*

Texture feature extraction and classification for CT image of Xinjiang local liver hydatid based on support vector machine

Fang Yang, Murat Hamit, Chuan Bo Yan, Abdugheni Kutluk, Wei Kang Yuan & Elzat Alip
Xinjiang Medical University, College of Public Health, Urumqi, China

Asat Matmusa
Xinjiang Medical University, College of Medical Engineering Technology, Urumqi, China

ABSTRACT: Texture feature is an important clue of image analysis. In this paper, we proposed the feature extraction based on gray-gradient co-occurrence matrix to characterize the normal liver and the mono-hydatid cyst. Firstly, the median filter had been applied to remove the image noise. The contrast was enhanced by adaptive histogram equalization. Secondly, the gray gradient co-occurrence matrix method was used to extract the texture features of the images. Finally, the classification ability of the texture features was evaluated by Lib-SVM classifier. The experiment was conducted repeatedly by adjusting the support vector machine kernel function parameters and penalty factor. Experiment results show that the classification accuracy varied from the different parameters. And the classification accuracy was the best when penalty factor was 1000 and the kernel function parameter was 0.4. This work proved the classification ability of support vector machine for Xinjiang local liver hydatid disease and verified the significance of parameter selection in classifier, which laid the foundation for the subsequent research.

KEYWORD: gray gradient co-occurrence matrix; support vector machine; Xinjiang local hepatic hydatid disease; feature extraction

1 INTRODUCTION

Hepatic hydatid disease, which often occurs in the pastoral area, is an infection of larval stage animal tapeworm. Xinjiang Uygur Autonomous Region, a multi-ethnic province in northwestern of China, is one of the most important foci of hepatic hydatid disease in the world [1]. CT examination is an effective method for early diagnosis and disease screening [2]. Texture, which is a basic property of the object surface, is widely spread in the nature world. It has important significance for the description and recognition of the object [3]. Texture feature is independent of the color and brightness to reflect the homogeneous phenomenon of visual characteristics [4]. It has been widely used in pattern recognition and computer vision. Fang DONG [5] proposed the texture feature extraction of liver CT images based on fractal dimension, which pointed that the fractal dimension value of the soft tissue in the same organs was just related to the property of organs. The fractal dimension value is different between the normal liver and

liver cancer, which can separate the normal liver from the cancer tissue. Stavroula G. M. defined an optimal performing computer-aided diagnosis architecture for the classification of liver tissue from non-enhanced CT images, five distinct sets of texture features were extracted using first order statistics, spatial gray level dependence matrix, gray level difference method, Laws' texture energy measures and fractal dimension measurements [6]. Doaa Mahmoud-Ghoneim investigated the accuracy of texture analysis results on three color spaces, conventional grey scale, RGB, and Hue-Saturation-Intensity (HSI), at different resolutions [7]. Zhang Gang extracted the texture features based on Gabor wavelet and investigated the classification of liver disease [8].

Gray gradient co-occurrence matrix combines the gradient histogram and edge gradieng histogram, through Gaussian Laplacian operator to calculate the gradient, which can obtain rotation invariant measure [9]. The Classification research of texture images has been widely applied in the filed of biomedical engineering, remote sensing and telemetry,

This work was supported by the National Natural Science Foundation of China, Xinjiang, 81160182 and 30960097.

industrial product testing. The common texture classification methods include: Markov random field model, Gibbs random field models, fractal models, neural networks, Bayes classifier, fuzzy C-means clustering and genetic algorithm. The methods mentioned above have several shortcomings, such as the large calculation, the poor classification accuracy and the bad antinoise capability etc. Support vector machine (SVM) can effectively avoid the dimension disaster when dealing with the high dimensional samples, it is also the small sample size required, which make it has the strong potential to classify the texture feature samples in the high dimension [10]. In this work, feature extraction based on gray gradient co-occurrence matrix was applied in CT images of Xinjiang local liver hydatid disease due to its particular pathological features. And the classification ability of the texture features was verified by Lib-SVM classifier.

2 METOLOGY

2.1 *Image preprocessing*

The preprocessing step aims to enhance the image or to reduce the effect of illumination changes, which can provide the better support for the doctor's diagnosis. For the liver CT images, converting RGB images to the grayscale intensity images, removing the image noise through median filter, using adaptive histogram equalization to enhance the contrast. The preprocessing results of normal liver CT images and mono-hydatid cyst CT image show in figure 1 and figure 2, respectively. As can be seen from the figures, the contrast of the images has been enhanced distinctly, which lay a good foundation for the subsequent image processing.

2.2 *Gray gradient co-occurrence matrix*

The element C_{ij} of gray gradient co-occurrence matrix can be defined as the total pixels, which gray value is i and gradient value is j in the normalized grayscale image $F(m, n)$ and the gradient image $G(m, n)$ [11]. Gray gradient co-occurrence matrix model reflects the relationship between the gray and the gradient of each pixel in the image. Grayscale is the basis of an image, while the gradient constitutes the edge contour. Therefore, gray gradient co-occurrence matrix not only describes the distribution of the gray and gradient of each pixel in the image, but also reflects the spatial relationship between the pixel and its neighboring pixels.

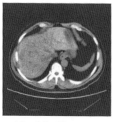 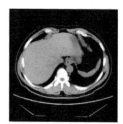 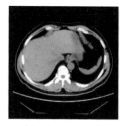 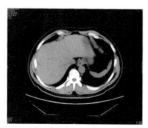

(a) original image   (b) scale normalized   (c) noise removed   (d) adaptive enhanced

Figure 1. Preprocessing results of normal liver CT image.

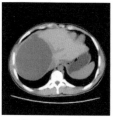 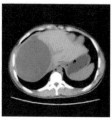 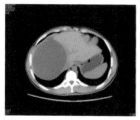 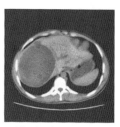

(a) original image   (b) scale normalized   (c) noise removed   (d) adaptive enhanced

Figure 2. Preprocessing results of mono-hydatid cyst CT image.

2.3 Normalization of gray gradient co-occurrence matrix

The purpose to normalize the gray gradient co-occurrence matrix is to reduce the amount of calculation through transforming the gray and gradient in the condition without affecting the texture feature of the images.

Gray array normalization:

$$F(K,L) = INT[f(K,L) \times N_f/f_m] + 1 \tag{1}$$

where INT represents the Integer arithmetic, f_m is the maximum gray value, N_f is the maximum gray value expected after the normalization.

Gradient array normalization:

$$G(K,L) = INT[g(K,L)N_g/g_m] + 1 \tag{2}$$

where g_m is the maximum gradient value, N_g is the maximum gradient value expected after the normalization. Then, the probability P_{ij} of the co-occurrence C_{ij} can be shown as follows[12]:

$$P_{ij} = c_{ij} \Big/ \Big(\sum_i \sum_j c_{ij} \Big) \quad (0 \le P_{ij} \le 1) \tag{3}$$

2.4 Feature extraction

In this study, the features based on gray gradient co-occurrence matrix were calculated for each image from the following equations:

Small gradient strength (T1):

$$T_1 = [\sum_{i=0}^{L-1}\sum_{j=0}^{L-1}\frac{H(i,j)}{j^2}] \Big/ H \quad , \text{where} \quad H = \sum_{i=0}^{L-1}\sum_{j=0}^{L-1}H(i,j) \tag{4}$$

Large gradient strength (T2):

$$T_2 = [\sum_{i=0}^{L-1}\sum_{j=0}^{L-1}j^2 H(i,j)] \Big/ H \tag{5}$$

Energy (T5):

$$T_5 = \sum_{i=0}^{L-1}\sum_{j=0}^{L-1}H(i,j)^2 \tag{6}$$

Average gradient (T7):

$$T_7 = \sum_{j=0}^{L-1}j[\sum_{i=0}^{L-1}H(i,j)] \tag{7}$$

Mean square gradient (T9):

$$T_9 = \{\sum_{j=0}^{L-1}(j-T_7)^2[\sum_{i=0}^{L-1}H(i,j)]\}^{1/2} \tag{8}$$

Gray entropy (T11):

$$T_{11} = -\sum_{i=0}^{L-1}\sum_{j=0}^{L-1}H(i,j)\log\sum_{j=0}^{L-1}H(i,j) \tag{9}$$

Gradient entropy (T12):

$$T_{12} = -\sum_{j=0}^{L-1}\sum_{i=0}^{L-1}H(i,j)\log\sum_{i=0}^{L-1}H(i,j) \tag{10}$$

Mix entropy (T13):

$$T_{13} = -\sum_{i=0}^{L-1}\sum_{j=0}^{L-1}H(i,j)\log H(i,j) \tag{11}$$

Inverse difference moment (T15):

$$T_{15} = \sum_{i=0}^{L-1}\sum_{j=0}^{L-1}\frac{1}{1+(i-j)^2}H(i,j) \tag{12}$$

3 EXPERIMENTAL RESULT

3.1 Feature extraction results of gray gradient co-occurrence matrix

All experiments were carried out using images acquired from the first affiliated hospital of Xinjiang medical university of China. The number of image for normal liver and mono-hydatid cyst is both 60. The features extracted of gray gradient co-occurrence matrix are shown in table1.

3.2 The results of SVM classification

SVM method is based on the VC dimension theory and the principle of structural risk minimization [13]. In the case of linear separable, according to the principle of structural risk minimization to design an ultra plane and make each kind of data has the maximum interval on the ultra plane, which can control the complexity of the classifier and achieve a better generalization ability. In the condition of non-linear or linear inseparable, mapping the sample set to a higher dimension space using the kernel function of SVM, which can achieve a linear separable sample set. Then the inner product operations in the high dimension space will be converted to the kernel function operations in the original space.

Lib-SVM classifier is flexible using, less input parameters, easy to generalize, and the operation speed significantly higher than normal SVM, which make it become the wider used classifier at present. In this study, the total amount of the experimental image feature data was 120 groups, of which 70 groups were training set and 50 groups were testing set, 9 features for each group. Selecting radial basis function

Table 1. Results of feature extraction by gray-gradient co-occurrence matrix method.

Image Type	T1	T2	T5	T7	T9	T11	T12	T13	T15
	0.81697	3.40812	0.49424	3.40812	8.46023	0.90467	0.42402	1.21101	0.00311
	0.81335	3.51718	0.49187	3.51718	8.58210	0.90699	0.43015	1.21977	0.00336
Normal liver	0.81631	3.45305	0.48989	3.45305	8.51449	0.90371	0.42543	1.21426	0.00319
	...	...	...	...	...	...	...	...	...
	0.80445	3.72381	0.4715	3.72381	8.80553	0.94385	0.44576	1.27174	0.00361
	0.77768	3.79564	0.36406	3.79564	8.62294	1.14611	0.50425	1.52713	0.00233
Mono-hydatid cyst	0.79013	3.07428	0.39497	3.07428	7.45403	1.10696	0.49291	1.47919	0.00185
	0.77793	3.2487	0.37418	3.2487	7.60468	1.14912	0.51489	1.53759	0.00178
	...	...	...	...	...	...	...	...	...
	0.77914	3.23104	0.36121	3.23104	7.62304	1.17246	0.51332	1.56128	0.00179

(RBF) to be the kernel function, and conducting the experiment repeatedly by adjusting the kernel function parameters and penalty factor in order to obtain the optimal parameters, the experimental results of the testing set are shown in table 2. As can be seen in the table, the classification accuracy is higher when the penalty factor is 1000 and kernel function parameter is 0.4.

Table 2. Results of classification by Lib-SVM.

Kernel function parameter (gama)	Penalty factor(cost)		
	10	100	1000
0.1	0.74	0.66	0.80
0.2	0.72	0.76	0.80
0.4	0.72	0.80	0.82

4 CONCLUSION

In this paper, we proposed the feature extraction based on gray-gradient co-occurrence matrix to characterizing the normal liver and the mono-hydatid cyst. The classification ability of the texture features were evaluated by Lib-SVM classifier. Experiment results show that the classification accuracy varied from the different parameters, the classification accuracy is the best when penalty factor is 1000 and the kernel function parameter is 0.4, which proved the classification ability of SVM for Xinjiang local liver hydatid disease, while verified the significance of parameter selection in classifier. Preliminary experimental study demonstrated the feasibility of the proposed technique in analysis the normal liver and mono-hydatid cyst, which laid the foundation for the subsequent research.

ACKNOWLEDGMENT

The authors would like to acknowledge the support of the first affiliated hospital of Xinjiang medical university of China.

REFERENCES

[1] Wen H, New RR, Craig PS. "Diagnosis and treatment of human hydatidosis". British Journal of Clinical Pharmacology, vol. 35, pp. 565–693, June 1993.
[2] Shao Xiong NIU, Zhen Zhong LIU, Xin Yu PENG, Jian Hua NIU, Shi Yue ZHANG. "Applicatic value of cystic echinococcosis biologic viability deternfmed in CT". Journal of medical imageing, vol. 17, pp. 478–481, May 2007.
[3] Cheng Cheng GAO, Xiao Wei HUI. "GLCM-Based Texture Feature Extraction". Journal of computer system and application, vol. 19, pp. 195–198, June 2010.
[4] Bao Yu HAO, Ren Li WANG, Jun MA, Bing SU, Jian Hua ZHEN. "Image retrieval method based on Tamura texture feature improved". Science of surveying and mapping, vol. 35, pp. 136–138, April 2010.
[5] Fang DONG, Bao Wei LI. "Texture characteristics of CT image for liver based on fractional dimension". Medical Journal of the Chinese People's Armed Police Forces, vol. 14, pp. 337–340, June 2003.
[6] Stavroula G. Mougiakakou, Ioannis K. ValavanisAlexandra Nikita, Konstantina S. Nikita. "Differential diagnosis of CT focal liver lesions using texture features, feature selection and ensemble driven classifiers". Artificial Intelligence in Medicine, vol. 41, pp. 25–37, 2007.
[7] Mahmoud Ghoneim D. "Optimizing automated characterization of liver fibrosis histological images by investigating color spaces at different resolutions". Theoretical biology and medical model, vol. 8, pp. 25–37, July 2011.
[8] Zhang Gang, Ma Zongmin. "Texture feature extraction and description using gabor wavelet in content-based

medical image retrieval". Beijing: Wavelet Analysis and Pattern Recognition, vol. 5, pp. 169–173, April 2007.

[9] Xie Qun ZHANG. "Texture classification method combining gray shade-gradient intergrowth with ring form gabor filtering". Technology and Economy in Areas of Communications, vol. 11, pp. 115–1173, April 2009.

[10] Kim K I, Jung K, Park S H et al. "Support vector machines for texture classification". IEEE Transactions on Pattern Analysis and Machine Intelligence, vol. 24, pp. 1542–1550, November 2002.

[11] Hong Jing ZHU. "The segmentation method of retinal blood vessels based on gray level-gradient co-occurrence matrix". Journal of shanghai jiaotong university, vol. 38, pp. 1485–1488, September 2004.

[12] Rui CONG, Guang Fu GAO, Rui Xiao FAN, Lei QIAO, Wei ZHANG. Vibration image recognition method based on gray gradient co-occurrence matrix and kernel based fuzzy clustering. Journal of vibration and shock, vol. 31, pp. 73–76, September 2012.

[13] Xiao Feng LI, Yi SHEN. "Support vector machines based on computer aided diagnosis system of breast tumor with ultrasound images". Journal of optoelectronics laser, vol. 19, pp. 115–119, January 2008.

Information Technology – Wan et al. (Eds)
© 2015 Taylor & Francis Group, London, ISBN 978-1-138-02785-5

The blind recognition of systematic convolutional code based on digital signal processor

Tong Li, Chenglin Miao & Jun Lv
Department of Information Engineering, Academy of Armored Forces Engineering, Beijing, China

ABSTRACT: This paper researches an algorithm about the blind recognition of systematic convolutional code. By building data matrix and analyzing the rank property of the matrix, the algorithm can realize the blind recognition of channel parameters, including the code length, code beginning, code rate, checking polynomial, as well as the generating polynomial. This paper also proposes the idea to transfer the algorithm to digital signal processor (DSP). After transferring the software codes onto the hardware platform, we optimize the algorithm in the software Software of Code Composer Studio (CCS) and then finish the simulation of the blind recognition of systematic convolutional code with code rate equals 1/2. It offers support to the fast implementation of the algorithm in actual projects.

KEYWORDS: convolutional codes, blind recognition, digital signal processor, rank characteristic

1 INTRODUCTION

The acquisition of information about channel coding sequences is the research emphasis and difficulties at present, and recognizing the channel coding parameters is the prerequisite for information acquisition. Parametric analysis about channel coding is an important component of intelligence communication, communication reconnaissance and computer network operations. Convolutional codes are widely used in many communication systems because of its advantages of simplicity implementation and powerful error correction. The blind recognition of channel codes is to quickly identify the coding system, method and parameters only by using received data under the condition of unknown encoding information. The fast blind recognition algorithm of convolution codes is a relatively new and professional field. With digital communication technology developing to the adaptive and intelligent direction, more and more areas will demand for the blind identification of channel coding technology. So the study is worthy of theoretical and practical significance.

2 DSP MODEL SELECTION

The TMS320F28335 DSP is a floating-point digital signal processor which is newly launched by TI company. Being developed based on existing DSP platform, it adds a kernel of floating point calculation. It not only keeps the original advantages of DSP chips, but also is able to operate complex floating

point calculation. At the same time, it also has advantages of high precision, low consumption, low cost, high peripheral integration, AD conversion precision, large storage of data and application as well as the save of code execution time and storage space. In this design, in view of lots of 0 or 1 symbols to be processed with, the requirement of storage space and the computing speed are demanding. The system frequency of TMS320F28335 is 150 MHz, and the Flash memory space is $256K \times 16$, and SARAM memory space is $34K \times 16$. It also supports for DSP/BIOS real time operating system on-debug simulation. These characteristics can meet the design requirements. Considering the price and performance, this design chooses TMS320F28335 DSP.

3 BLIND RECOGNITION ALGORITHM

3.1 Target parameters

The blind recognition of convolution codes aims at getting blind parameters, which are unknown and needed to be identified and analyzed. The main parameters of convolution codes are listed as follows: (a) the start point of code word i, (b) the length of code element n, (c) code rate k/n, (d) generating matrix G, (e) check matrix H, (f) generating polynomial $g(x)$ and so on. The generating matrix G is the research priority and the generating polynomial $g(x)$ is the goal of the whole recognition process. According to $CH^T = 0$, we mean to get the check matrix H via G which has been recognized.

3.2 Process of recognition algorithm

Based on reference, as to linear block codes, the paper establishes matrix model by data grouping and use the matrix rank characteristics to recognize blind parameters. Convolution codes are also a kind of linear block codes. In a similar way, we can also use the characteristics of matrix rank to recognize convolution codes blind parameters. Based on the algorithm proposed in reference, while synthesizing the characteristic of linear block codes, this paper employs the specific recognition process as shown in Figure 1.

To estimate n, considering that n is generally at most 10 in the practical application and the least common multiple of 2 to 10 is 2520, we can achieve that every code starting of each matrix row is less than

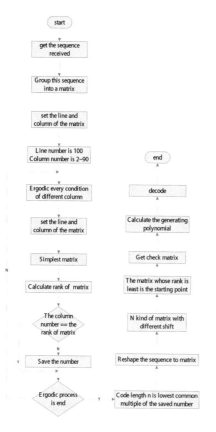

Figure 1. Recognition process.

$d(d=2520)$. We group the received code data into a matrix with x lines and y columns, and it is necessary that y is greater than the coding constraint length, which is $N=n(m+1)$, x>y. In the practical engineering, 64 is enough for y.

We identify the matrix in experiment as follows. As to the first line, if the start point of code word is i, then we take continuous y bits from the received data. As to the second line, the start point of code word is $i+d$, and also y bits should be taken. By such analogy, y bits will be taken in j line.

The next step is that we transform the matrix model into unitary matrix through elementary transformations. If the estimated value of n is greater than the actual value, after unitization, then the beginning of the unitary matrix will be a unit matrix whose rank is $(n-i)$. And the right of the unit matrix is all zeros. At last, from the data distribution characteristics of the submatrix we can get the information of n, k, i and check matrix H. If the convolution code is systematic, then we can directly get the generating matrix G by transforming H.

4 THE SIMULATION HARDWARE ENVIRONMENT

In this paper, the software and hardware design are implemented parallelly. The hardware circuit system contains the following several parts: external storage space expansion circuit, as shown in Figure 2; JTAG interface circuit, as shown in Figure 3; crystal oscillator circuit; reset and power circuit; filter circuit; the circuit related to communication with PC, as shown in Figure 4, in order to upgrade or improve the system. Among them, external memory expansion circuit adopts isl16lv25616 memory chips, whose storage is $256K \times 16$ bits. Therefore, it is necessary to write reasonable CMD files.

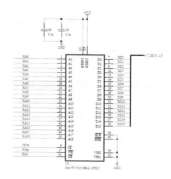

Figure 2. External storage space expansion circuit.

As to aspect of software, the relative algorithm is designed, optimized and simulated by CCS software. CCS is IDE for DSP chip design, which is developed by TI company. It supports online real-time

the number. After finding the least common multiple of these numbers, we get that the code length is 2. Then the relationship between the number of columns and the size of the corresponding unit matrix should be established. We then get a matrix, which is in the form of 64 lines, 65 columns and 4 bits data interval. The related code is as follows:

```
L=64;    // column numbers = L+1
d=4;     // data interval = d
for( k=0; k<L; k++)
{
       for( i =0; i <L+1; i ++)
       {
       data[i][k] =y[k+i*d]; //output matrix
       }
}
```

In DSP/BIOS tool, we call LOG_Printf function to output the data effectively, and get the reduced result of recognition matrix model as shown in Figure 5.

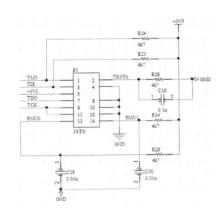

Figure 3. JTAG interface circuit.

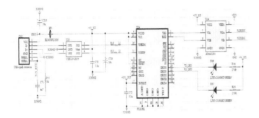

Figure 4. The circuit related to communication with PC.

simulation of JTAG interface based on IEEE1149.1 standard. And it supports DSP/BIOS real-time operating system. Therefore, experimental data is conveniently observed and analyzed.

Figure 5. The reduced result of recognition matrix model $i = 1$.

5 THE RESULTS OF SIMULATION ANALYSIS

This paper selects the $(2,1,6)$ convolution code as the research object. Its basic generating polynomial is $(11,01,11,11,00,10,11)$, $G(D) = [g_1(D) \ g_2(D)]$, $H(D) = [h_1(D) \ h_2(D)]$, and $g_1(D) = [1 \ 0 \ 1 \ 1 \ 0 \ 1 \ 1]$, $g_2(D) = [1 \ 1 \ 1 \ 1 \ 0 \ 0 \ 1]$. Because $G(D)H^T(D) = 0$, so $h_1(D) = g_2(D)$ and $h_2(D) = g_1(D)$. Also the starting point of captured sequences is $i = 1$.

First of all, encode $(2,1,6)$ convolution code. The convolution code sequence is simulated in CCS software, and the sequence is just the captured data, waiting to be analyzed.

Secondly, according to the blind recognition process shown in Figure.1, we transform the data sequence into matrix according to the number of columns. After being simplified, the rank of each matrix need to be calculated. Write down the rank which is not equal to the number of columns and save

From the distribution of sub-matrices, we can get that code rate k/n is 1/2. The top left corner of the matrix is all zero and its size is $(n-i) = 1$. As we can see from Figure 1, after deducting the code word starting point, a check sequence appears in 15th column. It indicates that the encode constraint length $N = 14$. So the memory unit number is 6. Considering that the code word starting point is 1, there should be check sequences in 3th, 5th, 7th, 9th, 11th, 13th columns. However, because of not satisfying the linear relation constraint, they finally didn't appear. Column 15 just meets the full linear constraint, and the column data contains information of the check sequence which we desire for.

As a result, the check sequence is 11010011111011. According to the code length $n = 2$, the extracted sequences are 1001111 and 1101101, in other words, $h_1(D) = [1 \ 1 \ 1 \ 1 \ 0 \ 0 \ 1]$ and $h_2(D) = [1 \ 0 \ 1 \ 1 \ 0 \ 1 \ 1]$. Thus we can get the generating polynomial matrix, and the result is the same with theory. Then generating polynomial G(x) can be obtained through the generating polynomial matrix. In this paper, the experiment doesn't decode the convolution codes. If we want to go on decoding code, other DSP, for example C5000, will be chosen. As to C5000 DSP, its price is higher and its performance is more powerful. The Spru553-Viterbi decoding arithmetic library can be implemented in C5000 DSP.

It is worth mentioning that the simplest transformations of each matrix is in $GF(2)$ according to the principle of modulo-2, which is different from general digital signal processing.

In addition, for another situation, we need modify the starting point and keep the rest unchanged. The available result is shown in Figure 6.

It thus can be seen that when $i = 0$, $(n - i) = 2$. Because $n = 2$, the starting point $i = 2$ is equivalent to $i = 0$. So all zero bits does not appear in the starting point.

is practical. But in this paper, the established model is suitable for the blind recognition of systematic code and 1/n convolution code. For the non-systematic code, it can only blindly recognize the code rate k/n as well as code length, and its recognition algorithm is the same. But due to the existence of error, some column numbers which should have been saved are not saved, and it is without prejudice that some columns which are likely to be saved will be saved. Then we can get the code length, code rate and starting point.

This paper builds the matrix model, uses features of matrix rank, and then gets the data matrix to be analyzed by iterating through all possible matrix forms. We get target parameters from these matrices. In practical application, for 1/n convolution code, if the code length n is greater than 4, then the coding efficiency is very low. So n is not too large in practical application. Due to characteristics of DSP, it is a key step that storage space is arranged reasonably with large data.

Figure 6. The reduced result of recognition matrix model $i = 0$.

6 THE RESULT

Only 300 bits data are used in this simulation, and the process is simple, and the result is accurate and easy to identify. Based on these advantages, this algorithm

REFERENCES

Chen Jin-jie, Yang Jun-an. Blind recognition for low code-linear block codes based on bit frequency detection, Journal of Electronic Measure and instrument, 2011, 25(7); 642–646.

Chen Jin-jie, Yang Jun-an. Blind recognition for low code-linear block codes based on bit frequency detection, Journal of Electronic Measure and instrument, 2011, 25(7); 642–646.

Zan Jun-jun, Li Yan-bin. Blind recognition of low code-rate binary linear block code, Radio Engineering, 2009, 39(1): 19–22.

Chai Xian-ming, Huang Zhi-tao, Wang Feng-hua, Zhou Yi-yu. On Blind Recognition of Channel Coding. Communication Countermeasures, 2008, 101(2).

Song Jing-ye. Study on the technology of channel coding recognition[D]. Xi'an: XiDian University, 2009.

WANG Feng-hua, HUANG Zhi-tao. A Method for Blind Recognition of Convolution Based on Euclidean Algorithm; IEEE Inter Conference on Wireless Com Networking and Mobile Computing, 2007.

JJ Chang, DJ Hwang, and MC Lin. Some Extend Results on the Search for Good Convolutional Codes, IEEE Trans. Inform.Theory, IT-43:1682–97, September 1997.

Information Technology – Wan et al. (Eds)
© *2015 Taylor & Francis Group, London, ISBN 978-1-138-02785-5*

An input sample selection method based on frequency domain filter design

Gaojian Zhang & Gang Yang
College of Information Engineering, Communication University of China, Beijing, China

ABSTRACT: In the paper, we introduce an algorithm called "frequency-domain least-square (FDLS)", which was proposed by Professor Greg Berchin in 1988. The algorithm can help to design the digital filter, especially suitable for the situation that we only know the frequency response curves of the filter. And based on the FDLS design method, we analyze the influence of the input samples for filter design. At last, we put forward a method called non-uniform sampling to select the input samples. The input samples obtained by this method can help to improve the computational efficiency and design the filter which is closer to the prototype.

1 INTRODUCTION

Impulse invariance method and bilinear transform method are the most common methods in the digital filter design.

The impulse invariance method is based on the idea that that we can design a digital filter whose time-domain impulse response is a sampled version of the impulse response of a continuous analog filter. Because the impulse responses of time-domain and frequency-domain are one-to-one correspondence, matching the impulse response can be matched to the frequency response. However, convert a time-domain impulse response to the frequency domain, it will become infinite bandwidth. So it's easy to cause spectrum aliasing if we use the method to design filter.

There's a popular filter design method known as the bilinear transform method. This method is no longer involved in the time domain. It matches the frequency response of the prototype filter directly, so the spectrum aliasing are avoided. We can design a filter which is very close to the standard if the parameters of prototype filter are known. However, designing digital filter in the method will cause the nonlinear transformation of phase in the process from the s plane to z plane. So we can't use this method to design linear phase fir digital filter.

However, the two methods still have their limitations in the practical application. It may require another type of filter design methods if we only know the frequency response curves of the filter. And one of the design methods is the frequency-domain least squares (FDLS) algorithm.

The FDLS method is proposed by Greg Berchin during the period of his graduate. It's quite flexible in that it can create transfer functions containing poles and zeros, only zeros, or only poles. The input to the algorithm is a set of magnitude and phase values of arbitrary frequencies. The algorithm's output is a set of transfer function coefficients. The more input samples we select, the more calculations we need to get the coefficients. Therefore, it's meaningful to find out a selection method to get the better input samples in practical applications.

2 THE ALGORITHMASE THEORY

Next, let's see the specific process of FDLS algorithm. The standard form of digital filter transfer function is

$$\frac{Y(z)}{U(z)} = \frac{b_0 + b_1 z^{-1} + \cdots + b_N z^{-N}}{1 + a_1 z^{-1} + \cdots + a_D z^{-D}} \quad (1)$$

where $U(z)$ is the z-transform of the input signal, $Y(z)$ is the z-transform of the output signal. Furthermore, we assume that the filter is causal, meaning that its response to an input does not begin until after the input is applied. Under these assumptions, to make the above equation becomes the time-domain difference equation is

$$y(k) = -a_1 y(k-1) - \cdots - a_D y(k-D) + b_0 u(k) + \cdots \quad (2)$$
$$+ b_N u(k-N)$$

where k is the time index, u(k) and y(k) are the current values of the input and output, $u(k-N)$ was the input value N samples in the past, and $y(k-D)$ was the output value D samples in the past. We can write the equation above in matrix form as

$$y(k) = [-y(k-1) \cdots -y(k-D) \, u(k) \cdots u(k-N)] \begin{bmatrix} a_1 \\ \vdots \\ a_D \\ b_0 \\ \vdots \\ b_N \end{bmatrix} \quad (3)$$

And the $a1...b_N$ column vector is the coefficient vector of transfer function.

Assuming at the k moment, we select a point with the frequency of ω_1 and give the filter an input signal $u_1(k) = \cos(k\omega_1 t_s)$. Then the output will be $y_1(k) = A_1\cos(k\omega_1 t_s + \phi_1)$. The input and output values at any sample time can be determined by the amplitude and phase. It means the frequency response of a filter at different frequencies, is portrayed by the amplitude A and phase ϕ. Based on the theory, we also know that the relationship between input value u and output value y at any sample time can be inferred from the amplitude A and phase ϕ at frequency ω.

In order to facilitate the calculation, we set the value of k to zero. Combining these two ideas, we obtain one equation in $D+N+1$ unknown

$$y_1(0) = [-y_1(-1)\cdots -y_1(-D)\ u_1(0)\cdots u_1(-N)]\begin{bmatrix}a_1\\ \vdots \\ a_D \\ b_0 \\ \vdots \\ b_N\end{bmatrix} \quad (4)$$

If we repeat the above process at a different frequency ω_2, we obtain a second equation in $D+N+1$ unknown. And if we repeat at many more different frequencies M than we have unknowns $D+N+1$, we can write the equation (5):

$$\begin{bmatrix}y_1(0)\\y_2(0)\\ \vdots \\ y_M(0)\end{bmatrix} =$$

$$\begin{bmatrix}-y_1(-1) & \cdots & -y_1(-D) & u_1(0) & \cdots & u_1(-N)\\ -y_2(-1) & \cdots & -y_2(-D) & u_2(0) & \cdots & u_2(-N)\\ \vdots & & \vdots & \vdots & & \vdots \\ -y_M(-1) & \cdots & -y_M(-D) & u_M(0) & \cdots & u_M(-N)\end{bmatrix}\begin{bmatrix}a_1\\ \vdots \\ a_D \\ b_0 \\ \vdots \\ b_N\end{bmatrix} \quad (5)$$

We can denote the $y_1(0)\ldots y_M(0)$ column vector above as Y, the matrix as X, and the $a_1\ldots b_N$ column vector as Θ.

Now, there's a matrix equationand $Y = X\Theta$ and our task is to solve out the Θ. The X and Y are composed by the u and y, they can be represented if we know the magnitude and phase values of input frequencies. However, X is not a square matrix, and we can't get the inverse matrix unless the M and $(N+D+1)$ are equal. Luckily, the linear algebra provides the "pseudoinverse" to determine the values of Θ.

How does the pseudoinverse work? For example, if we have the following matrix equation

$$\begin{bmatrix}5 & 2\\ -6 & 4\end{bmatrix}\begin{bmatrix}x\\y\end{bmatrix} = \begin{bmatrix}7\\9\end{bmatrix} \quad (6)$$

We can get the unique solutions for both x and y by the way of matrix inversion.

$$\begin{bmatrix}x\\y\end{bmatrix} = \begin{bmatrix}5 & 2\\ -6 & 4\end{bmatrix}^{-1}\begin{bmatrix}7\\9\end{bmatrix} = \begin{bmatrix}\frac{1}{8} & -\frac{1}{16}\\ \frac{3}{16} & \frac{5}{32}\end{bmatrix}\begin{bmatrix}7\\9\end{bmatrix} = \begin{bmatrix}\frac{5}{16}\\ \frac{87}{32}\end{bmatrix} \quad (7)$$

Let's consider what happens if we add another equation, $x+y = 5$, to the pair that we already have above. We can't get the inverse matrix of coefficient matrix because it's not a square matrix. There are no values of x and y that satisfy all three equations simultaneously.

$$\begin{bmatrix}5 & 2\\ -6 & 4\\ 1 & 1\end{bmatrix}\begin{bmatrix}x\\y\end{bmatrix} = \begin{bmatrix}7\\9\\5\end{bmatrix} \quad (8)$$

In the situation, we need pseudoinverse to determines the values of x and y that come, in the least-squares sense, as close as possible to satisfying all three equations.

It's unnecessary to detail the definition of pseudoinverse in the paper, and the solution is then given by

$$\begin{bmatrix}x\\y\end{bmatrix} \approx \left(\begin{bmatrix}5 & 2\\ -6 & 4\\ 1 & 1\end{bmatrix}^T\begin{bmatrix}5 & 2\\ -6 & 4\\ 1 & 1\end{bmatrix}\right)^{-1}\begin{bmatrix}5 & 2\\ -6 & 4\\ 1 & 1\end{bmatrix}^T\begin{bmatrix}7\\9\\5\end{bmatrix} \approx \begin{bmatrix}0.3716\\ 2.8491\end{bmatrix} \quad (11)$$

where T denotes the matrix transpose and -1 denotes the matrix inverse.

With $Y = X\Theta$, the pseudoinverse can solve the vector Θ that contains the desired filter coefficients by $\Theta \approx (X^TX)^{-1}X^TY$ and complete the filter design.

3 THE ANALYSIS OF THE INPUT SAMPLES

We need to select the input samples from the prototype response curves when the FDLS algorithm is used to design filters. The number and location of input samples will affect the filter design. So it's necessary for us to analysis the selection of input samples.

At first, we need to know the impact of the number of input samples because they determine the calculation of our design directly. And then we hope to find out a selection method of the input samples. We can design a digital filter which is closer to the prototype filter on the basis of the same calculation in this way.

The proximity of prototype filter and our designing digital filter can be expressed by their frequency response curve. The closer the two curves are, the more similar their performances are, and our design is more success.

Let's use an example to illustrate the judgment method. We can see the Figure 1. The curves of the two colors represent the prototype filter and designing digital filter. The above section says the amplitude, the following section shows phase. The red curve represents the prototype, black curve on behalf of the design of the filter.

In order to express the similarity of two curves accurately, we can set the amplitude curve of the prototype filter as l_{1A} and the phase curve as l_{1f}. The response curves of our designed filter can be represented as l_{2A} and l_{2f}. The difference curves $\Delta l_A = |l_{1A} - l_{2A}|$ and $\Delta l_\theta = |l_{1\theta} - l_{2\theta}|$ can be got through subtracting the response curves. We can determine how close the two filters by calculating the mean and variance of the difference curves.

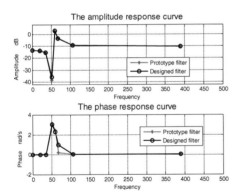

Figure 1.   The frequency response curves.

3.1   *The number of input samples*

From the equation (5) we know that, the more input samples we select, the more calculation we need. Supposing the number of input samples is M. It's helpful to reduce the calculation if we understand the impact of the M.

Selecting an analog filter as the prototype, the means and variances of Δl_A and Δl_θ can be got by the experiment. The data are shown in Figure 1.

Figure 2 shows the mean and variance of the difference curves at different values of M. The upper figure represents amplitude and the under represents phase. The blue line represents the change of means with M and the red represents the change of variances. Some values between 26 and 2000 are selected as M in the experiment, and we get the M input samples from the prototype curve with uniform sampling.

From the trend of the curve, the means and variances of Δl_A and Δl_θ are basically unchanged when M reaches a certain value. There's no too much difference between the digital filters we designed by 50 input samples and 2000 input samples.

For the further analysis, we can select more detailed values to experiment. There's the data figure

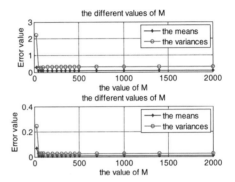

Figure 2.   The frequency response curves.

by choosing all integer values between 26 and 100 as M:

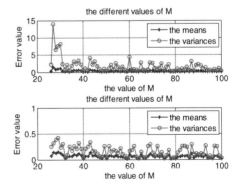

Figure 3.   The mean and variance of difference curves.

From the Figure 3, we can find the mean and variance are changed in a very small range when M is more than 30. Especially the means, has reached a low value.

In the case, the gap between the prototype filter and our designed filter can't shrink with the increasing of M.

For validating the universality of the above conclusion, we need more filter models. With low-pass and high-pass filters as the prototype respectively, we can get another two figures about the relationships between input numbers and differential data.

From Figure 4 and Figure 5, differential data does not have the big ups or downs with the increase of M. So when we design any filter with this method, it's useless to increase input samples if the number to a certain extent.

3.2 *The Non-uniform sampling method*

We analyze the impact of input sample numbers for the filter design before. Now we will introduce a

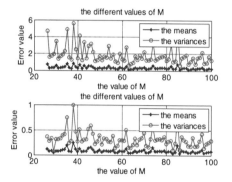

Figure 4.   The mean and variance of difference curves.

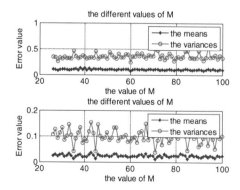

Figure 5.   The mean and variance of difference curves.

specific sampling method to get the input samples. The method can help us to design a digital filter

closer to the prototype when the calculation is the unchanged. The means and variances of the difference curves can reduce an order of magnitude if we get the input data in the way.

This efficient method of extracting samples is called the non-uniform sampling method because it gets the input samples on the curve by the way of non-uniform sampling. The principle of this method is to pick the extreme points on the curve when selecting the input samples. The substance of selecting the input samples is to describe the response curve of the prototype filter. It can help us to describe the places where the rate of change is big if we select as many extreme points as possible. And then get the better design.

Using the original prototype filter and set the value of M to 30(We choose the number because the design is not perfect with 30 input samples. It's help to compare the gaps under different input sample). We design the filters with uniform and non-uniform methods to extract input samples respectively. The frequency response curves of prototype filter and digital filters are shown in Figure 6.

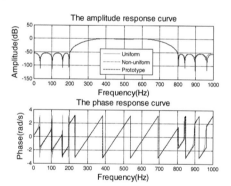

Figure 6.   The frequency response curves of the filters.

The black dash indicates the response curve of the prototype filter, and the red and blue curves show the uniform and non-uniform method. On the whole, the three curves are close to. But in a larger view of Figure 7, we can clearly see the gaps.

On the Figure 7, we amplify the part which the rate of change is big on the curves. The red curve and black have gaps obviously in the place of the turn. However, the blue curve and black are hard to tell. It means the method is effectual and can help to make a better design with the same number of input samples.

In order to validate the universality of this method, we select the low-pass, band-pass, and high-pass filter as the prototype filter respectively to carry out the experiment. Using uniform and non-uniform extracting methods to get the input samples under

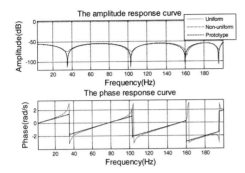

Figure 7.   The frequency response curves of the filters.

the premise that the number of samples is the same. This time we show the means and variances of the difference curves in the data tables for observing the distinction clearly.

Table 1.   The means and variances based on amplitude.

The types of filters	Sampling method	Amplitude	
		Means(dB)	Variances
low-pass	Uniform	1.3476	10.915
	Non-uniform	0.0665	0.6391
band-pass	Uniform	1.0463	8.1677
	Non-uniform	0.0698	0.2174
high-pass	Uniform	1.2881	9.5799
	Non-uniform	0.0542	0.5781

Table 2.   The means and variances based on phase.

The types of filters	Sampling method	Amplitude	
		Means(rad/s)	Variances
low-pass	Uniform	0.2731	0.8507
	Non-uniform	0.0307	0.0981
band-pass	Uniform	0.1243	0.2352
	Non-uniform	0.0103	0.0334
high-pass	Uniform	0.2348	0.6838
	Non-uniform	0.0263	0.1509

No matter which type the prototype filter is, the digital filter designed by the non-uniform sampling method is closer to the prototype filter than the uniform sampling. And from the two tables, we know the means and variances can be reduced at least an order of magnitude if we use this method.

4   CONCLUSION

In the paper, we introduce the FDLS algorithm and discuss the input samples selection of the algorithm. FDLS is a powerful method for designing digital filters. And it is most useful in cases where a specified frequency response must be duplicated to within tight tolerances over a wide frequency range or when the frequency response of an existing system is known but the coefficients of the system's transfer function are unknown. About the input samples, we analyze the impact of numbers at first. We know that it's meaningless to increase the number of input samples when it reaches a certain range. It's helpful to reduce the amount of calculation in the case of the filter performance is not affected. The next, we propose the non-uniform sampling method to select the input samples. This method can help to design the digital filter which is closer to the prototype on the premise of the same amount of calculation. And the method can also be used for the condition that calculations are restricted.

REFERENCES

G. Berchin, "A New Algorithm For System Identification From Frequency Response Information". Master's Thesis, University of California–Davis, 1988.

R. Lyons, "Understanding Digital Signal Processing", 2nd ed., Upper Saddle River, NJ:Prentice-Hall, pp. 232–240, 2004.

G. Strang, *Linear Algebra and Its Applications*,2nd ed., Orlando, FL: Academic, pp. 103–152, 1980.

G. Berchin and M.A. Soderstrand, "A transformdomain least-squares beamforming technique," in *Proc. IEEE Oceans '90 Conf.*, Arlington, VA,Sept.1990

M.A. Soderstrand, K.V. Rangarao and G. Berchin, "Enhanced FDLS Algorithm For Parameter-based System Identification For Automatic Testing".in Proc. IEEE Int. Conf, vol.1, 1990, pp. 96–99.

G. Berchin, "Precise Filter Design". DSP Tips & Tricks, Signal Processing Magazine, 24, 137–139, January 2007.

Information Technology – Wan et al. (Eds)
© 2015 Taylor & Francis Group, London, ISBN 978-1-138-02785-5

The method exploration to improve the frame rate of digital ultrasound imaging system

Rui Yang & Zhexu Li

University of Shanghai for Science and Technology, Shanghai Medical Instrumentation College, Shanghai, China

ABSTRACT: The time resolution of an ultrasound imaging system characterizes the ability to capture the motion changes of two adjacent phases. It can be described with the frame rate of ultrasound equipment. A high frame rate can capture the subtle information of motion changes. It is very important to scan the motion of the organ. On the basis of the multi-beam formation technique and the flying scanning technique which are two methods of improving the frame rate, a combination of the two methods is proposed. The method has a significant effect for improving the frame rate of an ultrasound imaging system.

KEYWORDS: Frame rate; Multi-beam formation; Flying scanning; Ultrasound imaging

1 INTRODUCTION

The medical ultrasound imaging technique and the X-ray imaging technique, magnetic resonance imaging technique and the nuclear medicine imaging technique are recognized as four major modern medical imaging techniques. The size of ultrasound wavelength is the same as a biological cell. When interacting with a biological tissue, it can obtain more information than the X-ray. The ultrasound imaging principle decides that the imaging can be dynamic and repeated. Using the physical characteristics of ultrasound that it spreads in human body, the instrument can show images of the inside organs or diseases and more diseases can be diagnosed accordingly. Compared to other imaging techniques, the medical ultrasound imaging is given widely attention and applications it is portable, relatively safe, real-time, non-invasive and a low price and so on, but it has also some weaknesses. For example, the low frame rate, significantly weaker than CT on the spatial resolution, susceptible to noise and so on. Thus the ultrasound imaging quality should be improved further.

A frame rate is the number of frames for per second image. A single rectangular image or scan is called a frame. It is an important parameter of describing the time resolution of an ultrasound imaging system. It reflects the ability of real-time imaging. When scanning an organ, which is moving with a high speed, a high frame rate can capture subtle changes in the motion information. The higher the frame rate is, the more timely organs motion displays. Today, the frame rate is about 30 frames/s which is still a lower frame rate for the real-time display of organs motion.

Therefore, improving the frame rate of an ultrasound imaging system has an important significance.

In terms of the B-mode, the frame rate can be improved by reducing the density of scanning lines, the depth of detecting and the range of imaging. For the method of reducing the density of scanning lines, the image resolution will be decreased. How to guarantee a certain image quality under the premise of improving the frame rate is a research focus now. This paper will explain the principles of the multi-beam formation technique and the flying scanning technique which can both improve the frame rate. A combination of the two methods is proposed on this basis which has a significant effect in improving the frame rate.

2 TWO METHODS OF IMPROVING THE FRAME RATE

2.1 *Digital multi-beam formation technique*

The digital beam forming is a most critical technique in a digital ultrasound imaging system. The beam former is where the action originates. It consists of a pulser, pulse delays, transmit/receive (T/R) switch, amplifiers, analog-to-digital (A/D) converters, echo delays, and a summer. The beam former is responsible for electronic beam scanning, steering, focusing, apodization, and aperture function with arrays.

The digital beam forming can be divided into two categories: the single beam forming and the multi-beam forming. In China, most of the B-mode devices are still the single beam forming. In simple terms, the single beam forming is that one pulse of ultrasound generates a single scan line. Multi-beam forming is that one pulse of ultrasound generates a plurality of scan lines. So the multi-beam formation is an effective method to improve the frame rate. However, it can also cause some other problems. For example, the process will generate a greater storage capacity of delay parameters in real time which is a key problem to be solved, especially the multi-beam formation with dynamic segmented focusing which has a relatively better image quality. Now we introduce the principle of the dual-beam formation which is an example of the digital multi-beam formation.

The transducers generate and receive ultrasound and form the connecting link between the ultrasound-tissue interactions and the instruments. These beams are focused and are automatically scanned through tissue by the transducers. Ultrasound transducers operate according to the principle of piezoelectricity. Piezoelectric elements convert electric voltages into ultrasound pulses and convert returning echoes back into voltages.

The sound produced by these transducers is confined in beams rather than traveling in all directions away from the source. To improve resolution, diagnostic transducers are focused. Focusing sound in the same manner as focusing light reduces beam width. Sound may be focused by using curved transducer elements, by using a lens, or by phasing. Phasing is that the phased array is operated by applying voltage pulses to most or all elements in the assembly, but with small time differences between them. Of course, phasing is also applied to some linear array to steer the beam from each element group in several directions by sending out several pulses from each group with different phasing. When focusing, the outer elements are pulsed slightly ahead of the inner ones. This produces a curved pulse that is focused at a depth determined by the delay between the firing of the outer and inner elements. These pulses are focused by phasing. When receiving, echoes can be generated from a specific anatomic location with several viewing angles. The echo information from these multiple views is processed to present an image of improved quality. The method of phased delays applied to each element group to focus the pulse is shown in Fig. 1.

Transmitting mode: the array contains a straight line of rectangular elements. Each element is about a wavelength wide and is operated by applying voltage pulses to groups of elements in succession. Each group of elements acts like a larger transducer element, providing a large enough aperture to confine the ultrasound to a fine enough beam for satisfactory resolution. As different groups are energized, the origin of the sound beam moves across the face of the transducer assembly from one end to the other. In the meantime, diagnostic transducers are focused and the resolution is improved.

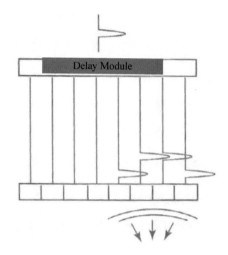

Figure 1.   Phased delays are applied to each element group to focus the pulse.

Receiving mode: a weak focusing can be achieved by increasing the focal length or decreasing the aperture in transmitting. The beam width is increased as a result. When an array is receiving echoes, it can continuously receive the echoes coming from two different beam receiving lines. And for every beam receiving line, the reception focus depth may be increased continually as the transmitted pulse travels through tissues and the echoes arrive from deeper and deeper locations. The continually changing reception focus is called dynamic focusing .Then the echo data are obtained by A/D sampling and sent into two synthesizers. In the end, we obtain two scanning lines by transmitting only once. The multi-beam formation is similar with the dual-beam formation. As shown in Fig. 2.

2.2   Flying scanning technique

The flying scanning is another method to increase the frame rate by changing the way of scanning. Sequential scanning and interval scanning are two scanning methods. Different scanning methods can obtain different number of scan lines, which finally affect the image quality. For example, the sequential scanning is a basic scanning mode which linear array and convex array can both use. Its characteristics are to achieve an electronic scanning, guarantee the power by groups transmitting, and improve the sensitivity and so on. But the image quality is not good, because it generates less scan lines that are acquired for one frame. The interval scanning can increase the scan line density and improves the image resolution. But with the increasing of the scan lines, more time is needed for one frame. The frame rate will be reduced and impacts the real-time display. Based on this, flying scanning technique is used to solve the problem.

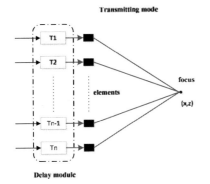

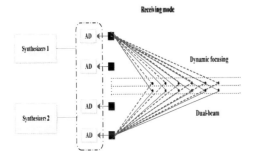

Figure 2. The schematic of dual-beam formation.

The process of flying scanning is that the elements of the transducer are divided into two parts A and B just like two transducers, and the two parts began scanning at the same time. In order to obtain a higher resolution, the two parts can use the interval scanning. The method can ensure that the frame rate can be improved twice the interval scanning. Assuming a 128-element array in group of 8 or 9, the scanning process is shown in Table 1:

Table 1. The process of flying scanning.

Number of scanning	Part A	Part B
1	1–8	65–72
2	1–9	65–73
3	2–9	66–73
4	2–10	66–74
5	3–10	67–74
6	3–11	67–75
...	...	...
...	...	...
113	57–64	121–128

For the two methods of improving the frame rate of a digital ultrasound imaging system, the multi-beam

formation technique is one that is transmitting once and two or more scan lines are got. The time required to complete one frame is shortened and the frame rate is increased. The flying scanning is that the transducer transmits two or more sets of ultrasonic beams at the same time every time to increase the number of scanning lines per unit time. The same result is that the time of completing one frame is shortened and the frame rate is also improved. The two different methods are effective in increasing the frame rate.

3 THE COMBINATION OF MULTI-BEAM FORMATION TECHNIQUE AND FLYING SCANNING TECHNIQUE

Based on the characteristics of multi-beam formation technique and flying scanning technique, we propose a new method to improve the frame rate which is a combination of the two techniques. The ways of transmitting and receiving are shown in Fig. 3:

Transmitting process: two parts elements of the transducer simultaneously transmit ultrasonic beams. During the transmitting, two ultrasonic beams transmitted simultaneously from the two parts should be always kept with a certain interval to prevent the mutual interference between them. Obviously, the way of transmitting is flying scanning.

Receiving process: the echoes will be received immediately after the transducer transmits two ultrasonic beams every time. Assuming the four-beam formation, 8 scanning lines can be obtained with the combination of two techniques and the frame rate will be twice of the four-beam formation and only four times the flying scanning. So the combination of both techniques can greatly improve the frame rate of the system. But the technology generates more delay parameters when improving the frame rate and the system needs higher requirement of hardware and software.

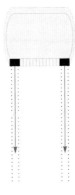

Figure 3. The combination of multi-beam formation technique and flying scanning technique.

4 CONCLUSION

This paper describes the principles and characteristics of multi-beam formation and flying scanning which can both improve the frame rate and the combination of both techniques is proposed on this basis. The method has a significant effect for increasing the frame rate of an ultrasound imaging system and has broad application prospects. It can support the real-time display of color Doppler flow and provide a strong guarantee to the moving organs imaging. It lays the foundation for the further study of generation algorithms and storage of the delay parameters and the realization of software and hardware.

REFERENCES

[1] Wang Wei-ming, "Study on Technologies and Optimization of Digital B-mode Ultrasound imaging," Thesis submitted to Chongqing University, 2010.

[2] Zhao Hai-long, Wang Yan, Fang Yan-hong, "Study and implementation of digital dual beam formation techniques in the ultrasound imaging system," *Journal of Southwest University of Science and Technology*, 2007, 22(3): 87–91.

[3] Gao Yang, "Study on Beamforming of Digital Ultrasound Imaging System," Thesis submitted to Chongqing University, 2013.

[4] Hedrick.Hykes. Starchman, Ultrasound Physics And Instrumentation, 4th ed., 2005.

[5] Frederic kW. Kremkau, Sonography Principles And Instruments, 8th ed, 2011.

[6] C.Passman,H.Ermert, "A 100 MHz ultrasound imaging system for dematologic and ophthalmologic diagnostics," *IEEETrans. Ultraon., Ferroelect. Freq. Contr.*, 1996, 43: 545–552.

[7] H.Djelouah, J.C. Baboux, M. Perdrix, "Theoretical and experimental study of the field radiated by ultrasonic focused transducers," *Ultrasonics*, 1991, 29: 188–200.

[8] Jensen J.A., Munk P.A., "New method for estimation of velocity," *IEEE Transactions on Ultrasonics, Ferroelectrics, and Frequenc Control*, 1998, 45(3):852–861.

[9] M. Karaman, P.-C. Li, M. Odonnell, "Synthetic aperture imaging for small scale systems," *IEEE Trans. Biomed. Eng.*, 1995, 42: 429–442.

[10] Ramazan Demirli, Jafar Saniie, "Model-based estimation of ultrasonic echoes part1:analysis and algorithms," *IEEE Transactions on Ultrasonics, Ferroelectrics, and Frequency Control*, 2001 48.

[11] Jensen J. A., Holm Ole, "Experimental ultrasound system for real-time synthetic imaging," *IEEE Ultrasonics Symposium Proceedings*, 1999, 1595–1599.

[12] S. Tamano, T. Kobayashi, S. Sano, K. Hara, J. Sakano, T. Azuma, "3D ultrosound imaging system using Fresnel ring array & high voltage multiplexer," *IEEE Ultrasonics Symposium*, 2004, 1:782–785.

[13] Wall Kieran, Lockwood, Geoffrey R.A, "New multi-beam approach to real-time 3D imaging," *IEEE Ultrasonics Symposium*, 2002, 2:1803–1806.

[14] Jin Hu-chang, "A new synthetic aperture focusing method to suppress the diffraction of ultrasound," *IEEE Transaction on Ultrasonics, Ferroelectrics and Frequency Control*, 2011, 58(2):327–337.

[15] T.Bjastad, S.A.Aase.Synthetic, "Transmit Beam Technique in an Aberrating Environment," IEEE Transactions onUltrasonics, Ferroelectrics and Frequency Control, 2009, 56(7):1340–1351.

[16] Martin Christian Hemmsen, Image processing in medical ultrasound, Technical University of Denmark, 2011.

[17] Thanassis X. Misaridis, Ultrasound imaging using coded signals, Technical University of Denmark, 2011.

[18] Jorgen Arendt Jensen, Users'guide for the field program, Techinical University of Denmark, 2001.

[19] Wang Shun-Li, Li Pai-Chi, "High frame rate adaptive imaging using coherence factor weighting and the MVDR method," IEEE Ultrasonics Symosium, 2009, 1175–1178.

[20] Burkholder R. J., Browne K. E., "coherence factor enhancement of through-wall radar image," *Antennas and Wireless Propagation Letters*, 2011, 9:842–845.

[21] Park S,.B.Karpiouk A,"Adaptive beamforming for photoacoustic imaging using linear array transducer,"IEEE Ultrasonics Symposium, 2010, 11:1924–1927.

[22] C.H. Frazier, W.D. Obrien, "Synthetic aperture techniques with a virtual source element," *IEEE Transactions on Ultrasonics, Ferrroelectrics and Frequency Control*, 1998, 45(1):196–207.

[23] Capon Jack, "High resolution frequency-wavenumber spectrum analysis," *Proceedings of the IEEE*, 1969, 57(8):1408–1418.

[24] Li Jian, Stoica P., Wang Zhi-Song, "On robust capon beamforming and diagonal loading,"*IEEE Transactions on Ultrasonics, Ferroelectrics and Frequency Control*, 2003, 51(7):1702–1714.

Information Technology – Wan et al. (Eds)
© 2015 Taylor & Francis Group, London, ISBN 978-1-138-02785-5

The query by humming method based on cluster analysis and dynamic time warping

Menglu Li, Zhijun Zhao & Ping Shi
Communication University of China, Beijing, China

ABSTRACT: Among many content-based music retrieval ways, query by humming (QBH) is considered as a retrieval method that is convenient and easily accepted by users, which is also a study of hotspots in recent years. This article divided WAV format music into sentences by using cluster analysis. By combining cluster analysis and dynamic time warping (DTW), this paper implemented WAV format QBH with hum fragments and music melody. The experiment results showed that the accuracy of the segmentation method based on cluster analysis was high. The final retrieval accuracy improved and the retrieval time declined considerably. Using cluster analysis to implement sentence segmentation of WAV format music provided a new method for improving the effect of QBH.

1 INSTRUCTION

In the modern information society, multimedia information has become a main component of the information superhighway. Users often query the multimedia information according to their need. Traditional multimedia retrieval based on text (such as Baidu, Google) has its inherent limitations. Because people are unable to retrieve multimedia files without key words, content-based retrieval was proposed in response to the proper time and conditions. Of many music retrieval methods, the music retrieval based on melody matching is a popular research direction in recent years. According to the musical characteristics such as melody and rhythm, we can efficiently achieve matching. Query by humming (QBH) is considered as a content-based music retrieval way that is convenient and easily accepted by users. Namely inquirers record a music melody through microphone and then can retrieve similar music on computer. Many research institutions such as MIT and University of Southern California have done some research about content-based audio retrieval, and have made great progress in many fields such as QBH, audio classification, and audio structured [1]. Currently, the Muscle Fish Company has developed a commercial audio retrieval engine based on audio perceptual characteristics. Siri voice recognition system in Apple products has a high utilization rate and a wide range of user [2]. A few Chinese web hum retrieval systems have come into use, but retrieval effect is not satisfactory. Current research results and applications show that QBH remains to be further perfected.

Content-based music retrieval has two core problems: melody feature extraction and similarity calculation. Current QBH systems can hardly be extended to massive database because most of them adopt the features extracted from MIDI files. The MIDI format is not widely used and the match method consumes much time [3]. Traditional music melody matching algorithm based on dynamic time warping (DTW) can realize similarity calculation and matching between hum fragments and music of WAV format. But this method has a high error detection rate and needs much time. This paper put forward a music melody matching way based on cluster analysis and DTW. The experiment result showed that the music sentence segmentation method using cluster analysis improved the detection accuracy and the retrieval speed.

2 INTRODUCTION OF MUSIC MELODY MATCHING ALGORITHM

The traditional melody feature matching algorithm can return the most similar results, which is based on the pitch vectors. By extracting the absolute or relative pitch value and calculating similarity with DTW, we can get the most similar music fragments.

Due to the large amount of data in a music pitch sequence, the music must be divided into sentences and achieve matching by sentences. The traditional matching algorithm has not solved the problem well. The fixed small energy amplitude was usually used for music segmentation in this algorithm. This method essentially does not distinguish sentence interval from word interval, causing a low melody similarity between hum fragments and music. The error judgment and error detection are easily caused and the processing time is longer.

3 IMPROVED MUSIC MELODY MATCHING ALGORITHM

An improved algorithm combining cluster analysis and DTW was proposed in this paper. This algorithm realized the accurate segmentation of music sentences, and improved the accuracy and the speed of QBH retrieval. In this paper, the QBH method consisted of three parts: cluster analysis, melody feature extraction, and similarity calculation.

3.1 Cluster analysis

QBH realizes melody matching sentence by sentences. It is significant of the accuracy of music sentence end-point detection. The foundation of data processing is dividing music into sentences and extracting features by sentences.

Sentence end-point detection was implemented by using voice energy analysis and cluster analysis in this paper. Through framing and short-time energy analysis, mute frames were detected and the adjacent mute frames were combined. Because of the randomness of hum fragments, a threshold should be defined through cluster analysis, by which word interval could be distinguished from sentence interval. The specific processing included framing, short-time energy calculation, and cluster analysis.

3.1.1 Framing

In the process of speech signal processing, to facilitate calculation, we usually take N voice signals for a consecutive frame. For the smoothness of data, there is a certain overlap between the adjacent frames. Frame length of 20 ms and the overlap of 1/2 frame were taken in this paper.

3.1.2 Short-time energy calculation

The energy of speech signals changes over time. For hum fragments, energy of punctuation is close to zero. For songs, the punctuation contains accompaniment only. Compared with the parts containing singer's voice, the energy of punctuation is obviously different. So according to the analysis of short-time energy, the change of the voice can be described accurately.

The short-time energy En is defined as follows, where N is the window length, $w(m)$ is the window function, $x(m)$ is the speech sample values.

$$En = \sum_{m=n-(N-1)}^{n} [x(m)w(n-m)]^2 \qquad (1)$$

The rectangular window was chosen to truncate the signal as follows:

$$En = \sum_{m=n-(N-1)}^{n} x^2(m) \qquad (2)$$

The short-time energy of each frame according to this formula is the basis of the mute frame judgment.

The short-time energy analysis of hum fragments and music fragments of the song "tornado" were shown in Figure 1.

As shown in Figure 1, because there is no accompaniment interference in hum fragments, we can extract the mute frames and merge them as the time interval, which is saved into the time interval vector [t1, t2, t3, ..., tn]. The time interval vector is the key to the cluster analysis. In the song clips, with the overlapping of vocals and accompaniments, a threshold needs to be defined. Accompaniment frames under the threshold need to be combined as mute frames, whose duration will be saved into the interval time vectors of music.

There are two categories of the interval vector [t1, t2, t3, ..., tn]: time interval between sentences and time interval between words. So the threshold needs to be determined to separate two categories through cluster analysis.

3.1.3 Cluster analysis

The purpose of clustering is to determine the threshold for classifying word intervals and sentence intervals. First, we set up an initial threshold for the classification and calculate the average of the two categories as the category center. Then calculate the Euclidean distance of each value and the category center in turn, and put each value into the closer category. After calculating all, there are the new threshold and new category centers. Repeat the iteration, until there is not any interval value transferred into the other category. Eventually determined threshold is the judgment of words and sentences intervals.

Music sentences were saved into database to facilitate melody matching.

3.2 Melody feature extraction

Relative pitch vectors were adopted to express the melody feature in this paper. The pitch values need to be converted into note numbers to get the absolute and relative pitch vectors.

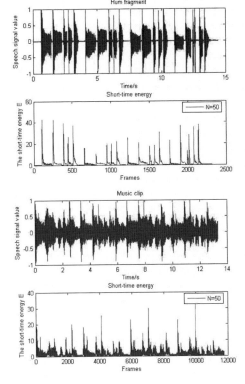

Figure 1.   Short-time energy analysis of the hum fragment and the actual music clip.

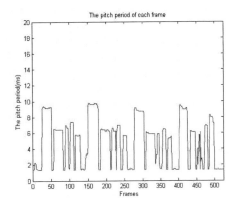

Figure 2.   The pitch period of the hum fragment in Figure 1.

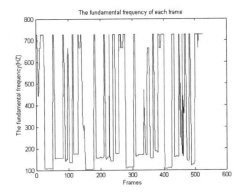

Figure 3.   The fundamental frequency of the hum fragment in Figure 1.

3.2.1   *The fundamental frequency extraction*

For the WAV format songs, the melody information of notes was got by detecting the fundamental frequency.

The algorithms of fundamental frequency detection include the autocorrelation function, the average magnitude difference function, cepstrum analysis, and so on. The autocorrelation function is simple and the results are more accurate [4]. So in this paper, autocorrelation function was used to extract the fundamental frequency. Autocorrelation function is defined as follows:

$$R(v) = \sum_{n=-\infty}^{n=\infty} x(n)x(n+v) \qquad (3)$$

Figure 2 shows the pitch period and fundamental frequency information of the hum fragment in Figure 1.

By extracting the period of autocorrelation function in each frame and eliminating the outliers to smooth signals, we can determine the pitch period.

3.2.2   *The transformation between the frequency and notes*

After extracting the fundamental frequency information in each frame, fundamental frequency information needs to be transformed into note information according to the comparison table of frequency and notes.

According to the principle of proximity, the fundamental frequency was mapped into note numbers of MIDI format. The note information was saved into the absolute pitch sequences to prepare for the matching.

3.2.3   *Relative pitch sequence*

Because reference tones of hum fragments may be different from songs, in this paper the absolute pitch sequences were transformed into relative pitch sequences as melody feature when matching.

The melodic contour of relative pitch sequence has the widespread application. For any note of a melody, it has three kinds of states: if the note number is higher than the previous note's, the state is represented by "U," "D" for lower than the previous note's, and "S" for same as the previous note's. According to this rule, any melody can be converted into a character sequence containing the letters U, S, D [5]. This paper improved the algorithm. By calculating the difference between adjacent notes' number, music is converted to relative pitch sequences to prepare for the similarity calculation.

3.3 Similarity calculation

DTW was adopted in this paper to calculate the similarity of music melody feature vectors. The songs were sorted according to the similarity, and the most similar songs were returned.

DTW is an algorithm translating the global optimization problem to the local optimum problem, which is a nonlinear warping technique combining time alignment with calculation of distance measures. Automatic searching for a path using local optimization to acquire the minimum distortion between two feature vectors can avoid the error introduced by different length effectively. For the isolated words recognition and pitch sequences matching by the sentence, DTW algorithm is relatively simple. And DTW can solve the problem of the inconsistency in the note amount of hum fragments and the actual music. So in this paper DTW algorithm was adopted to make similarity calculation.

For QBH, the melody feature vector of the hum fragment is considered X axis, while the melody feature vector of the standard music is Y axis. The shortest distance of two vectors was calculated by DTW. The formula is expressed by Equation (4), in which i is the index of the X axis, and j is the index of Y, $D(i, j)$ is the Euclidean distance between two vectors, and $dist(i, j)$ is the cumulative distance.

$$D(i, j) = \min \begin{cases} D(i-1, j-2) \\ D(i-1, j-1) \\ D(i-2, j-1) \end{cases} + dist(i, j) \qquad (4)$$

In this paper, the cumulative distance of hum fragments and songs were calculated by the sentence, and then the most similar songs were acquired by sorting according to the returned distance value.

4 THE EXPERIMENTAL RESULTS

4.1 Segmentation experiment based on cluster analysis

In this paper, cluster analysis experiments of 100 songs were carried out. Sentence intervals were detected in each song, and compared with the actual intervals. The percentage of detected intervals and actual intervals is shown in Figure 4. The results demonstrated that the highest percentage is 100%, lowest percentage is 90.18%, average percentage is 96.25%, and the average percentage is above 90%.

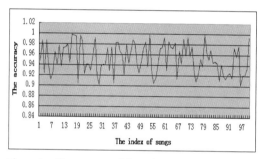

Figure 4. The accuracy of clustering segmentation.

In the traditional QBH algorithm, music is segmented by estimating a small energy amplitude values because there are small energy amplitude value between words' endpoint. At the beginning of a song or in the part with small accompaniment sounds, single words may be determined as sentences. Meanwhile, in the refrain of the music, the traditional algorithm may fail to detect endpoints, which would lead to a similarity calculation based on a long time music fragment. By the traditional algorithm, the average error detection rate is 78.6%. The low accuracy of the endpoint detection based on traditional algorithm would lead to the low accuracy of QBH.

4.2 QBH experiment

In this paper, 30 recorded WAV format hum fragments and 50 WAV format songs were used to calculate DTW distance. At the same time, an experiment was done by traditional music melody matching algorithm for comparison. Traditional music melody matching algorithm divided music into sentences by detecting note's energy amplitude. In this paper, the extraction results of improved algorithm were compared with that of the optimal traditional algorithm.

The cumulative distance of 30 hum fragments calculated by two methods were shown in Figure 5, and the maximum, minimum, average were shown in Table 1. It can be seen that the improved melody

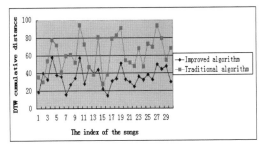

Figure 5.   DTW cumulative distance value.

Table 1.   The comparison table of DTW.

	Improved melody matching algorithm	Traditional melody matching algorithm
The highest DTW value	58.10	94.29
The lowest DTW value	14.67	28.17
The average value	35.62	61.55

matching algorithm greatly reduces the DTW cumulative distance, which means the similarity of hum fragments and songs was higher and the correct extraction results were higher.

The top three and the top five retrieval accuracy, the average time (the ratio of total retrieval time and the total number of music) using two methods were shown in Table 2, with the i5 professor of 4 Core and 4G memory.

Table 2.   The similar audio detection accuracy.

	Improved melody matching algorithm	Traditional melody matching algorithm
The top three retrieval accuracy	43.33%	26.66%
The top five retrieval accuracy	63.33%	53.33%
The average retrieval time/ms	1132	2175

Compared with traditional algorithm, the experiment results showed that the top three retrieval accuracy of improved algorithm increased by 14%, and the top five retrieval accuracy increased by 8%, and the average retrieval time declined by 1043 ms. Retrieval accuracy and efficiency of the improved extraction algorithm improved substantially.

5   THE CONCLUSION

In this paper, cluster analysis was applied to the music sentence segmentation and relative pitch information was extracted by the sentence to calculate similarity by DTW. 30 WAV format hum fragments and 50 WAV songs were tested, and the results were compared with the results of the traditional method punctuating music by the energy of amplitude. The results showed that cluster analysis adopted to segment the songs obviously improved the accuracy and effect of retrieval. This paper provided a new method that improved QBH effect.

Content-based audio retrieval technology still has some problems to be solved, such as the relevance of high-level concepts and low-level features to realize automatic computer relevance of media semantic, and the low retrieval speed of web content–based audio retrieval, and browsing, retrieving, and submitting of large-scale continuous audio media [6]. In short, to have good recognition accuracy and high detection speed, some parts of content-based audio retrieval technology still need to be improved.

REFERENCES

[1] Chaijong Song and SeobPil Lee, "The Contents Based Music Retrieval Method Using Audio Feature Analysis against Polyphonic Music," ISA 2011, CCIS 200, pp.257–263, 2011.
[2] Xuedong Huang, James Baker, and Raj Reddy, "A Historical Perspective of Speech Recognition," Communications Of The Acm, Jan, 2014.
[3] Guangchao Yao, Yao Zheng, and Limin Xiao, "GPU-Accelerated Query by Humming Using Modified SPRING Algorithm," 2013 IEEE 13th International Conference on Data Mining Workshops, pp.654–663, Dec, 2013.
[4] Jie Zhang and Ziye Long, "The review of the fundamental frequency extraction algorithm," Journal of university of electronic science and technology, vol. 39, no., pp.99–103, April, 2010.
[5] Yongqiang Shang, Linmei Zhang, and Dawei Xu, "Content-based audio retrieval algorithm," Journal of henan institute of science and technology, vol.37, no.3, pp. 69–72, Sep, 2009.
[6] Chen Li and Mingquan Zhou, "Audio retrieval technique research," The development of computer science and technology, vol.18, no.8, pp. 215–218, Aug, 2008.

Information Technology – Wan et al. (Eds)
© *2015 Taylor & Francis Group, London, ISBN 978-1-138-02785-5*

The recovery of sheared papers with regular boundaries based on the picture pixel

Mei Xiong, Longwei Chen & Heng Wang
Statistics and Mathematics College, Information School, Yunnan University of Finance and Economics, Kunming, China

ABSTRACT: For the stitching problem of sheared papers with regular boundaries after the square image cropping, Matlab software is used for the reconstruction. First, only the longitudinal case: We read the picture's stored pixels, get each picture's left and right sides' edge value vector, and go to convert them into binaryzation, the correlation coefficient of distances between the right boundary of a picture and the left boundary of another picture is calculated, the smaller the distance, the higher the similarity, goes on splicing of two images. After the most left picture is selected by artificial intervention, we can machine, automatic mosaic, get a complete restored image. Second, both longitudinal and transverse situation: We read the pictures' image pixels, in accordance with the conditions that the left column or right column of the pixels are all 0, combine it with artificial intervention that find the left and right's boundary fragments; all the pictures are to be horizontal projection and binaryzation, extract the row height characteristics of the fragments' texts, According to the characteristics and the left or right border fragments, the fragments are classified by the classification results; According to the method of the first step, the horizontal strip of papers are around top to bottom to splice, get the complete restored image. Third, because the picture is English text, we don't extract the row height characteristics for classification in accordance with the second methods, and the graph has positive and negative, therefore we define a new distance function, the positive and negative distance is the minimum, then construct the distance matrix of the two pieces of paper, and press the distance to sort, through artificial intervention to be around top to bottom to splice, and finally get the whole jigsaw puzzle.

KEYWORDS: Matlab the recovery of sheared papers; image binaryzation; the difference value of the pixel matching

1 INTRODUCTION

A recovery of sheared papers has its important application in the fields of legal evidences, historical documents, and the military information. Traditionally, the reassembling work needs to be finished artificially. However, the accuracy of artificial method has been proved very high in productive practice, and the efficiency is very low. Especially, under a huge number of sheared papers, an artificial recovery maybe spends a long time. With the development of the computer technique, we try to invent the recovery method automatically to improve the efficiency and to shorten the time. In this paper, we are concerned with a method of image auto-splicing coupled with artificial.

From a general point of view, two strategies exist to deal with the splicing of scrap papers, and each method has its own different object. One is to use the semi-automate stitching of scrap papers based on character of the text pattern. The other one is to

research on the edge detection algorithm for scrap papers with line segments scanning. These methods are reviewed [1–5]. Here, we want to discuss the application of the former in great detail. The text position algorithm is usually applied to deal with regular edge, such as Sobel, Roberts, Prewitt, Harris, and SUSAN [1], which have their characteristics and their scope of application, but they have a together shortcoming need to be overcome. In order to avoid the disadvantages of the stitch based on geometry characteristics, some effective methods are researched to recover scrapped papers, semi-automatically, such as characters in the rows of text, lines in a table, and other symbols in a sheared paper. A typical application in the field of the computer vision and the pattern recognition is just to complete the reassembling of paper fragments through the scan and image extraction techniques. It contains two steps: the image preprocessing and the information matching.

In our recovery technique, we employ the method based on the character information and manual intervention. These problems are as follows:

1. For 1×19=19 sheared single papers with regular edges (only rip cutting) from the same page cut by a paper shredder, we want to build a mathematical model and its corresponding algorithm to stitch these papers. If the process of the stitching needs to intervene artificially, we will give the intervening way and intervening time. The result is shown as a figure and a serial number table.
2. For the 11×19=209 sheared single papers with regular edges sheared by both vertical cutting and horizontal cutting from the same single paper and double paper cut by a paper shredder, we want to build a mathematical model and its corresponding algorithm to stitch the paper. Other dements is the same as (1).

2 SCRAPS OF PAPER CUT VERTICALLY

After pixels of 19 pictures cut vertically are read by the software R2011a, we program as follows:

```
for i=1:19
    imageName = strcat(num2str(i),'.bmp'); % read pictures
    I1= imread(imageName);% pictures are changed into pixels
    L=[L I1(:,1)];%take pixels the left boundary of these pictures
    R=[R I1(:,m)];%take pixels the right boundary of these pictures
end
```

We obtain pixel matrixes of left boundary and right boundary:

$$L = \begin{bmatrix} l_{11} & \cdots & l_{1n} \\ \vdots & \vdots & \vdots \\ l_{m1} & \cdots & l_{mn} \end{bmatrix}_{1980 \times 19}, R = \begin{bmatrix} r_{11} & \cdots & r_{1n} \\ \vdots & \vdots & \vdots \\ r_{m1} & \cdots & r_{mn} \end{bmatrix}_{1980 \times 19},$$

Which are converted to binary matrixes for the goal that there is an appreciable difference between two pictures' boundary?

$$l_{ij} \text{ or } r_{ij} = \begin{cases} 1 & 0 \leq x_{ij} < 255 \\ 0 & x_{ij} = 255 \end{cases}.$$

Where, l_{ij} or $r_{ij} = x_{ij}$ is pixels of initial picture, and matrixes of binary are l_{ij} and r_{ij} yet.

We assume that there are n-row in the pixel matrix of a picture, use the pixel value $a = (a_{i1}, a_{i2}, \cdots, a_{in})^T$ of i picture's pixel matrix on the right side of the last column, compared with the pixel value $b = (b_1, b_2, \cdots, b_n)^T$ of j picture's pixel matrix on the left of the first column.

Thus, define the similarity:

$$d_{ij} = \sqrt{\sum_{k=1}^{n} (a_{ik} - b_{jk})^2}$$

$(i = 1, 2, \cdots, N, j = 1, 2, \cdots, N, i \neq j)$.

Get the similarity matrix:
$D = [d_{ij}]_{19 \times 19}$,

When $d_{ij} = 0$, the similarity of the i picture's right boundary and the j picture's left field is very high, or take $\min\{d_{ij}\}$. For the j picture, we should put j on the right of i picture. Find out the left boundary pictures by artificial means, and swap sequence number with the first picture, stitching program automatically, and splice according to the following method.

```
E=[1 14 12 15 3 10 2 16 19 4 5 9 13 18 7 17 8 6];
G=[];
for i=E %according to the picture number, read pictures
    imageName=strcat(num2str(i),'.bmp');
    I= imread(imageName); % take picture pixels
    G=[G I]; %combine picture pixels
end
PicData=[G]; % If it's the horizontal splicing
imshow(PicData); %display the picture
```

Get the complete jigsaw puzzle, as shown in figure 2.

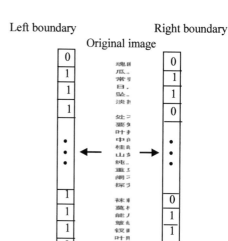

Figure 1. The binaryzation of boundaries of the original image.

Figure 2.   Whole jigsaw under with vertical cut.

3   SCRAPS OF PAPER CUT VERTICALLY AND HORIZONTALLY

3.1   *To look for the boundary*

Left and right boundary, the search results are as follows: the possible of left boundary is 7, 14, 29, 38, 49, 61, 62, 67, 71, 80, 89, 94, 125, 135, 143, 168。The possible of right boundary is 18, 36, 43, 59, 60, 65, 74, 76, 78, 123, 141, 145, 146, 176, and 196.

Through artificial intervention and judgment, then we delete those are not boundary figure number, the left and right side table are as follows:

Select the picture above, make the extraction of line features, i.e. to make the horizontal projection of the text line, we put the pictures are projected into a one-dimensional vector.

Suppose there are N picture, each image pixel matrix size is the same, is $n \times m$, the line feature vector of picture is defined as $t^{(i)} = (t_1^{(i)}, t_2^{(i)}, \cdots, t_n^{(i)})$, $t_l^{(i)} = \begin{cases} 1 & x_{ik}^{(l)} < 255 \\ 0 & x_{ik}^{(l)} = 255 \end{cases}$

$i = 1, 2, \cdots, N$, $l = 1, 2, \cdots, n$, $k = 1, 2, \cdots, m$ 。

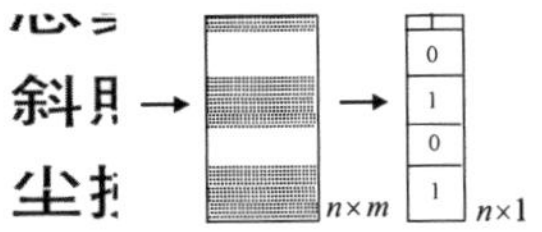

Figure 3.   Schematic diagram for line feature extraction.

We define the line feature similarity of the i picture and j picture is

$$d_{ij} = \sqrt{\sum_{l=1}^{n} (t_l^{(i)} - t_l^{(j)})^2}$$

$$(i = 1, 2, \cdots, N, j = 1, 2, \cdots, N, i \neq j)$$

So that we can get the similarity matrix of the i picture and the j picture is $D = (d_{ij})_{N \times N}$, among it suppose $d_{ii} = \inf$, the purpose is to eliminate the self similarity of images, the smaller d_{ij}, the higher the similarity of i picture and j picture, the possibility that they are in the same line is more.

Table 1.   The left boundary image sequence number.

The Left boundary	7	14	29	38	49	61	71	89	94	125	168

Table 2.   The right boundary image sequence number.

The Right boundary	18	36	36	43	59	60	74	123	141	145	176	196

Table 3.   The left and right boundary possible matching table.

The Left boundary	14	29	38	49	61	71	89	94	125	168	7
The 1st candidate	176	59	43	141	36	36	123	43	60	18	43
The 2nd candidate			74			60			145		196

We compare the table above number's pictures that have the left and right boundary, get the following results.

Since the feature vector height errors, so we need to be the first candidate and second candidate, and go on the artificial intervention. After artificial intervention, we get the correct results:

3.2 *For classification of all the pictures*

According to the method of (1) to classify all the pictures, which has similar features vector images into a class, and in accordance with the left boundary similarity from big to small in order, in the classification results the first 19 figure number:

The figures in the frame in the table are 164, 42, 77, 183, 34, 43, and although the similarity of the left boundary is the higher, but the actual splicing can be seen from the picture, they do not belong to the class, such as the puzzle of that the left boundary is seven is as follows:

Then, how to look for the lost 6 figures, the method is very simple, and finds out the highest similarity 6 graphics with the right boundary 196, the replacement

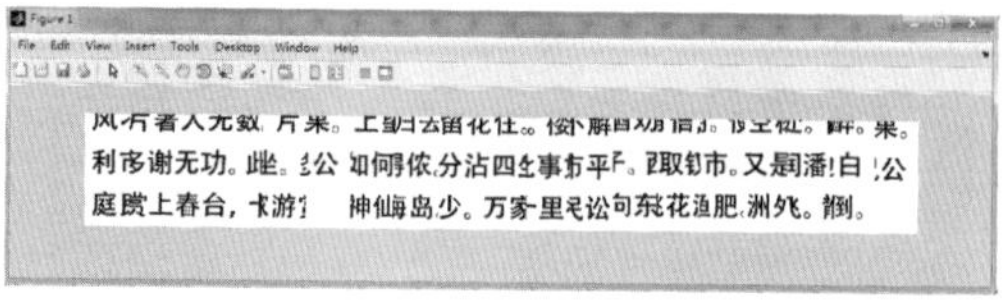

Figure 4. The left boundary of 7, mosaic image with the highest similarity.

method with the rest data in frame is the similar. The highest similarity first 18 graphic sequence number tables with the right boundary 196 as follows:

Framed images in the top 6 are just 6 images that are covered and replaced, the correct result, we see in the last complete puzzle table.

In the table, we transfer the program only longitudinal situation, transverse splicing of 11 categories, such as the image mosaic which left boundary is 49, right boundary is 141 is as follows:

The mosaic image of the left boundary is 89, right boundary is 123 is as follows:

As can be seen, the above two pictures are the first and last picture.

Table 4. The correct boundary matching table.

The Left boundary	7	14	29	38	49	61	71	89	94	125	168
The 1st candidate	196	176	59	74	141	36	60	123	43	145	18

Table 5. The left boundary similarity the first 19 graphics from big to small order table.

7	68	175	45	126	196	209	53	138	158	174	208	164	183	34	43	196
14	135	31	203	3	39	89	159	12	51	73	128	134	160	176	199	176
29	59	75	5	98	172	180	206	10	37	48	104	171	92	201	44	59
38	148	35	46	189	81	88	103	130	122	24	161	167	112	84	94	74
49	22	49	91	118	141	192	129	186	54	57	95	143	11	28	190	141
61	63	72	6	19	20	52	79	99	162	163	69	96	67	131	36	36
71	15	17	80	202	83	132	156	200	198	33	133	36	116	67	131	60
89	114	119	207	14	117	40	140	146	108	113	185	194	4	101	123	123
94	42	144	34	47	112	136	84	121	124	127	183	43	97	149	164	43
125	60	152	155	4	27	108	113	123	165	194	40	140	207	85	101	145
168	23	41	120	26	76	179	18	50	62	147	191	86	195	100	1	18

Table 6. The highest similarity first 18 graphic sequence number table with the right boundary is 196.

196	153	93	32	166	70	56	85	8	205	9	25	27	165	74	105	152	60

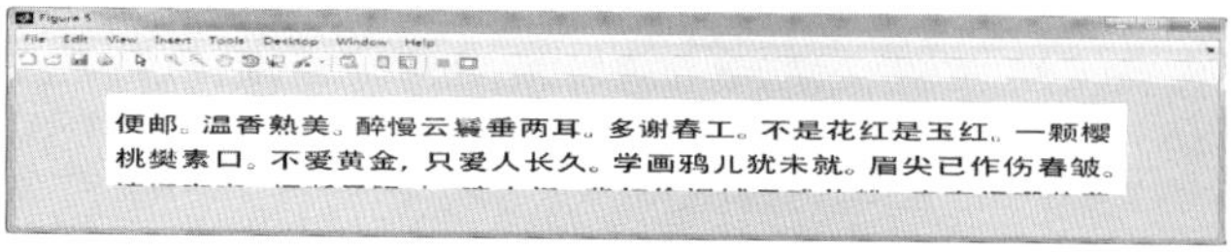

Figure 5. The mosaic image of the left boundary is 49, right boundary is 141.

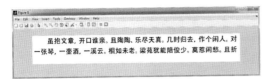

Figure 6. The mosaic image of the left boundary is 89, right boundary is 123.

Figure 7. Whole jigsaw with horizontal and vertical cut.

We have only the rip cutting's program slightly altered, it changed to only crosscutting situation. Then we performed longitudinal splicing of 11 cross cutting, the results are as follows:

The correct results for each class should be 19.

4 SCRAPS OF PAPER CUT VERTICALLY AND HORIZONTALLY(DOUBLE-EDGED)

With the same method of Q.3 to look for boundary, find out the left boundary in group a (the components of two valued vector are all zero vector)

The left boundary is group A:

9a,54a,89a,99a,100a,114a,136a,141a,143a,146a, 157a,160a.

The right boundary is group B:

3a,5a,13a,23a,35a,55a,78a,83a,88a,90a,105a, 160a,128a,153a,165a,172a,186a,199a.

Through artificial intervention, we can know the box number is not a boundary; they may be the end of natural paragraph. So a total of 22 left, we put the right border B group number and B end and A group together, just as positive or negative left border l, i.e.

Positive and negative left boundary group:

9a, 54a, 89a, 99a, 114a, 136a, 143a, 146a ,
3b,5b,13b,23b,35b,78b,83b,88b,90b,105b,165b, 172b,186b,199b.

Positive and negative right boundary group:

9b, 54b, 89b, 99b, 114b, 136b, 143b, 146b,
3a,5a,13a,23a,35a,78a,83a,88a,90a,105a,165a, 172a,186a,199a.

Because it could not know which is the front boundary, which is the opposite of the boundary, but if 9a and 35b are upper and lower splicing, the lower boundary of 9a and upper boundary of 35b's similarity should be higher, at the same time, in the opposite boundary, the lower boundary of 9a and upper boundary of 35b's similarity should be also higher.

Table 7. Correct and complete puzzle number table (11×19).

49	54	65	143	186	2	57	192	178	118	190	95	11	129	28	91	188	141
61	19	78	67	69	99	162	96	131	79	63	116	163	6	177	20	52	36
168	100	76	62	142	30	41	23	147	191	50	179	120	195	26	1	87	18
38	148	46	161	24	35	81	189	122	103	130	193	88	25	8	9	105	74
71	156	83	132	200	17	80	33	202	198	15	133	170	85	152	165	27	60
14	128	3	159	82	199	135	12	73	160	203	169	134	31	51	107	115	176
94	34	84	183	90	47	121	42	124	144	77	112	149	136	164	127	58	43
125	13	182	109	197	16	184	110	187	66	106	150	21	157	181	204	139	145
29	64	111	201	5	92	180	48	37	75	55	44	206	104	98	172	171	59
7	208	138	158	126	68	175	45	174	209	137	53	56	153	70	166	032	196
89	146	102	154	114	40	151	207	155	140	185	108	117	101	113	194	119	123

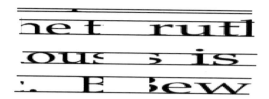

Figure 8. The feature of line height extraction diagram.

Because we cannot get the line height form English pictures, as shown in figure 8. Only according to the

left and right boundary in the face and the upper and lower boundary in the back of highly similar, to look for the matching graph.

Through the positive and negative's boundary distance and define two picture's distances:

$$d_{ij} = \sqrt{\sum_{l=1}^{n}(a_l^{(i)} - a_l^{(j)})^2} + \sqrt{\sum_{l=1}^{n}(b_l^{(i)} - b_l^{(j)})^2}$$

$$(i = 1, 2, \cdots, N, j = 1, 2, \cdots, N, i \neq j)$$

Table 8. The correct and complete jigsaw puzzle sequence number (11×19).

136a	47b	20b	164a	81a	189a	29b	18a	108b	66b	110b	174a	183a	150b	155b	140b	125b	111a	78a
5b	152b	147b	60a	59b	14b	79b	144b	120a	22b	124a	192b	25a	44b	178b	76a	36b	10a	89b
143a	200a	86a	187a	131a	56a	138b	45b	137a	61a	94a	98b	121b	38b	30b	42a	84a	153b	186a
83b	39a	97b	175b	72a	93b	132a	87b	198a	181a	34b	156b	206a	173a	194a	169a	161b	11a	199a
90b	203a	162a	2b	139a	70a	41b	170a	151a	1a	166a	115a	65a	191b	37a	180b	149a	107b	88a
13b	24b	57b	142b	208b	64a	102a	17a	12b	28a	154a	197b	158b	58b	207b	116a	179a	184a	114b
35b	159b	73a	193a	163b	130b	21a	202b	53a	177a	16a	19a	92a	190a	50b	201b	31b	171a	146b
172b	122b	182a	40b	127b	188b	68a	8a	117a	167b	75a	63a	67b	46b	168b	157b	128b	195b	165a
105b	204a	141b	135a	27b	80a	0a	185b	176b	126a	74a	32b	69b	4b	77b	148a	85a	7a	3a
9a	145b	82a	205b	15a	101b	118a	129a	62b	52b	71a	33a	119b	160a	95b	51a	48b	133b	23a
54a	196a	112b	103b	55a	100a	106a	91b	49a	26a	113b	134b	104b	6b	123b	109b	96a	43b	99b

We get the distance matrix: $D = (d_{ij})_{419 \times 419}$, according to the size of dij to be in ascending order, we take the top 10 ranking, the picture sequence is output to the excel table, through artificial intervention to find it, the results are in Table 7. In Table 7, we first determine the left and right boundary, press the left boundary to the right to find, identify the right boundary to the left to find (underlined). Framed figures in the tables for the top 10 ranking in both the left

and right cannot find digital, found through artificial intervention. The first bar, and then press the first question in the upper and lower splicing, get the complete puzzle table.

The back of the jigsaw puzzle with the positive sequence is the same, the corresponding digital a for B, B for a, then reverse to obtain the first column opposite number. The table is as follows, the rest are omitted.

……… ………

78b	111b	125a	140a	155a	150a	183b	174b	110a	66a	108a	18b	29a	189b	81b	164b	20a	47a	136b

5 CONCLUSION

For the stitching problem of Sheared Papers with regular boundaries after the square image cropping, Matlab software is used for the reconstruction. First, only the longitudinal case: We read the pictures' stored pixels, get each picture's left and right sides' edge value vector, and go to convert them into binaryzation, the correlation coefficient of distances between the right boundary of a picture and the left boundary of another picture is calculated, the smaller the distance, the higher the similarity, go on splicing of two images. After the most left picture is selected

by artificial intervention, we can machine automatic mosaic, get a complete restored image. Second, both longitudinal and transverse situation: We read the pictures' image pixels, in accordance with the conditions that the left column or right column of the pixels are all 0, combine it with artificial intervention that find the left and right's boundary fragments; all the pictures are to be horizontal projection and binaryzation, extract the row height characteristics of the fragments' texts, According to the characteristics and the left or right border fragments, the fragments are classified by the classification results; According to the method of the first step, the horizontal strip of papers are around top to bottom to splice, get the complete restored image. Third, because the picture is English text, we don't extract the row height characteristics for classification in accordance with the second methods, and the graph has positive and negative, therefore we define a new distance function, the positive and negative distance is the minimum, then construct the distance matrix of the two pieces of paper, and press the distance to sort, through artificial intervention to be around top to bottom to splice, and finally get the whole jigsaw puzzle.

Figure 9.   Whole jig saw with horizontal and vertical cut, and both sides.

In the third question, because the degree of manual intervention is very big, so the algorithm can be improved, such as the introduction of the Harris corner detection algorithm to detect the characteristic points of each picture, looking for the adjacent suited picture with the matching algorithm of RANSAC, sorting according to the matching results of pictures, then stitching to get the complete picture [9].

REFERENCES

[1] Zhang Guofeng. Research and Application of Background Filtering Technology Based on Text Detection. North China University of Technology, 2013, 5:I.

[2] Luo Zhizhong. Semi-auto Stitching of Scrapped Paper Based on Character Characteristic . Computer Engineering and Applications, 2012, 48(5):207–210.

[3] Luo Zhizhong. Reshearch on Edge Detection Algorithm for Scrapped Paper with Linc Segment Scanning. Chinese Journal of Scientific Instrument, 2011,32(2):289–294.

[4] Jia Haiyan. Research on The Key Technologies of Computer-aided Paper Fragments Reassembly. Graduate School of National University of Defense Technology, 2005,11.

[5] Jia Haiyan, Zhu Liangjia, Zhou Zongtan, etc.. A Shape Matching Method for Automatic Reassembly of Paper Frangments. Computer simulation, 2006, 23(11):180–183.

[6] National award: The recovery of sheared papers. file:///D:/ Mathematical Contest in modeling /2013/ 2013b title /b title information / The recovery of sheared papers %20-%20 Douding network.htm, 2013/11.

Information Technology – Wan et al. (Eds)
© *2015 Taylor & Francis Group, London, ISBN 978-1-138-02785-5*

The resource scheduling system of performance theaters based on the optimal path

Wei Jiang, Yuanyuan Huo, Yujian Jiang & Jingjing Zhang
School of Information Engineering, Communication University of China, Beijing, China

ABSTRACT: Facing the status quo of scattered resources and unreasonable scheduling, this topic firstly researches and analyzes the current situation and the development trend of performance industry gives the related conceptions of performance cinema, using this new idea to innovate performances resource management model. Then it researches the problem of resource scheduling under the mode of performance cinema, extracts the special requirements and the main problem of the existing performances resource scheduling, which raises demand for study of new resource scheduling model, on these bases it emphatically analyses the problem of single performance troupe scheduling based on the shortest path, which provides the shortest path route for single performance troupe traveling through many cities by building the corresponding mathematical model with the corresponding algorithm. Additionally, it also carries the research on the optimal cost path route by reference to the relevant calculating basis of transport costs and accommodation costs. Finally, it develops and designs the performance troupe resource scheduling system with ASP and SQL Server in view of the important role of scheduling routes for the troupe multiple city tour. Compared with other routes, it directly shows rationality of the system.

KEYWORDS: performance theaters; resource scheduling system; optimal path algorithm

1 INTRODUCTION

In recent years, the domestic theatrical performance market is becoming more prosperous [1]. Thus, cultural performance theaters came into being. Performance industry has formed the industrial chain, including performance troupes, theaters, related companies and audiences, etc. The resource scheduling problems of healthy development of China's performance market are concerned by relevant departments.

Theater's performances are a new cultural transmission channel. Its main mode is the innovation chain. It is an emerging cultural service which provides performances to the performance market. It bases theater troupes touring the country, to achieve the optimal allocation of performance resources. Performance theaters aim to change the status quo of theater poor utilization rates and a small number of theaters Performance, by mastering theater information and troupe information, in the form of troupe touring to connect the national theater. Theater's performance is to achieve win-win situation as the goal to promote the overall development of the performance industry chain. Healthy development of heaters performance needs reasonable performances resource scheduling. The mature performance theater should have a rich performance program effective flow in the theaters. Only reasonable resource scheduling can fully

mobilize the theater, troupe and other resources, can optimize the configuration optimize the allocation of performance resources, and through large-scale, intensive operations, It not only enriches the content of theater performance, but also reduces costs and attract more and better programs to enter the theater, in order to achieve the purpose of the performance theater system to activate the industry. Thus, the rapid development of cinematic performance has requested reasonable performance resource scheduling.

Reasonable cultural performance resource scheduling makes theaters and troupes in different cities optimal configuration in accordance with the performance time sequence. It ensures theaters and troupes reasonable match. Through the performance troupes and the performance facilities optimize theater corresponding configuration to achieve a troupes' high performance rate and theaters higher utilization. However, so far, China has not yet studied the performance of resource scheduling, and now there is no special system on performance resource scheduling [2]. Theaters and troupes resource scheduling regularly rely on excessive historical data and performance experience. Therefore, along with cultural performance theaters booming development, research on performance resource scheduling imminent.

In this paper, performance theater resource scheduling system based on the optimal path is designed.

The system uses the B / S framework to simplify the client's work, which combines the cost-optimal scheduling algorithms and shortest path scheduling algorithm, and designs the database containing data for each theater and information for troupes. The system can find suitable theaters for a troupe, and be able to achieve without considering the props transportation cases, a single performance troupe in many theatres meeting the performance conditions optimal tour route.

2 SYSTEM DESIGN AND IMPLEMENTATION

2.1 *System functions*

1 *System function modules:* Functions of the system include: system maintenance and management modules (including theatres information maintenance, adding theatres' coordinates maintenance, and theater weights maintenance), theatres match and query by a variety of conditions module, adding the target theaters to my plan table module, and the optimal route scheduling module (including the shortest path route scheduling and the minimum cost route scheduling). These features are distributed in the front and in the back system modules, the front-end system module is mainly for the majority of users, and the average user is able to enter the system through registered an account. The Backend system module is only opened for the system administrators. The administrators can modify and maintain theaters and Other operations in the system. Functions of the front end and backend is shown in Figure 1.

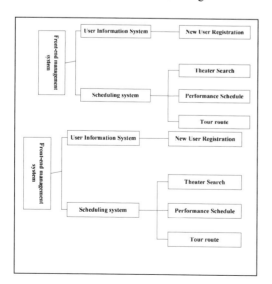

Figure 1. Front-end system functions diagram.

2 *Realization of system functions:* Figure 2 is a timing diagram of the system, which shows the time sequence of the transmission of information between the various modules. In the graph, each object arranges at the top of the timing diagram. At the bottom of each object, a dashed vertical line is drawn. When an object sends a message to another object, the message starts at the dotted line at the bottom of the sending object, and it is terminated at the dotted line at the bottom of the receiving object. These messages are indicated by arrows, placed horizontally arranged in the vertical direction. In the vertical direction, the messages near the top are sent earlier. When an object receives a message, the message is treated as a command to perform some action. So the timing diagram provides the users with a stream of events over time, clear and visualization track. Based on the above UML, a timing diagram of resource scheduling system is shown in Figure 2.

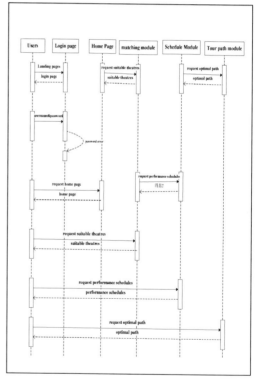

Figure 2.   System timing diagram.

2.2 *Key technologies*

1 *Layering processing framework:* The entire system uses Browser / Web / Database three-tier architecture. Presentation, business logic, and data service

is divided into three independent units. The system scripting language is ASP. Three-tier architecture is shown in Figure 3.

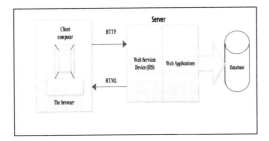

Figure 3. Three-tier architecture diagram.

a. The first layer (presentation): interactive interface (web browser) is located in the client. Its mission is to provide the service request by the Web browser to the Web server on the network, and use the HTTP protocol to transmit the required homepage to the client. The client accepts the homepage file, and displays it in a Web browser.
b. The second layer (business logic): a Web browser with application extended function. The business logic accepts the user's request, and makes data processing application to the database server, then database server processes the data, and submits the results to the Web server, finally, the results are transmitted to the client.
c. The third layer (data service): a database server. This layer contains data processing logic of the system. It is located in a database server. It accepts Web server requests for the database manipulation. It can query, modify, update the database, and submits the results to the Web server. The system uses Microsoft SQL Server 2000 [3]. Five aspects of database, including.
  • User data information: including user name, password, mobile phone number, e-mail and other basic data.
  • The data includes theatre: theatre, the city, rental price, pictures of theater, data evaluation information.
  • The theatre coordinates: including name of theatre, the city theatre, coordinates.
  • The performance plan: including the theater, theater, the market rent prices.
  • The weight allocation: includes describing weight configuration data information for the theatre.
  Data tables and their corresponding functions contained in the database are shown in Table I.
2 *Optimal path scheduling algorithm:* Optimal path scheduling algorithm [4] is divided into the shortest path scheduling algorithm and cost optimal

Table 1. Database tables.

Tables	Functions
Username	user information
Cinema	description of theaters
Cinema_zuobiao	geographical coordinates of theaters
Quanzhong	the weights of theaters
Jihuabiao	performance schedule of users

scheduling algorithm. Unlike other areas of scheduling, Performance scheduling resources are stricter requirements for time, weather, traffic and other conditions. In performance Theatres mode, the number of troupe performances will increase dramatically. The troupe will continuously tour across multiple cities. Troupe adapting to the venue and rehearsals are all made high demands of time. Only enough time can be able to guarantee the quality of the performances of troupe performance. The urgency of performance needs a reasonable performance scheduling path. According to experience, the shorter route, transport costs are also lower. The optimal path saves the troupes show time. Thus, it not only can guarantee the time required, but also provides troupes for additional performance opportunity. So the shortest path scheduling algorithm is the focus of the study. The essence of the shortest path scheduling algorithm is a single troupe touring among multiple cities. Each city can only tour only once, and get the minimum path among all the paths. Actually the problem is solving the shortest path of the global traverse in the case of the city's geographical coordinates are known. In this system, it uses the latitude and the longitude to calculate the distance between cities.

Dijkstra's algorithm [5] is one of the classic algorithms to calculate the shortest path. However, the calculation speed of the algorithm is not very ideal. So the system optimizes Dijkstra's algorithm. The optimization algorithm only calculates the shortest path' neighboring nodes, not all other nodes.

Assuming V_0 is the starting point of the city, and city V_1, V_2, V_3, V_4, V_5 are five destinations. According to the latitude and longitude of the city, the tour can be represented as a weighted model. The model is shown in Figure 4.

In Figure 4, V1 is short from V_0, so the tour path is $V_0 \rightarrow V_1$. City V_2, V_3, V_4 close to V_1, the distance of $V_1 \rightarrow V_2$ is shortest. Therefore, the shortest path V_0 to V_5 is $V_0 \rightarrow V_1 \rightarrow V_2 \rightarrow V_4 \rightarrow V_3 \rightarrow V_5$.

For cost optimal scheduling algorithm, when the troupes continue touring, the costs mainly consider transport costs and accommodation costs. Troupe performances in different time lead to a different

415

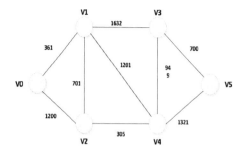

Figure 4. Weighted model diagram.

departure time. And economic levels in different cities led to very different performance cost of travel and accommodation. How to obtain real-time performance cost and how to properly plan the cost are the key the of resource scheduling.

In the accommodation cost estimates, the cost refers to economical three-star hotel. By querying the major cities of the three-star hotel prices, cost optimal scheduling algorithm calculate the cost of accommodation prices in different cities, according to the first-tier cities, second-tier cities in Class A, second-tier cities in Class B and third-tier cities. Therefore, it can be quickly estimated the cost of accommodation in different cities.

About the transportation cost estimation, this algorithm screened a number of important indicators to divide the level of cities. These indicators include the number of passengers, airport passenger throughput, GDP, international and domestic visitors and retail sales of social commodities. According to these indicators rank the cities; the cities are divided into four categories: first-tier cities, second-tier cities in Class A, second-tier cities in Class B and third-tier cities. According to the large number of ticket prices and distance, transport costs per kilometers in different cities are calculated. This factor can be calculated transport costs.

For the calculation of the city distance, latitude and longitude coordinates in different cities are acquired. According to the corresponding latitude and longitude, the distance between any two cities was calculated. So about cost optimal scheduling algorithm, Latitude and longitude coordinates between different cities are known. According to the latitude and longitude coordinates, it can get straight-line distance between any two cities. By querying the relational data the transportation cost factor and the price of three-star hotel chains in different cities are getting. Straight-line distance multiplied by the coefficient of transportation calculated, and plus the price of the hotel as the corresponding weights. So the system can calculate the different scheduling route.

2.3 System implementation

In the system, ordinary users have to log, the system can determine whether the user exists, if the user does not exist, users haves to register. The system displays the appropriate theaters based on user requirements. Users can add theaters meeting their requirements to the schedule. Then decide the start and end cities, the system will give the optimal route.

Along with more and more theaters adding system, on the one hand, administrators are able to add more information of new theaters, which includes geographical coordinates, performance types, and theater facilities; on the other hand system administrators also can maintain theater information. A flowchart is shown in Figures 5 and 6.

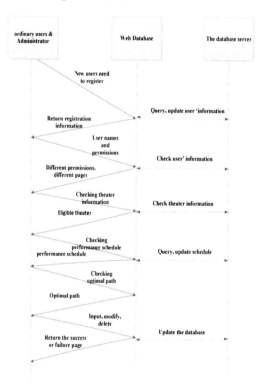

Figure 5. Implementation figure.

3 CONCLUSION

The resource scheduling system of performance theaters based on optimal path is easy to use. It can provide suitable theaters for users. And also be able to provide optimal path. It provides a practical reference for the actual performance theaters scheduling.

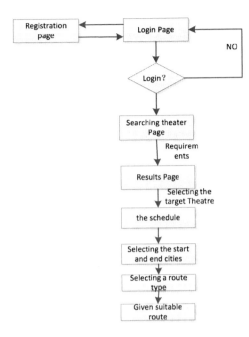

Figure 6.   Flow chart.

Compared with traditional methods, the rationality and practicality of the route have a great advantage. It saves a lot of manpower and time, and helps administrators manage the data.

ACKNOWLEDGMENT

We would like to thank the National Science and Technology Support Program (Foundation project: Research on They manage systems for entertainment performance chains and key technology of web collaboration services platform, Project number: 2012BAH02F04) and Engineer Project of CUC (Foundation item: Research on resources cooperative scheduling problem and optimization technique based on entertainment performance chains, Project number: 3132013XNG1305) for their help and support.

REFERENCES

[1] Jiang Wei, Qi Chao, Wang jin-yong, Jiang Yu-jian, Zhang Jing-jing and Zhou Bin. "Analysis of Operation mode about Our Artistic Show Courtyard" Beijing, China, September 2012.pp.38–42.
[2] QI Chao, "Research of performance theaters resource scheduling" unpublished.
[3] Yaosong Ye, Xiangfa Ruan,Yan Zhao. "The inventory management system based on SQL Server2000 database" Mechanical design and manufacturing, 3rd ed.,China, 2006, pp. 169–170.
[4] C. Gong, Y.J. Jiang.J.J. Zhang. C. Qi, W. Jiang. Research on path scheduling algorithm for performing touring, Proceedings of 2013 IEEE 4th ICEI and EC, Beijing,China.November, 2013.
[5] Li Luo,Feng Wang. "Shortest path improved Algorithm based Dijkstra's algorithm", Hubei Automotive Industries Institute, 2007.

Information Technology – Wan et al. (Eds)
© *2015 Taylor & Francis Group, London, ISBN 978-1-138-02785-5*

The situation analysis of biological students' self-regulated learning in network

Kai Deng
Institute of Eugenics and Genetics, Renmin Hospital, Hubei University of Medicine, Shiyan, China

Chenxi Wang
China Women's University, Beijing, China

Xiaoe Ouyang
Centre of Health Administration and Development Studies, Hubei University of Medicine, Shiyan, China

Changjun Zhang
Institute of Eugenics and Genetics, Renmin Hospital, Hubei University of Medicine, Shiyan, China

Jianguo Wang
Centre of Health Administration and Development Studies, Hubei University of Medicine, Shiyan, China

Xiaoyan Wang
Department of clinical Oncology, Taihe Hospital, Hubei University of Medicine, Shiyan, China

ABSTRACT: What are the effects of self-regulated learning online, what the problems are and how to solve these problems, have been the workers in network education theory and the teachers' attention and research important topic? With the questionnaire survey, this paper finds out the existence of Medical College Students' self-regulated online learning problems, analyzes the causes of these problems, and researches on strategies and methods to solve these problems

KEYWORDS: Medical, Biology, Computer Network, self-regulated learning

1 INTRODUCTION

In computer networks, especially the rapid development of the Internet today, the self-regulated learning online is an important form of extracurricular learning students in medical colleges. Compared with the traditional forms of learning, however, what are the effects of self-regulated learning online, what are the problems and how to solve these problems have been the workers in network education theory and the teachers' attention and important topic to research. With the questionnaire survey, this paper finds out the existence of Medical College Students' self-reg ulated learning online problems, analyzes the causes of these problems, researches on strategies and methods to solve these problems, giving enlightenment to people including the organizers of network, medical educators, the supervisors and managers

An important feature of education in Medical University is learning, "three class, seven class". Because of this, there will be knowledge in the "three class", in "seven class"; lapping in "three class", then "seven points in extracurricular"; inheritance in the "three class", innovation in "seven class" classical education[1]. It emphasizes the importance of classroom teaching, and explains the extracurricular learning's being significant in learning ability, innovation and culture, especially education in Medical University at the same time [2].

In recent years, computer network, especially the Internet develops rapidly. According to the twenty-sixth survey data of the CNNIC (China Internet Network Information Center), the number of Internet users in China exceeded 400000000[3]. According to the CNNIC standard, nearly 98% of students in Medical College are Internet users[4]. Namely, most of the students in medical college possess the hardware conditions of self-regulated learning by using the internet [5].

In general, the biology professional course teacher would get learning resources of extracurricular network ready for students. So, how to organize teaching resources of the network itself? Unfold extracurricular self-regulated learning online how to guide students, how to manage the students' self-regulated learning? How does it work out? What are the problems? How to solve these problems? Thus, it will be an important and very meaningful topic to research. Of course, for different disciplines, different majors, different class, different teachers, there may be different methods of implementation and results [6].

Taking an introductory course of professional biology science, for example, this paper discussed as teachers, self-regulated learning in the network organization, guidance and education and supervision – management issues of practice and thinking. Aiming at improving the network teaching resources and services in the future, striving to make students achieve a higher degree of satisfaction, better results corning self-regulated learning in the network.

2 THE SITUATION AND THE MAIN PROBLEMS OF STUDENTS' SELF-REGULATED LEARNING IN NETWORK

As types of review and guide in the course, because "Introduction" involves the intensive knowledge, making it difficult for students to master the essentials of curriculum relying on classroom teaching barely. At this time, the auxiliary of out-of-class in network is the essential solution.

2.1 Investigation on the effect of the extensive network of self-regulated learning

Also the early use of self-regulated learning mode, relatively simple networks is an FTP server. The teachers and students establish accounts, and then related learning resources are uploaded by teachers and downloaded by students themselves. This approach is too simple, almost no management, guidance, communication, computer network has become the temporary storage for self-regulated learning resources. How does it work out? We made a questionnaire survey of 116 people in 2 classes. According to the results of the survey, 116 college students in every subject will be more or less use computer, mobile phone and other communication tools on the internet. This fully shows that the college students have realized the important role of networks in their own development. Most of the information resources in the network are helpful to the study and life of college students, helping them solve part of problems in learning. So that they are willing to seek help of the network, and make the network become an indispensable part of learning life. The Internet

every day for more than 4 occupies 66.67% hours and 2-4 accounted occupies 20%, which proves that the Internet has become one of the main contents of students after school life. On that survey students access to the Internet, students mainly from the following several aspects: chat, watch movies, understand the various types of information, playing online games, learning, other, the network is applied to learn only 14.4%, students' access to a diversity of characteristics, and the entertainment and communication with friends become the main purpose of the Internet, but the use of online learning is not much.

When it comes to the question how does the network help with your study, 33.3% of the students think that "great"; 61.1% think that "so-so", 5.6% think that the "little", so most of the students have realized the function of network on learning. However, the survey found the number of students actively and frequently using network for learning is not much, only to 14.4%, and mostly passively finds some information to study. Especially when teachers assigned homework, they do not know how to do, mostly on the Internet to find a related data copy.

Investigation shows the main course leading to the difference in efforts of online learning is the subjective intention, bad independent ability and low interest in learning of learners. The key to succeed in network learning is its degree of own effort. Aware of network learning is highly self-regulated learning, only when the learner realizes his subject status and learning goals, relying on his own efforts, can he complete the learning tasks better, achieving successful online learning.

2.2 Summarize of the extensive network of self-regulated learning

The extensive network of self-regulated learning, as is shown in the survey, the result is unsatisfactory. But the data from this questionnaire reveal important information, which has great inspiration for our improving the organization, guidance and education and supervision - management of resource concerning self-regulated learning in the network.

3 STRATEGIES OF THE SELF-REGULATED LEARNING IN A NETWORK'S ORGANIZATION, GUIDE THE EDUCATION SUPERVISION AND MANAGEMENT

When organizing, self-regulated learning, teachers must consider the difference between its form and content and classroom learning. Each of the students uses their computer, self-regulated learning in a virtual space. From the form, the network study is a kind of individual learning, whose features are as follows: learners can choose learning content according to their own need, can select learning methods adapt to themselves, can arrange learning progress according to their own time, can choose the depth of [11] learning content according to their ability. Therefore, the traditional classroom teaching, teachers usually have no time to allow for diversity of different individuals' ability to accept new things, making it difficult for each student to make great progress actually. But during the organization of network learning process resources, we can give full attention to differences in student's foundation and learning ability, guide the good basic, help the less basic.

3.1 Problems of self-regulated learning resource organization

The questionnaire mainly tells us the following information:

1 The vast majority of students (94%) is on the Internet every day.
2 Due to various activities, real learning is less than 2 hours, which warns us, whether quality - education, guidance, supervision and management of self-regulated learning in network resources is good or bad, will directly affect, even change the students' online activities. So we need to talk to the network game, QQ chat activities "for student's market share, "rob"".
3 Most of the students (95%) confirm network help with learning, which forms a huge contrast to then second conclusions above. Therefore, our teacher should reflect on the responsibility.
4 70% of the students actively or passively use network for learning, as high as 30% students rarely use web-based for self-regulated learning. We must figure out the reason why they do not use the network to learn, guide and help students get habits of using network for learning and like it sincerely.
5 Students little rely on teachers and other students. If the teacher can make a good organization about learning resources, then key problems of self-regulated learning can be solved.

3.2 Resource guidance and management strategies of self-regulated learning in network

According to these problems we should take corresponding measures to deal with:

1 Abandoning the FTP lack of a model of interactive web-based self-regulated learning resource, start using model of group website or blog +QQ courses, which takes into account the characteristics of students' love for QQ and being accessible to the Internet.
2 According to the characteristics of curriculum and content, set a limit to the content and time of extra-curricular self-regulated Learning that must be completed by the students. In addition, introduce an incentive mechanism and reward the students according to their recording and the residence time on the system of web-based self-regulated learning, and link up with usually grades of the curriculum.
3 Because most of the students recognized the form of network learning, therefore, only when we make efforts in the selection of materials, can we attract the students to the course of web-based self-regu lated learning. For example, in the "Introduction" of biological science, our approach is to hang what is taught in the class on the Internet or the shared folder. Prerequisite foundation - increase the background and basic content of knowledge, letting the slow learners be able to understand; improve application - offer the quick a play.
4 As for the 30% who don't like the learning in the network, which depends on the teacher to guide and help the students to recognize the unique advantages of Web-based self-regulated learning. For example, making full use of computer's good interactivity, can reallocate the better teaching material in the resources of traditional learning, such as text, animation, video, pictures and so on, drawing students' attention to the learning in the network and join the team. Through the creation of scenarios, teachers can arouse students' interest and guide the questioning and motivate students' self-regulated exploration.Of course, it needs more exchange feedback between teachers and students.
5 University has been emphasizing that college should cultivate the independent learning ability of students. But the teacher's necessary guidance can let students avoid detours, obtaining faster growth. Although most of the students are getting the habit of independent learning, but other students are eager to communicate with teachers and classmates, opening QQ courses develops the instant communication in order to facilitate the teachers and students. Some people say that self-regulated learning system with no interactivity is a cold-blooded system, not the real self-regulated learning system.However, considering the workload of teachers, QQ robot that can substitute for teachers in many occasions may be the better choice to perfect the interaction.

4 ETHICAL PROBLEMS OF THE SELF- REGULATED LEARNING IN THE NETWORK WAITS TO BE DISCUSSED

1 Encouraging students' self-regulated learning in the network will make the students addicted to the network every day, contacting with some bad information, and causing political and ideological and moral quality problems.

2 When students depend too much on the network, the school will lose some of the functions, and status of teachers in the minds of the students will also reduce. For a long time, it is difficult for a teacher to educate students, according to the teaching plan.

3 The network is a robot without life. Relying on the network for a long time will make students tired from the exchange of ideas and communication and drift apart the distance between teachers and students in the mentality. As a result, teachers' course teaching and experimental demonstration are also possibly a contemptuous disregard for a student.

4 On one hand web-based self-regulated learning needs to have some economic basis, but most of the students are from the mountains, rural and city with low income. So it is currently not possible for everyone to own a computer. On the other hand, if the school lets students completely rely on the course of self-regulated learning in the network. Naturally, students and parents will have cast doubt on the school, leading to various social problems.

5 CONCLUSION

Self-regulated learning in the network is a new form for students' study, where the teacher can break through the constraints of the teaching plan or the 45 minutes in the classroom to build his ideal curriculum learning space, allowing students to roam and stroll freely. Both the mode group of website and blog +QQ courses are easy to be accepted and loved by students. Because it meets the requirements of the times and the students' growth individuality.

The university not only concentrates on learning, and attaches more importance to training students' ability to learn new things. With the fast development of the Internet today, self-regulated in the network is a strong supplement to classroom teaching and provides a more broad stage for students to learn more. It can really help to improve the effectiveness of classroom teaching, improving the quality of teaching courses and greatly cultivating the students' self-regulated learning ability, which is what the human society advocates, namely ability to receive lifelong education. "Introduction to computer science" as the introductory courses of freshman at the first grade, students can benefit from such a learning experience for five years and so far as to the whole life, which is the goal of self-regulated learning online.

ACKNOWLEDGMENT

This work was supported by the Key Research Center for Humanities and Social Sciences in Hubei Province (Hubei University of medicine). This work was also supported by the Key Discipline Project of Hubei Province (2014XKJSSJ08) and the grants from Nature Science Foundation of Hubei province(2013CFB479), Teaching Research Program of Hubei University of medicine (2013018), National Undergraduate Training Programs for Innovation and Entrepreneurship (201310929017), Scientific Research Fund for National Project, Hubei University of Medicine(2013GPY10/2013GPY04), Shiyan science and technology research and development projects(2013069), Scientific Research Foundation for the Graduate, Hubei University of Medicine(2012QDJ03/2012QDJ06).

REFERENCES

[1] Barrera, A. (2001). Students' conceptions of intelligence, emotions, and achievement in university statistics courses. Unpublished master's thesis, University of Munich, Germany.

[2] Goetz, T., Pekrun, R., Perry, R. P., & Hladkyi, S. (2001). Academic Emotions Questionnaire: Codebook for English-language scale versions (Tech. Rep.). Munich, Germany: University of Munich, Department of Psychology.

[3] Olafson, K. M., & Ferraro, F. R. (2001). Effects of emotional state on lexical decision performance. Brain and Cognition, 45, 15–20.

[4] Pekrun, R. (2000). A social cognitive, control–value theory of achievement emotions. In J. Heckhausen (Ed.), Motivational psychology of human development (pp. 143–163). Oxford, England: Elsevier.

[5] Perry, R. P., Hladkyj, S., Pekrun, R., & Clifton, R. (2001, April). Self-regulation in college students' scholastic attainment: A three-year field study involving academic control and action control. Paper presented at the annual meeting of the American Educational Research Association, Seattle, WA.

[6] Pape, S., & Wang, C. (2003). Middle School Children Strategic Behaviour: Classification and Relation to Academic Achievement and Mathematical Problem Solving. Instructional Science 31, 419–449.

Information Technology – Wan et al. (Eds)
© *2015 Taylor & Francis Group, London, ISBN 978-1-138-02785-5*

The studies and implementation for conversion of image file format

Yu Zhang
Shanghai University of Engineering Science, Shanghai, China

ABSTRACT: With BMP and GIF file format being analyzed, we studied compression algorithm by comparing different color model between the two file formats, then designed conversion flow, and implemented conversion from BMP format to GIF format finally. According to the test, the obtained image quality was satisfied after conversion.

1   INTRODUCTION

BMP (Bitmap-File) format is one of the most commonly used image formats, which is the standard image formats for its Windows environment equipment by Microsoft. The default extension name of bitmap file is ".BMP". Sometimes it would be instead of ".DIB ". The BMP file is not compressed, so it's size is usually larger than the same image file but compressed.

The BMP file format is capable of storing 2D di-gital images of arbitrary width, height, and resolution, both monochrome and color, in various color depths, and optionally with data compression, alpha channels, and color profiles. Microsoft has defined a particular representation of color bitmaps of diffe-rent color depths, as an aid to exchanging bitmaps between devices and applications with a variety of internal representations. They called these device-independent bitmaps or DIBs, and the file format for them is called DIB file format or BMP image file format [1].

According to Microsoft support [3], a device-independent bitmap (DIB) is a format used to define device-independent bitmaps in various color resolutions. The main purpose of DIBs is to allow bitmaps to be moved from one device to another (hence, the device-independent part of the name). A DIB is an external format, in contrast to a device-dependent bitmap, which appears in the system as a bitmap object (created by an application...). A DIB is normally transported in metafiles (usually using the StretchDIBits() function), BMP files, and the Clipboard (CF_DIB data format). If the images were stored with BMP format, it can be read by almost all of the software.

The full name of GIF format is the "Graphics Interchange Format." In the 1980s, .It was image formats developed for bandwidth limitations by CompuServe Network Company in the U.S. Because it was permitted to open use right by the CompuServe Company, and higher compression ratio and less disk space in GIF format, so this image format had been widely applied quickly, and was supported by a variety of application environments.

The GIF format files supports up to 8 bits per pixel for each image, allowing a single image to reference its own palette of up to 256 different colors chosen from the 24-bit RGB color space. It also supports animations and allows a separate palette of up to 256 colors for each frame. These palette limitations make the GIF format unsuitable for reprodu-cing color photographs and other images with continuous color, but it is well-suited for simpler images such as graphics or logos with solid areas of color [4].

The GIF format can be used to display animation. An animated GIF file comprises a number of frames that are displayed in succession, each introduced by its own GCE (Graphics Control Extension), which gives the time delay to wait after the frame is drawn. Global information at the start of the file is applied by default to all frames.

With the research and development of the conversion of BMP and GIF file formats, there is important significance for the use and processing of digital images, and also it is a foundation for the Synesthesia studies of digital audio and images and the ultimate realization of the synesthesia-based conversion for Digital Image and Digital Audio.

2   BMP IMAGE FILE FORMAT PARSING [1]

There are four kinds of BMP image file format, mon-ochrome, 16 colors, 256 colors and true-color image data. The arranged order of data is different from the general file, which is stored the image from the lower left corner as the starting point instead of the top-left corner on the image. In addition, there is a distinctive characteristic in the BMP image file format, that is, its data structure used in palette (color table) data, the order of red, green, and blue color also issued contrary to other image file formats.

The bitmap image file consists of fixed-size structures (headers) as well as variable-size structures appearing in a predetermined sequence. The bitmap file composed of structures is shown in the Table 1:

Table 1. Composition of BMP file.

Structur Name	Optional	Purpose
Bitmap File Header	No	To store general information
DIB Header	No	To store detailed information
Extra bit masks	Yes	To define the pixel format
Color Table	Semi-optional	To define colors)
Gap1	Yes	Structure alignment
Pixel Array	No	To define the actual values
Gap2	Yes	Structure alignment
ICC Color Profile	Yes	To define the color profile

2.1 *Bitmap file header*

This block of bytes is at the start of the file and is used to identify the file. A typical application reads this block first to ensure that the file is actually a BMP file and that it is not damaged. The first two bytes of the BMP file format are the character 'B' then the character 'M' in 1-byte ASCII encoding. All of the integer values are stored in little-endian format (i.e. least-significant byte first).

2.2 *DIB header (bitmap information header)*

This block of bytes tells the application detailed information about the image, which will be used to display the image on the screen. The block also matches the header used internally by Windows and OS/2 and has several different variants. All of them contain a Dword (32 bit) field, specifying their size, so that an application can easily determine which header is used in the image. In order to resolve the ambiguity of which bits define which samples, the DIB Headers provides certain defaults as well as specific BITFIELDS which are bit masks that define the membership of particular group of bits in a pixel to a particular channel.

2.3 *Color table*

The color table (palette) occurs in the BMP image file directly after the BMP file header, the DIB

header (and after optional three red, green and blue bitmasks if the BITMAPINFOHEADER header with BI_BITFIELDS option is used). Therefore, its offset is the size of the BITMAPFILEHEADER plus the size of the DIB header (plus optional 12 bytes for the three bit masks) [2].

2.4 *Pixel array (bitmap data)*

The pixel array is a block of 32 bit DWORDs that describes the image pixel by pixel. Normally pixels are stored "upside-down" with respect to normal image raster scan order, starting in the lower left corner, going from left to right, and then row by row from the bottom to the top of the image.[5] Unless BITMAPCOREHEADER is used, uncompressed Windows bitmaps also can be stored from the top to bottom, when the Image Height value is negative.

3 GIF IMAGE FILE FORMAT PARSING [7]

In GIF image, there are five major components with a fixed order. All the components are composed one or more data blocks. GIF data blocks are the basic units of data storage. Each block is represented by the code or signature in the first byte. The appearing order of five components in GIF is: file header block (Header), logical screen descriptor block, optional global palette, the image data block and the end of file blocks (Trailer); which logical screen descriptor block, optional global palette and image data blocks together are known as GIF data stream. All the control information and data blocks in data stream must be between the head and the end of file blocks in the file. Image data block is also a complex structure, which includes a graphical depiction block, local color palette, the compressed image data, and a plurality of extents. A GIF file can contain more than one image data block. Fig. 1 is a GIF file structure:

GIF files start with a fixed-length header ("GIF87a" or "GIF89a") giving the version, followed by a fixed-length Logical Screen Descriptor giving the size and other characteristics of the logical screen. The screen descriptor may also specify the presence and size of a Global Color Table, which follows next if present [5] [6].

Although the Graphics Control Extension block declares color index 16 (hexadecimal 10) to be transparent, that index is not used in the image. The only color indexes appearing in the image data are decimal 40 and 255, which the Global Color Table maps to black and white, respectively.

The image pixel data, scanned horizontally from top left, are converted by LZW encoding to codes

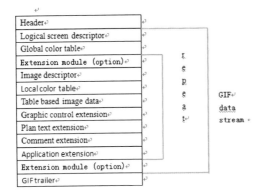

Header
Logical screen descriptor
Global color table
Extension module (option)
Image descriptor
Local color table
Table based image data
Graphic control extension
Plan text extension
Comment extension
Application extension
Extension module (option)
GIF trailer

GIF data stream

Figure 1. The structure of GIF file.

that are then mapped into bytes for storing in the file. The pixel codes typically do not match the 8-bit size of the bytes, so the codes are packed into bytes by a "little-Endian" scheme: the least significant bit of the first code is stored in the least significant bit of the first byte, higher order bits of the code into higher order bits of the byte, spilling over into the low order bits of the next byte as necessary. Each subsequent code is stored starting at the least significant bit not already be used.

4 THE IMPLEMENTATION OF ALGORITHM ON CONVERTING OF IMAGE FILE FORMAT

BMP file format is converted into GIF file format in two steps, the first step is read binary data from BMP format file, the second step is format conversion, and save the converted files in GIF format. There are two key technologies in converting BMP file format into GIF file format, the first is color conversion, as GIF supports only 256 colors, when BMP is converted into GIF, it will need color conversing in many cases, while the second is the use of LZW algorithm in compression encoding

4.1 Color conversion [8][9][10]

It is read out binary data in BMP format file, then completed that more than 256-color image file to be converted into 256-color image file by color conversion function SaveAs256 (const char * pszDibFileName, int x, int y, int nWidth, int nHeight)

Here, const char * pszDibFileName is pointed bitmap path, int x is saved a bitmap starting coordinates x, int y is saved a bitmap starting coordinates y, int nWidth is saved bitmapwidth, int nHeight is saved bitmap highly.

SaveAs256 function first converts the data into 24 bit device-Related bitmap, and then gets the quantized data to be filled with bitmap- related data and saved.

Quantization should be called QuantizeColor-function, QuantizeColor(LPBYTE lpbyDdbBits24, int nScanWidth, int nScanHeight, LPBYTE lpbyDdbBits8, CPalette * pPalette) function is to quantify the color, to count the color distribution, to select the highest frequency of occurrence 256 kinds of color allocate memory to LOGOALETTE, and to fill them, using color quantization result to rewrite 256 located in the middle and to create a mapping table so as to create a logical palette Palette, and also to map to neighboring color for 16 kinds of abandoned colors, to re-map for 8 -bit data and to stored in lpbyDdbBits8.

4.2 Format conversion [11]

After converted 256-color image data, it will be begun to encode BMP file into GIF file, BOOL WINAPI DIBToGIF (LPSTR lpDIB, CFile & file, BOOL bInterlace, int NUM) function is a key function for conversion BMP file into GIF file. When encoding, it will be called a EncodeGIF_LZW (LPSTR lpDIBBits, CFile & file, LPGIFC_VAR lpGIFCVar, WORD wWidthBytes, BOOL bInterlace) function, which is encoded BMP file into LZW file for the specified image. Here, parameterLPSTR lpDIBBits is pointed to source DIB image pointer, CFile & file is the file to be saved, while LPGIFC_VAR lpGIFCVar is pointed to GIFC_VAR structure pointer, WORD wWidthBytes represents the bytes per rowimage, BOOL bInterlace decide whether to be saved in a interleave manner respectively.

First, it is declared the structure of GIF file format in GIFAPI.h. Then variables are defined which are included: DIB height: WORD wHeight; DIB width: WORD wWidth; define DIB pixel pointer: LPSTR lpDIBBits; GIF file header: GIFHEADER GIFH; GIF logical screen descriptor block: GIFSCRDESC GIFS; GIF image description block: GIFIMAGE GIFI ; GIF encoding parameters: GIFC_VAR GIFCVar; GIF image extension block: GIFCONTROL GIFCON; number of colors: WORD wColors;

number of bytes per row: WORD wWidthBytes; palette: BYTE byGIF_Pal [768]; byte variable: BYTE byChar; pointing BITMAPINFO pointer to a structure (Win3.0): LPBITMAPINFO lpbmi; pointing BITMAPCOREINFO structure pointer: LPBITMAPCOREINFO lpbmc; indicate whether Win3.0 DIB tag: BOOL bWinStyleDIB, and so on. Then convert the image format. The conversion process is shown in Figure 2.

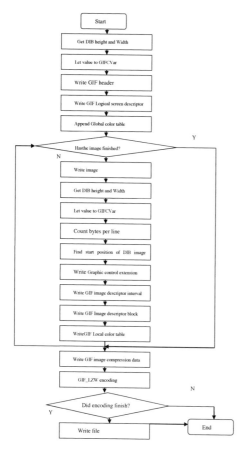

Figure 2.   The conversion process of DIBToGIF.

5   CONCLUSION

In this paper, the image file formats and the method of conversion have been studied. Both BMP, GIF image file formats are analyzed. As it is only providing support for BMP format by the VC itself, through the file structure analysis, GIF's structure is defined in the program. By comparing the color difference between the two models of different file formats, researched two compressions for BMP, GIF format, designed conversion step process, and implemented

the image file format conversion from BMP to GIF. After testing, the satisfactory results are obtained in quality of converted image.

Image file format conversion algorithm is not only one. Their efficiency and quality are not the same. With the rapid development of computer technology, the image compression algorithms are improved in continuous and the image file formats are constantly modified, such as the GIF 87a and 89a edition version. This study is a foundation for the study on other image file format conversion. Consistency of style is very important.

ACKNOWLEDGMENT

This work is supported by Shanghai University of Engineering Science disciplines construction project (No: XKCZ1212) and the course construction project of SUES (No. S201402001)

REFERENCES

[1] "DIBs and Their Uses". Microsoft Help and Support. 2005.
[2] "[MS-WMF]: Windows Metafile Format". MSDN. 2014.
[3] BITMAPINFOHEADER (Windows CE 5.0): BI_ALPHABITFIELDS in biCompression member.
[4] Royal Frazier. "All About GIF89a". Retrieved 7 January 2013.
[5] "Graphics Interchange Format, Version 87a". W3C. 15 June 1987. Retrieved 13 October 2012.
[6] "Graphics Interchange Format, Version 89a". W3C. 31 July 1990. Retrieved 6 March 2009.
[7] Jiang Nan, Wang Jian. *Multimedia File Formats and Compression Standards Analysis*. Beijing: Electronic Industry Press, 2005.
[8] Li Wei, Zhang Lihua. Res image mplementation of BMP file decoding. Publishing and Printing, 2006.
[9] Rafael C. Gonzalez, Richard E. Woods. Digital Image Processing. International of 3rd revised. Pearson Education (US), 2008.
[10] Kenneth R. *Digital Image Processing*. Prentive Hall Inc a Simon & Schuster Company, 2000.
[11] Xue peiding, Xu Guoding Editor: (Japan) Bo Harashima. *Image Information Compression*. Beijing: Science Press, 2004.

Information Technology – Wan et al. (Eds)
© 2015 Taylor & Francis Group, London, ISBN 978-1-138-02785-5

The two-level analysis of disclosure of profit forecasts—empirical evidence from IPO companies

Hui-yun Li

School of Management and Economics, Beijing Institute of Technology & Beijing, China

Qiu-bo Zhao & Hua-chao Qu

School of Management and Economics, Beijing Institute of Technology & Beijing, China

ABSTRACT: As voluntary disclosure is applied to profit forecasts of IPO companies in China, the proportion of companies disclosing such information has been declining generally. From this perspective, the two-level analysis is conducted on the disclosure of profit forecasts of 683 listed companies with initial public offering of A shares in 2007–2011. It finally comes out that there are significantly positive correlations respectively between the companies' financial condition and the disclosure of the profit forecasts information; the corporate governance structure and the disclosure of the profit forecasts information. Meanwhile, there are also positive correlations respectively between the industry types and the level of the disclosure of the profit forecasts; whether audited by the Big Four accounting firms and the level of the disclosure of the profit forecasts, but unremarkable.

KEYWORDS: IPO companies; Disclosure of profit forecasts; Two-level analysis

1 INTRODUCTION

On March 15, 2001, *the Rules on the Content and Format of Information Disclosure of Companies that Publicly Offer Securities No. 1—Prospectus* released by CSRC marked the change of disclosure method for profit forecasts of IPO companies from mandatory to voluntary disclosure in China. Nevertheless, the proportion of companies disclosing profit forecasts in total IPO companies fell significantly after the change of disclosure method in 2001. Now the questions are why those IPO companies are increasingly reluctant to disclose the profit forecasts information? Which kind of factors influences the level of the disclosure? This paper chooses some IPO reformatory companies of A shares from 2007 to 2011 as samples to conduct the two-level analysis to its disclosure of the profit forecasts information. The first level analysis uses the logistic regression to study the influencing factors of the disclosure of the profit forecasts information of those IPO companies. On the basis of building the evaluation index system of the disclosed level on its own, the second level analysis probes into the influencing factors of the level of the disclosure of those IPO companies by means of the multiple linear regression models. The ultimate empirical demonstrations will offer some suggestions to the decision makers and regulators of the IPO companies to improve the present environment of the profit forecasts disclosure.

2 HYPOTHESES

2.1 Financial position and disclosure of profit forecasts

Three indicators including net return on assets, asset-liability ratio, and company size are used in this article to study the relationship between company's financial position and disclosure of profit forecasts.

- Net return on assets. The contract theory holds the view that the firm managers will be more active to disclose more information to the outside world when the company has high profitability. In this way, on one hand, they can achieve better personal reputation and higher salaries. On the other hand, they gain the reliance of the investors and attract more funds, which is benefit for the improvement of the firm's value. Luo Wei and Zhu Chunyan

Supported by National Natural Science Foundation of China (71140006);
Major Project of National Social Science Foundation of China (13ATJ003)

(2012) found that, with the rise of profits, the company's information disclosure will be improved correspondingly.

- Asset-liability ratio. In accordance with the agency theory, when the firm has a high asset-liability, creditors' potential wealth is more likely to be lost. According to the study on companies listed in Kuala Lumpur Stock Exchange (Nan Xu [2012]), when a company presents high asset-liability ratio, potential wealth will be more likely to be transferred from creditors to shareholders and the management.
- Company size. As large-scale companies are confident enough with their development prospect and pay more attention to social image and reputation, they are willing to strengthen the communication with investors by means of disclosure of profit forecasts.

In summary, the first hypothesis is proposed:

- H1a: Positive correlation between the IPO companies' financial condition and the disclosure of the profit forecasts information.
- H1b: Positive correlation between the IPO companies' financial condition and the level of the disclosure of the profit forecasts information.

2.2 Corporate governance structure and disclosure of profit forecasts

In general, corporate governance structure includes three parts, that is, shareholding structure (shareholding concentration, shareholding ratio of directors and supervisors, and percentage of state-owned shares), independence of the Board of Directors (percentage of independent directors), and internal supervision (size of the Board of Supervisors and whether the Audit Committee is established). As all sample companies selected in this article have established an audit committee, the indicator of "internal supervision" can be ignored.

- Shareholding concentration. If companies' shareholders reach a certain number, there will be more requirements on the aspect of information disclosure, which leads to the firms' voluntary disclosure and higher level of the information disclosure. Gerald and Sidney (2011) make a study of 62 industry IPO companies located in H. K and Singapore, and the results show that there is a positive correlation between the shareholding concentration and the level of the information disclosure.

- Percentage of independent directors. High percentage of independent directors can enhance the supervisory role of financial information and reduce the necessity of concealing information by the management.
- Shareholding ratio of directors and supervisors. The higher the shareholding ratio of directors and supervisors is, the lower the demands for public disclosure of profit forecasts are.
- Percentage of state-owned shares. Listed companies in China have a special shareholding structure. Most state-owned enterprises are restructured, with a big percentage of state-owned shares. The empirical study of Li Hao (2010) shows that under such circumstances, insiders are often reluctant to voluntarily disclose more information to stakeholders, which, to a large extent, weakens the overall voluntary information disclosure of listed companies.

To make consistent affecting direction of the above mentioned four factors on disclosure of profit forecasts, the second hypothesis is proposed based on the reciprocal of the three variables including shareholding concentration, shareholding ratio of directors and supervisors, and percentage of state-owned shares:

- H2a: Positive correlation between the IPO companies' corporate governance structure and the disclosure of the profit forecasts information.
- H2b: Positive correlation between the IPO companies' corporate governance structure and the level of the disclosure of the profit forecasts information.

2.3 External factors and disclosure of profit forecasts

According to behavioral science theory, no enterprise exists in isolation, and the existence and evolvement of an enterprise are always affected and restricted by its surrounding environment. External factors in this article include industry type and whether being audited by the Big Four accounting firms.

- Industry type. Different industries face different environments, with diverse asset allocation, thus leading to varied motives for disclosure of profit forecasts. Numerous researches indicate that the phenomenon of "industry herding behavior" is remarkable when companies conduct the voluntary disclosure. Industries in China can be generally divided into two categories: industry and nonindustry.

- Whether audited by the Big Four accounting firms. To keep their good reputation and independent audit, large-scale accounting firms normally disclose more information and details than those small accounting firms.

3 THE FIRST LEVEL ANALYSIS OF THE DISCLOSURE OF THE PROFIT FORECASTS INFORMATION

3.1 Selection of samples and data collection

The samples selected for empirical study in this article include 638 IPO companies listed on the main board and SME board of stock exchanges in Shenzhen and Shanghai from January 1, 2007 to December 31, 2011. Some sample data are extracted from the annual reports of companies listed on the A share main board and SME board downloaded from the websites of Shanghai Stock Exchange and Shenzhen Stock Exchange, and other sample data are sourced from GTA Economic and Financial Research Database (www.gtarsc.com).

3.2 Model design

First, this paper will conduct the factor analysis to the variables firs. Second, according to the outcomes of the analysis, common factors will be collected as the independent variable to conduct the regression analysis.

1 Factor analysis
   a. KMP statistical quantity and Bartlett sphericity test
   On testing, the KMO value of sample data is 0.862, higher than the minimum standard at 0.6. In the meantime, Bartlett sphericity test P value

is 0.05, which rejects the original hypothesis that related matrices are identity matrices, but supports factor analysis. Consequently, factor analysis is suitable for the data in this article.

   b. Determination of factor number
   The loading factors after rotation show that the first common factor has big loading on the four variables including shareholding concentration, percentage of independent directors, shareholding ratio of directors and supervisors, and percentage of state-owned shares, indicating strong correlation among the four indicators, thus, it could be classified into the first category of factor—corporate governance structure (CGS); the second common factor has big loading on three variables including company size, asset-liability ratio, and net return on assets, which could be classified into the second category of factor—company's financial position (CFS).

2 The Logistic regression model
   Expression of the logistic regression model:

$$\ln\left(\frac{p}{1-p}\right) = \beta_0 + \beta_1 CGS + \beta_2 CFS + \beta_3 Industry + \beta_4 Type + \varepsilon$$

3 Variables description

   Table 1 Selection of Variables and Definitions, see Table 1.

3.3 Empirical test and analysis of results

1 Descriptive statistics
   See Table 2, among 638 IPO companies during the period from 2007 to 2011, only 36 companies disclosed profit forecasts, accounting for 5.64% in average, suggesting that the disclosure of profit forecasts by IPO companies stayed at a low level in China.

Table 1. Selection of variables and definitions.

Variable Type	Name of Variable	Definition of Variables and Descriptions
Explanatory variable	Company's financial position (CFS)	Cannot be measured directly, which can be measured by three indicators including net return on assets, asset-liability ratio, and company size
	Corporate governance structure (CGS)	Cannot be measured directly, which can be measured by four indicators including shareholding concentration, percentage of independent directors, shareholding ratio of directors and supervisors, and percentage of state-owned shares
Controlled variable	External environment	Measured by two indicators including industry type and whether being audited by the Big Four accounting firms
Dependent variable	P	Disclose, $P=1$; NOT Disclose, $P=0$

(Continued)

Table 1. Selection of variables and definitions. (*Continued*)

Variable Type	Name of Variable	Definition of Variables and Descriptions
Independent variable	Company size	Measured by the logarithm of company's total assets
	Net return on assets	Measured by company's net profit/total assets
	Asset-liability ratio	Measured by company's total liabilities/total assets
	Shareholding concentration	Measured by CR-5 indicator (sum of shareholding ratios of top five shareholders)
	Percentage of independent directors	Proportion of independent directors in total directors
	Shareholding ratio of directors and supervisors	Proportion of company's common shares held by the board of directors and board of supervisors in the company's total common shares
	Percentage of state-owned shares	Proportion of state-owned shares in the company's total shares
	Industry type	1 for industry, and 0 for non-industry
	Whether being audited by the Big Four accounting firms	1 for listed companies audited by the Big Four accounting firms, i.e. KPMG, PwC, Deloitte and Ernst & Young; 0 for other companies

Table 2. Number of IPO companies disclosing profit forecasts.

Year	Total number of IPO companies	Number of companies disclosing profit forecasts	Percentage (%)
2007	119	14	11.76
2008	76	12	15.79
2009	64	2	3.13
2010	228	3	1.32
2011	151	5	3.31
Total	638	36	5.64

Table 3. Table type styles.

R	Cox & Snell R2	Nagelkerke R2
0.839a	0.724	0.715

2 Outcomes of the Logistic regression

See Table 3, the index of Cox & Snell R2 and Nagelkerke R2 are respectively, 0.724 and 0.715, which indicates that the model is available and has certain ability to explain the data.

See Table 4, from the significant influence that the common factors have on the dependent variable, the Sig index of the firms' financial condition and the corporate governance show that there is a positive correlation between these two factors and the IPO companies' disclosure

Table 4. Logistic regression coefficients and significance table.

	B	S.E,	Wals	df	Sig.
FAC1_1	0.038	0.005	52.445	1	0.001***
FAC2_1	0.532	0.266	4.012	1	0.045**
Industry type	0.469	0.500	0.879	1	0.348
Whether being audited by the Big Four accounting firms	0.003	0.008	0.140	1	0.708

of the profit forecasts information. Similarly, positive correlation exists between the industry types, whether being audited by the Big Four accounting firms and the IPO companies' disclosure of the profit forecasts information, but is not remarkable.

4 THE FIRST LEVEL ANALYSIS OF THE DISCLOSURE OF THE PROFIT FORECASTS INFORMATION

Among the sample of 638 companies, there are only 36 of them that are willing to conduct the voluntary disclosure, accounting only 5.64% for the total. In that case, what is the level of these IPO companies' disclosure? And which kind of factors impacts the disclosed level?

4.1 Selection of samples and data collection

638 IPO companies of A share located in Shanghai and Shenzhen are chosen first, then 36 of them who conduct the voluntary disclosure are selected to be the sample of the study.

4.2 Model design

1 Preparation of profit forecast disclosure indicator
The profit forecasts issued as required by CSRC are prepared according to the new standards on voluntary disclosure, Guidelines for Preparation of Profit Forecast Report of Companies Offering New Shares and China's national conditions. 14 profit forecast disclosure indicators are determined: operating income, operating costs, business tax and surcharges, sales expenses, management expenses, financial expenses, asset impairment losses, profit/loss from changes in fair value, investment income, subsidy income, nonoperating expenses, income tax payable, net profit, earnings per share. Next, we separately give score to the14 disclosure index of each sample company. The grading principle is as following: If the sample company is willing to do the voluntary disclosure, it would get 1 point. If not, it will get 0.

2 The multiple linear regression model
Expression of the multiple linear regression model:

$$VDI = \beta_0 + \beta_1 CGS + \beta_2 CFS + \beta_3 Industry + \beta_4 Type + \varepsilon$$

4.3 Empirical test and analysis of results

See Table 5, the index of Cox & Snell R2 is 0.501. The result is acceptable in consideration of its small sample size—only 36 IPO ample companies.

Table 5. Multiple linear regression model parameter table.

R	0.633[a]
Cox & Snell R^2	0.501
Nagelkerke R^2	0.426
Standard error of estimate	0.022
D-W	2.074

See Table 6, the symbols of the regression results are identifying with the predicted positive correlation. On one hand, the Sig index of the financial condition and the corporate governance are lower than 0.05, indicating that they have a high degree of correlation with the dependent variable. That is to say, there is positive correlation respectively

between the companies' financial condition and the disclosure of the profit forecasts information; the corporate governance structure and the disclosure of the profit forecasts information. On the other hand, the Sig index of the industry type and whether being audited by the Big Four accounting firms are greater than 0.1, which explains that these two factors do not have a significant impact to the level of the IPO companies' disclosure of profit forecasts information.

Table 6. Multiple linear regression coefficients and significance table.

Variable	B	Standard error	Standard coefficient	t	Sig.
FAC1_1	4.258	5.445	0.048	0.276	0.012**
FAC2_1	1.745	1.809	0.345	0.965	0.034**
Industry type	0.564	1.831	0.056	1.574	0.185
Whether being audited by the Big Four accounting firms	0.086	0.152	0.136	1.522	0.141

5 CONCLUSIONS AND SUGGESTIONS

Through the relevant data collection and processing, the first level analysis of this paper concludes that there are significant positive correlations, respectively, between the companies' financial condition and the disclosure of the profit forecasts information; the corporate governance structure and the disclosure of the profit forecasts information; the industry types and the disclosure of the profit forecasts; whether audited by the Big Four accounting firms and the disclosure of the profit forecast. Meanwhile, there is unremarkable positive correlation, respectively, between the industry types and the level of the disclosure of the profit forecasts; whether audited by the Big Four accounting firms and the level of the disclosure of the profit forecasts.

As suggested by empirical results, in terms of corporate structure, shareholding structure should be optimized, standardized board of directors and independent and efficient board of supervisors should be set up, so as to enhance the disclosure of profit forecasts by IPO companies in China. With respect to external factors, information disclosure quality of other domestic accounting firms should be improved, legal responsibilities of CPAs for review should be intensified and implemented, and a set of complete, scientific, and standardized profit forecast system and exemption system should be established to ameliorate the profit forecast disclosure system of IPO companies.

REFERENCES

[1] Luo Wei and Zhu Chunyan. Agency Costs and Company's Voluntary Disclosure [J]. Economic Research: 2010, (10): 142–155.

[2] Nan Xu, Study on Management Characteristics and Earnings Forecast Disclosure Willingness Based on Logistic Regression Model [J]. Journal of Computers, 2012, 7 (8): 1951–1958.

[3] Gerald K. Chau, Sidney J. Gray. Ownership Structure and Corporate Voluntary Disclosure in Hong Kong and Singapore [J]. International Journal of Accounting, 2011(2): 247–265.

[4] Li Hao. Shareholding Structure, Voluntary Disclosure of IPO Profit Forecasts and Its Economic Consequences [J]. On Economic Problems, 2010, (6): 98–102.

[5] Peng Yuanyuan. "Comparison of Chinese and American voluntary disclosure" [J]. Auditing and Finance, 2008, (2), pp. 56–57. (in Chinese)

[6] Timo Niemi, Marko Junkkari, Kalervo Järvelin. Concept-based query language approach to enterprise information systems [J]. Enterprise Information Systems, 2011(1): 26–66.

[7] Wang Kemin, Lian Peng. The Regulation On IPO Earnings Forecasts and Earnings Management of Listed Companies [J]. Accounting Research, 2013, (3): 72–77.

Information Technology – Wan et al. (Eds)
© 2015 Taylor & Francis Group, London, ISBN 978-1-138-02785-5

Track and recognize unresolved target and decoy with monopulse radar

Jian Su, Zhi-yong Song & Qiang Fu
ATR Laboratory, National University of Defense Technology, Changsha, China

ABSTRACT: The air-launched radar decoy (MALD) flied cooperatively with the true target and imitated the radar characteristics to form the unresolved targets within the radar beam. The unresolved target and decoy result in echoes aliasing and observation merging, and then led the measuring error and made the traditional track process failed. This paper analyzed the pattern and principle of the jamming, and proposed the signal mode of unresolved targets. Through combing the detection of jamming with the characteristics of echoes observation and likelihood function, the particle filter was employed to approximate the joint conditional state probability density and realize the joint tracking for unresolved target and decoy. Utilizing the joint tracking information with the jamming characteristic, the DOA information was adopted to realize the identity recognition of the target and decoy. The simulation experiment was conducted to verify the effectiveness.

1 INTRODUCTION

The electromagnetic environment that the homing radar counters becomes more and more complex with the advancement in the ECM and ECCM. The air-launched radar decoy (MALD) is a kind of new off-board jamming, through flying cooperatively to the true target and imitating the radar feature to form the unresolved multiple targets within radar beam. The unresolved targets cause the echoes aliasing and lead observation merging, and thus make a serious error of monopulse measuring (Zhang 2005). There is an obvious bias between the true angular deviations from the target and the measuring angle deviations that obtained from the monopulse ratio estimation. By this inaccurate measure information, radar will be guided to track the decoy, while the true target will be lost (Zhang 2014). The jamming principle of the MALD is the typical two-point source interference. The unresolved target and decoy disturb the natural process of target search, capture and track, and make the boresight of radar point to the centroid of two sources. With echoes merging and identity uncertainty, the tracking process that based on the observation abstraction and parameter measurement is unavailable to initiate and retain the steady tracking for true target.

Many technologies such as statistical detection, MIMO radar, probability hypothesis density (PHD) filter, etc., were employed to detect and track targets in multiple target scenario. However, for monopulse radar in terminal guidance, these methods need too many additional conditions, and the availability and reliability are uncertain.

In order to realize the steady tracking of true target under jamming condition, the joint tracking for unresolved target and decoy is the precondition. Combine the tracking information with jamming pattern, attack state, geometry format, etc., monopulse radar will be likely to distinguish the target and the decoy with DOA diversity, and realize the identity recognition. In this paper, the scene of MALD jamming is chosen to give a demonstration, the signal mode of unresolved targets was proposed. With the detection information of jamming and the echoes observation, this paper operates directly on the monopulse sum/difference data, bypasses the measurement extraction and parameter measuring, utilizes the particles propagating in state space to obtain the conditional state probability density of the unresolved target and decoy, and realizes the joint tracking of the target and decoy. After obtaining the tracking state and trajectory, the DOA information that derived from the joint tracking are adopted to distinguish the target and decoy.

2 JAMMING PATTERN AND SIGNAL MODE

2.1 Jamming pattern and characteristics

The jamming essence of the MALD is through signal imitating and cooperative flying to compose the non-coherent two-point sources within radar beam. The unresolved sources beguile the radar boresight into pointing to decoy, and resulting in target lost and tracking failed. Set the received power ratio between the jamming echo and target echo as PR

$$PR = \frac{P_{Decoy}}{P_{T\arg et}} \quad (1)$$

So the angular deviation of the radar boresight that relative to the geometric center of the target and decoy is (Song 2011)

$$\theta = \frac{\theta}{2}\frac{PR^2-1}{PR^2+1} \qquad (2)$$

where $\Delta\theta$ is the angle interval between target and decoy relative to radar. (2) shows that the boresight points to the power centroid of unresolved targets.

The typical flight mode of MALD can be described as figure 1. The target launches the decoy when it is exposed to the radar, and the decoy intercepts the radar signal, then imitates and retransmits to monopulse radar. In the initial stage of launching, it is easy for the decoy to capture the radar tracking gate. The decoy imitates the flight envelope of target through cooperative flying, and induces the monopulse to regard the decoy as the true target to measure and track. With the increasing of the θ, the true target will move to the edge of the radar beam, and at last escape from the beam.

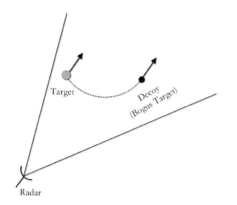

Figure 1. The flight mode and jamming course of MALD.

In the course of jamming, the positions of the target and decoy in different coordinates are able to translate into the corresponding parameter vector reciprocally with bellow equations (Isaac 2008).

$$x_k^j = \rho_k^j \cos\left(\varsigma_{e,k}^j\right)\cos\left(\varsigma_{a,k}^j\right) \qquad y_k^j = \rho_k^j \cos\left(\varsigma_{e,k}^j\right)\sin\left(\varsigma_{a,k}^j\right) \qquad (3)$$
$$z_k^j = \rho_k^j \sin\left(\varsigma_{e,k}^j\right)$$

$$\rho_k^j = \sqrt{\left(x_k^j\right)^2+\left(y_k^j\right)^2+\left(z_k^j\right)^2} \qquad \varsigma_{a,k}^j = \arctan\left(\frac{y_k^j}{x_k^j}\right)$$

$$\varsigma_{e,k}^j = \arctan\left(\frac{z_k^j}{\rho_k^j}\right) \qquad (4)$$

$$\eta_{a,k}^j = \frac{\left(\varsigma_{a,k}^j-\theta_{a,k}\right)}{B_a} \qquad \eta_{e,k}^j = \frac{\left(\varsigma_{e,k}^j-\theta_{e,k}\right)}{B_e}$$
$$\alpha_k^j = \frac{\mod\left(\rho_k^j,BW\right)}{BW} \qquad (5)$$

$$\rho_k^j = \left(N_k+\alpha_k^j\right)BW \qquad \varsigma_{a,k}^j = \theta_{a,k}+\left(\eta_{a,k}^j B_a\right)$$
$$\varsigma_{e,k}^j = \theta_{e,k}+\left(\eta_{e,k}^j B_e\right) \qquad (6)$$

where $\left(\theta_{a,k},\theta_{e,k}\right)$ is taken as the boresight azimuth and elevation angle. $P_k^j = \left(\rho_k^j, \varsigma_{a,k}^j, \varsigma_{e,k}^j\right)^T (j=T,D)$ is the position in spherical coordinates, which denote the range, azimuth and elevation of the target and decoy, where T denotes the target and D denotes the decoy. $X_k^j = \left(x_k^j, y_k^j, z_k^j\right)^T$ is the position in Cartesian coordinates. $\lambda_k^j = \left(\alpha_k^j, \eta_{a,k}^j, \eta_{e,k}^j\right)(j=T,D)$ is the parameter vector of the target and the decoy, which denote sub-bin range and azimuth/elevation electronic-angle respectively.

2.2 Echoes modew

In typical monopulse radar, the in-phase and quadrature parts of the sum, horizontal, and vertical difference signals for a target are (Levanon,1988)

$$s_i = \beta\cos\phi + n_{si}$$
$$s_q = \beta\sin\phi + n_{sq}$$
$$d_{ai} = \beta\eta_a\cos\phi + n_{dai}$$
$$d_{aq} = \beta\eta_a\sin\phi + n_{daq} \qquad (7)$$
$$d_{ei} = \beta\eta_e\cos\phi + n_{dei}$$
$$d_{eq} = \beta\eta_e\sin\phi + n_{deq}$$

where β,ϕ,η_a,η_e are the amplitude, phase, horizontal and vertical DOA of the target respectively, $n_{si},n_{sq}\sim N\left(0,\sigma_s^2\right)$ are the sum channel noise, $n_{dai},n_{daq},n_{dei},n_{deq}\sim N\left(0,\sigma_d^2\right)$ are the difference channel noise. Assume β has a Rayleigh distribution, therefore $x,y\sim N\left(0,\sigma_0^2\right)$.

In the actual sampling of matched filter for the monopulse radar, the energy of the target will spill over to the consecutive sampling points, that is to say the position of the target will not just locate on some single sampling points but fall between two or more consecutive sampling points (Zhang 2009). As for the unresolved target and decoy, they are both within the same resolution cell. Set the matched filter sampling rate is once per pulse-length, the target and the decoy will locate between two consecutive sampling points. The signal mode is shown in Figure 2.

In Figure 2, define the time offset T_T and T_D of the target and the decoy respect to the leading sampling point. x_T and x_D denote the amplitude of the

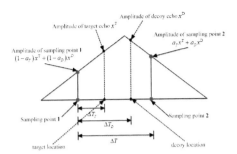

Figure 2.   The signal mode of unresolved targets.

target echo and decoy jamming respectively. Define $\alpha = T'/T$, thus the amplitude of sampling point 1 is $(1-\alpha_T)x_T+(1-\alpha_D)x_D$, and that of sampling point 2 is $\alpha_T x_T + \alpha_D x_D$.

At the discrete time instance k, when radar emits the mth pulse, the in-phase parts of the observation vector $z_k(m)$ is

$$\mathbf{z_k}(\mathbf{m}) = \left[s_{i1,k}(m), s_{i2,k}(m), d_{ai1,k}(m), d_{ai2,k}(m), d_{ei1,k}(m), d_{ei2,k}(m)\right]^T \quad (8)$$

$$s_{i1,k}(m) = \sum_{j=T,D}^{2}\left(1-\alpha_k^j\right)x_k^j(m)+n_{si1,k}(m)$$

$$s_{i2,k}(m) = \sum_{j=T,D}^{2}\alpha_k^j x_k^j(m)+n_{si2,k}(m)$$

$$d_{ai1,k}(m) = \sum_{j=T,D}^{2}\left(1-\alpha_k^j\right)\eta_{a,k}^j x_k^j(m)+n_{dai1,k}(m) \quad (9)$$

$$d_{ai2,k}(m) = \sum_{j=T,D}^{2}\alpha_k^j \eta_{a,k}^j x_k^j(m)+n_{dai2,k}(m)$$

$$d_{ei1,k}(m) = \sum_{j=T,D}^{2}\left(1-\alpha_k^j\right)\eta_{e,k}^j x_k^j(m)+n_{dei1,k}(m)$$

$$d_{ei2,k}(m) = \sum_{j=T,D}^{2}\alpha_k^j \eta_{e,k}^j x_k^j(m)+n_{dei2,k}(m)$$

The parameter vector of the target is $\lambda_k^T = \left(\alpha_k^T, \eta_{a,k}^T, \eta_{e,k}^T\right)$, and that of the decoy is $\lambda_k^D = \left(\alpha_k^D, \eta_{a,k}^D, \eta_{e,k}^D\right)$. Thus the joint state vector of the unresolved targets is

$$\lambda_k^J = \left[\lambda_k^T \ , \ \lambda_k^D\right] \quad (10)$$

Therefore, the observation vector Z_k follows the Gaussian distribution with zero mean and covariance matrix R.

$$L\left(\lambda_k^J\right) = p\left(\mathbf{Z_k} \mid \lambda_k^J\right)$$
$$= \prod_{m=1}^{M}\left(\frac{1}{|2\pi\mathbf{R_k}|^{1/2}}\exp\left(-\frac{1}{2}\mathbf{z_k}(\mathbf{m})^T \mathbf{R_k^{-1}z_k}(\mathbf{m})\right)\right) \quad (11)$$

The signal-to noise ratio(SNR) of the target and the decoy is defined as

$$SNR^j = 10\log\left(\frac{M}{\sigma_s^2}\sum\left(\sigma^j\right)^2\right)(j=T,D) \quad (12)$$

where $\left(\sigma^j\right)^2$ denote the power of the target echo or decoy jamming.

3   JOINT TRACKING FOR UNRESOLVED TARGETS

The unresolved target and decoy cause the echoes aliasing and observation merging, thus the parameter measuring based on observation can only obtain the parameter and state of the power centroid. The nonlinear characteristic of observation and the unknown covariance matrix deteriorate the state estimation problem. Particle filter (Arulampalam, 2002) is a strong tool to resolve complex nonlinear problem, and state particles of both target and decoy can be utilized to approximate the conditional state probability density of unresolved target and decoy.

Without loss of generality, assume the motion of both target and decoy can be expressed as a constant velocity linear motion model, therefore the state equation and observation equation are as follows (Monakov 2012)

$$s^j(k+1) = Fs^j(k)+v^j(k) \quad (13)$$

$$z^j(k) = H\left(s^j(k)\right)+w^j(k) \quad (14)$$

where F and H are the state transition and input-output matrices. $s^j(k)$ is the state at time k, a six-dimensional vector with the first to third elements representing the positional coordinates, and the others representing the corresponding velocities along these coordinates.

When monopulse radar detects the presence of decoy (Song, 2012), based on echo observation and kinematic mode, the radar operates directly on the monopulse sum/difference, utilizes the particle filter to realize the joint tracking for unresolved targets. The process is shown in Figure 3.

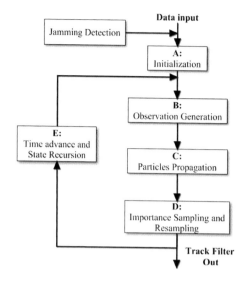

Figure 3. The joint tracking process of unresolved targets.

Stage A: Initialization

Step 1: Define I as the total number of particles among the joint tracking of the unresolved group. For $i = 1, 2, \cdots, I$ and $j = 1, 2$ (denote target and decoy respectively), according to the definition and characteristics of parameter vector, set the initial particles follow the corresponding uniform distribution, and $\left[\alpha_0^j\right]_i \sim U(0,1)$, $\left[\eta_{a,0}^j\right]_i \sim U(-1,1)$ and $\left[\eta_{e,0}^j\right]_i \sim U(-1,1)$;

Step 2: Get the initial integrated Cartesian positional coordinates $[\chi_0]_i = \left[\ \left[\chi_0^1\right]_i^{\mathrm{T}}\ \left[\chi_0^2\right]_i^{\mathrm{T}}\ \right]^{\mathrm{T}}$ and the initial integrated spherical positional coordinates $P_{M(0)}$ using (3)~(6)based on the above initial parameter particles and measurement information about the unresolved group targets at the moment, where the $\left[\chi_0^1\right]_i^{\mathrm{T}}$ and $\left[\chi_0^2\right]_i^{\mathrm{T}}$ denote the initial positional coordinates of the target and decoy respectively;

Step 3: Get the joint space position of the target and decoy using $\chi_0 = \sum_{i=1}^{I} [\chi_0]_i / I$ with the initial positional particles $[\chi_0]_i$;

Step 4: Construct the integrated state vector particles of the target and decoy under the Cartesian positional coordinates as $\left[s(0)\right]_{i=1}^{I} = \left[\ \left[\chi_0^T\right]_{i=1}^{I}\ \dot{\chi}_0^{\ \mathrm{T}}\ \right]^{\mathrm{T}}$, where the $\dot{\chi}_0^{\ T}$ is the velocity vector and can be obtained by the initial assumption or the two-point method;

Step 5: Predict the spherical coordinates $P_{1/0}$ pertaining to P_0 with (13) using χ_0 and $\dot{\chi}_0^{\ \mathrm{T}}$, and get the radar position coordinate $P_{M(1/0)}$ using the proportion guidance law, and set $k = 1$;

Stage B: Observation generation

Step 6: With the monopulse radar coordinates $P_{M(k/k-1)}$, the power relationship of the target and decoy echoes and the orientation part of $P_{k/k-1}^1$ and $P_{k/k-1}^2$, the radar boresight $(\theta_{a,k}, \theta_{e,k})$ is obtained with the power ratio and the two point source principle;

Step 7: Move the coordinate original position to the homing radar position $P_{M(k/k-1)}$, and get the M-sub-pulse radar observations $\mathbf{z}_k(m) = \{z_k(m)\}_{m=1}^{M}$ with (8) and (9);

Stage C: Particles propagation

Step 8: The integrated state particle is propagated on step in time, such that the conditional density function $p\left([\chi_k]_i \mid [\chi_{k-1}]_i\right)$ satisfies (13);

Step 9: Set $[s(k \mid k-1)]_i = F[s(k-1)]_i$, get the corresponding integrated state vector $\left[\chi_{k|k-1}\right]_i$ from the positional coordinates particles $[s(k \mid k-1)]_i$;

Step 10: Add a disorder to the propagation of each particle, and set $\left[\chi_{k|k-1}^u\right]_i \sim N\left(\left[\chi_{k|k-1}\right]_i, r\right)$, use the congregation method and make sure the position particles $\left[\chi_{k|k-1}^u\right]_i$ always lie within the radar beam width and bin width;

Step 11: Convert the position particles $\left[\chi_{k|k-1}^u\right]_i$ to the corresponding joint parameter vector $\{\chi_k^u\}_i$;

Stage D: Importance Sampling and Resampling

Step 12: Accord to (11) to evaluate the importance weights of each particle $W^i = p\left(\mathbf{z}_k(m) \mid \left[\lambda_k^u\right]_i\right)$, and normalize the importance weights $w^i = W^i / \sum_{i=1}^{I} W^i$;

Step 13: Assign the normalized weight w^i to particle $[\lambda^u_k]_i$, and resample I particles $\left\{\left[\lambda_k^{sampled}\right]_i\right\}_{i=1}^{I}$ with replacement from the set $\{[\lambda^u_k]_i\}_{i=1}^{I}$ according to the latter set's weights w^i (Saha 2012);

Step 14: Convert $\left\{\left[\lambda_k^{sampled}\right]_i\right\}_{i=1}^{I}$ to the corresponding position vector $[\chi_k]_i$ according to (6) and (3);

Step 15: Get the joint position vector $\chi_k = \frac{1}{I} \sum_{i=1}^{I} [\chi_k]_i$, and predict the joint position coordinates $P_{k+1|k}$ pertaining to P_k using (13), at the same time get the homing radar position $P_{M(k+1/k)}$ according to guidance law;

Stage E: Time advance and State Recursion

Step 16: Set $k+1 \to k$ and go to *step 6*.

4 IDENTITY RECOGNITION

After realizing the joint tracking of the target and decoy, the estimation of the parameter vector of them are obtained. The identifying information of the target and decoy are embodied in their parameter vectors. According to the jamming principle of the radar decoy, the power of jamming echo is usually larger than that of the target echo, so that the boresight of monopulse radar is near the target and apart from the decoy. Therefore, under the serious jamming condition, the azimuth and elevation DOA of the decoy is gathering together with the zero, and the absolute value of DOA is small, while the azimuth and elevation DOA of the target is apart from the zero, and the absolute value DOA is large. That is to say, when PR is large, the big DOA denotes the target, and the small DOA denotes the decoy.

According to analysis, with joint tracking results, we can obtain two groups of DOA parameters $\{\eta_{a1}, \eta_{e1}\}$ and $\{\eta_{a2}, \eta_{e2}\}$ via (4)~(5). Through the DOA matching with the jamming characteristic, the identity information about $\{\eta_{a1}, \eta_{e1}\}$ and $\{\eta_{a2}, \eta_{e2}\}$ can be resolved. The identity recognition based on DOA matching is shown in Figure 4.

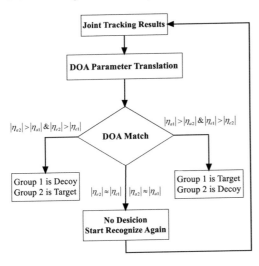

Figure 4. The identity recognition process of target.

5 SIMULATION EXPERIMENTS

This part adopts the Monte Carlo simulation to validate the performance of joint tracking and recognition algorithm under jamming conditions. In the experiment, assume the target and the decoy are located within the radar beam. The root mean square error (RMSE) of tracking trajectory comparing to the actual trajectory is used to evaluate the tracking performance, and the correct recognition probability is used to evaluate the performance of recognition method. The SNR of the target echo is set as $\Re_T = 13$dB, and that of decoy is educed by PR. The typical condition is $PR = 1, 2, 4, 8$. The parameter setting of the experiment is shown in Table 1.

Table 1. Simulation setting.

Parameter Style	Parameter Value
Initial Position(m)	Radar: (0,0,0) Target: (10830;10;2240) Decoy: (10900;80;2290)
Velocity (m/s)	Target v_T: (300;30;0) Decoy v_D: (300;30;0)
Other Parameters	Radar velocity v_M: 1000 m/s Beamform width B: 3° Bin range width BW: 100m Time step T: 0.1s Proportion Navigation coefficient: 3 Total particles number I: 2000 Pulses number M: 10 Monte Carlo experiments: 200

The flight of the radar, target and decoy under the simulation setting are shown in Figure 5.

Experiments 1: Select the $PR = 1, 2, 4, 8$ one by one, utilize process in Figure 3 to deal with the echoes, the joint track trajectory is shown in Figure 6.

Carry out 200 times Monte Carlo experiments for the joint tracking, the tracking RMSE of the target and the decoy are shown in Figure 7.

The tracking trajectory and RMSE in Figure 6 and Figure 7 show that the radar can realize joint tracking of the unresolved target and decoy under 4 kinds of jamming conditions and obtain corresponding state and trajectory information. In Figure 6(a), $PR = 1$ means the power of target and that of decoy are equal, and the boresight points to the geometric center, thus the tracking performance of both is equivalent. With the increasing of PR, radar boresight gradually

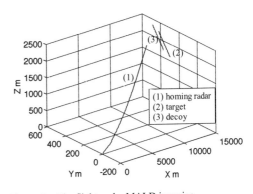

Figure 5. The flight under MALD jamming.

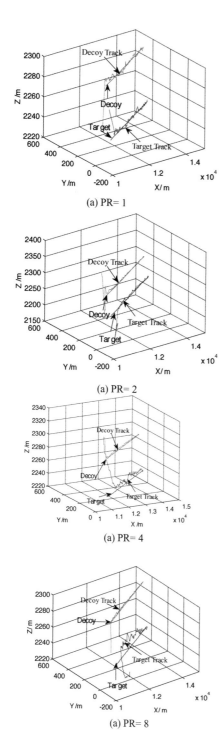

(a) PR= 1

(a) PR= 2

(a) PR= 4

(a) PR= 8

Figure 6. Joint tracking trajectory for unresolved target and decoy.

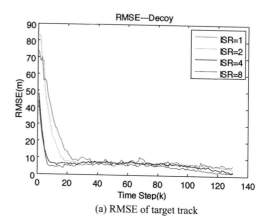

(a) RMSE of target track

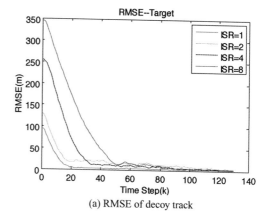

(a) RMSE of decoy track

Figure 7. Tracking RMSE of the target and decoy.

deviates from the geometric center and move to decoy. Therefore, in Figure 6(b), (c) and (d), the tracking trajectory of the target becomes more and more smooth, and that of target undulates more and more heavily. This conclusion is consistent with the result in Figure 7. In the initial stage of tracking, due to the insufficiency of particle iteration, the tracking dose not converges, so the RMSE of target and decoy is poor. With the iteration advances, the tracking gradually converges and the corresponding RMSE decreases. Although the PR affects the track performance of the target, when $k > 40$, the RMSE of decoy tracking is about 10m, while that of the target is larger a little. When $k > 60$, the RMSE of target tracking also decreases to 10m, and satisfies the requirement of steady tracking for aerial target.

Experiments 2: Translate the joint tracking state of the target and decoy to the parameter vector via (4)~(5). In order to evaluate the recognition performance under severe jamming conditions, take $PR = 2, 4, 8$. Based on the two groups of DOA information, utilize the identity recognition method in Fig 4, the recognition probability of the target under $PR = 2, 4, 8$ among the 200 experiments is shown in Figure 8.

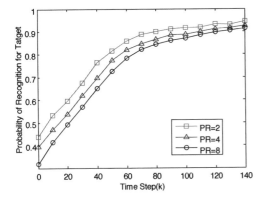

Figure 8. The recognition probability of target under the MALD jamming.

The recognition probability at the discrete time shows that in the initial stage of joint tracking, the track RMSE is large, so that the correct recognition probability of target is low. It means that monopulse radar is difficult to distinguish the target and decoy in this stage. With the process of joint tracking, the tracking RMSE decreases, and the recognition probability improves at the same time. When $k > 60$, the average recognition probability of the target can reach to 80%. At the same time, with the increasing of PR, the recognition probability of the target decreases in some sort, which due to the undulation of the target trajectory.

6 CONCLUSION

In this paper, a novel method to track and recognize the unresolved targets within the radar beam was proposed. Combing the detection information with the echo characteristic, the joint tracking of the unresolved target and decoy was realized, and their tracking state and trajectory were all obtained. Through utilizing the DOA information and jamming characteristic information, the identity recognition of target was achieved. Experimental results indicate that the tracking RMSE of target can decrease to 10m, and the recognition probability of target can reach to above 80%.

REFERENCES

Arulampalam, M. S., Maskell, S. & Gordon, N. 2002. A Tutorial on particle filters for online nonlinear/non-Gaussian Bayesian tracking. *IEEE Transaction on Signal Processing* 50(2): 174–188.

Isaac, A, Willett, P. K. & Bar-Shalom, Y. 2008. Quickest detection and tracking of spawning targets using monopulse radar channel signals. *IEEE Transaction on Signal Processing* 56(3): 1302–1308.

Levanon, N. 1988. *Radar Principle.* New York: Wiley.

Monakov, A. 2012. Maximum-likelihood estimation of parameters of an extended target in tracking monopulse radars. *IEEE Transaction on Aerospace and Electronic Systems* 48(3): 2653–2665.

Saha, S. & Gustafsson, F. 2012. Particle filtering with dependent noise processes. *IEEE Transaction on Signal Processing* 60(9):4497–4508.

Song, Z. Y. & Xiao, H. T. 2011. Detection of presence of towed radar active decoys based on angle glint. *Signal Process* 27(4):522–528.

Song, Z. Y. Xiao, H. T. & Zhu. Y. L. 2012. A novel approach to detect the unresolved towed decoy in terminal guidance. *Chinese Journal of Electronics* 21(2):367–373.

Zhang. T.,Yu, L. & Zhou, Z. L. 2014. Coordinated engagement strategy for MALD and fighter in beyond-visual-range air combat. *Journal of Projectiles, Rockets, Missile and Guidance* 34(1): 60–65.

Zhang, X. , Willett, P. K. & Bar-shalom, Y.2005. Monopulse Radar detection and localization of multiple unresolved targets via joint bin Processing. *IEEE Transaction on Signal Processing* 45(2): 455–472.

Zhang, X. , Willett, P. K. & Bar-shalom, Y. 2009. Detection and Localization of Multiple Unresolved Extended Targets via Monopulse Radar Signal Processing. *IEEE Transaction on Signal Processing* 53(4): 1225–1236.

Information Technology – Wan et al. (Eds)
© 2015 Taylor & Francis Group, London, ISBN 978-1-138-02785-5

An uncertain multiple attribute decision making method based on the ideal point with four-point iinterval numbers

Lei Zhong, Yuejin Lv & Yanhua Yang
School of Mathematics and Information Science, Guangxi University, Nanning, China

ABSTRACT: Based on the three-point interval number, a kind of four-point interval number is proposed whose most probable value is the interval number. The probability of the probable value which falls into the most probable interval is greater than the others. At the same time, some algorithms of it are putted forward. What is more, the similarity degree of two four-point interval numbers is defined to measure their close degree. The technique for order preference by similarity to an ideal solution (TOPSIS) which achieves the alternative to approximate the positive ideal point based on similarity degree is used to solve the uncertain multiple attribute decision making problem whose attribute values are four-point interval numbers. When a relative closeness degree of each alternative for the ideal point (the positive one and the negative one) is identified by the similarity, alternatives are ranked. A demonstration is given at last in the paper, and it verifies the method validity.

1 INTRODUCTION

The multiple attribute decision making is a limited multi-objective decision (Hwang & Yoon, 1981). The essence of it is to use the existing decision-making information to sort and choose the preferred options.

It is an undeniable fact that it plays a very important role in the economy, management, social, military and scientific research. With the development of the society, it cannot already meet the need of the present social reality. Because of the complexity and uncertainty of the decision problem, there still exists the decision information given by some uncertain parameters, such as fuzzy number by Jain (1978), Buckley (2001), Dubois & Prade (1978) and Zhu & Jing (1999), interval number by Saaty (1980), Wang & Yang et al. (2005),Zhu & Liu et al. (2005) and Xu & Da (2003),triangle fuzzy number by Jiang & Fan (2002) and Kao & Liu (2001), trapezoidal fuzzy number by Li (2002) and Gong & Liang (2008), interval-valued intuitionistic fuzzy number by Atanasov & Gargov (1989) and Xu & Chen (2007), random variables by Basak (1998), interval rough number by Zeng & Zeng (2010), Qian & Zeng (2013) and Jin & Guo (2013), and three-point interval number by Tian & Zhu (2008) and Hu et al. (2007). The uncertain theory derived from uncertain number gives a valid tool for solving uncertain multiple attribute decision making issues.

Tian & Zhu (2008) has proposed the three-point interval number, using a new preferred approach to express decider's uncertain preference. The main idea is that it adopts three-point interval number in the form of $[a^L, a^U, a^M]$ to express decider's preference.

First, estimates the range of interval as $[a^L, a^U]$. At the same time gets the probability of all of possible values. After that find the corresponding probable value which is marked as a^M whose probability gets the maximal point and exceeds a specific value.

Obviously, it is very hard to give the probability of other judgment values in the interval $[a^L, a^U]$ and is prone to error.

This paper firstly proposes a kind of four-point interval judgment method. Decider gives a judgment interval $[a^L, a^U]$ and estimates the range of most probable value marked as an interval $[a^-, a^+]$ (It's confidence level should exceed a specific value.)

Finally decider's judgment is obtained in the form of $[a^L, [a^-, a^+], a^U]$. Specifically, the probability of possible values in the interval $[a^-, a^+]$ is greater than in the interval $[a^L, a^-]$, as well as $[a^+, a^U]$. For an instance, decider expects to reach investment benefit of a project in a four-point interval number [3, [4, 5], 9]. This means that the project benefit range is in an interval [3, 9]. But the possible benefit is more likely to fall into the most probable interval [4, 5].Obviously, this decision is effective and conforms to the decider's judgment.

In the multiple attribute decision making method, the technique for order preference by similarity to an ideal solution (TOPSIS) is a decision method to approximate the positive ideal point. This approach has been popularized broadly. When getting closer to the positive ideal point and farther to the negative ideal point (the main work of it is to define the distance), the alternative is obviously more optimal. This paper defines the similarity degree of four-point interval numbers to measure the distance between them and uses TOPSIS to achieve alternatives to approximate the positive ideal point. When getting a relative closeness degree of each alternative for the ideal point, alternatives are ranked.

2 FOUR-POINT INTERVAL NUMBERS

Generally, the four-point interval number is defined in the form of $[a^L, [a^-, a^+], a^U]$ and the common sorting is $a^L < a^- < a^+ < a^U$. a^L denotes the lower limit of the possible value. $[a^-, a^+]$ denotes the interval of the most probable value. a^U denotes the upper limit of the possible value. The probability of possible values falling into $[a^-, a^+]$ is greater and exceeds a specific value. All four-point interval numbers are recorded as $FEI(R)$.

Specifically, considering the probability distribution of all possible values in the interval when $a^- = a^+$, $a^L \neq a^U$, the four-point interval number degrades into the three-point interval number. Also, when $a^L = a^U$, the four-point interval number degrades into the real number. Next, calculations of the four-point interval number are introduced to make decision.

Firstly, Let $a_1 = [a_1^L, [a_1^-, a_1^+], a_1^U]$, $a_2 = [a_2^L, [a_2^-, a_2^+], a_2^U]$. ($a_1, a_2 \in FEI(R)$). When $\lambda \in R^+$, $\lambda > R$ and $0 \notin a_2$, there are some calculations about four-point interval number:

$$a_1 + a_2 = [a_1^L + a_2^L, [a_1^- + a_2^-, a_1^+ + a_2^+], a_2^U] \quad (1)$$

$$a_1 - a_2 = [a_1^L - a_2^U, [a_1^- - a_2^+, a_1^+ - a_2^+], a_1^U - a_2^L] \quad (2)$$

$$\lambda a_1 = [\lambda a_1^L, [\lambda a_1^-, \lambda a_1^+], \lambda a_1^U] \quad (3)$$

$$a_1 - \lambda = [a_1^L - \lambda, [a_1^- - \lambda, a_1^+ - \lambda], a_1^U - \lambda] \quad (4)$$

$$a_1 a_2 = [\min(a_1^L a_2^U, a_1^L a_2^L, a_1^U a_2^U, a_1^U a_2^L), \\ [\min(a_1^- a_2^+, a_1^- a_2^-, a_1^+ a_2^+, a_1^+ a_2^-), \\ \max(a_1^- a_2^+, a_1^- a_2^-, a_1^+ a_2^+, a_1^+ a_2^-)], \\ \max(a_1^L a_2^U, a_1^L a_2^L, a_1^U a_2^U, a_1^U a_2^L)] \quad (5)$$

$$a_1/a_2 = [\min(a_1^L/a_2^U, a_1^L/a_2^L, a_1^U/a_2^U, a_1^U/a_2^L), \\ [\min(a_1^-/a_2^+, a_1^-/a_2^-, a_1^+/a_2^+, a_1^+/a_2^-), \\ \max(a_1^-/a_2^+, a_1^-/a_2^-, a_1^+/a_2^+, a_1^+/a_2^-)], \\ \max(a_1^L/a_2^U, a_1^L/a_2^L, a_1^U/a_2^U, a_1^U/a_2^L)] \quad (6)$$

specifically, if $a_1, a_2 > 0$, then:

$$a_1 a_2 = [a_1^L a_2^L, [a_1^- a_2^-, a_1^+ a_2^+], a_1^U a_2^U] \quad (7)$$

$$a_1/a_2 = [a_1^L/a_2^U, [a_1^-/a_2^+, a_1^+ a_2^-], a_1^U/a_2^L] \quad (8)$$

Obviously, the form of four-point interval number has not changed according to related operations.

By reference to the definition of similarity about two interval numbers (Chen & Qin 2009), this paper also puts forward the following similarity degree of four-point interval number:

Let $a_1 = [a_1^L, [a_1^-, a_1^+], a_1^U]$ and $a_2 = [a_2^L, [a_2^-, a_2^+], a_2^U]$ the similarity degree ofis defined by:

$$S(a_1, a_2) = \frac{\beta}{2} \frac{\left| [a_1^L, a_1^-] \cap [a_2^L, a_2^-] \right|}{\left| [a_1^L, a_1^-] \cup [a_2^L, a_2^-] \right|} + \alpha \frac{\left| [a_1^-, a_1^+] \cap [a_2^-, a_2^+] \right|}{\left| [a_1^-, a_1^+] \cup [a_2^-, a_2^+] \right|}$$
$$+ \frac{\beta}{2} \frac{\left| [a_1^+, a_1^U] \cap [a_2^+, a_2^U] \right|}{\left| [a_1^+, a_1^U] \cup [a_2^+, a_2^U] \right|} \quad (9)$$

among them, we set $\alpha + \beta = 1$ and $\alpha > 0.5$. $|*|$ denotes the length of an interval. Its value is equal or greater than zero. Obviously, The length of the empty set and single point set is zero.

Then, the similarity degree has the following properties:

a) It is easy to know $S(a_1, a_2) \in [0, 1]$, then obviously when $\|[a_1^L, a_1^-] \cap [a_2^L, a_2^-]\| = 0$, $\|[a_1^-, a_1^+] \cap [a_2^-, a_2^+]\| = 0$ and also $\|[a_1^+, a_1^U] \cap [a_2^+, a_2^U]\| = 0$, $S(a_1, a_2) = 1$. What's more, When $a_1^L = a_2^L$, $a_1^U = a_2^U$, $a_1^- = a_2^-$ and $a_1^+ = a_2^+$, $S(a_1, a_2) = 1$.

b) $S(a_1, a_2) = S(a_2, a_1)$

The proposition is tenable according to the definition of the similarity degree.

Zeng & Zeng(2010) proposed the definition of interval rough number whose form is similar to the four-point interval number. But there are great differences in practical significance among them.

Firstly, We defined the interval rough number as $\xi = ([a, b], [c, d])$ whose lower approximation and upper approximation are intervals. Among them, $[a, b]$ denotes the lower approximation and $[c, d]$ denotes the upper approximation (Zeng & Zeng 2010).

For instance, set the project investment to be expressed as interval rough number ([4,6], [3,7]).It is very definite that the project investment is between 4 and 6 (the range of lower approximation). But it is possibly between 3 and 7 (the range of upper approximation) by the paper of Zeng & Zeng(2010).

Zeng proposed an explanation about the practical significance of interval rough number which completely goes against the essence of interval inclusion relations.It is supposed to be a defect. But if we use four-point interval number ([3,[4,6],7])to denotes investment, it can be explained that investment certainly is between 3 and 7. But, more likely it is between 4 and 6. Four-point interval numbers decisively abandon the concept of the upper and lower approximation in interval rough numbers. It satisfies the essence of interval inclusion relations compared with interval rough numbers. It reflects decider's judgment by the upper and lower limit of the possible value as well as the most possible value. Specifically, the most possible value of it is an interval which clearly shows the decider's hesitation. The results are fair and conform to reality.

3 UNCERTAIN MULTIPLE ATTRIBUTE DECISION MAKING BASED ON FOUR-POINT INTERVAL NUMBERS

In this chapter, an uncertain multiple attribute decision making problem based on the four-point interval number is discussed by: $s = \{s_1, s_2, \dots, s_m\}(m \geq 2)$ denotes a solution set. And then attribute set is $u = \{u_1, u_2, \dots, u_n\}(n \geq 2)$. And the property weight is $w = \{w_1, w_2, \dots, w_n\}^T(n \geq 2)$ (Given). The value of solution s_i is a four-point interval number $a_{ij} = [a_{ij}{}^L, [a_{ij}{}^-, a_{ij}{}^+], a_{ij}{}^U]$ under the attribute $u_j(i \in M = \{1, 2, \dots m\}, j \in N = \{1, 2, \dots n\}, a_{ij}{}^L \geq 0)$.

This paper proposes the TOPSIS decision making method of a four-point interval number problem based on the similarity degree defined. Specific steps are as follows:

Step 1 Decider uses four-point interval numbers to make decision for the related issues and gets the decision matrix in the form of $A = (a_{ij})_{m \times n}$.

Step 2 Typically, attribute types are classified as a benefit-type or cost-type. In order to eliminate the influence of different physical dimensions to the result of the decision, it is very necessary to normalize decision matrix $A = (a_{ij})_{m \times n}$. As a result, the normalized decision matrix is obtained in the form of $R = (r_{ij})_{m \times n}$ and r_{ij} is still the four-point interval number. Conversion formulas are as follows:

To attribute values of benefit-type:

$$r_{ij} = \frac{a_{ij} - \min_i\{a_{ij}{}^L\}}{\max_i\{a_{ij}{}^U\} - \min_i\{a_{ij}{}^L\}} \tag{10}$$

To attribute values of cost-type:

$$r_{ij} = \frac{\max_i\{a_{ij}{}^U\} - a_{ij}}{\max_i\{a_{ij}{}^U\} - \min_i\{a_{ij}{}^L\}} \tag{11}$$

Step 3 To construct a weighted normalized decision matrix $Z = (z_{ij})_{m \times n}$ according the known attribute weight vector w and the normalized decision matrix $R = (r_{ij})_{m \times n}$. Among them, $z_{ij} = [z_{ij}{}^L, [z_{ij}{}^-, z_{ij}{}^+], z_{ij}{}^U]$:

$$z_{ij} = z_{ij}w_j \tag{12}$$

Step 4 To calculate the positive ideal point $x^+ = (x_1{}^+, x_2{}^+, \dots, x_n{}^+)$ as well as the negative ideal point

$x^- = (x_1{}^-, x_2{}^-, \dots, x_n{}^-)$ of each scheme given. Formulas are as follows:

$$x_j^+ = [\max_i z_{ij}^L, [\max_i z_{ij}^-, \max_i z_{ij}^+], \max_i z_{ij}^U] \tag{13}$$

$$x_j^- = [\min_i z_{ij}^L, [\min_i z_{ij}^-, \min_i z_{ij}^+], \min_i z_{ij}^U] \tag{14}$$

Step 5 To Use the (9) to calculate the similarity degree between each scheme and the positive ideal point s_i^+ and negative ideal point s_i^-:

$$S_i^+ = \sum_{j=1}^n S(z_{ij}, x_j^+) \tag{15}$$

$$S_i^- = \sum_{j=1}^n S(z_{ij}, x_j^-) \tag{16}$$

Step 6 To calculate the relative closeness degree of each alternative for the ideal point:

$$D_i = \frac{S_i^+}{S_i^+ + S_i^-} \tag{17}$$

Step 7 To sort each alternative by the relative closeness degree D_i. The larger the value itself is, the more optimal the solution is.

4 INSTANCE ANALYSIS

A company plans to execute an investment project about the estate. The initial investment schemes are x_1, x_2, x_3, x_4, x_5, and then $\mu_1, \mu_2, \mu_3, \mu_4$ are evaluation indicators. Among them, μ_2 and μ_3 denotes attribute values of benefit-type. μ_1 and μ_4 denotes attribute values of cost-type. Attribute values of each scheme are proposed by the form of four-point interval numbers. Therefore, according to the decider's uncertain judgment, $A = (a_{ij})_{5 \times 4}$ denotes a decision matrix. In order to calculate, we set $\alpha = 0.6$, $\beta = 0.4$ to be the known conditions.

$$A = \begin{pmatrix} [15,[18,20],22] & [5,[7,8],10] & [10,[11,12],13] & [0.5,[0.7,0.8],1.0] \\ [10,[11,12],13] & [5,[6,7],8] & [14,[16,18],20] & [1.5,[1.6,1.7],1.8] \\ [9,[11,12],13] & [2,[4,5],7] & [8,[9,11],13] & [0.2,[0.3,0.4],0.6] \\ [6,[7,9],11] & [1,[2,3],5] & [11,[12,13],14] & [1.2,[1.3,1.4],1.6] \\ [7,[8,9],10] & [2,[3,4],6] & [1,[2,3],4] & [0.3,[0.4,0.5],0.7] \end{pmatrix}$$

At the same time, the attribute weights are given as $w = \{0.23, 0.31, 0.31, 0.15\}^T$.

According to the previous talking, the solving process is as follows:

Step 1 To normalize decision matrix by (10) and (11).The normalized decision matrix is obtained.

Obviously, r_{ij} is still the four-point interval number.

$R = (r_{ij})_{m \times n}$ is as follows:

$$R = \begin{bmatrix} [0.00,[0.13,0.25]0.44] & [0.45,[0.67,0.78]1.00] & [0.48,[0.53,0.58]0.63] & [0.50,[0.63,0.69]0.81] \\ [0.56,[0.63,0.69]0.75] & [0.45,[0.56,0.67]0.78] & [0.68,[0.79,0.89]1.00] & [0.00,[0.06,0.13]0.19] \\ [0.56,[0.63,0.69]0.81] & [0.11,[0.34,0.45]0.67] & [0.37,[0.42,0.53]0.63] & [0.75,[0.88,0.94]1.00] \\ [0.69,[0.81,0.94]1.00] & [0.00,[0.11,0.22]0.45] & [0.53,[0.58,0.63]0.68] & [0.13,[0.25,0.31]0.38] \\ [0.75,[0.81,0.89]0.94] & [0.11,[0.22,0.34]0.56] & [0.00,[0.05,0.11]0.16] & [0.69,[0.81,0.88]0.94] \end{bmatrix}$$

Step 2 To calculate the weighted normalized decision matrix $Z = (z_{ij})_{m \times n}$ by (3) and (12), and we get:

$$Z = \begin{bmatrix} [0.00,[0.03,0.06]0.10] & [0.14,[0.21,0.24]0.31] & [0.15,[0.16,0.18]0.20] & [0.08,[0.09,0.10]0.12] \\ [0.13,[0.14,0.16]0.17] & [0.14,[0.17,0.21]0.24] & [0.21,[0.24,0.28]0.31] & [0.00,[0.01,0.02]0.03] \\ [0.13,[0.14,0.16]0.19] & [0.03,[0.11,0.14]0.21] & [0.11,[0.13,0.16]0.20] & [0.11,[0.13,0.14]0.15] \\ [0.16,[0.19,0.22]0.23] & [0.00,[0.03,0.07]0.14] & [0.16,[0.18,0.20]0.21] & [0.02,[0.04,0.05]0.06] \\ [0.17,[0.19,0.20]0.22] & [0.03,[0.07,0.11]0.17] & [0.00,[0.02,0.03]0.05] & [0.10,[0.12,0.13]0.14] \end{bmatrix}$$

Step 3 To calculate the positive ideal point and negative ideal point by (13) and (14), and we get:

$$x^+ = [0.17,[0.19,0.22]0.23],[0.14,[0.21,0.24]0.31],[0.21,[0.24,0.28]0.31],[0.11,[0.13,0.14],0.15]$$
$$x^- = [0.00,[0.03,0.06]0.10],[0.00,[0.03,0.07]0.14],[0.00,[0.02,0.03]0.05],[0.00,[0.01,0.02],0.03]$$

Step 4 To use the (9), (15) and (16) to calculate the similarity degree between each scheme and the positive ideal point and negative ideal point. Among them $\alpha = 0.6$, $\beta = 0.4$. the similarity degrees between each scheme and the positive ideal point are as follows:

$$S_1^+ = \sum_{j=1}^{n} S(z_{1j}, x_j^+) = 0+1+0+0 = 1$$

$$S_2^+ = \sum_{j=1}^{n} S(z_{2j}, x_j^+) = 0+0.1+1+0 = 1.1$$

$$S_3^+ = \sum_{j=1}^{n} S(z_{3j}, x_j^+) = 0+0+0+1 = 1$$

$$S_4^+ = \sum_{j=1}^{n} S(z_{4j}, x_j^+) = 0.9+0+0+0 = 0.9$$

$$S_5^+ = \sum_{j=1}^{n} S(z_{5j}, x_j^+) = 0.57+0+0+0.06 = 0.63$$

the similarity degrees between each scheme and the negative ideal point are as follows:

$$S_1^- = \sum_{j=1}^{n} S(z_{1j}, x_j^-) = 0+0.06+1+0 = 1.06$$

$$S_2^- = \sum_{j=1}^{n} S(z_{2j}, x_j^-) = 0+1+0+0 = 1$$

$$S_3^- = \sum_{j=1}^{n} S(z_{3j}, x_j^-) = 0+0+0+0 = 0$$

$$S_4^- = \sum_{j=1}^{n} S(z_{4j}, x_j^-) = 0+0+0+1 = 1$$

$$S_5^- = \sum_{j=1}^{n} S(z_{5j}, x_j^-) = 1+0+0+0 = 1$$

Step 5 To calculate the relative closeness degree of each alternative for the ideal point by (17):

$$D_1 = \frac{S_1^+}{S_1^+ + S_1^-} = \frac{1}{1+1.06} = 0.49$$

$$D_2 = \frac{S_2^+}{S_2^+ + S_2^-} = \frac{1.1}{1.1+1} = 0.52$$

$$D_3 = \frac{S_3^+}{S_3^+ + S_3^-} = \frac{1}{1+0} = 1$$

$$D_4 = \frac{S_4^+}{S_4^+ + S_4^-} = \frac{0.9}{0.9+1} = 0.47$$

$$D_5 = \frac{S_5^+}{S_5^+ + S_5^-} = \frac{0.63}{0.63+1} = 0.39$$

Step 6 The result of calculate process above is $D_3 > D_2 > D_1 > D_4 > D_5$. Obviously, the ranking of alternatives is $x_3 > x_2 > x_1 > x_4 > x_5$. The optimal solution is x_3.

5 CONCLUSIONS

Based on the three-point interval number, the paper proposes the four-point interval number whose most possible value is interval number. This achievement can effectively show the uncertain judgment made by decider. Besides, the paper provides some algorithms, and redefines the similarity degree of four-point interval numbers to measure the distances between these numbers. What's more, this paper proves the advantages of four-point interval number by comparing it with three-point interval number and interval rough number. Then the paper provides the solution of a kind of uncertain multiple attribute decision making problem whose property value are four-point interval numbers. Finally, the success of the sample testifies shows that the solution of the article is easy to handle and it does work.

However, there are still some shortages remained in the definition of the four-point interval number similarity degree. So, How to define a concise and functional similarity degree is the goal of improving. At the same time, the author is going to do some deep-going research on the attribute reduction of the information system on the basis of four-point interval numbers.

ACKNOWLEDGMENT

This work is supported by the National Natural Science Foundation of China (No. 71361002) and the Natural Science Foundation of Guangxi (2013GXNSFAA No. 019016).

REFERENCES

Atanasov, K. & Gargov, G. Interval-valued intuitionistic fuzzy sets. *Fuzzy Sets and Systems*, 1989, 31(3):343–349.

Buckley, J. & Feuring, T. & Hayashi, Y. Fuzzy Hierarchical Analysis Revisited. *European Journal of Operational Research*, 2001, 129(1):48–64.

Basak, I. Probabilistic Judgments Specified Partially in The Analytic Hierarchy Process. *European Journal of Operational Research*, 1998, 108(1):153–164.

Chen, Z.C. & Qin, K.Y. Attribute reduction of interval-valued information system based on variable precision tolerance relation. *Computer Science*, 2009, 36(3):163–166.

Dubois, D. & Prade, H. Operations on fuzzy numbers. *International Journal of Systems Science*, 1978, 9(6):613–626.

Gong, Y.B. & Liang, X.C. Fuzzy multi-attribute decision making method based on fuzzy C-OWA operator. *Systems Engineering and Electronics*, 2008, 30(7):1478–1480.

Hu, Q.Z. & Zhang, W.H. & Yu, Li. The research and application of interval numbers of three parameters *Engineering Science*, 2007, 9(3):47–51.

Hwang, C.L. & Yoon, K. (1981). *Multiple Attribute Decision Making*. London:Springer-Verlag Berlin Heidelberg New York.

Jain, R. A procedure for multi-aspect decision making using fuzzy sets. *International Journal of Systems Science*, 1978, (8):1–7.

Jiang, Y.P. & Fan, Z.P. A practical ranking method for reciprocal judgement matrix with triangular fuzzy Numbers. *Systems Engineering*, 2002,20(2):89–92.

Jin, Z.W. & Guo, H. Research on multiple attribute decision making method based on ideal point with interval rough numbers. *Journal of Chongqing University of technology(natural science)* 27(5):113–117.

Kao, C. & Liu, S.T. (2001).Fractional programming approach to fuzzy weighted average.Fuzzy sets and Systems 120(3): 435 444.

Li, R.J. (2002). *The Theory and Application of Fuzzy Multiple Attribute Decision Making*.Beijing:Science Press.

Qian, W.Y. & Zeng, Z. (2013).Method for Ranking Interval Rough Numbers Based on Possibility Degree.*operations research and management science* 22(1):71–76.

Saaty, T. (1980).*The Analytic Hierarchy Process*. New York: McGraw-Hill.

Tian, F. & Zhu, J.J. & Yao, D.P. & Zhu, H.Y. (2008). Consistency and weight estimation of novel three-point interval number complementary judgment matrix. *Systems Engineering-theory & Practice* 28(10):108–113.

Wang, Y.M. & Yang, J.B. & Xu, D.L. (2005).A Two–stage Logarithmic Goal Programming Methods for Generating Weights from Interval Comparison Matrices. *Fuzzy Sets and Systems* 152(1):475–498.

Xu, Z.S. & Da, Q.L. (2003).Possibility degree method for ranking interval numbers and its application. *Journal of Systems Engineering* 18(1):67–70.

Xu, Z.S. & Chen, J. (2007).An approach to group decision making based on interval-valued intuitionistic judgment matrices. *Systems Engineering-theory & Practice* 27(4):126–133.

Zhu, K.J. & Jing, Y.U. (1999).A Discussion on Extent Analysis Method and Applications of Fuzzy AHP. *European Journal of Operational Research* 116(2):450–456.

Zhu, J.J. & Liu,S.X. & Wang, M.G. (2005).Integration Of Weights Model of Interval Numbers Comparison Matrix. *Acta Automatica Sinica* 31(3):434–439.

Zeng, L. & Zeng, X.Y. (2010).Research on a class of multiple attribute decision making method with interval rough numbers. *Control and Decision* 25(11):1757–1760.

Information Technology – Wan et al. (Eds)
© *2015 Taylor & Francis Group, London, ISBN 978-1-138-02785-5*

Using imputation method to improve whole exome sequencing analysis with reference panel reconstruction

Zhiyong Pei, Xuenan Li, Jiapeng Zhou, Xiaoke Gui, Lijie Zhang & Yubao Chen
Beijing Computing Center, Beijing Academy of Science and Technology
Beijing, China

Zhiyong Pei
Beijing Institute of Genomics, Chinese Academy of Sciences, Beijing, China

ABSTRACT: With the advances of Next Generation Sequencing technology, the application of Bioinformatics to solve genomics problem become a focus of recent research. The accumulation of human genetic variation data makes it feasible to utilize computational method to improve the analysis, such as using imputation miss data in Genome wide association studies. In this work, in order to study how imputation method perform when it's coupled with whole exome sequencing result analysis. We carried on comparative analysis using with and without imputation data, using a strategy by reconstructing the reference panels by a custom-built way. The results indicated that more genetic markers were identified, especially one in intergenic and upstream/downstream regions. Meanwhile, the computing time cost was reduced and it is acceptable for further application.

KEYWORDS: Bioinformatics; Imputation; Whole exome sequencing; 1000 Genomes Project; Reference panel

1 INTRODUCTION

The Whole exome sequencing (WES) technology is wildly used in genetic research to identify the causes of diseases such as monogenic and clinical disorders in recent years [1–2]. Next-generation sequencing coupled with efficient DNA capture enables the use of WES as an effective way to study the genetic variations and phenotypes in human genome [3]. With other technological advances like genome wide association studies (GWAS), enormous amounts of genotype data are being generated, mostly from genome-wide SNP microarrays. While many studies often augmented of untyped variants by imputation methods. Pasaniuc [4] et al's work shows the imputation method increases the power of GWAS based on extremely low-coverage sequencing. This work provided a novel strategy to improve the sequencing result analysis with imputation, which is a statistical inference of unobserved genotypes[5]. There are several software have been published such as MaCH[6], IMPUTE2[7], BEAGLE[8] and so on. Most imputation achieved by using known haplotypes in a population, for instance from the HapMap or the 1000 Genomes Project in humans using HapMap or 1000 Genomes Project (1000G) reference [9–12].

However, it is time-consuming work for imputing from large reference panels, which imposes a high computational burden. Howie et al's have carried on a study with a pre-phasing strategy for fast and accurate genotype imputation. The method could maintains the accuracy of leading methods while reducing computational costs[13]. Meanwhile, the result of Zheng et al's work shows that the v variants with lower minor allele frequency (MAF) are more difficult to impute, especially for the very rare variants when using HapMap2 and 1000 Genomes pilot as reference panels[14]. In the data from the 1000 Genomes Project, there are 1092 individual with genetic variations. It might not be a suitable strategy to use all the dataset when carrying on particular imputation, such as imputing the results from WES. Besides, considering the divergence of allele frequency in different populations. In this case work, the reference panel was particularly reconstructed according to the test data set before imputing. The results indicate that this strategy could decrease the computing time, and the performance of result analysis protocol as well as when identifies the correlated variation cause with phenotypes.

2 RESULT AND DISCUSSION

2.1 *The accuracy and run-time performance with simulated data*

Firstly, we build the test data set for imputation from 1000G data, in which 286 individuals from Asian were derived and randomly missed by 50% and 80% respectively. Based on the data, two sets of simulated data were constructed as sequencing data with 0.5X and 0.2X coverage which is considered as a simulation of WES data. To test the computational advantages of strategy with different reference panel, all data with genetic variation of 1092 individuals and ASN data of 286 Asian individuals were respectively used as reference for imputing analysis. Then, the imputation was carried on using BEAGLE software. The result of accuracy and the time cost of chromosome 22 of the human genome is comparatively shown in Fig. 1 as pre-test evaluation. We measured imputation accuracy at these SNPs as the mean R^2, average squared correlation, between masked genotypes and imputed allele dosages as Howie et al' work [13]. The Running times are shown as the central processing unit (CPU) minutes needed to impute across one chromosome for all 286 individuals.

It is shown in Fig. 1 that, the accuracy of imputation in ASN data as a reference panel maintains the R2 compared to using the all data in 1000G.

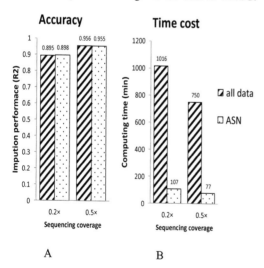

Figure 1. The comparative imputation results of simulated data with 0.2× and 0.5× sequencing coverage in chromosome 22 of the human genome for all data and ASN data respectively. A. The imputation accuracy with R2. B. The computing time with CPU minute as unit. The result values are indicated at the top of bars in the figure.

For 0.5× simulated data, the average R2 of all data and ASN result are 0.956 and 0.955 respectively, and for 0.2× simulated data, are 0.895 and 0.898 which are lower on the whole than 0.5× simulated data. On the other hand, the running times with ASN data are much less used compared with using all data as a reference panel in imputing. The computing time is only one tenth when using ASN reference than use all data. It is indicated that the computational costs could be reduced by adjusting the reference panel in imputing calculation. The general performance is acceptable for low coverage sequencing data with an approximate 0.9 in average R2 in 0.2× simulated data. From the result in Fig. 1B we can see the running time is longer for lower coverage.

It should be noted as that the distribution of genetic markers (SNP, indel, etc.) in simulated data is relatively less of bias, which might not be the same as real sequencing data. For instance, in WES data, the distances between the locations of markers could differ from each other dramatically. Then, the data in the reference panel far of the markers on unsurpassed test data might play much less roles in imputation. Then to reconstruct the reference panel according to test data as a custom-built way might be a suit method to improve the performance and analysis.

2.2 *Improvement of the WES analysis result by imputation*

To improve the analysis of analysis of WES such SNP and indel calling, imputation was carried on for increasing the amount of genetic variation numbers before annotation, data filtration and candidate variation identification. The test data set of WES was from Chinese individual, that only ASN individuals in 1000G were used for imputing. MAF of the reference panel were also be calculated. Any markers with MAF larger than 0.05 as a threshold were omitted as there were considered as common variation. Only those rare variations with MAF < 0.05 were used for imputation considering they might be more likely be the correlated with disease phenotypes. After that, the reference panels were particularly generated for each test data set, according to which the location of markers in test data sets was enlarged to a certain distance window. These could be useful to reduce the run-time in following imputation.

The results show that the amount of genetic variation, including SNP and Indel, has different extents of growth in all chromosomes (Table 1). The WES variation calling result of all chromosome data

This work is supported by BNSF7120001, BCCMY-2013-01, OTP-2011-010 and OTP-2012-011.

was carried on imputation with the reconstructed reference panel. It is shown in Table 1 that some chromosome has a relatively large increasing percent of data, such as chromosome 5, 6, 8, 10 and 12, whose achieve more than 30 percent increasing. While in some chromosome like 2, 17, 18 and 21, very limited increasing had been observed. Averagely, 16.6±12.2 percent of genetic variation had been increased by using the imputation method.

The run-time of imputing for every chromosome of one individual data is shown in Table 1 as well. The total time of imputing for a genome is about 4.5 hours, which is acceptable for practically applied to high throughput sequencing analysis. In the actual calculating process, every date from each chromosome submitted to server cluster to separate cores, then the imputing tasks could carry on parallel computation. This parallel processing computation could speed up the total process when large scale data to be dealt with. That result indicated the performance of this method is useful for following analysis. More candidate genetic variation might be found out to be the cause of particular phenotypes.

Table 1. The result WES analysis improved by imputation for each chromosome.

| Chromsome | Genetic variation amounts | | | |
	Before imputation	After imputation	Increased percent (%)	Run-time (min)
1	2470	3122	26.4	9.5
2	2107	2203	4.6	18.4
3	1774	2105	18.7	16.2
4	1246	1310	5.1	15.4
5	1428	1922	34.6	15.7
6	2111	2745	30.0	18.8
7	1695	1972	16.3	15.3
8	1170	1561	33.4	12.5
9	1433	1563	9.1	12.9
10	1510	2133	41.3	5.4
11	2229	2457	10.2	14.1
12	1631	2268	39.1	14.2
13	621	748	20.5	8.9
14	1251	1402	12.1	10.3
15	1162	1380	18.8	10.9
16	1442	1686	16.9	7.8
17	2061	2131	3.4	9.3
18	494	509	3.0	7.7
19	2556	2712	6.1	9.1
20	907	1013	11.7	7.1
21	521	527	1.2	2.4
22	957	1073	12.1	5.9
X	438	466	6.4	26.0
			Average	Total
			16.6±12.2	273.8

Table 2. The comparative result of genetic markers after annotation.

Genetic marker numbers	Before imputation	After imputation	Increased percent (%)
upstream	100	161	61.00
downstream	64	155	142.19
exonic	19863	19900	0.19
intergenic	1161	3387	191.73
intronic	2724	5684	108.66
ncRNA_exonic	2420	2445	1.03
ncRNA_intronic	250	536	114.40
splicing	35	38	8.57
exonic_noxious	2203	2206	0.14

Analysis of a series of annotation was carried on for variation calling results without and with imputation respectively. The annotation results for both analysis hand been compared with each other, in order to evaluate the practical improvement in WES study. It is shown in Table 2 that, the exonic markers that located in exonic regions are increased very slightly. Meanwhile, the improvement by using after-imputation data achieve a good performance in intragenic or intrinsic regions rather than exonic regions. Much more markers in upstream/downstream in genome had been identified by doubly increased.

3 CONCLUSION

The imputation by using the custom-built reference panel for WES analysis could improve the result with the genetic marker number increased. This strategy could also be useful for reducing the computational time by minimizing the reference panel size. Much more genetic markers that likely to be correlated with phenotypes could be found in upstream/downstream and intergenic/intronic, which could be helpful for further research on those neighborhood regions of genes and exome.

4 MATERIALS AND METHODS

The raw data we reproduced by 30X whole exome sequencing of the sample from a Chinese male individual. The WES sequencing instrument was Hiseq2000. About 30G raw reads were produced and analyzed with a process to get genetic variation data in VCF format. Then imputation was carried on by using the latest beagle (version 4) software, which could be taken the VCF files as input conveniently.

Before imputation, a key step is to prepare the reference panel according to the unphased test data to be imputed. The locations of variation marker in test data have been extracted and the marker blocks edges were identified. Based on that, larger target blocks were defined by enlarging the edges of an approximately 300k window. Then the ASN reference data in 1000G data were reconstructed, that only variation marker within the target blocks were kept while omitting the others. The purpose to so is to reduce the size of the reference panel while maintains the relatively effective data preserved for imputing. As the computational cost grows quickly with reference panel size, it can be useful to time-consuming limit to run a high throughput data by NGS.

REFERENCES

[1] S. B. Ng, et al., Targeted capture and massively parallel sequencing of 12 human exomes. Nature, 2009. 461(7261): 272–6.

[2] B. A. Schuler, S. Z. Prisco, and H. J. Jacob, Using Whole Exome Sequencing to Walk From Clinical Practice to Research and Back Again. Circulation, 2013. 127(9): 968–970.

[3] A. Kiezun, et al., Exome sequencing and the genetic basis of complex traits. Nat Genet, 2012. 44(6): 623–30.

[4] B. Pasaniuc, et al., Extremely low-coverage sequencing and imputation increases power for genome-wide association studies. Nat Genet, 2012. 44(6): 631–5.

[5] P. Scheet and M. Stephens, A fast and flexible statistical model for large-scale population genotype data: applications to inferring missing genotypes and haplotypic phase. Am J Hum Genet, 2006. 78(4): 629–44.

[6] Y. Li, et al., MaCH: using sequence and genotype data to estimate haplotypes and unobserved genotypes. Genet Epidemiol, 2010. 34(8): 816–34.

[7] B. N. Howie, P. Donnelly, and J. Marchini, A flexible and accurate genotype imputation method for the next generation of genome-wide association studies. PLoS genetic, 2009. 5(6): e1000529.

[8] B. L. Browning and S. R. Browning, A fast, powerful method for detecting identity by descent. Am J Hum Genet, 2011. 88(2): 173–82.

[9] C. Genomes Project, et al., A map of human genome variation from population-scale sequencing. Nature, 2010. 467(7319): 1061–73.

[10] C. International HapMap, The International HapMap Project. Nature, 2003. 426(6968): 789–96.

[11] N. Siva, 1000 Genomes project. Nat Biotechnol, 2008. 26(3): 256.

[12] J. Marchini and B. Howie, Genotype imputation for genome-wide association studies. Nat Rev Genet, 2010. 11(7): 499–511.

[13] B. Howie, et al., Fast and accurate genotype imputation in genome-wide association studies through pre-phasing. Nat Genet, 2012. 44(8): 955–9.

[14] H. F. Zheng, et al., Effect of genome-wide genotyping and reference panels on rare variants imputation. J Genet Genomics, 2012. 39(10): 545–50.

Information Technology – Wan et al. (Eds)
© 2015 Taylor & Francis Group, London, ISBN 978-1-138-02785-5

Vehicle identification based on feature point matching and epipolar geometry constraint

Sai Liu
Beijing Lab of Intelligent Information Technology
School of Computer Science Beijing Institute of Technology
Beijing, CHINA

Mingtao Pei
Beijing Lab of Intelligent Information Technology
School of Computer Science Beijing Institute of Technology
Beijing, CHINA

ABSTRACT: Vehicle identification has been extensively researched in recent years. Usually the license plate is used to identify the vehicle. However when the vehicle does not have a license plate or the license plate is occluded, other features of the vehicle have to be employed. In this paper, we propose a method to identify vehicles by feature point matching and epipolar geometry constraint. Affine-sift (Affine Scale-Invariant Feature Transform), which is invariant to affine transformation, is used to generate feature points, and a collection of matched feature point pairs is obtained which contains many false matched pairs. Then epipolar geometry constraint is employed to eliminate false matched point pairs. And the vehicle is identified by the number of corrected matched point pairs. Our method is motivated by the insight that same vehicle in different backgrounds has plenty of matched point pairs on the vehicle, and the matched point pairs should coincide with epipolar geometry constraint, while false matched point pairs in different backgrounds and on different vehicles will not coincide with epipolar geometry constraint. Experimental results show that our method is able to identify vehicles in different places under different view points.

1 INTRODUCTION

Vehicle identification across non-overlapping cameras has wide applications such as parking management, speed estimation and vehicle tracking. In these applications license plate plays an important role, [13, 14, 15] use license plate characters to identify vehicles by exploiting image processing technology. However, when the vehicle does not have a license plate or the license plate is occluded, other features of the vehicle have to be employed. Shan et al. [3] propose a novel solution converting problem of identifying vehicles across non-overlapping cameras into a same-different classification problem without direct feature matching by computing the same-different probabilities. Ferencz et al. [18] propose an on-line algorithm building an efficient same-different classification cascade by predicating the most discriminative feature set for vehicles, not only estimates the saliency and scoring function for each candidate feature, but also models the dependency between features. Wang et al. [17] tackle

vehicle identification by reconstructing vehicles with multiple linear regression models and sparse coding which has been successfully used in multi-samples classification and identification. Wang et al. [6] introduced an inter period adjusting technique based on the exponential smoothing to define an appropriate time-window constraint to identify vehicles. Matching vehicles which are subjected to drastic pose change and extreme illumination variation is conducted in [16], Hou explicitly estimates pose and illumination effectively for vehicles in reference image and target image respectively using approximated 3D vehicle models and albedos estimation, then re-render vehicles in reference image according to vehicles' pose and illumination in target image to generate relit image, finally comparisons is made between the relit image and the re-rendered target image to determinate whether the vehicle in the reference image is identical to the vehicle in target image. Tian et al. [4] use multiple sensors to accomplish vehicle identification; vehicle status and correct signature segmentation can be determinated by the matching result of one vehicle's signature obtained by different sensors. The co-relationship between signatures can be obtained, and then the time offset is corrected depends on such a co-relationship. Sanchez et al. [19] propose a method based on matching electromagnetic vehicle signatures, which are obtained from wireless magnetic sensors. Jazayeri et al. [21] apply HMM (Hidden Markov Model) to separate background and moving vehicles in the temporal domain. Their identification process is based on vehicle and background motions.

Almost all above-mentioned methods have to detect the position of vehicles first before process vehicle identification while vehicle detection itself is a hard problem. We propose a method to identify vehicles, which does not need to know the location of the vehicle in the image. Our method is based on the idea that same vehicle in different backgrounds has many same feature points on the vehicle and different vehicles in different backgrounds possess almost zero same feature point. The goal of vehicle identification now is to find a vehicle image that has been selected in one view (probe image) in all the images from another view (gallery images). This is achieved by calculating the number of matched feature point pairs between the probe image and all gallery images. But in real application two main problems usually occur: view point change and false matched feature point pairs. Two additional improvements are made to deal with these two problems. First we employ Affine-sift [5] instead of using traditional method like SIFT (Scale-Invariant Feature Transform) [1] or SURF (Speeded-up Robust Features) [20], which are perform well at image scaling, rotation and shift but bad at view point change, to generate feature points. Affine-sift, which is invariant to affine transformation, simulates view point changes by varying longitudes and latitudes defined in [5]. And we apply epipolar geometry constraint originals from photogrammetry to eliminate the false matched point pairs. Fundamental matrix can be obtained from the two feature point sets formed from two images. Point pairs do not satisfy the constraint that are eliminated as outliers while preserving inliers as true matched point pairs. Depending on the number of matched feature point pairs, whether the vehicle is the same one can be judged.

The rest of the paper is organized as follows: Section 2 gives a detail description of our method including affine-sift feature point generation and matching between the feature points. Epipolar geometry constraint is explained as well. Two experimental results are presented in Section 3. Section 4 gives a conclusion of the paper.

2 VEHICLEIDENTIFICATION

2.1 *Extraction and matching of feature points*

SIFT is wildly used because it is invariant to image scale, rotation, shift, etc. However our dataset are taken under two view points, sift cannot deal with this situation. Affine-SIFT is applied to extract feature points which is invariant to affine transformation by simulating all view point changes. Yu et al. [5] point out that image distortions arising from view point change can be locally modeled by affine planar transformation and it gives the local image deformation model under a camera motion:

$$u(x, y) \rightarrow u(\alpha x + by + e, cx + dy + f) \qquad (1)$$

Any linear planar map $\mathbf{A} = \begin{bmatrix} a & b \\ c & d \end{bmatrix}$ with positive determinant and can be decomposed as:

$$\mathbf{A} = \lambda R(\psi) T, R(\varphi) = \lambda \begin{bmatrix} \cos\psi & -\sin\psi \\ \sin\psi & \cos\psi \end{bmatrix} \begin{bmatrix} t & 0 \\ 0 & 1 \end{bmatrix} \begin{bmatrix} \cos\varphi & -\sin\varphi \\ \sin\varphi & \cos\varphi \end{bmatrix}$$

$$(2)$$

where R is matrix of rotation, $\lambda > 0$, and λt is the determinant of A, $\varphi \in [0, \pi)$, $\psi \in [0, 2\pi)$. A can be viewed in a geometric interpretation in Fig. 1.

As Fig. 1 implies, angle φ is called longitude, the optical axis deviates θ from the perpendicular line of the image plane u is called latitude. In the A's expression, t is defined by $t\cos(\theta) = 1$ and is called tilt ($t \geq 1$). ψ indicates the angle of camera rotation around its optical axis. Image deformation implemented by (1) can be represented by A. Then

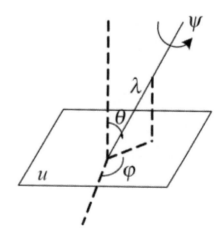

Figure 1. Geometric interpretation of A.

simulated images can be obtained through A by varying longitudes and latitudes.

Once simulated images of probe image are generated, a collection of sift features points and descriptors of the probe image are obtained for each simulated images, then all the feature point descriptors from the probe image will be matched one by one with other feature point descriptors from the gallery images. A collection of matched point pairs for the probe image and each gallery image is generated consisting of both true matched point pairs and false matched point pairs.

2.2 *Epipolar geometry constraint*

Method proposed in this paper is based on the idea that one vehicle in different places share plenty of true feature point pairs while different vehicles in different places share almost zero true point pair. However, in real application there are many false matched feature point pairs, we apply epipolar geometry constraint to eliminate the false matched pairs. Fig. 2 reveals the geometry relationship between the two points in two camera planes subjecting to epipolar geometry constraint.

In Fig. 2 P is a 3D scene point, P_l is the projection of P in the left camera's image plane π_l, P_r is the

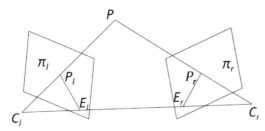

Figure 2.   Epipolar geometry constraint.

projection of P in the right camera's image plane π_r. Let C_l and C_r be optical centers of the two cameras. Intersection points between line C_lC_r and π_l, π_r are E_l, E_r. which are called epipoles. Plane constructed by P, C_l and C_r is epipolar plane. According to epipolar geometry constraint, relationship between P_l and P_r can be formulized as:

$$P_r^T F P_l = 0 \tag{3}$$

As the vehicle is a rigid body, feature point pairs, which are generated from two images on the same vehicle will satisfy (3), while those point pairs in the different backgrounds or on different vehicles will not.

The F in (3) is fundamental matrix. Here P is point on the vehicle, and P_l, P_r are projections of P

in the two view points. Once F is calculated, same point projection P_l, P_r of P will make (3) established while different point projections will not conform to (3). A common solution to solve F is RANSAC [7, 8] (Random Sample Consensus), which is an iterative paradigm to fit a model to experimental data while the data can be contaminated. RANSAC which is wildly used in [9, 10, 11, 22] can estimate model parameters, but fails when data contains a significant percentage of "outliers" and cannot always find an optimal solution.

We use a robust method to calculate the fundamental matrix which is introduced by Moisan et al. in [2].

Figure 3.   Before using epipolar geometry constraint.

Moisan points out the epipolar geometry constraint relies on a rigidity constraint, changing the view point amounts to applying linear transform on object. Thus estimating fundamental matrix is equivalent to rigid motion estimation. Moisan proposes a probabilistic criterion to detect the existence of a rigid motion between two sets of point matches that permits to decide whether these points are independent or if they are correlated by a rigid motion. A large number of point correspondences between the two images cannot be explained by such a rigid motion, so that true point correspondences can be detected as large deviations from randomness.

Top row in Figs. 3 and shows the same vehicle, while bottom row in Figs. 3 and 4 shows different

Figure 5(b).   False matched pairs.

pairs on different vehicles and in different backgrounds. Fig. 4 exhibits feature point pairs result that are filtered by epipolar geometry constraint. Fig. 4 illuminates the necessity of epipolar geometry constraint. It can be seen that false matched point pairs are eliminated by epipolar geometry constraint.

3 EXPERIMENT

105 image pairs are taken for 105 vehicles. Thus only one sample per vehicle is given. License plates are erased manually to demonstrate that our method can handle situations that a vehicle does not have a license plate or license plate is obscured. As images are taken in different places from two different view points, our dataset bears illumination change, shadow change and perspective change, all of above making our dataset a challenging one. We divide dataset vehicles images into two datasets according to the two view points.

Figs. 3 and 4 show the left column and right column which represent two view points respectively. After feature points are detected by affine-sift and matched feature point pairs are generated. We use the number of corresponding point pairs filtered by epipolar geometry constraint to determinate whether the two vehicles are identical. We consider a pair of vehicles is identical when the number of corresponding point pairs between them is more than others and exceeds a given threshold. Results show that 9 pairs out of 105 are false matched achieving accuracy of 91.44%. Fig. 5(a) and (b) shows some of the true matched pairs and a false matched pair. From Fig. 5(b), we can see that the false matched pairs are due to high contrast of light, shadows on the windshield and similar backgrounds.

To show the advantage of affine-sift, we apply SIFT instead of affine-sift to generate feature points and keep the remained steps the same. According to the judgment criterion mentioned above, result shows

Figure 4.   After using epipolar geometry constraint.

vehicles. Fig. 3 shows feature points matching result before using epipolar geometry constraint. There are a lot true matched feature point pairs in the same vehicle, so as to a lot of false matched feature point

Figure 5(a).   True matched pairs.

Figure 6. SIFT instead of affine-sift.

that 47 pairs out of 105 are false matched achieving accuracy of 55.24%. Fig. 6 shows the same vehicle pairs' matching results corresponds to Fig. 5.

It can be seen from the two experiments that affine-sift performs much better than sift for the sake of its invariant to affine transformation by simulating images, more feature points can be generated by affine-sift and the final results are more reliable.

4 CONCLUSION

Vehicle identification has been a challenging topic, especially when only one sample is given. Methods designed for multi-samples to deal with classification and identification will be infeasible. We propose a method to identify vehicles by feature points matching. Affine-sift is employed to detect feature points and epipolar geometry constraint is used to filter false matched feature point pairs.

Our method does not need to detect the location of vehicle in the image. The vehicle's location can be roughly inferred from the matched points' position as matched feature point pairs are detected on the vehicle's body. Our future work is to exploit more features such as histogram of oriented gradient [23], color and vehicle logo to identify vehicles.

REFERENCES

Lowe D G. Distinctive image features from scale-invarant keypoints. *International Journal of Computer Vision*, 2004, 60(2): 91–110.

Moisan L, Stival B. A probabilistic criterion to detect rigid point matches between two images and estimate the funda-mental matrix, *International Journal of Computer Vision,* 2004, 57(3): 201–218.

Shan Y, Sawhney H S, Kumar R. Vehicle identification between non-overlapping cameras without direct feature matching, Computer Vision, ICCV. Tenth IEEE Interna- tional Conference on. IEEE, 2005, vol. 1, pp. 378–385.

Tian Y, Dong H, Jia L, et al. A vehicle re-identification algorithm based on multi-sensor correlation, *Journal of Zhejiang University Science C,* 2014, 15(5): 372–382.

Yu G, Morel J M. A fully affine invariant image comparison method, Acoustics, IEEE International Conference on Speech and Signal Processing, ICASSP 2009, IEEE, 2009, pp. 1597–1600.

Wang J, Indra-Payoong N, Sumalee A, et al. Vehicle reidentification with self-adaptive time windows for real-time travel time estimation, *IEEE Transactions on In telligent Transportation Systems,* 2014, 15: 540–552.

Fischler M A, Bolles R C.1981.Random sample consensus: a paradigm for model fitting with applications to image analy- sis and automated cartography, *Communications of the ACM,* 1981, 24(6): 381–395.

Hartley R, Zisserman A. *Multiple View Geometry in Computer Vision,* Cambridge, Cambridge University Press, 2003.

Yang A Y, Rao S R, Ma Y.2006. Robust statistical estimation and segmentation of multiple subspaces. *Conference on Computer Vision and Pattern Recognition Workshop, CVPRW'06,* IEEE, 2006, pp. 99–99.

Borkar, A., Hayes, M., Smith, M. Robust lane detection and tracking with RANSAC and Kalman filter, In: *Interna- tional Conference on Image Processing,* 2009, pp. 3261–3264.

Borkar A, Hayes M, Smith M T. A novel lane detection system with efficient ground truth generation, *IEEE Transactions on Intelligent Transportation Systems* 2012, 13(1): 365–374.

Sur F, Noury N, Berger M O.2008. Computing the uncertainty of the 8 point algorithm for fundamental matrix estimation, *19th British Machine Vision Conference-BMVC,* pp. 96.1– 96.10.

Qadri M T, Asif M.2009. Automatic number plate recognition system for vehicle identification using optical character recognition, *International Conference on Education Technology and Computer,* ICETC'09,. IEEE, 2009, pp. 335–338.

Deb K, Le M H, Woo B S, et al.2011. Automatic vehicle iden- tification by plate recognition for intelligent transportation system applications, *Modern Approaches in Applied Intelli- gence.* Springer Berlin Heidelberg, pp: 163–172.

Ozbay S, Ercelebi E. Automatic vehicle identification by plate recognition, *World Academy of Science, Engineering and Technology*, 2005, 9(41): 222–225.

Hou T, Wang S, Qin H. Vehicle matching and recognition under large variations of pose and illumination, *IEEE Computer Society Conference on Computer Vision and Pattern Recognition Workshops, CVPR Workshops,* 2009, pp. IEEE, pp, 24–29.Wang S, Cui L, Liu D, et al.2012.Vehicle identification via sparse representation, *IEEE Transactions on Intelligent Transportation Systems*, 2012, 13(2): 955–962.

Ferencz A, Learned-Miller E G, Malik J.2005. Building a classification cascade for visual identification from one example, Tenth IEEE International Conference on Computer Vision, ICCV 2005. IEEE, 2005, 1, pp. 286–293.

Sanchez R O, Flores C, Horowitz R, et al.2011.Vehicle re-identification using wireless magnetic sensors: algorithm revision, modifications and performance analysis, 2011 *IEEE International Conference on. IEEE Vehicular Electronics and Safety (ICVES)*, 2011, pp. 226–231.

Bay H, Ess A, Tuytelaars T, et al.2008.Speeded-up robust features, Computer vision and image understanding, vol.110(3), pp. 346–359.

Jazayeri A, Cai H, Zheng J Y, et al.2010. Motion based vehicle identification in car video, Intelligent Vehicles Symposium (IV), IEEE. pp. 493–499.

Lipman Y, Yagev S, Poranne R, et al.2014. Feature Matching with Bounded Distortion, *ACM Transactions on Graphics (TOG)*, 2014, 33(3): 26.

Dalal N, Triggs B.2005.Histograms of oriented gradients for human detection, *IEEE Computer Society Conference on Computer Vision and Pattern Recognition, CVPR*, 2005. IEEE, 2005, 1: 886–893.

Information Technology – Wan et al. (Eds)
© 2015 Taylor & Francis Group, London, ISBN 978-1-138-02785-5

Author index